INTRODUCTION TO CHEMISTRY

INTRODUCTION TO CHEMISTRY

AMOS TURK □ HERBERT MEISLICH

FRANK BRESCIA □ JOHN ARENTS

Department of Chemistry, The City College of the City University of New York

ACADEMIC PRESS NEW YORK and LONDON

ACADEMIC PRESS INC.
111 Fifth Avenue, New York, New York 10003

United Kingdom Edition published by
ACADEMIC PRESS INC. (LONDON) LTD.
Berkeley Square House, London W.1

Library of Congress Catalog Card Number: 68-14653

Printed in the United States of America

PREFACE

The pursuit of chemistry as a science and a technology is one of the most productive of man's activities today. Concepts, techniques, tools, and the instruction of the student are all undergoing very rapid advances.

An introductory textbook of chemistry should catch the spirit of vitality, of intellectual challenge and excitement, of professional tradition and style, and of great utility for man's needs that pervades the world of the chemist.

This text is designed to serve in the one-year course in general chemistry. It has been our objective to hold the interest of the student who does not have vocational objectives in chemistry, and also to provide the necessary preparation for students who do plan to continue their studies in the advanced courses.

The approaches used in first-year chemistry texts have been classified* as the "high road" and the "low road" methods. The "high road" begins with the study of atoms and proceeds from the properties of these chemical microcosms to evolve subject matter like stoichiometry, and molecular structure and transformations. The "low road" proceeds from macroscopic beginnings—the densities of vapors, the composition of materials, and the like—and thus lays the basis for our atomistic concepts.

The approach of this text does not fit well into either category; it is, perhaps, a "middle road." After an introduction to atomic structure, the text proceeds to chemical periodicity, and to a classical consideration of atomic weights. Thus, by the time that stoichiometry is introduced, the student has been exposed to the study of the atom from both macroscopic and microscopic viewpoints.

Having considered atoms and the arithmetic and energetics of their combination into molecules (Chapters 1–6), the next phase (Chapters 7–9) considers the nature of the interactions among atoms—in other words, the nature of chemical bonding. After this introduction to the question of how atoms aggregate to form molecules, it seems reasonable to proceed to the question of how molecules aggregate to form materials. This phase is covered in Chapters 10–13, which deal with the nature of intermolecular forces, and the study of the states of matter. Chapters 14 to 19 are concerned with the statics and dynamics

* Leonard K. Nash, *Stoichiometry*. New York: Addison-Wesley Publishing Company, 1966.

of chemistry—equilibrium and kinetics. Included in these chapters are ionic equilibrium, acids and bases, and a study of galvanic cells. The final portions of the book, Chapters 20–26, apply the previously developed principles to a descriptive study of chemistry—representative and transition elements, organic chemistry, metals, nuclear chemistry, polymers, and biochemistry. These descriptive sections do not abandon the approach of the first two thirds of the text. There is no question of whether "principles" or "facts" take precedence, for the two do not separate easily. It is important to introduce the descriptive matter of chemistry in such a way that the student learns the facts as his understanding of principles is reinforced.

Many of the features of our more comprehensive text, *Fundamentals of Chemistry: A Modern Introduction*, are retained here. There are two sets of problems for almost every chapter. The first set can be assigned *in toto* to cover the essential ideas of the chapter. The second set, "additional problems," provides additional drill, extends the contents of the chapter, or, in some cases, poses a greater challenge. Answers are provided for about one-half of the problems requiring computation.

One other matter. A chemistry textbook usually serves in part as a reference for the experiments in the general chemistry laboratory. We are very much concerned with safety, and have taken every reasonable opportunity to instill the philosophy of safe practice into these pages, even though this is not a laboratory book. The laboratory manual that is an optional accompaniment of this text, *Fundamentals of Chemistry: Laboratory Studies*, emphasizes safety along with critical selection of experimental procedures and the use of techniques of quantitative chemistry.

We express our gratitude to our wives, our students, and our colleagues for their many kinds of help. We also thank Mrs. Coleman London for research assistance and Mrs. Evelyn Manacek who typed most of the manuscript. We particularly wish to acknowledge our appreciation to Professor Darrell Eyman of the University of Iowa, whose contribution to the concept of the "middle road" did much to set the approach of this book, and whose critical reading of the entire manuscript provided helpful insights and suggestions.

A. T., H. M., F. B., J. A.

New York,
February, 1968

CONTENTS

21 ▪ THE CHEMISTRY OF THE TRANSITION ELEMENTS

22 ▪ ORGANIC CHEMISTRY

23 ▪ METALS AND METALLURGY

24 ▪ NUCLEAR CHEMISTRY

25 ▪ POLYMERS

INTRODUCTION TO CHEMISTRY

1 ■ INTRODUCTION

1.1 □ GENERAL AND HISTORICAL

Chemistry deals with the properties and transformations of materials. Materials are samples of matter; they include the secretion from a pituitary gland, the varnish on a table, the odorant from a flower, cobra venom, a flame, a helium atom, and a proton. Chemists observe how materials change; they coordinate their observations into useful concepts, and they predict conditions under which specific changes will occur—often to produce materials never previously observed. Chemistry's origins are ancient. Metallurgy, leather tanning, fermentation, and the manufacture of soap, glass, and pigments were all devised before man acquired the concepts that now underlie his understanding of chemical change. Progress in chemical theory and technology during the past two centuries, however, has been spectacular compared with that in all previous history. The most valuable group of ideas held by chemists has proved to be the atomic and molecular hypotheses.

What we now consider to be modern chemistry began with investigations of gases under pressure and vacuum (a discipline then called "pneumatics") in the seventeenth century by Evangelista Torricelli, Blaise Pascal, Otto von Guericke, and Robert Boyle. These studies led to improvements in laboratory techniques that helped another generation of scientists (notably Karl Wilhelm Scheele, Henry Cavendish, Joseph Priestley, and Antoine Lavoisier), about a century later, to formulate a quantitative basis for chemical changes, especially combustion and other reactions involving oxygen. These advances, in turn, set the stage for the chemical pioneers of the nineteenth century (John Dalton, Amadeo Avogadro, Jons Jakob Berzelius, and Stanislao

Cannizzaro) to interpret chemical changes in terms of atoms and molecules, and to devise rational systems of atomic and molecular weights. The latter half of that century witnessed a very fruitful growth of systematizing concepts—the periodic table, the structural theory of organic chemistry, and stereochemistry (the geometry of molecules). In 1896 Henri Becquerel discovered radioactivity, thus initiating a new chain of discoveries that led to a great refinement of our ideas about the atom, and to new understanding of chemical processes. It is to these ideas that we shall turn to provide the major conceptual framework for our study of chemistry.

Refer to Appendix I for a review of physical concepts, measurement scales, and significant figures. This material is fundamental to discussions throughout the book. Some of the problems at the end of this chapter are based on material in the Appendix.

1.2 □ DEFINITIONS OF SOME CHEMICAL TERMS

We shall define and discuss briefly some terms which, over a period of many years, have become part of the language of chemists.

The properties of a material are its distinguishing characteristics. **Accidental** or **extensive properties** depend on the amount of matter present in a sample of the material; **specific,** or **intensive, properties,** on the other hand, do not depend on the amount of matter in the sample. Thus, the white color of a piece of chalk is a specific property; its length is an accidental one.

A **substance** is any variety of matter of recognizably definite composition and specific properties. The term is used in distinction to **body,** or **object,** which refers to a particular item of matter. Thus, a chair (object) is made of wood (substance). The **composition** of a substance is its makeup of constituent substances, usually expressed in terms of percent or fraction by weight.

Some substances have precisely fixed compositions which are associated with their properties; they are said to be **pure substances.** For example, red iron rust can be obtained as a pure substance comprising 69.94% iron and 30.06% oxygen. Coal, on the other hand, is not a pure substance; its carbon content ranges from 35 to 84%. Of course, a pure substance may be contaminated by admixture of foreign matter. The important point, however, is that the pure substance, when it is recovered from such a **mixture,** retains its definite composition and specific properties.

It is believed that the constant compositions associated with pure substances are maintained by linkages among elementary units of matter; such linkages are called **chemical bonds.** A transformation accompanied by the making or breaking of chemical bonds is called a **chemical change,** or **chemical reaction.** Examples are combustion, corrosion, photosynthesis, and digestion.

A **physical change** of a substance does not involve change of definite composition or specific properties. Alterations in the dimensions of objects, or in the states of aggregation of their constituents, are considered to be physical changes. Examples are fracture, deformation, pulverizing, drawing (as of a metal wire), thermal expansion or contraction, melting, boiling, and freezing.

The types of behavior that a substance exhibits in chemical reactions are called its **chemical properties;** other intensive characteristics of a substance are called its **physical properties.**

Decomposition is a chemical reaction in which the constituent entities of a substance break down into simpler forms. Most of the many substances known to man can undergo decompositions that involve net energy changes up to about 2×10^3 cal/g (released) or 3×10^4 cal/g (absorbed) and which yield two or more decomposition products. A relatively few substances (somewhat over one hundred) do not decompose at all within these ranges of energy change or, if they do, give only one ultimate product (for example, ozone → oxygen). Such substances are considered to be the stuff of which all other substances are made, and are called **elements.** The fundamental unit of the element is the **atom.**

A nonelemental pure substance is called a **compound substance** or a **compound.**

Electrically neutral individual particles of ordinary matter whose atoms are linked together by chemical bonds are called **molecules.** Individual unbonded atoms are also considered to be molecules; for example, the molecules in helium gas are individual atoms. There is no absolute upper limit of size for molecules. Viruses, which reach dimensions of hundreds of Ångström units,* are sometimes called "giant molecules." Molecules of substances that are gaseous in ordinary terrestrial environments are usually less than 10 Å; small molecules like those of water and hydrogen chloride are around 2 to 4 Å.

Electrically charged atoms or groups of atoms are called **ions.** Positive ions are **cations,** negative ones **anions.** The smallest ions are individual charged atoms (for example, sodium ion, Na^+, or fluoride ion, F^-). Ions may also be groups of relatively few atoms, such as sulfate ion, SO_4^{2-}. Scientists concerned with air use the term "atmospheric ions" to denote charged dust particles. Ions can be arranged in some closely packed pattern, as in a crystal. In such cases, it is sometimes conventional to designate as a molecule the smallest group of ions

* One Ångström unit = 10^{-8} cm; see Appendix 1.

whose charges just neutralize each other—for example, a molecule of calcium chloride consisting of one Ca^{2+} and two Cl^- ions. We shall, however, use the word "molecule" to denote only individual entities as described above.

1.3 □ THE PURITY OF COMPOUNDS

A pure substance can be defined as one with a precisely fixed composition that is associated with its properties. We usually think of it as being entirely composed of like molecules. But the situation is really not so simple. When a given constituent has been separated from a mixture, and no further efforts succeed in separating that constituent into additional components, we must *assume* that it is pure. As far as laboratory operations are concerned, there is no way of showing that the assumption is false except by developing a more effective method to separate the "pure" substance into new components. In laboratory practice, then, the criterion of purity of a substance is simply the inability of the experimenter to isolate or otherwise detect foreign material.

Purification involves physical separation, which occurs as a consequence of the differences in properties among the components to be separated. For example, bricks and grains of sand differ from each other in size; a sieve will retain the bricks, and allow the sand to fall through under the influence of gravity. Materials may differ from each other in many attributes other than particle size—electrical, magnetic, and solubility properties are among those which frequently serve as bases for purification procedures. The important thing to remember is that the chemical investigator does not take the viewpoint that purification ceases when the compound is pure, rather that the compound must be called pure when purification ceases.

1.4 □ CHEMICAL SYMBOLS, FORMULAS, AND EQUATIONS

Atoms or elements are denoted by symbols of one or two letters, like H, U, W, Ba, At, and Zn. (See inside back cover for names.)

Compounds or molecules are represented by formulas that consist of symbols and subscripts, sometimes with parentheses. The subscript denotes the number of atoms represented by the symbols to which it is attached. Thus $(COOH)_2$ is a formula that represents a molecule of oxalic acid or the substance oxalic acid. The molecule consists of 2 atoms each of carbon and hydrogen, and 4 atoms of oxygen. The substance consists of matter that is an aggregate of such molecules. The formula for phosphorus vapor is P_4; this tells us that the molecules consist of 4 atoms each. Sometimes formulas are written so as to give other information in addition to the numbers or ratios of atoms; such representations will be taken up in later chapters. In the nineteenth century it was customary to use dots to separate portions of a formula that correspond to formulas of simpler substances. For example, calcium sulfate, $CaSO_4$, was written as $CaO \cdot SO_3$. Today such usage persists in some formulas of substances, called **hydrates,** that produce water when they decompose, like $CuSO_4 \cdot 5H_2O$. We prefer to write $CuSO_4(H_2O)_5$.

Chemical transformations are represented by chemical equations, which tell us what molecules or substances react and what ones are produced, and in what molecular ratios. The equation for the burning of methane in oxygen to produce carbon dioxide and water is

$$CH_4 + 2O_2 \longrightarrow CO_2 + 2H_2O$$

Each coefficient applies to the entire formula that follows it. Thus $2H_2O$ means $2(H_2O)$. This gives the following molecular ratios: reacting materials, 2 molecules of oxygen to 1 of methane; products, 2 molecules of water to 1 of carbon dioxide. The equation written above is balanced because the same number and kinds of atoms, one of carbon and four each of hydrogen and oxygen, appear on each side of the arrow. The balancing of chemical equations is taken up in more detail in Chapters 5 and 15.

PROBLEMS

1. Chemical and physical change Identify the chemical and the physical changes in the following sequences.

(a) Carrots for an ulcer patient are scraped, cooked, mashed, sieved, swallowed, digested, absorbed, and utilized metabolically for coenzyme manufacture.

(b) Oxygen is liquefied, poured into rocket storage tanks, used as a fuel component during flight, used to pressurize the capsule, inhaled by the astronauts, and converted during respiration to CO_2 and H_2O.

(c) A lump of sugar is ground to a powder and then heated in air. It melts, then darkens, and finally bursts into flame and burns completely.

2. Definitions of terms Discuss the inadequacies and limitations of the following definitions. Supply better ones. (a) A molecule is the smallest particle of matter that retains the specific properties of the matter. (b) An element is a substance that cannot be decomposed by chemical means. (c) An ion is any electrically charged body. (d) A compound is a substance that contains more than one element.

3. Purification Black gunpowder consists of sulfur, charcoal, and potassium nitrate. Sulfur dissolves in the volatile solvent carbon bisulfide; potassium nitrate dissolves in water; charcoal is insoluble in both. Outline a procedure for separation and isolation of the components, and suggest methods to identify them.

4. Purification For each of the following purification methods, point out what physical separation is occurring, and what differences in the properties of the substances involved make the separation possible.

(a) Scrap iron is recovered from junk with the aid of a magnet.
(b) Moving ionized material is deflected in a magnetic field; the lighter particles deflect to the greater extent. The separation is recorded on a photographic plate.
(c) A perfume chemist places a drop of an essential oil on a strip of blotting paper, which he leaves open to evaporation. He sniffs the paper each hour, until a constant, characteristic odor is noted.

5. Purity of matter In 1868 a sample of water, after treatment by distillation and other methods, resisted further attempts at purification and was called "pure." Techniques available in 1968 made it possible to separate the "pure" sample into several components, including protium deuterium oxide and dideuterium oxide. Has the purity of the unchanged sample deteriorated because of advance in technique? Was the sample pure in 1868? Is it possible for future advances in methods to show that components separated today are "impure"?

6. Chemical symbols, formulas For the following formulas, give the number of atoms of each element found in a molecule of the compound; give the total number of atoms in one molecule of the compound: $(NH_4)_2HPO_4$; $Al(OH)(CH_3CO_2)_2$; $K_3Fe(C_2O_4)_3(H_2O)_3$.

The following problems are related to material in Appendix 1.

7. Matter and energy Explain the fallacy in each of the following statements.

(a) Smoke pollution from power plants could be profitably controlled by collecting all the stack emissions and chemically reconverting them to fuel.
(b) A window air conditioner will also provide cooling if it is placed inside the room with the windows and doors closed.
(c) Energy used to pulverize coal is lost when the coal burns.

8. Significant figures How many significant figures are there in each of the following quantities? (a) 6.0342 g; (b) 2.00×10^{-3} ml; (c) 1000 dollars; (d) 0.002°C; (e) 62.9%.

9. Units; significant figures It has been said that a number as large as 10^{100} is never needed to express any magnitude that has physical significance. What is the ratio between the apparent mass of a neutron star, estimated density 6×10^{13} g/cm^3 and diameter 1.6×10^4 m, and that of the electron, approximate rest mass 9.11×10^{-28} g?

10. Density Copper pellets are poured into a measuring cylinder up to the 50-ml mark. The cylinder is then filled with water to the same mark. Data are

Weight of empty cylinder	104 g
Weight of cylinder + copper	371 g
Weight of cylinder + copper + water	391 g
Temperature of system	20°C
Density of water at 20°C	0.998 g/ml

Calculate the density of copper at 20°C.

11. Pressure The density of mercury at 0°C is 13.595 g/cm^3. Prove that 1 torr = pressure exerted by a mass of 1.3595 g on an area of 1 cm^2.

ANSWER 10. 8.9 g/ml.

2 ■ ELECTRONIC STRUCTURES OF ATOMS

2.1 □ INTRODUCTION

Our present concepts concerning the nature of the atom and its role as the unit of chemical transformations may be traced to three groups of sources:

(a) The speculations of the Greek physical philosophers Democritus (fifth century B.C.), adopting his ideas in part from the doctrines of Leucippus, reasoned that things consist of atoms that are eternal, absolutely small so that they cannot be further diminished (*A-tomos*, indivisible), absolutely full and incompressible, and all identical in quality but differing in shape—the atoms of water, for example, being smooth and globular and able to roll over each other, and the atoms of iron being rough and jagged, clinging together to form a solid body.

(b) The assumptions of John Dalton (1803) and others who conceived that atoms were the fundamental units that uniquely characterized the elements Thus, for example, the element tin consists only of tin atoms. Chemical changes were understood to involve only the rearrangements of linkages between atoms, not their creation, destruction, or other alteration. These "Daltonian atoms" were assumed to be ultimate and immutable particles—Dalton himself thought of them as spheres—whose intimate structures would defy any examination.

(c) Experiments on electricity that began about 150 years ago and culminated at the turn of the century in the discovery of x rays and radioactivity The spontaneous disintegration of naturally radioactive atoms into smaller particles contradicted the Daltonian hypothesis that atoms are unalterable. In turn, these discoveries inaugurated the nuclear theory of the atom, and reaffirmed the atom as the unit of chemical changes.

We therefore direct our attention first to some fundamentals concerning the nature of electricity.

2.2 □ QUANTIZATION OF ELECTRICITY

When a property of matter exists in discrete (separate, distinct) amounts, we say the property is "quantized." For example, in accord with the atomic theory of matter, mass is quantized. Solid iron, liquid water, gaseous hydrogen are not continuous; they are composed of atoms or molecules and therefore are quantized.

Electrolysis That matter is electrical in nature is evidenced by the fact that electricity can effect chemical change, a process called **electrolysis.** Michael Faraday (1833) showed that a relationship exists between the mass of products produced in an electrolysis and the quantity of electricity that passes through the sample. The practical unit of charge or quantity of electricity is the coulomb (C); one coulomb is the amount of charge obtained from a steady current of one ampere (A) in one second (sec).

From these results, George Johnstone Stoney in 1874 concluded that electricity, like matter, is atomic in nature and consists of particles. He called the single particle of electricity the **electron.** Various experiments based on the work of Faraday yield 1.60×10^{-19} C or 4.80×10^{-10} electrostatic unit (esu) for the electronic charge.

Discharge Tubes The passage of the electric current through a gas is called an electric discharge. This phenomenon was concurrently studied by Faraday, Johann W. Hittorf, Sir William Crookes, and many other investigators. The results of these studies are best explained in terms of the electron particle suggested by Stoney.

A sealed glass tube with two electrodes attached to a high-voltage source is a discharge tube (see Fig. 2.1). The negative electrode is called the **cathode;** the positive one is the **anode.*** When the gas pressure in the tube is reduced to about 10 torr, the gas becomes a conductor, current flows, and the gas fluoresces (emits light); when the pressure is reduced below about 10^{-3} torr, emission of light by the gas ceases. The current, however, still flows between the electrodes and the glass at the anode end of the tube fluoresces. If this end of the tube is coated with a luminescent material, such as impure zinc sulfide, it glows brilliantly. It thus appears that a radiation capable of inducing

* The subject of electrical conductance and electrolysis is more extensively covered in Chapter 15.

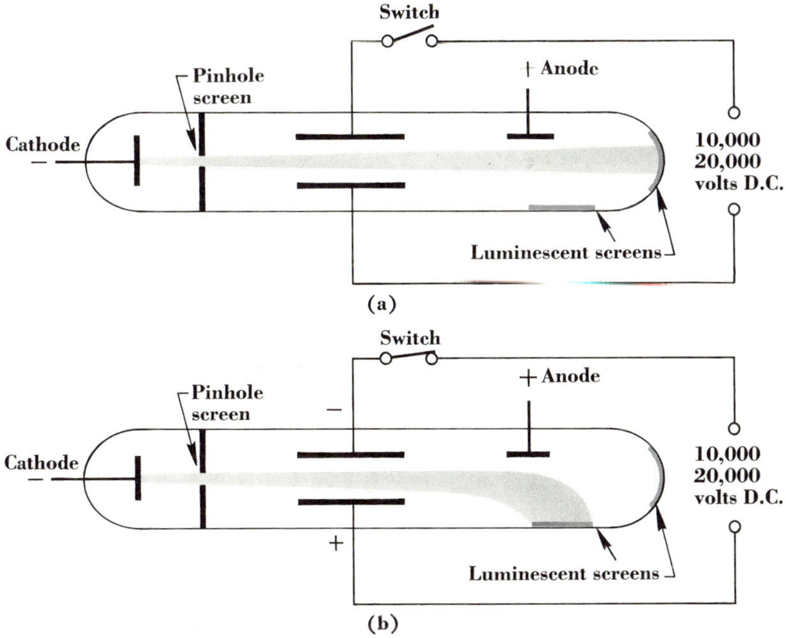

Fig. 2.1 A discharge tube. Experiment showing that the cathode rays possess a negative charge: (a) position of ray in the absence of electric field; (b) position of ray in an electric field. The pressure in the tube is about 10^{-3} torr.

luminescence is emitted from the cathode; this radiation, because of its origin, is named the **"cathode ray."**

These rays possess momentum, a characteristic property of matter in motion, and they appear also to possess mass. They are deflected in magnetic and electric fields in a manner predicted for negatively charged particles (see Fig. 2.1). They also darken a photographic plate. Their properties are independent of the nature of the residual gas in the tube. These experiments demonstrated the particle character of the cathode rays. Moreover, Sir J. J. Thomson in 1897 showed the particles to be electrons by measuring the ratio of the charge of the particle to its mass, symbolized by e/m. The currently accepted value of e/m is 1.76×10^8 C/g. From the electronic charge calculated from Faraday's experiments and other data, and later confirmed and greatly refined by Robert Millikan, it is possible to calculate the (rest) mass of the electron on the following basis:

$$\frac{e}{e/m} = m$$

The currently accepted value is 9.11×10^{-28} g per electron.

The ratio e/m is independent of the nature of the gas in the discharge tube and of the materials of which the tube is composed. Also, electrons from other sources, such as a hot cathode filament or photoelectrons, yield the same ratio. This, with the observation that the mass of the electron is about 2×10^3 times smaller than the mass of the lightest atom known, implies that the electron is a universal constituent of matter.

Originally referred to as "rays" because they were believed to be light waves, cathode rays would be more accurately named "electron beams." Electromagnetic radiation (light), in contrast, is not deflected in electric or magnetic fields.

Wilhelm Röntgen, while performing experiments on the fluorescence produced by cathode rays, discovered in 1895 a radiation that penetrates glass and metal plates opaque to light and cathode rays. This radiation, named the **x ray,** is produced when high-speed electrons strike an object. X rays darken a photographic plate, exhibit the phenomena of diffraction, refraction, and reflection, but are not deflected in electric or magnetic fields. They are therefore electromagnetic waves of the same nature as light, radio waves, and ultraviolet radiation. Their wavelengths are of the order of 1 Å, compared to $4\text{–}7 \times 10^3$ Å for visible light. In addition, they cause luminescence and ionize gases. (The conversion of neutral particles into ions is called **ionization.**)

2.3 □ THE POSITIVE IONS (POSITIVE RAYS)

In the discharge tube, neutral particles—atoms or molecules—are ionized into electrons and positive ions. These charged particles acquire kinetic energy under the influence of the electric field. Many of the positive ions collide with the cathode. Evidently, if a perforated cathode is used, some positive ions will emerge behind it. Eugen Goldstein performed such an experiment in 1886 and detected what he called the positive "rays." Their behavior in electric and magnetic fields conforms to that predicted for positively charged particles but, unlike the cathode ray, the nature of the positive ions is characteristic of the gas in the tube. The mass of the positive ion depends on the atomic or molecular weight of the gas in the tube. Of all the positive ions whose masses have been determined, the *lightest ion found* is obtained from hydrogen. This ion, named the **proton,** designated H^+, bears a charge equal to that of an electron, but opposite in sign. The removal of two electrons from a neutral helium atom produces an **α (alpha) particle,** He^{2+}.

2.4 □ THE RUTHERFORD-BOHR NUCLEAR THEORY OF THE ATOM

The existence of nuclei within atoms was suggested by Lord Rutherford in 1911 to explain his experimental results on the scattering of alpha particles (see below), and confirmed by Niels Bohr in 1913 in his attempt to explain the origin of spectral lines (p. 19). These ideas, however, evolved over a period of years in which other scientists contributed significantly.

The problem was viewed in the following terms: If matter is electrical in nature and if the atom is the unit of matter, how is electrical charge distributed in the atom? Are the electrons, which are negative, distributed in a uniform sphere of positive electricity? If so, a heavy positive particle traveling at high velocity through such atoms should be repelled equally in all directions by the positive charge surrounding it, and thus the repulsive forces should cancel each other and the particle should proceed in a straight line.

To carry out such an experiment, Lord Rutherford, Hans Geiger, and Ernest Marsden, in 1909, allowed a beam of α particles, which have masses over 7000 times that of electrons, to impinge upon a thin metal foil (Fig. 2.2). Practically all the particles passed through the film without deviation from the original path but, much to the amazement of the experimenters, a small fraction of the particles was deflected through large angles; a minute fraction, in fact, was almost completely reversed in direction. In Lord Rutherford's own words, "It was almost as incredible as if you fired a 15-inch shell at a piece of tissue paper and it came back and hit you."

To explain these results, Rutherford in 1911 assumed that the atom must be almost completely empty space; this would account for the passage of most of the particles in a straight line through the foil. To account for the repulsive force required to produce large deflections, he further assumed that the positive electricity of the atom, and therefore the mass of the atom, must be concentrated in a very small volume, called the **nucleus** of the atom. When an α particle approaches a nucleus, it will be repelled, the angle of deflection being dependent on

Fig. 2.2 Alpha scattering.

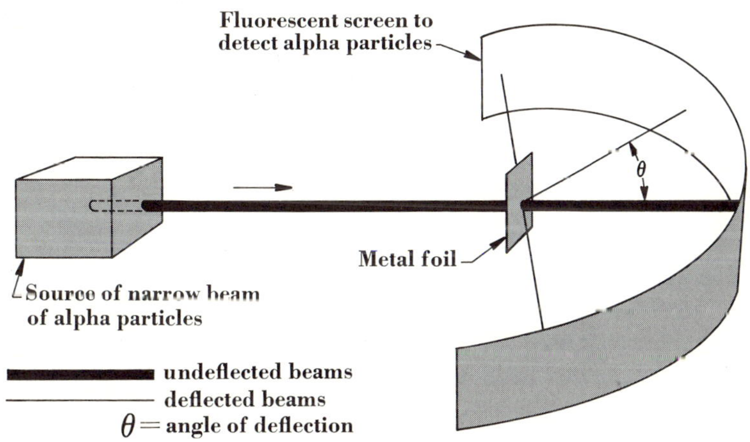

the closeness of approach to the nucleus. Since the atom is electrically neutral, electrons, sufficient in number to equalize the positive charge of the nucleus, reside around the nucleus and determine the effective volume of the atom.

Assuming a nuclear type of atom and the applicability of Coulomb's law,* Rutherford calculated a maximum value for the diameter of the nucleus, $\approx 10^{-13}$ cm, and found that the *number of unit positive charges on a nucleus* is numerically equal to the **atomic number,** the ordinal number of the atom in the periodic classification (see Chapter 3). Measurements of the nuclear charge from alpha particle deflection experiments by Geiger and Marsden, and later by Sir James Chadwick, verified this prediction.

Composition of the Nucleus A hydrogen atom, atomic number 1, consists of a proton, charge +1, as its nucleus, surrounded by one electron, charge −1. It is therefore inferred that the proton is a common constituent of atomic nuclei, sufficient in number to give a nucleus its characteristic positive charge. For example, the nucleus of any atom of neon, atomic number 10, contains 10 protons regardless of the mass of the atom.

These protons account for only about one-half the mass of a neon atom. The currently accepted view attributes the remainder of the mass to the presence of neutrons in the nucleus.† The neutron, an electrically neutral particle with a mass very slightly greater than that of a proton, was discovered in 1932 by Chadwick. The number of neutrons in a nucleus, N, together with the number of protons, Z, make up the **mass number** of the atom, A.

$$A \text{ (mass number)} = N \text{ (number of neutrons)} + Z \text{ (atomic number, or number of protons)}$$

An **element** is a substance all of whose atoms have the same atomic number, Z. Atoms of the same element may have different numbers of neutrons, and hence different mass numbers; such atoms are called **isotopes.** Naturally occurring hydrogen is a mixture of three isotopes: protium, deuterium, and tritium:

Isotope	Symbol	Z	$+ N$	$= A$
Protium	H	1	0	1
Deuterium	D	1	1	2
Tritium	T	1	2	3

* See Appendix 1.

† There is, however, a slight, but significant, difference between the mass of an atom and the sum of the masses of its constituent particles (Chapter 24).

When it is unnecessary to distinguish between protons and neutrons, they are referred to as **nucleons.** The mass number thus represents the total number of nucleons in the nucleus.

A nucleus having a particular number of protons and neutrons is called a **nuclide.** An example is $^{14}_{6}C$, where the subscript is the atomic number, the superscript the mass number, and the letter the symbol for the element.

2.5 □ THE NATURE OF LIGHT

The discussion of the electronic structure of an atom is so closely interwoven with an understanding of the nature of light that we shall pause here for a brief review of the nature and propagation of light.

James Clerk Maxwell in 1864 predicted that an alternating current of high frequency in a circuit would radiate energy in the form of waves —"electromagnetic waves"—traveling into space with the speed of light. A **wave,** by definition, *transmits energy without the transmission of matter.** This prediction led to the development of wireless telegraphy and the radio. The number of waves passing a given point in unit time is known as the **frequency** of the wave, ν; the unit of frequency is sec^{-1}. The distance between two crests of consecutive waves is called the **wavelength,** λ, illustrated in Fig. 2.3; the unit of wavelength is the centimeter. The speed of the wave, c, usually expressed in centimeters per second, is given by the equation

$$c = \lambda\nu$$

The quantity $+a$ or $-a$ is called the **amplitude** of the wave. The **intensity** of a wave is related to the amplitude; for a given frequency, the larger the amplitude, the greater the intensity.

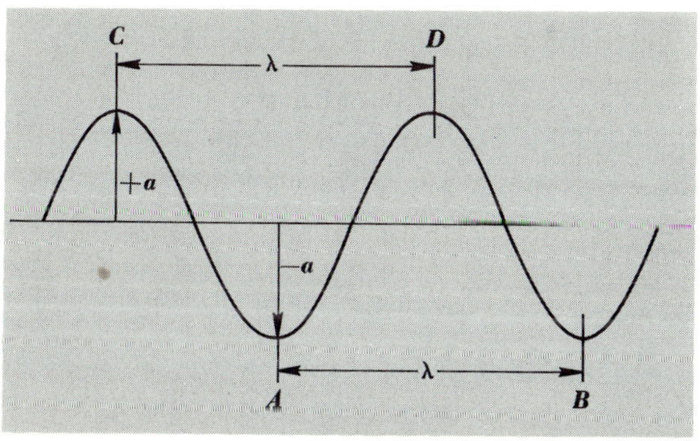

Fig. 2.3 Illustrating the wavelength and the amplitude, $+a$ or $-a$, of the wave set up in a vibrating wire. The distance between two consecutive troughs AB is the same as the distance between two consecutive crests CD.

* Although the word "wave" connotes a visible shape like that of a water wave or a vibrating wire (Fig. 2.3), the generalized concept of wave motion is simply that of energy in pulsating or periodically varying form. No material form or shape or sequence of waves need be imagined.

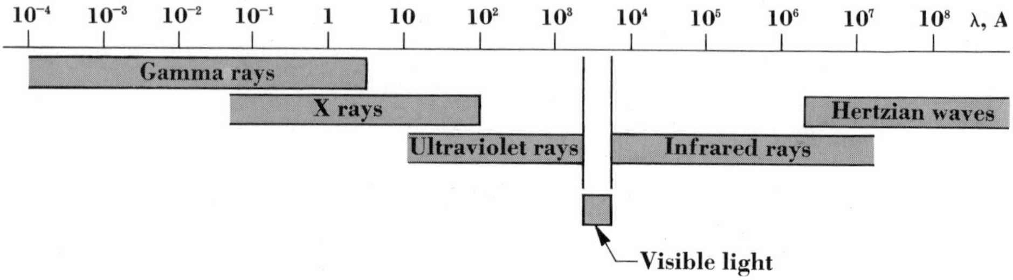

Fig. 2.4 Electromagnetic spectrum

Gamma rays	**(radiation from atomic nuclei)**
X rays	**(radiation from electrons striking a target)**
Ultraviolet rays	**(radiation from arcs and gas discharge)**
Visible light	**(radiation from stars, hot objects, or gases)**
Infrared rays	**(radiation from warm objects)**
Hertzian waves	**(radiation from electric currents: radio waves, television, microwaves, radar)**

Light is radiant energy that produces a visual effect. Light waves are regarded as electromagnetic waves. Electromagnetic waves emitted by a radio transmitter, by a hot tungsten wire (electric light bulb), or by a fluorescent lamp differ in frequency, not in speed. A portion of the electromagnetic wavelength range is shown in Fig. 2.4. Light is diffracted when it passes through a hole that is not too large compared to its wavelength. The resulting pattern, seen on a screen, consists of a series of dark and bright rings. This experiment shows that light is *propagated* as a wave motion (see Fig. 2.5).

Although the electromagnetic wave theory of light is successful in explaining the results dealing with the propagation of light, many phenomena can be explained only by attributing particle properties to light. Some of these properties are described in the following paragraphs.

(1) Light or other electromagnetic radiation falling upon a body exerts a pressure in the same direction as the motion of the radiation. The pressure produced by a gas is attributed to the collisions of its molecules with a body. Similarly, the radiation pressure may be attributed to "atoms of light."* The "atom of light" was named the **photon** in 1926 by the chemist Gilbert N. Lewis, who wrote, "I therefore take the liberty of proposing for this hypothetical new atom, which is not light but plays an essential part in every process of radiation, the name photon."

(2) In the late nineteenth century, many scientists were involved in the study of the distribution of the wavelengths and the intensities (the energies) of the light waves emitted by hot bodies, such as a hot tungsten filament. Science, at that time, recognized the quantization

* Although this explanation is simpler, Maxwell's wave theory also predicts the existence of a radiation pressure.

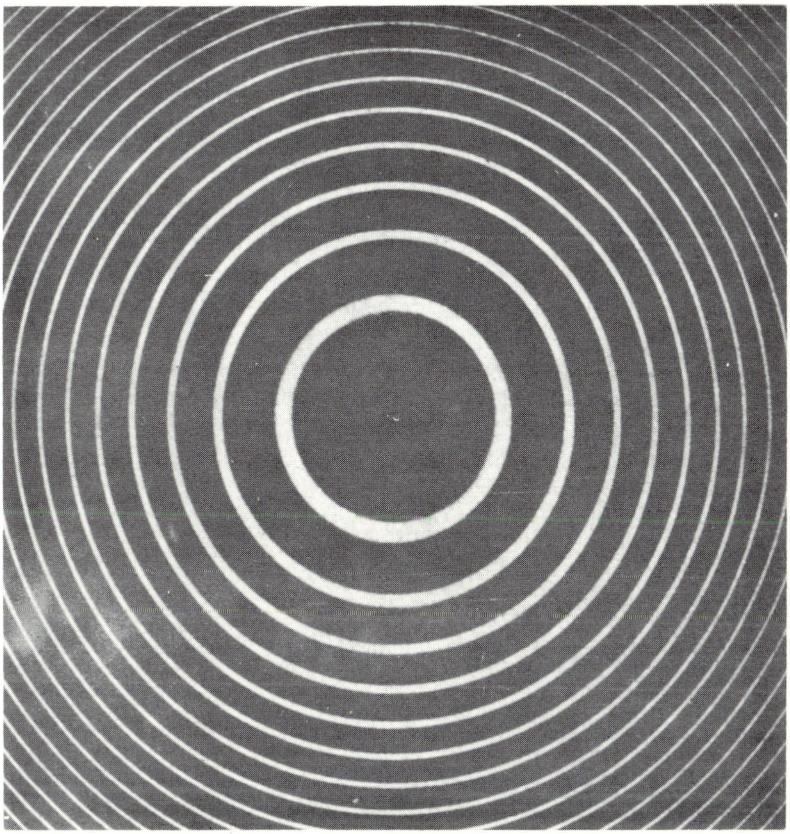

(a)

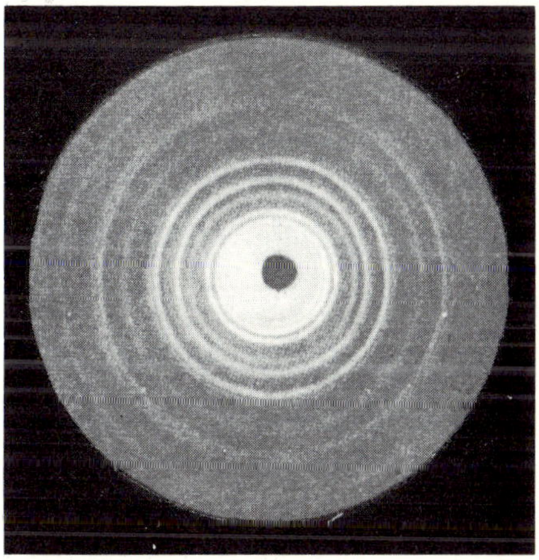

(b)

Fig. 2.5 Diffraction patterns showing the wave nature of electromagnetic waves: (a) visible light; (b) x-rays through a silver foil.

of matter and of electricity, but radiant energy was believed to be continuous. That is, the electromagnetic wave theory assumed that the energy of a radiation may be *any* value, from infinitely small to infinitely large. But all arguments based on this assumption failed miserably to account for the energy distribution found experimentally. To obtain complete agreement with experiment, Max Planck, in 1900, found it necessary to invent what is now known as the **quantum theory.** Namely, in interactions between radiation and matter, radiant energy behaves as if it is composed of "atoms" or "quanta." Further, the energy of a quantum, E, is given by

$$E = h\nu = \frac{hc}{\lambda} \tag{1}$$

where ν is the frequency of the radiation absorbed or emitted, and h is a constant, 6.625×10^{-27} erg sec per particle, known as Planck's constant. Finally, only a whole number of quanta may be absorbed or emitted by a body. As previously mentioned, the "quantum" has been renamed the "photon."

In terms of the quantum theory, the intensity of electromagnetic radiation is determined by the number of photons striking a unit area in unit time.

(3) When light falls upon a metal, electrons (**photoelectrons**) are ejected from its surface. Classical physics predicts that as the intensity of light of a given single frequency (**monochromatic light**) is increased, the kinetic energy of the ejected electron should increase. However, experiments show that the maximum kinetic energy of the ejected electrons is *independent* of the intensity, and is dependent only on the frequency of the monochromatic light. These results were explained by Einstein in terms of the quantum theory of radiation: the energies of all photons for a given frequency, ν, are identical. Each photon has the energy $h\nu$. For each photon absorbed, one electron is ejected. Then, the energy, E_s, required to remove the electron from the surface plus the kinetic energy of the ejected electron must equal the energy of the photon,

$$h\nu = \tfrac{1}{2}mv^2 + E_s$$

Here, m is the mass of an electron, and v is the measured maximum velocity. But, increasing the intensity of the light increases the number of photons and *not the energy* of an individual photon. Then, the kinetic energy of ejected electrons should remain constant but the number of electrons ejected should increase. This conclusion is in perfect agreement with experimental results.

Thus, while the wave character of radiation (electromagnetic waves) is verified by experiments involving diffraction and interference phenomena, the particle character of radiation is verified by experiments involving energy exchanges between radiation and matter. We visualize

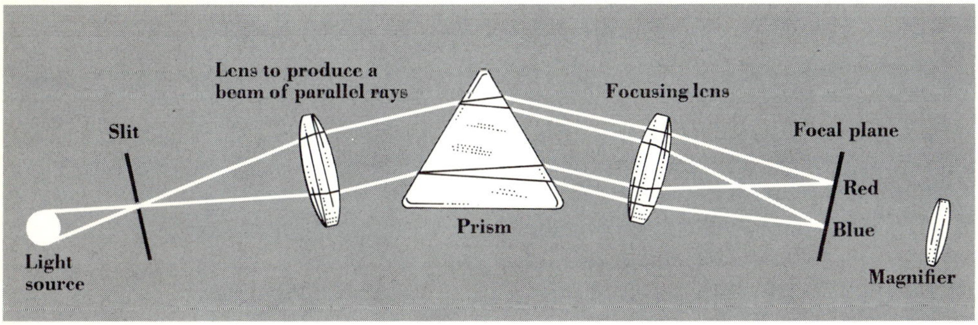

Fig. 2.6 The dispersion of light composed of two wavelengths by a prism is illustrated. The focusing lens and magnifier constitute a telescope.

a beam of light as either a stream of photons or a wave train only for our educational convenience.

2.6 □ SPECTRA OF ELEMENTS

A **spectrum** (Latin, *image*) is a separated group of components arranged in the sequence of their separation. The word usually applies to the separation of electromagnetic energies, which can be analyzed by an instrument called a **spectroscope.** When a beam of parallel rays of light enters the face of a triangular glass prism, it is dispersed into its component wavelengths. This dispersion of light is the basis of the spectroscope. The essential parts of a spectroscope are illustrated in Fig. 2.6.

Solids, liquids, and dense gases become incandescent at high temperatures. If the light is examined with a spectroscope, it will consist of a *continuous band of colors* as observed in a rainbow. An incandescent substance thus emits a **continuous spectrum.**

When the light emitted from a source is analyzed, the spectrum obtained is called an **emission spectrum.** On the other hand, the spectrum obtained *after light from some source has passed through a substance* is known as an **absorption spectrum.**

If the light of a neon lamp is examined with a spectroscope, a **line spectrum** is observed. A rainbow will not appear; rather, several isolated colored lines appear, such as those in Fig. 2.7. If the light from a flame in which a volatile salt is placed is examined in a spectroscope, again, a line spectrum is observed. Very significantly, the line spectrum of an element is characteristic of that element regardless of its source.

Fig. 2.7 Emission spectra in the visible range. (Note: 4101 should read 4102 in the hydrogen spectrum.) [See color plate used, by permission, from *Chemistry* by M. J. Sienko and R. A. Plane (2nd ed.) Copyright. 1961, McGraw-Hill, New York.]

Sodium and its compounds, for example, always yield the same sodium line spectrum, differing from that of all other elements. It is thus the *atoms* of the element that emit the line spectrum. The study of the spectra of elements is therefore an important source of information in the determination of the structure of atoms. The lines in the hydrogen spectrum, for example, can be represented by a relation *involving whole numbers*,

$$\nu = 3.289 \times 10^{15}\ \text{sec}^{-1}\left(\frac{1}{n_1^2} - \frac{1}{n_2^2}\right) \tag{2}$$

where ν is the frequency of the line in sec^{-1}, and $3.289 \times 10^{15}\ \text{sec}^{-1}$ is a constant for hydrogen, known as the Rydberg constant.* n_1 is a dimensionless integer, and n_2 is another dimensionless integer greater than n_1. Thus when $n_1 = 1$, n_2 may be any integer greater than 1 and each value of n_2 yields a frequency, ν, corresponding to a line in the spectrum of hydrogen. When $n_1 = 2$, n_2 is any integer greater than 2, and so on. According to Equation 1 (p. 16), the radiant energy emitted or absorbed per atom depends on the frequency. Thus each spectral line corresponds to a definite energy. For example, the 4102 Å line in the spectrum of hydrogen corresponds to 4.842×10^{-12} erg/atom,

$$E = h\frac{c}{\lambda} = 6.625 \times 10^{-27}\ \frac{\text{erg}\ \cancel{\text{sec}}}{\text{atom}} \times \frac{2.998 \times 10^{10}\ \cancel{\text{cm}}/\cancel{\text{sec}} \times 10^{8}\ \text{Å}/\cancel{\text{cm}}}{4102\ \text{Å}}$$

$$= 4.842 \times 10^{-12}\ \frac{\text{erg}}{\text{atom}}$$

Since only certain definite frequencies of radiation occur in the spectrum (Equation 2), atoms therefore emit or absorb only photons of the corresponding definite energies.

Other atoms have spectra more complicated than that of hydrogen, but their emitted or absorbed photons also correspond only to definite energies.

2.7 □ QUANTIZATION OF THE ENERGY OF AN ELECTRON ASSOCIATED WITH A NUCLEUS

An atom consists of a positively charged nucleus surrounded by electrons within a volume roughly 10^{-8} cm in radius. The arrangement of these electrons, we further believe, determines the physical and chemical properties of the elements.

What prevents the electrons from being attracted into the nucleus? All attempts to answer this question on the basis of classical physics fail. Let us consider the simplest of atomic systems, the hydrogen atom.

* Named for Johannes Robert Rydberg, who discovered this empirical relation in 1890 after a study of a large mass of data.

If the electron were stationary, the coulombic attractive force* would cause it to fall into the nucleus. Rutherford then suggested that the electron revolves about the proton (nucleus) in a manner analogous to the revolution of the moon about the earth. The coulombic attractive force acting on the electron keeps it in circular motion and gives it just the centripetal acceleration needed to keep it from leaving its orbit and flying on in a straight line.

However, the electric charge of the particles that compose the hydrogen atom introduces a new factor. The Maxwell electromagnetic theory of light predicts that the revolving electron will emit radiant energy. The revolving electron should lose energy, resulting in a gradual decrease in its kinetic energy, and should therefore spiral toward the proton, emitting a rainbow of colors and falling finally into the proton. This theory thus incorrectly predicts a *continuous spectrum and a collapsed atom.*

This dilemma was boldly resolved by Niels Bohr in 1913 by applying the concepts of the quantum theory, as exemplified by Equation 1 (p. 16), to Rutherford's nuclear theory of the atom:

(1) In his theory of the hydrogen atom, Bohr assumed that the electron does revolve with a definite fixed energy, without the emission or absorption of energy. He determined the energy of the electron by calculating how much work must be done against the attractive force to remove the electron from its normal position near the proton to a distance so far that the attraction becomes practically zero. The answer is 21.79×10^{-12} erg. An energy value of zero is arbitrarily assigned to the isolated electron. When the electron is near the proton, it has *less* energy than when it is isolated because work must be done to move the electron away from the positive charge to which it is attracted. Therefore the electron near the nucleus is assigned a *negative* energy. Thus, Bohr pictured a normal hydrogen atom as a stationary proton and a revolving electron whose energy is -21.79×10^{-12} erg.

(2) We come now to a crucial question. Can the energy of the electron assume some other value? If so, what are these other values? Bohr postulated that the energy of the electron is quantized and can take only the values given by

$$E = -\frac{21.79 \times 10^{-12}\ \text{erg}}{n^2} \tag{3}$$

in which n is a dimensionless integer, known as the **principal quantum number.** We speak of these energy values as "the principal energy levels" of the electron or of the hydrogen atom. So, the energy levels of the electron (hydrogen atom) are restricted to certain definite values. The first and lowest energy level, when the principal quantum number of the electron is 1, is known as the "normal" or **"ground" state** of the

* See Appendix 1.

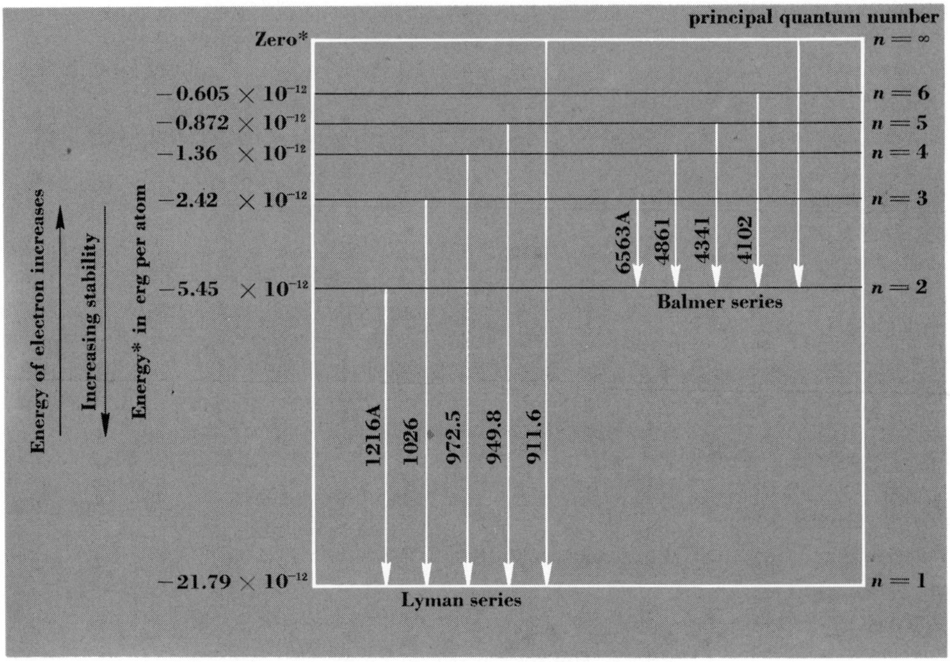

Fig. 2.8 Energy level diagram of the hydrogen atom. Wavelengths are in angstrom units. Compare the predicted lines in the "Balmer series" with the line emission spectrum of hydrogen, Fig. 2.7.

hydrogen atom; all other energy levels are known as **"excited" states.** A convenient method of representing the ground and excited states of the hydrogen atom is illustrated in Fig. 2.8, which shows the energy levels as lines. Such a figure is known as an **energy level diagram.**

(3) Bohr next proposed a mechanism for the absorption and emission of energy. The absorption of radiation will correspond to a transition—"an electron jump"—between any two of the possible energy levels, from a lower to a higher level. The emission of radiation will correspond to a transition between any two of the possible energy levels, but from a higher to a lower level. Thus, in absorption, the energy of the electron is increased and, in emission, the energy of the electron is decreased, but only in accord with the difference between any two energy levels. The energy lost appears as a photon. Since the energy of the electron is restricted (or quantized), *the hydrogen atom can emit or absorb only photons whose energy is equal to the difference between two*

* The reader is reminded that the energy is expressed as the work necessary to separate the electron from the hydrogen atom against the attractive force of the nucleus. Zero energy then means that no work is required to remove the electron when the distance between the proton and the electron is infinite ($n = \infty$).

energy levels. A spectral line thus originates when the electron "drops" from one energy level to a lower level,

$$E \text{ (energy of photon emitted)} = E_{n_2} - E_{n_1} \tag{4}$$

where E_{n_2} and E_{n_1} are the energy levels with principal quantum numbers n_2 and n_1, n_2 being larger than n_1.

(4) Finally, using Equation 4, Bohr predicted the energy of the photons that should be emitted by the hydrogen atom. For an electron transition between any two levels (Fig. 2.8), the difference in the energy of the two levels equals the energy of the radiation absorbed or emitted by a hydrogen atom.

EXAMPLE 1 A particle "strikes" a normal hydrogen atom, exciting the electron to its 6th energy level ($n = 6$); this electron then "drops" to its 2nd energy level ($n = 2$). What is the energy of the photon emitted?

ANSWER According to the Bohr theory, the energy of the photon emitted must be the difference between the higher and the lower energy levels of the electron; hence, from Fig. 2.8,

$$E \text{ (energy of the photon)} = E_6 - E_2$$

$$= (-0.605 \times 10^{-12}) \frac{\text{erg}}{\text{atom}} - (-5.45 \times 10^{-12}) \frac{\text{erg}}{\text{atom}}$$

$$= 4.84 \times 10^{-12} \frac{\text{erg}}{\text{atom}}$$

This photon corresponds to a radiation of wavelength 4102 Å, in perfect agreement with one of the lines (Fig. 2.7) in the spectrum of hydrogen.

It is significant that the "Lyman series" of lines in the ultraviolet range, given in Fig. 2.8, as well as other series of lines, were discovered after Bohr predicted their existence.

The dimensionless integers discovered empirically (Equation 2) are now known as principal quantum numbers (Equation 3). Also, Equation 3 may be shown to be in practically perfect agreement with the empirical equation 2 as follows. According to the Bohr theory,

$$E \text{ (energy of the photon)} = h\nu = E_{n_2} - E_{n_1}$$

$$= -\frac{21.79 \times 10^{-12}}{n_2^2} - \left(\frac{-21.79 \times 10^{-12}}{n_1^2}\right)$$

$$= 21.79 \times 10^{-12} \frac{\text{erg}}{\text{atom}} \left(\frac{1}{n_1^2} - \frac{1}{n_2^2}\right) \tag{5}$$

Now, if we multiply Equation 2 by Planck's constant, h, to convert the frequency of the radiated line to its corresponding energy,

$$E \text{ (energy of the radiated line)} = h\nu$$
$$= 6.625 \times 10^{-27} \frac{\text{erg } \cancel{\text{sec}}}{\text{atom}} \times 3.289 \times 10^{15} \, \cancel{\text{sec}}^{-1} \left(\frac{1}{n_1^2} - \frac{1}{n_2^2}\right)$$
$$= 21.79 \times 10^{-12} \frac{\text{erg}}{\text{atom}} \left(\frac{1}{n_1^2} - \frac{1}{n_2^2}\right) \quad (6)$$

The agreement between Equations 5 and 6 is indeed remarkable. However, only the general features, and not the quantitative treatment, of the Bohr theory may be extended to atomic systems containing more than one electron.

To recapitulate, the assumptions of the Bohr theory, advanced primarily to explain the origin of spectral lines, are as follows:

(1) Spectral lines are produced by atoms.
(2) The Rutherford (nuclear) model of the atom is acceptable.
(3) The energy of the hydrogen atom is quantized and is determined by the square of an integer, the quantum number n.
(4) In any one energy level, the electron does not radiate or absorb energy.
(5) An electron transition between two energy levels, E_1 and E_2, determines the energy absorbed or emitted by a hydrogen atom.
(6) The Planck equation, $E = h\nu$, is valid for the absorption or the emission of the radiation.
(7) The frequency of the radiation absorbed or radiated is determined only by the energy difference of the two levels between which an electron "jumps" or "drops": $h\nu = E_2 - E_1$.

2.8 □ MODIFICATION OF THE BOHR THEORY

The extraordinary success of Bohr's celebrated atomic theory in explaining the origin of the spectra of hydrogenlike particles does not, of course, make it infallible. In fact, the agreement with experimental results may be regarded as a special case, and the Bohr theory has been superseded by a radically different concept, as discussed in the ensuing sections.

Thus, the Bohr theory predicts that one and only one spectral line can originate between any two principal energy levels. For example (see Fig. 2.8), for the transition $n = 3 \rightarrow n = 2$, only one line, 6563 Å, can be radiated. But if a spectroscope of high resolution is used, *two very slightly different lines*, 6562.8 and 6562.7 Å, *are observed* in place of the 6563 Å line. This is a typical and not an exceptional characteristic of atomic spectra. When the light source—the atoms emitting radiation—is placed in a strong magnetic field, each spectral line is further split into a number of spectral lines. To account for these additional spectral lines, it was assumed that each principal energy level was subdivided into a number of sublevels, some of which possess slightly different energies. Increasing the number of energy levels increases the number of spectral lines that may be radiated.

2.9 □ THE ELECTRON—HOW SHALL WE THINK ABOUT IT?

Thus far, we have discussed the electron as a particle of matter. However, Louis de Broglie in 1923 reasoned that since radiation possesses attributes both of waves and of particles, the two sets of properties must be associated with each other. Thus particles, including electrons, may also show wave characteristics, such as diffraction. This hypothesis was fully verified by experiment; the electron microscope is a practical application of the wave properties of electrons.

The ensuing decade was a period of great intellectual excitement and ferment regarding our concepts about the electron. In 1926 Erwin Schrödinger began to publish his papers on wave mechanics, in which the electron was treated not as a particle but as a mathematical function of the distribution of negative charge in space. The energy of the electron depends on the particular spatial distribution of charge that the electron assumes. The charge density function is quantized and the energy values that the electron can assume are correspondingly quantized. This concept of electron charge density leads to the abandonment of the idea of electrons moving in fixed orbits. What constraints do such mathematical ideas impose on our visualization of the electron? For one possibility, we might well choose not to visualize it at all, but to accept the electron as a mathematical wave function whose implications concerning the behavior of matter can supply a conceptual framework for our experiments. Most of us are not satisfied with such a choice, however, because we cannot imagine it. Instead, we "see" the distribution of charge as a cloud in space, denser in one region, more tenuous in another.

During this period (the 1920's) the group of scientists that included Max Born, Werner Heisenberg, and others, in view of the success of experiments on atomic and molecular collisions, could not bring themselves to "abolish" particles. In Born's own words, "A way had to be found for reconciling particles and waves. I saw the connecting link in the idea of probability." By this concept, an electron is indeed a particle, but we cannot specify both its position and velocity with certainty at the same time. The best we can do, referring to any given location with respect to the nucleus, is to specify the probability that the electron may occupy that location at any given time. If we specify such probabilities for all locations, we will engender a probability distribution in space that is mathematically identical with Schrödinger's wave function. Some of us may prefer to think of a "probability cloud" rather than a "charge density cloud." In any case (a) the position and velocity of the electron cannot be specified at any given time, (b) the distribution of the electron can be represented as a cloud with shape and density in space, (c) the possible energy values of the electron are quantized, and (d) the lowest possible energy of the electron decreases as the number of atoms with which it is associated increases.

The last point is of great significance for chemists, and its consequences will become apparent from time to time during our study. Consider one electron whose oscillations are localized in the vicinity of a single atom. By contrast, consider another electron oscillating in the vicinity of, say, four such atoms. (Chemists describe such an electron's relative freedom from constraint by saying that it is "delocalized.") The delocalized electron has a lower energy than the localized one. A lower energy content of a system implies a greater stability. Thus, we say that the delocalization of an electron enhances the stability of the chemical system of which it is a part. We shall refer to this concept in our studies of the chemical bond, the acidity of compounds, and the metallic state.

2.10 □ THE DISTRIBUTION OF ELECTRONS IN QUANTUM LEVELS

The distribution of the electron cloud in space is described by a mathematical function which satisfies a certain differential equation, the **wave equation.** This function is called a **wave function.** A mathematical solution of the wave equation in simple form has been obtained only for the simplest chemical system—the hydrogen atom. The permissible energies of the single electron in the hydrogen atom can be characterized in terms of a system of quantum numbers, by a procedure that involves only the following assumptions: (1) The wave equation is applicable; (2) the energy of the hydrogen atom is the sum of the kinetic and potential energies of the electron and of the proton; (3) the law describing the forces between charged macrobodies is applicable to electrons and protons; and (4) the special theory of relativity is applicable.

The permissible states of the electron in hydrogen are described by wave functions that are called **atomic orbitals.**

Let us now consider atoms other than hydrogen, which may have up to about 100 electrons. Each electron is repelled by all the others, as well as attracted to the nucleus; this leads to a very complex situation. We make the simplifying assumption, however, that each electron is affected only by the average (not the instantaneous) conditions of the others; this assumption has the effect of treating each electron approximately as though it were the lone electron of a hydrogen atom. In this way we assign all the electrons to orbitals that are qualitatively similar to the orbitals in hydrogen.

The energies of atomic orbitals are determined by the quantum numbers which identify them. In the absence of an external magnetic field, the orbitals are classified as occurring in **shells** and **subshells.** All of the orbitals in a given subshell have the same energy. In the presence of an external magnetic field, the orbitals within a given subshell acquire different energy values.

We now ask a crucial question. How many electrons in the same atom may occupy a given orbital? The answer is provided by the famous hypothesis known as the **Pauli* exclusion principle,** which tells us that as a consequence of the fact that each electron in an atom can be uniquely described by its quantum numbers, no more that *two* electrons can occupy the same orbital.

We summarize the assignment of quantized energy levels of electrons in atoms as follows:

Principal quantum levels (also called **shells**) These are designated by the principal quantum numbers 1,2,3, ... The corresponding shell designations are K, L, M, ...

Subordinate quantum levels or **sublevels** (also called **subshells**) The number of sublevels for each principal level is the same as the quantum number of the principal level. The sublevels are designated s, p, d, and f.† These are shown in Table 2.1.

TABLE 2.1 ▪ QUANTUM LEVELS

Principal levels		Sublevels	
Quantum number, n	Shell	How many?	Designation
1	K	1	s
2	L	2	s
			p
3	M	3	s
			p
			d
4	N	4	s
			p
			d
			f

* Wolfgang Pauli discovered this fundamental law in 1925 from a study of atomic spectra.

† Based on observations of the line spectra of elements such as sodium, certain lines were observed to belong to a "sharp" series (s), these were associated with energy transitions of the s subshell; other lines were classified as belonging to "principal" (p), diffuse (d), or "fundamental" (f) series.

Orbitals The number of orbitals for each sublevel or subshell is

Sublevel	Number of orbitals per sublevel
s	1
p	3
d	5
f	7

Electrons There are 2 at most electrons per orbital. The distribution of electrons in the first four principal quantum levels is summarized in Table 2.2.

TABLE 2.2 DISTRIBUTION OF ELECTRONS IN ORBITALS

Principal quantum level					
Quantum number, n	Shell	Sublevels	Orbitals per sublevel	Max. no. of electrons per orbital	Max. no. of electrons per shell
1	K	s	1	2	2
2	L	s	1	2	2 } 8
		p	3	2	6
3	M	s	1	2	2 } 18
		p	3	2	6
		d	5	2	10
4	N	s	1	2	2 } 32
		p	3	2	6
		d	5	2	10
		f	7	2	14

2.11 □ THE ENERGIES OF THE QUANTUM LEVELS

The energies of electrons in principal quantum levels for hydrogen were given by Equation 3 (p. 19) and in Fig. 2.8 (p. 20). The relative energies corresponding to the principal quantum levels of an electron in any atom, which are approximations based on the assumptions discussed in Section 2.10 (p. 24), are shown in Fig. 2.9. The energies of the subshells of a given shell (except K) are not all the same. Relative subshell energies for the first four shells are shown in Fig. 2.10. Note that as a result of the differences in energy among the subshells, the $4s$ level is lower than the $3d$ level, even though its principal quantum number is higher. Figure 2.11 shows the progression in energy levels of the subshells.

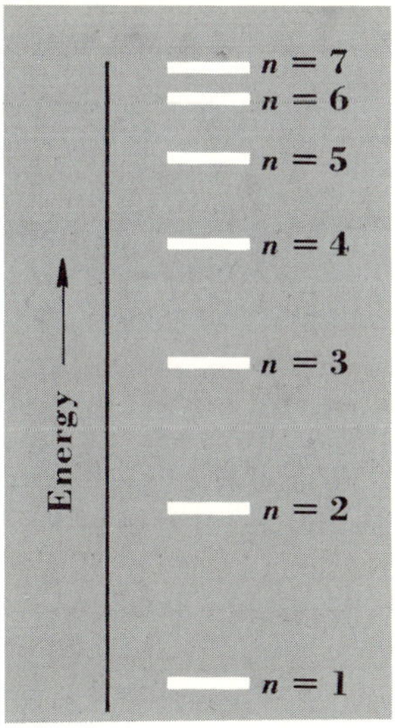

Fig. 2.9 Principal energy levels of electrons in an atom.

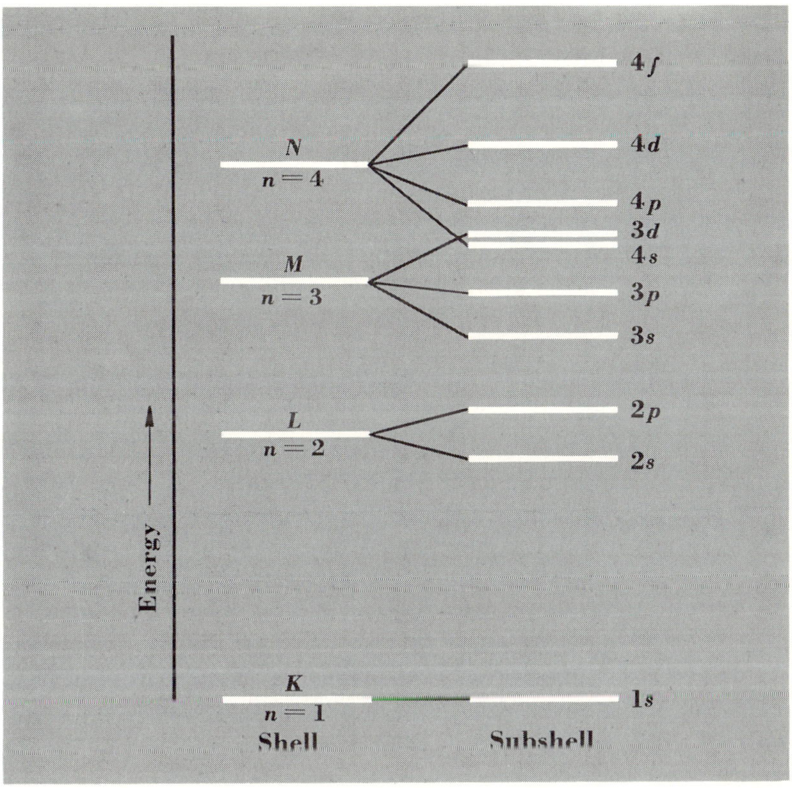

Fig. 2.10 Energy levels of electrons in the principal shells and subshells.

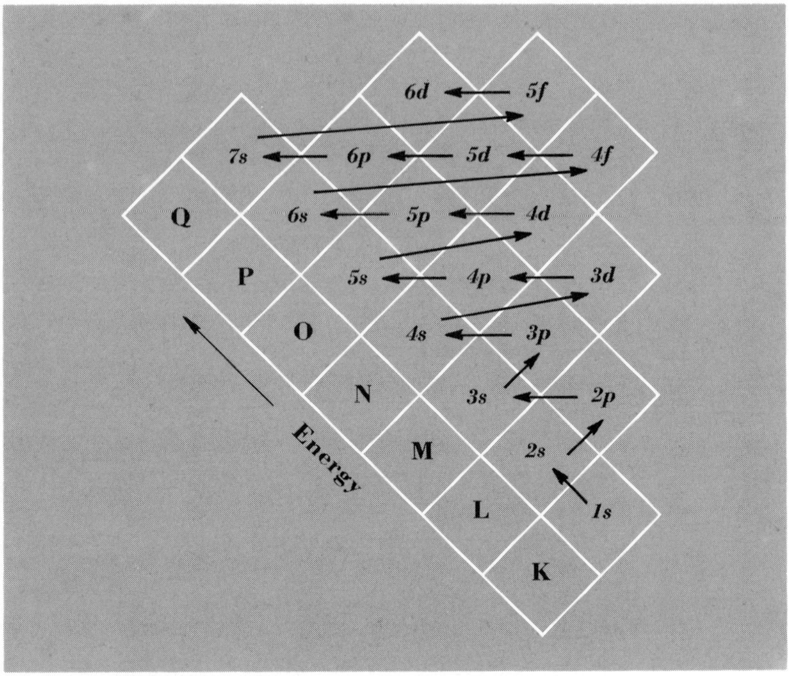

Fig. 2.11 Sequence of energy levels of the subshells.

2.12 ▫ ELECTRON SPIN

It was mentioned on p. 22 that when atoms emitting radiation are placed in a magnetic field, the number of spectral lines increases. In 1925 George Uhlenbeck and Samuel Goudsmit accounted for this increase by postulating that electrons were magnets by virtue of their "spinning"; the direction of revolution determined the orientation of the magnetic field engendered by the electron. Now if we say that a macroscopic object such as a top is spinning, we imply that its surface has some discontinuity, like a speck of dirt, whose revolving motion we can observe. There are no specks to observe on electrons, and it is therefore meaningless to ask whether the electron is *really* spinning; the electron exhibits a magnetic moment, and that is how spinning charged bodies behave, and we therefore assign a "spin" to electrons by inference.

In representing the spin conditions of electrons in orbitals, we use a dash to denote the orbital, an arrow pointing down (↓) to denote one spin orientation, and up (↑) to denote the other spin orientation. Thus an empty orbital is __ , a half-filled orbital (one electron) is $\underline{\downarrow}$, and a filled orbital (two electrons) is $\underline{\downarrow\uparrow}$. The two electrons in the filled orbital (Pauli exclusion principle, p. 25) must have *opposite* spin orientations.

2.13 □ AUFBAU

The "aufbau," or buildup, principle, referring to the progressive accretion of electrons in the energy levels of atoms as their atomic numbers increase, states that *each successive electron goes into the lowest energy level available*. The following rules are generally applicable:

(a) The usual order of increasing subshell energy *for neutral atoms* is (see Fig. 2.11), 1*s*, 2*s*, 2*p*, 3*s*, 3*p*, 4*s*, 3*d*, 4*p*, 5*s*, 4*d*, 5*p*, 6*s*, 4*f*, 5*d*, 6*p*, 7*s*, 5*f*, 6*d*. This order does not apply to positive ions.

(b) An orbital may contain no more than two electrons (Pauli principle).

(c) Electrons, as a consequence of their mutual repulsions, enter empty orbitals of a given subshell in preference to orbitals that are already occupied by an electron (Hund's rule).*

(d) A subshell that is completely filled (each orbital containing two electrons) is an especially stable arrangement, and may exist in violation of rule (a). The reason for this is that if an electron is to be added it must enter a higher energy level (difficult) and if one is to be removed it must leave a relatively low energy level (also difficult).

(e) A subshell that is just half-filled (each orbital containing one electron) is also a relatively stable arrangement that can exist in violation of rule (a). The reason for this is that if an electron is to be added it must enter an orbital already occupied by an electron (difficult because of repulsion), if one is to be removed it must leave an orbital without the aid of a repulsive fellow occupant.

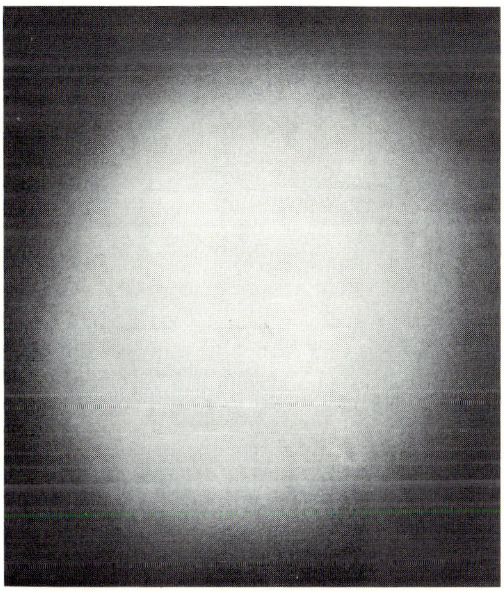

Fig. 2.12 The electron cloud of the hydrogen atom in the ground state (Is orbital) according to the predictions of wave mechanics. This may be visualized as a 3-dimensional cloud surrounding the nucleus.

* Named in honor of Friedrich Hund for his work with atomic and molecular spectra.

TABLE 2.3 AUFBAU

Element	Electron distributions K: 1s	L: 2s	L: 2p			Abbreviated designations	
$_1$H	↓	—	—	—	—	$1s^1$	
$_2$He	↓↑	—	—	—	—	$1s^2$	*K* shell full
$_3$Li	↓↑	↓	—	—	—	$K\ 2s^1$	
$_4$Be	↓↑	↓↑	—	—	—	$K\ 2s^2$	2*s* subshell full
$_5$B	↓↑	↓↑	↓	—	—	$K\ 2s^22p^1$	
$_6$C	↓↑	↓↑	↓	↓	—	$K\ 2s^22p^2$	
$_7$N	↓↑	↓↑	↓	↓	↓	$K\ 2s^22p^3$	
$_8$O	↓↑	↓↑	↓↑	↓	↓	$K\ 2s^22p^4$	
$_9$F	↓↑	↓↑	↓↑	↓↑	↓	$K\ 2s^22p^5$	
$_{10}$Ne	↓↑	↓↑	↓↑	↓↑	↓↑	$K\ 2s^22p^6$	*KL* shells full

Element	Core	M: 3s	M: 3p			M: 3d					Abbreviated designations	
$_{11}$Na	Ne core	↓	—	—	—	—	—	—	—	—	$KL\ 3s^1$	Ne core includes full *KL* shells
$_{12}$Mg	Ne core	↓↑	—	—	—	—	—	—	—	—	$KL\ 3s^2$	3*s* subshell full
$_{13}$Al	Ne core	↓↑	↓	—	—	—	—	—	—	—	$KL\ 3s^23p^1$	
$_{14}$Si	Ne core	↓↑	↓	↓	—	—	—	—	—	—	$KL\ 3s^23p^2$	
$_{15}$P	Ne core	↓↑	↓	↓	↓	—	—	—	—	—	$KL\ 3s^23p^3$	
$_{16}$S	Ne core	↓↑	↓↑	↓	↓	—	—	—	—	—	$KL\ 3s^23p^4$	
$_{17}$Cl	Ne core	↓↑	↓↑	↓↑	↓	—	—	—	—	—	$KL\ 3s^23p^5$	
$_{18}$Ar	Ne core	↓↑	↓↑	↓↑	↓↑	—	—	—	—	—	$KL\ 3s^23p^6$	3*s*, 3*p* subshells full

Element	Core	M: 3d	N: 4s	N: 4p	N: 4d	N: 4f	Abbreviated designations	
$_{19}$K	Ar core	— — — — —	↓	— — —	— — — — —	— — — — — — —	$KL\ 3s^23p^64s^1$	Ar core includes full *KL* shells, 3*s*3*p* subshells.
$_{20}$Ca	Ar core	— — — — —	↓↑	— — —	— — — — —	— — — — — — —	$KL\ 3s^23p^64s^2$	
$_{21}$Sc	Ar core	↓	↓↑				$KL\ 3s^23p^63d^14s^2$	

Element	Core	3d	4s	4p	4d	4f	Configuration	Remarks
$_{23}$V	Ar core	↑ ↑ ↑ — —	↑↓	— — —	— — — — —	— — — — — — —	$KL\ 3s^23p^63d^34s^2$	
$_{24}$Cr	Ar core	↑ ↑ ↑ ↑ ↑	↑	— — —	— — — — —	— — — — — — —	$KL\ 3s^23p^63d^54s^1$	3d subshell half-filled
$_{25}$Mn	Ar core	↑ ↑ ↑ ↑ ↑	↑↓	— — —	— — — — —	— — — — — — —	$KL\ 3s^23p^63d^54s^2$	3d subshell half-filled
$_{26}$Fe	Ar core	↑↓ ↑ ↑ ↑ ↑	↑↓	— — —	— — — — —	— — — — — — —	$KL\ 3s^23p^63d^64s^2$	
$_{27}$Co	Ar core	↑↓ ↑↓ ↑ ↑ ↑	↑↓	— — —	— — — — —	— — — — — — —	$KL\ 3s^23p^63d^74s^2$	
$_{28}$Ni	Ar core	↑↓ ↑↓ ↑↓ ↑ ↑	↑↓	— — —	— — — — —	— — — — — — —	$KL\ 3s^23p^63d^84s^2$	
$_{29}$Cu	Ar core	↑↓ ↑↓ ↑↓ ↑↓ ↑↓	↑	— — —	— — — — —	— — — — — — —	$KL\ 3s^23p^63d^{10}4s^1$	*M* shell full
$_{30}$Zn	Ar core	↑↓ ↑↓ ↑↓ ↑↓ ↑↓	↑↓	— — —	— — — — —	— — — — — — —	$KLM\ 4s^2$	4s subshell full
$_{31}$Ga	Ar core	↑↓ ↑↓ ↑↓ ↑↓ ↑↓	↑↓	↑ — —	— — — — —	— — — — — — —	$KLM\ 4s^24p^1$	
$_{32}$Ge	Ar core	↑↓ ↑↓ ↑↓ ↑↓ ↑↓	↑↓	↑ ↑ —	— — — — —	— — — — — — —	$KLM\ 4s^24p^2$	
$_{33}$As	Ar core	↑↓ ↑↓ ↑↓ ↑↓ ↑↓	↑↓	↑ ↑ ↑	— — — — —	— — — — — — —	$KLM\ 4s^24p^3$	
$_{34}$Se	Ar core	↑↓ ↑↓ ↑↓ ↑↓ ↑↓	↑↓	↑↓ ↑ ↑	— — — — —	— — — — — — —	$KLM\ 4s^24p^4$	
$_{35}$Br	Ar core	↑↓ ↑↓ ↑↓ ↑↓ ↑↓	↑↓	↑↓ ↑↓ ↑	— — — — —	— — — — — — —	$KLM\ 4s^24p^5$	
$_{36}$Kr	Ar core	↑↓ ↑↓ ↑↓ ↑↓ ↑↓	↑↓	↑↓ ↑↓ ↑↓	— — — — —	— — — — — — —	$KLM\ 4s^24p^6$	4p subshell full

Element	Core	4d	4f	*O*: 5s	*O*: 5p	*O*: 5d	Configuration	Remarks
$_{37}$Rb	Kr core	— — — — —	— — — — — — —	↑	— — —	— — — — —	$KLM\ 4s^24p^65s^1$	Kr core includes full *KLM* shells, 4s,4p subshells
$_{38}$Sr	Kr core	— — — — —	— — — — — — —	↑↓	— — —	— — — — —	$KLM\ 4s^24p^65s^2$	
$_{39}$Y	Kr core	↑ — — — —	— — — — — — —	↑↓	— — —	— — — — —	$KLM\ 4s^24p^64d^15s^2$	
$_{40}$Zr	Kr core	↑ ↑ — — —	— — — — — — —	↑↓	— — —	— — — — —	$KLM\ 4s^24p^64d^25s^2$	
$_{41}$Nb	Kr core	↑ ↑ ↑ ↑ —	— — — — — — —	↑	— — —	— — — — —	$KLM\ 4s^24p^64d^45s^1$	
$_{42}$Mo	Kr core	↑ ↑ ↑ ↑ ↑	— — — — — — —	↑	— — —	— — — — —	$KLM\ 4s^24p^64d^55s^1$	4d subshell half-filled
$_{43}$Tc	Kr core	↑ ↑ ↑ ↑ ↑	— — — — — — —	↑↓	— — —	— — — — —	$KLM\ 4s^24p^64d^55s^2$	4d subshell half-filled
$_{44}$Ru	Kr core	↑↓ ↑↓ ↑ ↑ ↑	— — — — — — —	↑	— — —	— — — — —	$KLM\ 4s^24p^64d^75s^1$	
$_{45}$Rh	Kr core	↑↓ ↑↓ ↑↓ ↑ ↑	— — — — — — —	↑	— — —	— — — — —	$KLM\ 4s^24p^64d^85s^1$	
$_{46}$Pd	Kr core	↑↓ ↑↓ ↑↓ ↑↓ ↑↓	— — — — — — —	—	— — —	— — — — —	$KLM\ 4s^24p^64d^{10}$	4d subshell full
$_{47}$Ag	Kr core	↑↓ ↑↓ ↑↓ ↑↓ ↑↓	— — — — — — —	↑	— — —	— — — — —	$KLM\ 4s^24p^64d^{10}5s^1$	
$_{48}$Cd	Kr core	↑↓ ↑↓ ↑↓ ↑↓ ↑↓	— — — — — — —	↑↓	— — —	— — — — —	$KLM\ 4s^24p^64d^{10}5s^2$	5s subshell full
$_{49}$In	Kr core	↑↓ ↑↓ ↑↓ ↑↓ ↑↓	— — — — — — —	↑↓	↑ — —	— — — — —	$KLM\ 4s^24p^64d^{10}5s^25p^1$	
$_{50}$Sn	Kr core	↑↓ ↑↓ ↑↓ ↑↓ ↑↓	— — — — — — —	↑↓	↑ ↑ —	— — — — —	$KLM\ 4s^24p^64d^{10}5s^25p^2$	
$_{51}$Sb	Kr core	↑↓ ↑↓ ↑↓ ↑↓ ↑↓	— — — — — — —	↑↓	↑ ↑ ↑	— — — — —	$KLM\ 4s^24p^64d^{10}5s^25p^3$	
$_{52}$Te	Kr core	↑↓ ↑↓ ↑↓ ↑↓ ↑↓	— — — — — — —	↑↓	↑↓ ↑ ↑	— — — — —	$KLM\ 4s^24p^64d^{10}5s^25p^4$	
$_{53}$I	Kr core	↑↓ ↑↓ ↑↓ ↑↓ ↑↓	— — — — — — —	↑↓	↑↓ ↑↓ ↑	— — — — —	$KLM\ 4s^24p^64d^{10}5s^25p^5$	
$_{54}$Xe	Kr core	↑↓ ↑↓ ↑↓ ↑↓ ↑↓	— — — — — — —	↑↓	↑↓ ↑↓ ↑↓	— — — — —	$KLM\ 4s^24p^64d^{10}5s^25p^6$	5p subshell full

Element	Core	*N*: 4f	*O*: 5d	*O*: 5f	*P*: 5s	*P*: 6p	Configuration	Remarks
$_{55}$Cs	Xe core	— — — — — — —	— — — — —	— — — — — — —	↑	— — —	$KLM\ 4s^24p^64d^{10}5s^25p^66s^1$	Xe core includes full *KLM* shells, 5s,5p subshells

TABLE 2.3 AUFBAU (*Continued*)

Element		Electron distributions: $4f$	$5d$	$5f$	$6s$	$6p$	Abbreviated designations	
$_{56}$Ba	Xe core	_ _ _ _ _ _ _	_ _ _ _ _	_ _ _ _ _ _ _	↓↑	_ _ _	$KLM\ 4s^24p^64d^{10}5s^25p^66s^2$	$6s$ subshell full
$_{57}$La	Xe core	_ _ _ _ _ _ _	↓ _ _ _ _	_ _ _ _ _ _ _	↓↑	_ _ _	$KLM\ 4s^24p^64d^{10}5s^25p^65d^16s^2$	Elements 57–71 comprise lanthanide series (rare earths)
$_{58}$Ce	Xe core	↓ ↓ _ _ _ _ _	_ _ _ _ _	_ _ _ _ _ _ _	↓↑	_ _ _	$KLM\ 4s^24p^64d^{10}4f^25s^25p^66s^2$	Uncertain
$_{59}$Pr	Xe core	↓ ↓ ↓ _ _ _ _	_ _ _ _ _	_ _ _ _ _ _ _	↓↑	_ _ _	$KLM\ 4s^24p^64d^{10}4f^35s^25p^66s^2$	Uncertain
$_{60}$Nd	Xe core	↓ ↓ ↓ ↓ _ _ _	_ _ _ _ _	_ _ _ _ _ _ _	↓↑	_ _ _	$KLM\ 4s^24p^64d^{10}4f^45s^25p^66s^2$	
$_{61}$Pm	Xe core	↓ ↓ ↓ ↓ ↓ _ _	_ _ _ _ _	_ _ _ _ _ _ _	↓↑	_ _ _	$KLM\ 4s^24p^64d^{10}4f^55s^25p^66s^2$	Uncertain
$_{62}$Sm	Xe core	↓ ↓ ↓ ↓ ↓ ↓ _	_ _ _ _ _	_ _ _ _ _ _ _	↓↑	_ _ _	$KLM\ 4s^24p^64d^{10}4f^65s^25p^66s^2$	
$_{63}$Eu	Xe core	↓ ↓ ↓ ↓ ↓ ↓ ↓	_ _ _ _ _	_ _ _ _ _ _ _	↓↑	_ _ _	$KLM\ 4s^24p^64d^{10}4f^75s^25p^66s^2$	$4f$ subshell half filled
$_{64}$Gd	Xe core	↓ ↓ ↓ ↓ ↓ ↓ ↓	↓ _ _ _ _	_ _ _ _ _ _ _	↓↑	_ _ _	$KLM\ 4s^24p^64d^{10}4f^75s^25d^15p^66s^2$	
$_{65}$Tb	Xe core	↓↑ ↓↑ ↓ ↓ ↓ ↓ ↓	_ _ _ _ _	_ _ _ _ _ _ _	↓↑	_ _ _	$KLM\ 4s^24p^64d^{10}4f^95s^25p^66s^2$	Uncertain
$_{66}$Dy	Xe core	↓↑ ↓↑ ↓↑ ↓ ↓ ↓ ↓	_ _ _ _ _	_ _ _ _ _ _ _	↓↑	_ _ _	$KLM\ 4s^24p^64d^{10}4f^{10}5s^25p^66s^2$	Uncertain
$_{67}$Ho	Xe core	↓↑ ↓↑ ↓↑ ↓↑ ↓ ↓ ↓	_ _ _ _ _	_ _ _ _ _ _ _	↓↑	_ _ _	$KLM\ 4s^24p^64d^{10}4f^{11}5s^25p^66s^2$	Uncertain
$_{68}$Er	Xe core	↓↑ ↓↑ ↓↑ ↓↑ ↓↑ ↓ ↓	_ _ _ _ _	_ _ _ _ _ _ _	↓↑	_ _ _	$KLM\ 4s^24p^64d^{10}4f^{12}5s^25p^66s^2$	
$_{69}$Tm	Xe core	↓↑ ↓↑ ↓↑ ↓↑ ↓↑ ↓↑ ↓	_ _ _ _ _	_ _ _ _ _ _ _	↓↑	_ _ _	$KLM\ 4s^24p^64d^{10}4f^{13}5s^25p^66s^2$	
$_{70}$Yb	Xe core	↓↑ ↓↑ ↓↑ ↓↑ ↓↑ ↓↑ ↓↑	_ _ _ _ _	_ _ _ _ _ _ _	↓↑	_ _ _	$KLM\ 4s^24p^64d^{10}4f^{14}5s^25p^66s^2$	N shell full
$_{71}$Lu	Xe core	↓↑ ↓↑ ↓↑ ↓↑ ↓↑ ↓↑ ↓↑	↓ _ _ _ _	_ _ _ _ _ _ _	↓↑	_ _ _	$KLMN\ 5s^25p^65d^16s^2$	End of lanthanide series

Element		N: $4f$	O: $5d$	O: $5f$	P: $6s$	P: $6p$	Abbreviated designations	
$_{72}$Hf	Xe core	↓↑ ↓↑ ↓↑ ↓↑ ↓↑ ↓↑ ↓↑	↓ ↓ _ _ _	_ _ _ _ _ _ _	↓↑	_ _ _	$KLMN\ 5s^25p^65d^26s^2$	
$_{73}$Ta	Xe core	↓↑ ↓↑ ↓↑ ↓↑ ↓↑ ↓↑ ↓↑	↓ ↓ ↓ _ _	_ _ _ _ _ _ _	↓↑	_ _ _	$KLMN\ 5s^25p^65d^36s^2$	
$_{74}$W	Xe core	↓↑ ↓↑ ↓↑ ↓↑ ↓↑ ↓↑ ↓↑	↓ ↓ ↓ ↓ _	_ _ _ _ _ _ _	↓↑	_ _ _	$KLMN\ 5s^25p^65d^46s^2$	
$_{75}$Re	Xe core	↓↑ ↓↑ ↓↑ ↓↑ ↓↑ ↓↑ ↓↑	↓ ↓ ↓ ↓ ↓	_ _ _ _ _ _ _	↓↑	_ _ _	$KLMN\ 5s^25p^65d^56s^2$	$5d$ subshell half full
$_{76}$Os	Xe core	↓↑ ↓↑ ↓↑ ↓↑ ↓↑ ↓↑ ↓↑	↓↑ ↓ ↓ ↓ ↓	_ _ _ _ _ _ _	↓↑	_ _ _	$KLMN\ 5s^25p^65d^66s^2$	
$_{77}$Ir	Xe core	↓↑ ↓↑ ↓↑ ↓↑ ↓↑ ↓↑ ↓↑	↓↑ ↓↑ ↓ ↓ ↓	_ _ _ _ _ _ _	↓↑	_ _ _	$KLMN\ 5s^25p^65d^76s^2$	
$_{78}$Pt	Xe core	↓↑ ↓↑ ↓↑ ↓↑ ↓↑ ↓↑ ↓↑	↓↑ ↓↑ ↓↑ ↓↑ ↓	_ _ _ _ _ _ _	↓	_ _ _	$KLMN\ 5s^25p^65d^96s^1$	
$_{79}$Au	Xe core	↓↑ ↓↑ ↓↑ ↓↑ ↓↑ ↓↑ ↓↑	↓↑ ↓↑ ↓↑ ↓↑ ↓↑	_ _ _ _ _ _ _	↓	_ _ _	$KLMN\ 5s^25p^65d^{10}6s^1$	$5d$ subshell full
$_{80}$Hg	Xe core	↓↑ ↓↑ ↓↑ ↓↑ ↓↑ ↓↑ ↓↑	↓↑ ↓↑ ↓↑ ↓↑ ↓↑	_ _ _ _ _ _ _	↓↑	_ _ _	$KLMN\ 5s^25p^65d^{10}6s^2$	
$_{81}$Te	Xe core	↓↑ ↓↑ ↓↑ ↓↑ ↓↑ ↓↑ ↓↑	↓↑ ↓↑ ↓↑ ↓↑ ↓↑	_ _ _ _ _ _ _	↓↑	↓ _ _	$KLMN\ 5s^25p^65d^{10}6s^26p^1$	
$_{82}$Pb	Xe core	↓↑ ↓↑ ↓↑ ↓↑ ↓↑ ↓↑ ↓↑	↓↑ ↓↑ ↓↑ ↓↑ ↓↑	_ _ _ _ _ _ _	↓↑	↓ ↓ _	$KLMN\ 5s^25p^65d^{10}6s^26p^2$	
$_{83}$Bi	Xe core	↓↑ ↓↑ ↓↑ ↓↑ ↓↑ ↓↑ ↓↑	↓↑ ↓↑ ↓↑ ↓↑ ↓↑	_ _ _ _ _ _ _	↓↑	↓ ↓ _	$KLMN\ 5s^25p^65d^{10}6s^26p^3$	
$_{84}$Po	Xe core	↓↑ ↓↑ ↓↑ ↓↑ ↓↑ ↓↑ ↓↑	↓↑ ↓↑ ↓↑ ↓↑ ↓↑	_ _ _ _ _ _ _	↓↑	↓↑ ↓ ↓	$KLMN\ 5s^25p^65d^{10}6s^26p^4$	
$_{85}$At	Xe core	↓↑ ↓↑ ↓↑ ↓↑ ↓↑ ↓↑ ↓↑	↓↑ ↓↑ ↓↑ ↓↑ ↓↑	_ _ _ _ _ _ _	↓↑	↓↑ ↓↑ ↓	$KLMN\ 5s^25p^55d^{10}6s^26p^5$	

	O	P				
	5*f*	6*d*	6*f*	7*s*		
$_{87}$Fr Rn core				↓	*KLMN* $5s^2 5p^6 5d^{10} 6s^2 6p^6 7s^1$	Rn core includes full *KLMN* shell, 5*s*, 5*p*, 5*d*, 6*s*, 6*p* subshells
$_{88}$Ra Rn core				⇅	*KLMN* $5s^2 5p^6 5d^{10} 6s^2 6p^6 7s^2$	7*s* subshell full.
$_{89}$Ac Rn core		↓		⇅	*KLMN* $5s^2 5p^6 5d^{10} 6s^2 6p^6 6d^1 7s^2$	Elements 89–103 comprise actinide series
$_{90}$Th Rn core		↓ ↓		⇅	*KLMN* $5s^2 5p^6 5d^{10} 6s^2 6p^6 6d^2 7s^2$	
$_{91}$Pa Rn core	↓ ↓	↓		⇅	*KLMN* $5s^2 5p^6 5d^{10} 5f^2 6s^2 6p^6 6d^1 7s^2$	Uncertain
$_{92}$U Rn core	↓ ↓ ↓	↓		⇅	*KLMN* $5s^2 5p^6 5d^{10} 5f^3 6s^2 6p^6 6d^1 7s^2$	
$_{93}$Np Rn core	↓ ↓ ↓ ↓	↓		⇅	*KLMN* $5s^2 5p^6 5d^{10} 5f^4 6s^2 6p^6 6d^1 7s^2$	Uncertain
$_{94}$Pu Rn core	↓ ↓ ↓ ↓ ↓	↓		⇅	*KLMN* $5s^2 5p^6 5d^{10} 5f^5 6s^2 6p^6 6d^1 7s^2$	Uncertain
$_{95}$Am Rn core	↓ ↓ ↓ ↓ ↓ ↓ ↓			⇅	*KLMN* $5s^2 5p^6 5d^{10} 5f^7 6s^2 6p^6 7s^2$	5*f* subshell half full
$_{96}$Cm Rn core	↓ ↓ ↓ ↓ ↓ ↓ ↓	↓		⇅	*KLMN* $5s^2 5p^6 5d^{10} 5f^7 6s^2 6p^6 6d^1 7s^2$	Uncertain
$_{97}$Bk Rn core	⇅ ⇅ ↓ ↓ ↓ ↓ ↓			⇅	*KLMN* $5s^2 5p^6 5d^{10} 5f^9 6s^2 6p^6 7s^2$	Uncertain
$_{98}$Cf Rn core	⇅ ⇅ ⇅ ↓ ↓ ↓ ↓			⇅	*KLMN* $5s^2 5p^6 5d^{10} 5f^{10} 6s^2 6p^6 7s^2$	Uncertain
$_{99}$Es Rn core	⇅ ⇅ ⇅ ⇅ ↓ ↓ ↓			⇅	*KLMN* $5s^2 5p^6 5d^{10} 5f^{11} 6s^2 6p^6 7s^2$	Uncertain
$_{100}$Fm Rn core	⇅ ⇅ ⇅ ⇅ ⇅ ↓ ↓			⇅	*KLMN* $5s^2 5p^6 5d^{10} 5f^{12} 6s^2 6p^6 7s^2$	Uncertain
$_{101}$Md Rn core	⇅ ⇅ ⇅ ⇅ ⇅ ⇅ ↓			⇅	*KLMN* $5s^2 5p^6 5d^{10} 5f^{13} 6s^2 6p^6 7s^2$	Uncertain
$_{102}$No Rn core	⇅ ⇅ ⇅ ⇅ ⇅ ⇅ ⇅			⇅	*KLMN* $5s^2 5p^6 5d^{10} 5f^{14} 6s^2 6p^6 7s^2$	Uncertain
$_{103}$Lw Rn core	⇅ ⇅ ⇅ ⇅ ⇅ ⇅ ⇅	↓		⇅	*KLMN* $5s^2 5p^6 5d^{10} 5f^{14} 6s^2 6p^6 6d^1 7s^2$	Uncertain. End of actinide series.

Representative elements, atomic numbers 1–20, 30–38, 48–56, 80–88: electron enters an *s* or *p* subshell.
Transition elements, atomic numbers 21–29, 39–47, 72–79: electron enters a *d* subshell.
Inner transition elements, atomic numbers 57–71, 89–103: electron enters an *f* subshell.

(f) For atoms of high atomic numbers, the differences between energy levels of subshells are small, and some assignments of electrons to orbitals are irregular and uncertain.

(g) In the abbreviated nomenclature exemplified by $4d^3$, the first number is the principal quantum level, the letter is the sublevel, and the superscript is the total number of electrons in the sublevel.

The distribution of electrons in the orbitals of the atoms in their most stable conditions (called the "ground states") is given in Table 2.3.

2.14 ▫ WHAT DO THE QUANTUM NUMBERS REPRESENT?

(a) The principal quantum level, n, is related to the most probable distance between the electron and the nucleus, or to the distance between the region of maximum electron density and the nucleus. In simplest terms, n may be considered to be related to *the size of the orbital.*

(b) The sublevels or subshells, s, p, d, f, are related to the charge density contours of the electron clouds. The sublevels may therefore be considered to represent the *shapes of the orbitals.* Different sublevels, as explained before, differ in energy.

(c) The orbitals in a given subshell, for example, the three $4p$ orbitals, are equivalent in energy in the absence of a magnetic field, but are split into different energy levels in a magnetic field. From these and other circumstances, we consider that *the orbitals in a given subshell differ among each other in their orientation in space.*

(d) The electrons in a given orbital differ between each other in *spin orientation*, as mentioned on p. 28.

2.15 ▫ REPRESENTATIONS OF THE SHAPES OF ATOMIC ORBITALS

Figure 2.12 shows the $1s$ electron cloud diagram. Figure 2.13 presents the cloud diagrams for several orbitals. These figures are drawings that correspond to equations; they are not photographs of electrons. Observe the profound differences between orbitals in the same shell. Observe also that the s orbitals are spherically symmetrical; this means that the electron density is uniform on the surface of any imaginary sphere whose center is the nucleus.

Representations of the electronic angular distributions, as given in Fig. 2.14 for s, p, and d orbitals, are also very useful. The shaded areas represent surfaces within which most of the electron density is found. Thus the diagrams show both the shapes of the orbitals and their orientations in space. *Notice that unlike the s orbitals, the p and d orbitals have preferred directions in space.*

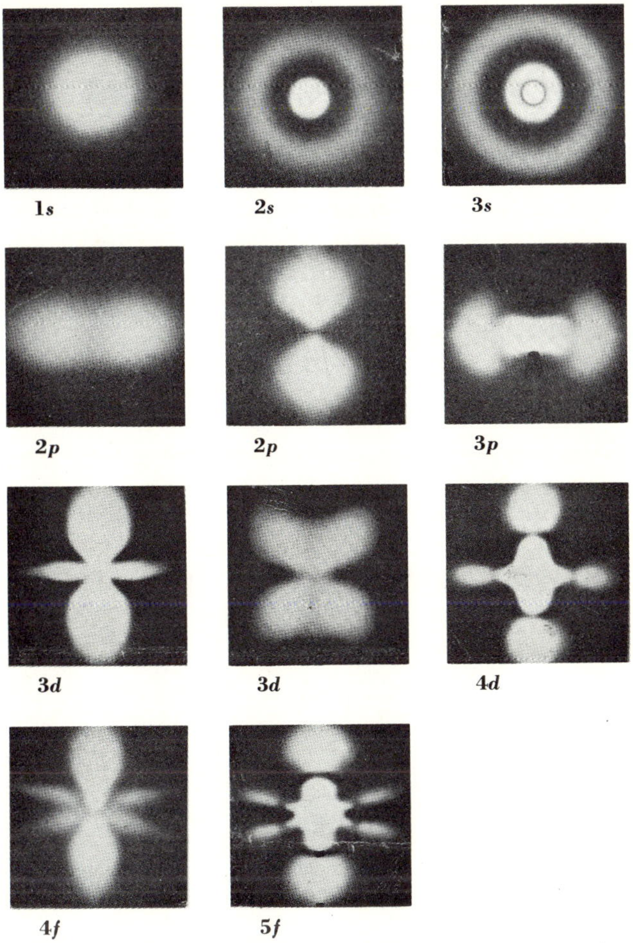

Fig. 2.13 Electron cloud diagrams of several states of the hydrogen atom according to the predictions of wave mechanics. These should be visualized as 3-dimensional clouds surrounding the nucleus of the hydrogen atom. [H. E. White, *Phys. Rev.* 37, 1416 (1931)].

One of the five d orbitals (d_{z^2}) is p-like in character. The other four d orbitals resemble four-leaf clovers. Picture one "clover" ($d_{x^2y^2}$) along the x and y axes; rotate the "clover" 45 degrees without tilting and you produce the d_{xy} orbital. Picture another "clover" (d_{yz}) upright between the y and z axes. Rotate this "clover" 90 degrees so that it now faces you; you are looking at the d_{xz} orbital.

The shapes of orbitals (electron cloud diagrams), we emphasize, are pictorial representations of the mathematical solutions of the wave equation; they are not pictures of electric charges or of matter.

2.16 □ PARAMAGNETISM

Atomic sodium, atomic hydrogen, molecular oxygen, and iron are attracted into magnetic fields. They are typical paramagnetic substances; they possess permanent magnetic moments. If a beam of

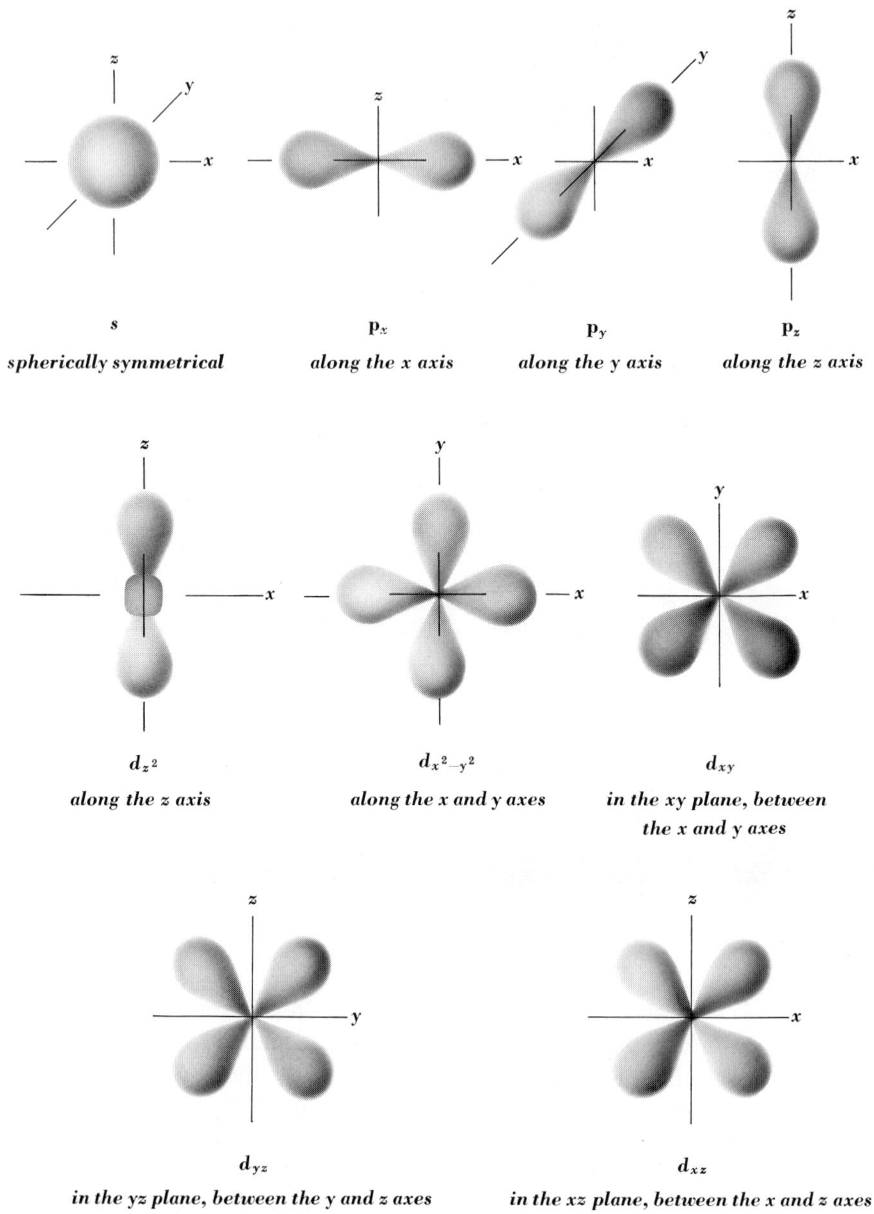

Fig. 2.14 The angular distributions of the s, the three p, and the five d electrons (orbitals). The nucleus resides at the origin of the axes.

hydrogen atoms, H $\underset{1s}{\underline{\downarrow}}$, is directed between the poles of a magnet, the deflection of the atoms will depend on the relative orientations of the two magnetic fields—that of the external magnet and that of the electron. But *only two spin orientations* are permitted for an electron in a magnetic field. These orientations are arbitrarily "pictured" as $\overset{S}{\underset{N}{\downarrow}}$ or $\overset{N}{\underset{S}{\uparrow}}$.

Thus the atoms with spins $\overset{S}{\underset{N}{\downarrow}}$ will be deflected in one direction, and those atoms with spins $\overset{N}{\underset{S}{\uparrow}}$ will be deflected in the opposite direction. The initial beam of hydrogen atoms should therefore emerge from the magnetic field as *two separate beams*. The actual experiment shows that a beam of monatomic hydrogen is split into two and only two parts. Thus **paramagnetism,** *a property of substances with permanent magnetic moments, is associated with* **unpaired electrons** *in an atom, ion, or molecule*. These permanent micromagnets are attracted into a magnetic field and align with the field, similarly to the magnetic needle of a compass.

The first experiment of this type was performed in 1921 by Otto Stern and Walther Gerlach, using silver (vapor) atoms. A silver atom contains one unpaired electron, as predicted by its electron configuration,

$_{47}$Ag	*KLM*	⇃↾	⇃↾	⇃↾	⇃↾	⇃↾	⇃↾	⇃↾	⇃↾	⇃↾	↓	
		4*s*		4*p*				4*d*			5*s*	

It is evident that the filled orbitals make no contribution to paramagnetic properties. Also, in silver, there is no question of the order of occupancy of the five 4*d* orbitals since 10 electrons are available. If, however, as in the case of $_{25}$Mn, 5 electrons are available for the five 3*d* orbitals, the order of occupancy is significant:

$_{25}$Mn	*KL*	⇃↾	⇃↾	⇃↾	⇃↾	↓	↓	↓	↓	↓	⇂↿	(correct)
		3*s*		3*p*				3*d*			4*s*	
or												
$_{25}$Mn	*KL*	⇃↾	⇃↾	⇃↾	⇃↾	⇃↾	⇃↾	↓	—	—	⇃↾	(incorrect)
		3*s*		3*p*				3*d*			4*s*	

Experimentally, the magnetic moment of Mn corresponds to five, not one, unpaired electrons. This is in accord with Hund's rule (see p. 29), which correctly predicts the number of unpaired electrons in atoms, ions, and molecules: *electrons remain unpaired with parallel spins in orbitals of equal energy until each such orbital has at least one electron in it*. For example, see the order of occupancy of the three 2*p* orbitals in the series from $_{3}$B to $_{10}$Ne in Table 2.3 (p. 30).

PROBLEMS

1. Quantization Is it possible to select a weight that cannot be counterpoised (within the errors of observation) by a quantity of pennies? Of water? Explain.

2. Atomic structure (a) In 1895, P. Lenard detected electrons outside a thin glass tube in which they were produced, and concluded that the atoms in the glass must have a very open structure. Explain the results of this experiment in terms of the presently accepted theory of matter. (b) Describe the fundamental differences among the Dalton, Rutherford-Bohr, and Schrödinger models of the atom.

3. Nuclear atom (a) When an α particle strikes an electron in an atom, what should occur? (Think of this encounter as you would a massive high-velocity body striking a relatively puny body.) (b) A fine stream consisting of 8×10^5 α particles impinges upon a gold foil of thickness 4.0×10^{-4} mm; 100 α particles are detected between the angles of deflection 180 degrees and 92 degrees. When the experiment is repeated with identical conditions except that the thickness of the foil is increased to 5.0×10^{-4} mm, 124 α particles are detected. Use these results to argue against the hypothesis that these deflections are caused by a surface reflection (similar to the action of a wall upon a handball).

4. Line spectra (a) Referring to Equation 2 (p. 18), would you predict the existence of a spectral line for atomic hydrogen corresponding to the transition $(n = 3) \rightarrow (n = 3.5)$? (b) Referring to Fig. 2.8 (p. 20), state which is greater, the energy difference between the ground state, $n = 1$, and the principal quantum level $n = 2$, or the energy involved in the transition $(n = 2) \rightarrow (n = 5)$.

5. Line spectra Calculate the energies of the hydrogen atom for (a) $n = 1$; (b) $n = 3$; (c) $n = 5$.

6. Bohr atom The electrons in a sample of hydrogen are excited to a maximum principal quantum level of $n = 4$. Assuming that, as this energy is lost, all possible transitions between $n = 4$ and the ground state occur, how many spectral lines are predicted by the original Bohr theory?

7. Bohr atom (a) 1.00 g of hydrogen contains 6.02×10^{23} atoms. How much work in kilocalories is required to remove the electrons in 1 gram of hydrogen atoms from the energy level, $n = 2$, to infinity? (b) What is the magnitude of the attractive force between the electron and the nucleus when n = infinity?

8. Nuclear composition Complete the following table by substituting the numerical value where an x appears and the symbol for the element where a question mark appears.

Isotope	Mass number	Z	N
(a) $^{65}_{x}$Cu	x	x	x
(b) $^{x}_{x}$?	23	x	11
(c) $^{x}_{x}$Ge	74	x	x
(d) $^{x}_{x}$?	x	47	62
(e) $^{x}_{x}$?	30	14	x
(f) $^{x}_{x}$P	x	x	16
(g) $^{x}_{x}$Be	9	x	x

9. Trajectory Do electrons travel around the nucleus in definite orbits? Justify your answer.

10. Paramagnetism Using Table 2.3 (p. 30-33), select three atoms you believe would possess paramagnetic properties.

11. Paramagnetism (a) Assume that for positive ions the 4*s* sublevel is *higher* in energy than the 3*d* sublevel. On the basis of this assumption, predict the electron distribution and the paramagnetic properties of Cu^+. (b) Calculate the number of unpaired electrons in $_{25}Mn^{2+}$, $_{25}Mn^{3+}$, $_{25}Mn^{4+}$, $_{25}Mn^{7+}$. See Table 2.3 (p. 31), and Fig. 2.11 (p. 28), and make the same assumption as in part (a).

12. 4*s* versus 3*d* The electron configuration of $_{20}Ca$ is $KL\ 3s^2 3p^6 4s^2$. Imagine that two protons are added to the nucleus. What is the chemical symbol of the ion produced? What is the electron configuration of this ion?

ADDITIONAL PROBLEMS

13. Nuclear atom Account for the fact that the density of matter, as exhibited in "white dwarf" stars, can be greater than 8×10^4 g/cm^3.

14. Line spectra (a) Using Equation 6 (p. 22), calculate the energy in ergs per atom for the $(n = 2) \rightarrow (n = 5)$ transition. (b) To what wavelength in the visible spectrum of hydrogen does this energy correspond?

15. Bohr atom In a "collision," a high-velocity electron transfers energy to a hydrogen atom, causing the transition $(n = 1) \rightarrow (n = 4)$; then, spontaneously, the transitions $(n = 4) \rightarrow (n = 2)$ and $(n = 2) \rightarrow$ ground state occur. (a) How many photons are emitted by the atom? (b) What are the energies of the different photons? (c) Which photon has the shortest wavelength?

16. Bohr atom Assume that a certain atom has the following electronic energy levels:

$$E_1 = -6.0 \times 10^{-11} \text{ erg/atom}$$
$$E_2 = -4.0 \times 10^{-11} \text{ erg/atom}$$
$$E_3 = -1.0 \times 10^{-11} \text{ erg/atom}$$

(a) Calculate the frequency of the radiation that would be emitted or absorbed in a transition from level E_1 to level E_3. (b) Calculate the frequency for a transition from level 3 to level 2. (c) State whether each process is absorption or emission of energy.

17. Bohr atom (a) Calculate the energy of the hydrogen atom when the principal quantum number of the electron is 8. (b) In what ways can we represent the distribution of an electron at this energy level? Can we locate it precisely in space in relation to the proton? (c) What does the minus sign in your answer to (a) signify?

18. Bohr atom In a discharge tube, hydrogen is subjected to collisions with electrons having a maximum kinetic energy of 21.1×10^{-12} erg/electron. On the basis of the original Bohr theory (Fig. 2.8, p. 20), how many lines are in the spectrum emitted by the bombarded H atoms?

19. Bohr atom Explain the statement, "If we could observe the 4861 Å line radiated by a few hydrogen atoms, the line would appear only intermittently."

20. Quantum theory Predict the minimum energy that an electron in a discharge tube must possess so that, upon collision with a mercury atom, it causes the appearance of the mercury line, $\lambda = 2537 \times 10^{-8}$ cm. This prediction was verified in 1913 by the experiment of James Franck and Gustav Herz in which electrons of precisely known energy passed through mercury vapor at low pressure.

21. 4*s* versus 3*d* Experimentally, the magnetic moment of ${}_{26}Fe^{3+}$ in the gaseous state corresponds to 5 unpaired electrons. Write the electron configuration of ${}_{26}Fe^{3+}$.

22. Electron spin (a) Assuming that for cations, the 4*s* electrons are higher in energy than the 3*d*, pick the ion having 5 unpaired electrons: ${}_{24}Cr^{2+}$; ${}_{25}Mn^{2+}$; ${}_{26}Fe^{2+}$. (b) Give the number of unpaired electrons in each of the other two ions named.

23. Nature of matter The Roman Lucretius concluded, "Yet things that are the tiniest will be composed of infinite parts just the same." The famous theoretical physicist Dirac wrote in 1947, "In a classical explanation of the constitution of matter, one would assume it to be made up of a large number of small constituent parts and one would postulate laws for the behavior of these parts, from which the laws of the matter in bulk could be deduced. This would not complete the explanation, however, since the question of the structure and stability of the constituent parts is left untouched. To go into this question, it becomes necessary to postulate that each constituent part is itself made up of smaller parts, in terms of which its behavior is to be explained. There is clearly no end to this procedure, so that one can never arrive at the ultimate structure of matter on these lines." Explain these statements in your own words.

ANSWERS

5. See Fig. 2.8, p. 20.

6. 6 lines.

15. (b) 4.09×10^{-12} ergs/atom for $(n = 4) \rightarrow (n = 2)$.

18. 10 lines.

20. 7.829×10^{-12} erg.

3 ■ CHEMICAL PERIODICITY

3.1 □ FORMULAS AND VALENCE

The examination of the formulas of such compounds as HCl (hydrogen chloride), H_2O (water), NH_3 (ammonia), NaCl (sodium chloride), $CaCl_2$ (calcium chloride), H_2SO_4 (sulfuric acid), and H_3PO_4 (phosphoric acid) immediately reveals that the **combining capacity** of atoms is not uniform. The combining capacity of an atom is commonly referred to as its **valence.** The formula HCl shows that the atoms of hydrogen and chlorine have the same combining capacity and therefore the same valence. The formula H_2O, however, shows that the combining capacity of an oxygen atom is twice that of the hydrogen atom and therefore the valence of oxygen is twice that of hydrogen. If we assign a valence of 1 to the hydrogen atom, then the valence of the chlorine atom becomes 1 and that of the oxygen atom 2. It also follows from the formula $CaCl_2$ that the combining capacity of the calcium atom is twice that of the chlorine atom; the valence of calcium is therefore 2.

Certain combinations of elements, known as **groups** or **polyatomic ions,** remain combined and behave as characteristic units during many chemical reactions. Such a combination is the sulfate group. Thus from the formula for sulfuric acid, H_2SO_4, a valence of 2 may be assigned to the sulfate (SO_4) group.

In Chapter 7, the combining capacities of elements will be discussed in terms of the broader concept of oxidation state.

3.2 □ NOMENCLATURE

Following are some rules for naming compounds.

(a) The name of a **binary compound,** a compound composed of two elements, is usually derived from the names of the elements (see back cover); most commonly the metallic or electropositive element is written first and the second element is given the suffix **-ide.** Although frequently omitted, a Greek prefix should be used to indicate the number of atoms when more than one atom is involved.

(b) When an element exhibits more than one valence, the higher state is denoted by the suffix **-ic** and the lower by **-ous;** however, the use of Roman numbers placed after the name of the element to denote its valence state is preferable. Examples:

BN	Li_3N	$SnCl_2$	$SnCl_4$	HCl
boron nitride	*trilithium nitride*	*tin (II) chloride*	*tin (IV) chloride*	*hydrogen chloride*
	lithium nitride	*tin dichloride*	*tin tetrachloride*	
		stannous chloride	stannic chloride	

Preferred names are in italics. Traditional names are used for very common compounds such as water (dihydrogen oxide) and ammonia (trihydrogen nitride).

(c) **Ternary compounds,** composed of three elements, are most frequently characterized by the presence of a group; the name is therefore usually derived from the name of the element and the group. Groups containing oxygen are named relative to the number of oxygen atoms, as shown in Table 3.1.

TABLE 3.1 ▪ COMMON NOMENCLATURE OF TERNARY COMPOUNDS CONTAINING OXYGEN

Prefix	Suffix	Number of oxygen atoms in group	Examples	
	-ate	Not specified	Na_2SO_4	Sodium sulfate
			$FeSO_4$	Iron(II) sulfate, or ferrous sulfate
			$Fe_2(SO_4)_3$	Iron(III) sulfate, or ferric sulfate
			Na_3PO_4	Trisodium phosphate
			$NaNO_3$	Sodium nitrate
			$NaClO_3$	Sodium chlorate
	-ite	One less O atom than *-ate*	Na_2SO_3	Disodium sulfite, or sodium sulfite
			$NaClO_2$	Sodium chlorite
hypo-	*-ite*	One less O atom than *-ite*	$NaClO$	Sodium hypochlorite
per-	*-ate*	One more O atom than *-ate*	$NaClO_4$	Sodium perchlorate
			$Mg(ClO_4)_2$	Magnesium perchlorate

3.3 □ CHEMICAL PERIODICITY BEFORE MENDELEEV

Elements can be grouped by properties into various classes or families, somewhat as living things are classified by biologists. For example, the metals iron, copper, and lead differ as a class from the nonmetals sulfur, phosphorus, and carbon. Within the metallic class, copper, silver, and gold (coinage metals) differ as a group from calcium, strontium, and barium (alkaline earth metals). These facts suggest that atoms are themselves aggregates of other fundamental entities, a notion that foreshadowed more refined concepts of atomic structure.

Hence, the effort of chemists to set up rational systems for classifying the elements was one of the necessary first steps leading to our current progress in understanding the complexities of the atom. Like other fundamental scientific concepts, the idea of chemical periodicity developed at an accelerating rate from quite unsophisticated beginnings to its current status as a very broad conceptual system. The first modern classification systems were based on relating chemical properties (especially valence) to the atomic weights of the elements; such systems therefore could not be formulated until methods for determining valences and atomic weights began to become available in the early part of the nineteenth century. In 1829, Johann Wolfgang Döbereiner reported several instances in which elements with closely related chemical properties, when arranged in order of atomic weight, showed approximately constant increments of atomic weight. These groups consisted of three elements each, and he therefore called them **triads.** Examples are given in Table 3.2.

TABLE 3.2 ▪ DÖBEREINER'S TRIADS

Triad	Properties	Atomic weight (approximate)	Atomic weight increment
Chlorine		35	
			⟶ 45
Bromine	Corrosive colored vapors	80	
			⟶ 47
Iodine		127	
Sulfur		32	
	Solids whose hydrides		⟶ 47
Selenium	(H_2S, H_2Se, H_2Te)	79	
	are stenches		⟶ 49
Tellurium		128	
Lithium		7	
			⟶ 16
Sodium	Very reactive metals	23	
			⟶ 16
Potassium		39	

In 1864, John Newlands made a great step forward by relating the properties of elements, not to a plot or scale of atomic weights, but simply to the **ordinal numbers*** *of the elements when they are arranged in an increasing series of atomic weights.* Newlands found that, in such a list, elements with similar properties recur at intervals of seven. He called this relationship, by a rather inapt analogy to the musical scale, the **law of octaves.** Thus, lithium (thought then to be the second heaviest element) and sodium (thought to be the ninth) are an "octave" apart.† This serialization was the beginning of our idea of "atomic number"—a concept vital to our present expressions of chemical periodicity and the electronic structures of atoms. Newlands' contribution, ridiculed at first, was later given recognition (Davy Medal, 1887).

3.4 □ MENDELEEV; MEYER

In 1869 Dmitrii Ivanovich Mendeleev and several months later Julius Lothar Meyer published independent versions of a periodic system of the elements. Credit for the "periodic law" is given to those two scientists (especially Mendeleev) not because they were the first to describe periodic classifications of the elements, but more particularly for the following reasons:

(a) The significance of periodicity was more fully appreciated, the system of classification was more elaborate, and attention was given to a broader range of physical and chemical properties than was examined by earlier investigators. Table 3.3 is an adaptation of Mendeleev's periodic table of 1871.

(b) Mendeleev recognized that, when the atomic weight order demanded a chemically incongruous location of elements in the periodic system, adherence to atomic weight order must be sacrificed. For example, if iodine (atomic weight in 1869: 127) is placed before tellurium (atomic weight in 1869: 128), iodine appears in a group with sulfur and selenium, and tellurium finds itself in the company of chlorine and bromine. This is obvious chemical mismatching and must mean that the atomic weights either are wrong or are not the fundamental basis of chemical periodicity. Mendeleev imagined that the weights were in error; actually they were not far from todays's accepted values.

(c) Mendeleev avoided what would otherwise be additional chemical inconsistencies by boldly leaving gaps in his table, predicting that elements would be discovered to fill the gaps, and describing the properties that the to-be-discovered elements would have. For example, the element after zinc was, in 1871, arsenic. But arsenic does not

* "Ordinal numbers" express order or succession in a series: first, second, third, and so on, or 1st, 2nd, 3rd, and so on. They are distinguished from **cardinal numbers,** the numbers used in counting: one, two, three, and so on.

† Helium and neon had not yet been discovered.

TABLE 3.3 ▪ PERIODIC TABLE OF THE ELEMENTS (ADAPTED FROM MENDELEEV, 1871)

Groups:	*I*		*II*		*III*		*IV*		*V*		*VI*		*VII*		*VIII*
Type formulas:	R_2O RCl		RO RCl_2		R_2O_3 RCl_3		RH_4 RO_2		RH_3 R_2O_5		RH_2 RO_3		RH R_2O_7		RO_4
Subgroup or family:	*A*	*B*	*A*	*B*	*A*	*B*	*A*	*B*	*A*	*B*	*A*	*B*	*A*	*B*	
Series 1		**H**													
2	**Li**		**Be**		**B**		**C**		**N**		**O**		**F**		
3		**Na**		**Mg**		**Al**		**Si**		**P**		**S**		**Cl**	
4	**K**		**Ca**		—		**Ti**		**V**		**Cr**		**Mn**		**Fe, Co, Ni, (Cu)**
5		**(Cu)**		**Zn**		—		—		**As**		**Se**		**Br**	
6	**Rb**		**Sr**		**Yt?**		**Zr**		**Nb**		**Mo**		—		**Ru, Rh, Pd, (Ag)**
7		**(Ag)**		**Cd**		**In**		**Sn**		**Sb**		**Te**		**I**	
8	**Cs**		**Ba**		**Di?**		**Ce?**		—		—		—		— — — —
9		—		—		—		—		—		—		—	
10	—		—		**Er?**		**La?**		**Ta**		**W**		—		**Os, Ir, Pt, (Au)**
11		**(Au)**		**Hg**		**Tl**		**Pb**		**Bi**					
12	—		—		—		**Th**		—		**U**		—		— — — —

belong in the same chemical group with aluminum or with silicon (see Table 3.3, groups III and IV). Arsenic is chemically like phosphorus, and therefore belongs in group V. Succeeding elements (selenium, bromine, and so on) also fall into reasonable locations if they follow arsenic located in group V. This means that two elements, one like aluminum and one like silicon, are missing.

For testing a theory, there is nothing like a prediction. The more spectacular the confrontation between predicted and observed data, the better the test. Table 3.4 shows the degree to which such predictions were successful for *eka*-silicon* (germanium). *Eka*-aluminum

* *Eka* (Greek) means "first after" or "first beyond."

TABLE 3.4 PREDICTION OF THE PROPERTIES OF GERMANIUM

	"Eka-silicon" (Es)	Germanium (Ge)	
Property	**Predicted in 1871 by Mendeleev**	**Reported in 1886 by Clemens Alexander Winkler**	**Currently accepted**
Atomic weight	72	72.32	72.59
Specific gravity	5.5	5.47	5.35
Melting point	High	—	947°C
Specific heat (cal/g deg)	0.073	0.076	0.074
Gram-atomic volume* (cm^3)	13	13.22	13.5
Color	Dark gray	Grayish white	Grayish white
Valence	4	4	4
Reaction with acids and alkalies	Es will be slightly attacked by such acids as HCl, but will resist attack by such alkalies as NaOH	Ge is dissolved by neither HCl nor NaOH, but is dissolved by concentrated NaOH	Ge is dissolved by neither HCl nor NaOH, but is dissolved by concentrated NaOH
Boiling point of the tetraethyl derivative	160°C	160°C	185°–187°C
Specific gravity of the dioxide	4.7	4.703	4.228
Specific gravity of the tetrachloride	1.9	1.887	1.8443
Boiling point of the tetrachloride	100°C	86°C	84°C

* The volume occupied by 1 gram-atomic weight. See Chapter 4.

(gallium) and *eka*-boron (scandium) are other examples of Mendeleev's successful predictions.

3.5 □ THE PERIODIC LAW AND THE PERIODIC TABLE. TYPES OF ELEMENTS

The properties of the elements are periodic functions of their atomic numbers. This oft-cited statement does not sound like a "law" at all, in the sense that it is not a quantitative expression. One must not, however, confuse precision with reliability. The periodic law is a broad, imprecise statement that grew out of early efforts, such as those described in the preceding section, and that still functions as a much-used framework on which comparisons and generalizations of chemical behavior are based. Chemists everywhere agree that the periodic law expresses deep-seated relationships in chemical properties and in the structures of atoms, but no two chemists are likely to offer the same detailed interpretation of the law.

Starting with Mendeleev and Meyer, and continuing to the present, enough periodic tables have been published to fill a book.* We shall examine the Mendeleev table of 1871 (Table 3.3), because some of its usages and conventions still persist. We shall then look at modern periodic tables.

Mendeleev's table After earlier efforts with other arrangements, Mendeleev in 1871 divided the elements into eight vertical columns, called **groups.** The horizontal rows thus formed were called **series** (now called **periods**). The most important characteristic common to elements in the same group is their valence—the numerical capacity of atoms of the element to combine with other atoms. If the other atoms are taken to be H, Cl, or O, it is possible to write "type formulas" that characterize the valence of elements within a group. Such formulas appear in Table 3.3, R denoting any element in the group. It was apparent to Mendeleev, however, that the elements within a given valence group exhibited not one, but two more or less distinct sets of properties. Accordingly, he divided each group into two **subgroups** or **families,** to correspond to this chemical duality. In group I, for example, all the metals form univalent oxides. Of these, Cu_2O, Ag_2O, and Au_2O lose oxygen on relatively mild heating, while Li_2O, K_2O, Rb_2O, and Cs_2O retain oxygen even under much more stringent conditions. Similar subgrouping continues through group VII. The makeup of group VIII is atypical—it consists of sets of chemically similar metals, only some of which exhibit octavalence (RuO_4, OsO_4).

Mendeleev noted that the assignments of elements to families alternated with increasing atomic number within each group, with confusing exceptions in some of the higher series. The uncertain assignments of Cu, Ag, and Au, and the misplacements of Na, O, and F are some consequences of this confusion. We are now aware that the subgroup classifications foreshadowed our modern understanding of the differences among elements based on electronic structure, and that Mendeleev's anomalies resulted from the fact that some of his series represented accretions of electrons in s and p subshells, and others in d subshells.

The important characteristic of elements in a series is *regular progression of chemical properties.* In series 2, 3, 5, and 7 this progression is accompanied by a marked shift from metallic to nonmetallic character with increasing group number.

A modern periodic table: the long-period form Modern periodic tables attempt to depict both chemical and electronic relationships. In all modern tables, the horizontal rows are called **"periods."** The so-called "short-period form," most popular up to about 1950, is a direct adaptation of Mendeleev's table, with inclusion of "group 0" to accom-

* Literally. See Edward Mazurs, "Types of Graphic Representations of the Periodic System of Chemical Elements," La Grange, Illinois, 1957.

TABLE 3.5[a] PERIODIC TABLE, LONG FORM

Periods	I A	II A	III B	IV B	V B	VI B	VII B	VIII	VIII	VIII	I B	II B	III A	IV A	V A	VI A	VII A	0
			Transition Elements															
1	H 1																H 1	He 2
2	Li 3	Be 4											B 5	C 6	N 7	O 8	F 9	Ne 10
3	Na 11	Mg 12											Al 13	Si 14	P 15	S 16	Cl 17	Ar 18
4	K 19	Ca 20	Sc 21	Ti 22	V 23	Cr 24	Mn 25	Fe 26	Co 27	Ni 28	Cu 29	Zn 30	Ga 31	Ge 32	As 33	Se 34	Br 35	Kr 36
5	Rb 37	Sr 38	Y 39	Zr 40	Nb 41	Mo 42	Tc 43	Ru 44	Rh 45	Pd 46	Ag 47	Cd 48	In 49	Sn 50	Sb 51	Te 52	I 53	Xe 54
6	Cs 55	Ba 56	La 57	Hf 72	Ta 73	W 74	Re 75	Os 76	Ir 77	Pt 78	Au 79	Hg 80	Tl 81	Pb 82	Bi 83	Po 84	At 85	Rn 86
7	Fr 87	Ra 88	Ac 89															

Lanthanide Series	Ce 58	Pr 59	Nd 60	Pm 61	Sm 62	Eu 63	Gd 64	Tb 65	Dy 66	Ho 67	Er 68	Tm 69	Yb 70	Lu 71
Actinide Series	Th 90	Pa 91	U 92	Np 93	Pu 94	Am 95	Cm 96	Bk 97	Cf 98	Es 99	Fm 100	Md 101	No 102	Lw 103

[a] The heavy black line separates the metallic from the nonmetallic elements; the distinction, however, is not so sharp as shown.

modate the noble gases, once thought to have "zero valence." The "long-period form" (Table 3.5), now widely used, has 18 vertical columns instead of eight, and uses this expansion to separate the A subgroups from the B subgroups. In the second and third periods, where no subgroup distinction exists, this wide separation, occurring between groups II and III, has no meaning at all. More serious, and common to both long and short forms, is the obscuration of sequences of properties (including valence) that are exhibited by certain elements in a given period (for example, K, Ca, Ga, Ge, As, Se, Br, Kr).

The interposition of lanthanide (rare earth) and actinide (heavy rare earth) elements into the body of the table would impose strains on the dimensions of charts and books, and is therefore not usually attempted.

Another modern table: the separated form; types of elements Another modern form of the periodic table is shown in Table 3.6. This arrangement has the advantages of preserving traditional group, subgroup, and period designations, of presenting meaningful separations based on electronic structures, and of facilitating the examination of many physical and chemical relationships that are the basis of the periodic law.

Note, particularly, the following characteristics of Table 3.6:

(a) The upper section of Table 3.6 comprises the elements that represent the regular progression of chemical properties on which all original periodic systems were based. They are therefore called **representative elements,** and the section in which they appear is a sort of "ideal" periodic table. The concept of "representative elements" is related to the progressive accretion of electrons in the s and p subshells of the atoms. The subgroup designation A is here reserved for representative elements.

(b) The middle section of Table 3.6 comprises the elements that interrupt the representative series. Interruption by ten elements first occurs in the 4th period between groups II and III, or between calcium and gallium. Ten elements also constitute the transition between groups II and III in the 5th and in the 6th periods. These groups of elements are called **transition elements.*** In comparison with the representative elements, the progression of chemical properties among the transition elements is not so marked. The transition elements thus constitute, chemically, a much more homogeneous group than the representative ones. The concept of "transition elements" is related to the progressive accretion of electrons in the d subshells of the atoms. The subgroup designation B is here reserved for transition elements.

(c) The lower section of Table 3.6 comprises the elements that interrupt the transition elements. This section originated as a reluctant

* The zinc subgroup elements (Zn, Cd, and Hg) have many properties of representative elements and therefore could be so classified.

TABLE 3.6 ■ NOMENCLATURE OF THE PERIODIC TABLE OF THE ELEMENTS

Periods	*I A*	*II A*	*III A*	*IV A*	*V A*	*VI A*	*VII A*	*0*
1	**H** 1							**He** 2
2	**Li** 3	**Be** 4	**B** 5	**C** 6	**N** 7	**O** 8	**F** 9	**Ne** 10
3	**Na** 11	**Mg** 12	**Al** 13	**Si** 14	**P** 15	**S** 16	**Cl** 17	**Ar** 18
4	**K** 19	**Ca** 20	**Ga** 31	**Ge** 32	**As** 33	**Se** 34	**Br** 35	**Kr** 36
5	**Rb** 37	**Sr** 38	**In** 49	**Sn** 50	**Sb** 51	**Te** 52	**I** 53	**Xe** 54
6	**Cs** 55	**Ba** 56	**Tl** 81	**Pb** 82	**Bi** 83	**Po** 84	**At** 85	**Rn** 86
7	**Fr** 87	**Ra** 88						

The Alkali Metals (I A); **The Alkaline-Earth Metals** (II A); **The Chalcones or Chalcogens** (VI A); **The Halogens** (VII A); **The Aerogens, or Noble Gases (formerly called "inert")** (0)

Other groups of elements are named after their Group Number (for example *The Group IV Elements,* C, Si, etc.) or after their first member (for example, *The Zinc Subgroup,* Zn, Cd, and Hg).

Periods	*III B*	*IV B*	*V B*	*VI B*	*VII B*	*VIII*			*I B*	*II B*
						The Iron Triad				
4	**Sc** 21	**Ti** 22	**V** 23	**Cr** 24	**Mn** 25	**Fe** 26	**Co** 27	**Ni** 28	**Cu** 29	**Zn** 30
						The Palladium Triad				
5	**Y** 39	**Zr** 40	**Nb** 41	**Mo** 42	**Tc** 43	**Ru** 44	**Rh** 45	**Pd** 46	**Ag** 47	**Cd** 48
						The Platinum Triad				
6	**La** 57	**Hf** 72	**Ta** 73	**W** 74	**Re** 75	**Os** 76	**Ir** 77	**Pt** 78	**Au** 79	**Hg** 80
7	**Ac** 89									

The Coinage Metals (I B)

No groups designated

Periods														
	The Rare Earth Elements, or the Lanthanides													
6	**Ce** 58	**Pr** 59	**Nd** 60	**Pm** 61	**Sm** 62	**Eu** 63	**Gd** 64	**Tb** 65	**Dy** 66	**Ho** 67	**Er** 68	**Tm** 69	**Yb** 70	**Lu** 71
	The Heavy Rare Earth Elements, or the Actinides													
7	**Th** 90	**Pa** 91	**U** 92	**Np** 93	**Pu** 94	**Am** 95	**Cm** 96	**Bk** 97	**Cf** 98	**Es** 99	**Fm** 100	**Md** 101	**No** 102	**Lw** 103

footnote to accommodate a frustrating group of 14 elements that all seemed to want to crowd into the one space in the 6th period in group III occupied by lanthanum. The elements were frustrating because they all seemed to be alike; it was difficult to characterize them, to separate them, to find out how many there were, to be sure indeed that some of the "elements" would not prove to be mixtures after further attempts at purification. This series came to be called the **"rare earth elements,"** or the **"lanthanides"*** (the word "earth" was used in the sense of "oxide"). That the 14 elements following actinium in the 7th period constitute an analogous series was learned much later; accordingly its members are called the **"heavy rare earth elements,"** or the **"actinides."*** Taken together, the two series are called the **"inner transition elements."** The actinides are *presumed* to be an interruption in the 7th period transition elements, although the elements that would continue after this interruption (starting with atomic number 104) have not yet been prepared (1967). The concept of "inner transition elements" is related to the progressive accretion of electrons in the f sublevels of the atoms.

(d) The slashes in the upper and middle sections show the interruptions in atomic number sequences that are filled by the transition and inner transition elements.

(e) The designations of energy levels in the listing to the left of the table afford, except for certain irregularities, a quick count of the number of electrons in the various sublevels.

EXAMPLE 1 How many $5f$ electrons and how many $6d$ electrons does $_{93}$Np have?

ANSWER Table 3.6 shows that Np is the fourth element in the $5f$ series; therefore it is $5f^4$. This entire period is taken from the position just after $_{89}$Ac; therefore it is $6d^1$. It must be recognized that the convenience of this method does not obviate the uncertainty of some of these electronic assignments (p. 33).

3.6 □ THE PERIODICITY OF CHEMICAL PROPERTIES

It will be helpful here to survey briefly some of the chemical properties of the elements in relation to their classification in the periodic system.

In writing chemical equations (see Chapter 5), we insert the following symbols after a formula to denote the state of matter of the substance under the conditions of the reaction: (c) crystalline solid; ($amorph$) amorphous or noncrystalline solid; (l) liquid; (g) gas; (sol) solution; (aq) aqueous (water) solution.

* Lanthanum and actinium, by virtue of their properties, are included in the classification of "lanthanide" and "actinide," respectively. Thus the space corresponding to group IIIB, period 6, may be assigned to all the elements $_{57}$La → $_{71}$Lu, and the corresponding space in period 7 to $_{89}$Ac → $_{103}$Lw.

The alkali metals These elements are soft, lustrous, highly reactive metals. They react vigorously with water, even in the cold, forming a solution of the hydroxide and liberating hydrogen gas and heat,

$$Li(c) + H_2O(l) \longrightarrow Li^+(aq) + OH^-(aq) + \tfrac{1}{2}H_2(g)$$

The alkali metals also react very vigorously with nonmetallic elements like oxygen and the halogens; in such reactions the alkali metals are converted to their 1+ ions.

The alkaline earth metals These elements resemble the alkali metals but are not quite so soft, lustrous, or reactive as the corresponding alkali metals. The alkaline earth metals, when they combine with nonmetals, are converted to their 2+ ions,

$$Ba(c) + Cl_2(g) \longrightarrow BaCl_2(c), \quad \textit{also written } Ba^{2+} + 2Cl^-$$

They react with water, but more slowly than the alkali metals,

$$Ca(c) + 2H_2O(l) \longrightarrow Ca^{2+}(aq) + 2OH^-(aq) + H_2(g)$$

The group III elements Boron is not classified as a metal; the other group III elements are. Boron is a hard, brittle, dull, rather inert element. It forms a series of hydrides and a series of complex borates (borate ion is BO_3^{3-}). Aluminum is a light, soft, abundant metal that protects itself from oxidation in air by forming a film of the oxide, Al_2O_3. Gallium, indium, and thallium are lustrous, moderately reactive metals.

The group IV elements These elements show a progression of metallic character from carbon (nonmetallic) to silicon and germanium (some metallic properties) to tin and lead (metals). Elemental carbon occurs in the forms of graphite and diamond. Such different forms of the same element are called **allotropic forms.** Carbon occurs in chemical combination in all living matter and in its derivatives such as coal and petroleum, in the atmosphere as CO_2 (carbon dioxide), in the toxic oxide CO (carbon monoxide), and in carbonate (CO_3^{2-}) minerals such as limestone, $CaCO_3$. Silicon is a dull solid, somewhat metallic in appearance. It forms hydrides which burn in oxygen to yield SiO_2 and water. A typical reaction is

$$SiH_4(g) + 2O_2(g) \longrightarrow SiO_2(c) + 2H_2O(g)$$

Silicon is widely distributed in the earth's crust as silicate rocks. Germanium is an element roughly midway between metallic and nonmetallic in character; the subject of semimetals in general will be discussed in Chapter 23. Tin and lead are relatively inert metals, widely used in the home and in industry. Tin dissolves in acid solutions to form stannous (2+) salts,

$$Sn(c) + \underbrace{2H^+(aq) + 2Cl^-(aq)}_{\textit{hydrochloric acid}} \longrightarrow Sn^{2+}(aq) + 2Cl^-(aq) + H_2(g)$$

Most of the salts of lead are insoluble in water; as a result, the action of acids on lead deposits an impervious layer, enabling the lead to resist further attack. Lead is therefore used widely as an "acidproof" material of construction.

The group V elements These elements, like those of group IV, also show a marked progression from nonmetallic (nitrogen, phosphorus, and arsenic) to metallic character (antimony and bismuth). Nitrogen is an odorless, colorless, relatively inert gas that constitutes about 80% of the earth's atmosphere. In combustion reactions in air, atmospheric nitrogen combines with oxygen to form nitric oxide and nitrogen dioxide,

$$N_2(g) + O_2(g) \longrightarrow 2NO(g)$$
$$2NO(g) + O_2(g) \longrightarrow 2NO_2(g)$$

Nitrogen also combines with metals to form **nitrides,** such as Mg_3N_2. The hydrides of nitrogen include ammonia (NH_3) and hydrazine (N_2H_4). Phosphorus is a soft solid that occurs in white and red allotropic forms. The white form, dangerous to handle, sublimes to give a vapor whose formula is P_4. This vapor ignites spontaneously in air or oxygen to form an oxide in which phosphorus is pentavalent,

$$P_4(g) + 5O_2(g) \longrightarrow P_4O_{10}(c)$$

The hydride of phosphorus is **phosphine,** PH_3, a very toxic gas. Arsenic exists in nonmetallic (yellow) and metallic (gray) modifications. It forms a hydride (arsine, AsH_3) and both trivalent and pentavalent oxides, As_2O_3 and As_2O_5, all very toxic compounds. Antimony and bismuth are both metals not useful in applications requiring great structural strength. Antimony forms the unstable hydride **stibine,** SbH_3. Both metals form the trivalent oxide (Sb_2O_3 and Bi_2O_3) and the pentavalent oxide (Sb_2O_5 and Bi_2O_5). All the group V elements form the trifluoride and the trichloride.

The group VI elements These elements are nonmetals except for the highest member, polonium. Gaseous oxygen is colorless and odorless; liquid oxygen is blue. Oxygen is a reactive element that combines with all the elements except some noble gases. In a number of combinations, oxygen atoms are bonded to each other; such compounds are called **peroxides.** A typical example is hydrogen peroxide, HOOH. Oxygen, when energized by ultraviolet irradiation or other means, produces ozone, an irritating, chemically reactive gas,

$$3O_2(g) \xrightarrow{\text{energy}} 2O_3(g)$$

Sulfur is a pale yellow solid obtained from underground beds of the element by melting and forcing it up with a mixture of steam, superheated water (about 175°C), and compressed air (**Frasch process**). The dioxide, SO_2, and trioxide, SO_3, are both well known. Sulfur combines with practically all metals to form sulfides. Its hydride is the foul-smelling gas H_2S. Selenium and tellurium are both semimetallic. Their hydrides, H_2Se and H_2Te, have even more evil odors than H_2S, and the stench, toxicity, and absorption and retention by the human body of these compounds and their derivatives have discouraged the investigation of their properties. Polonium is highly radioactive and the investigation of its chemistry is limited.

The group VII elements, the halogens These elements are nonmetallic, toxic, reactive, and gaseous or volatile at ordinary temperatures. Fluorine is a pale yellow gas of very high reactivity. It reacts violently with hydrogen,

$$H_2(g) + F_2(g) \longrightarrow H_2F_2(g)$$

Chlorine, a greenish-yellow gas, is also very reactive. It will replace hydrogen from hydrocarbons,

$$CH_4(g) + 4Cl_2(g) \longrightarrow CCl_4(l) + 4HCl(g)$$

Its hydride is HCl. Bromine at room temperature is a dense, dark red, volatile liquid. It is sufficiently reactive to combine rapidly with most metals,

$$Zn(c) + Br_2(l) \longrightarrow ZnBr_2(c), \quad \textit{also written } Zn^{2+} + 2Br^-$$

Its hydride is HBr. Iodine is a black solid that evaporates to yield a violet vapor. Its hydride is HI. The chemistry of astatine has not been extensively investigated, but it forms compounds in which it exists as astatide ion, At^-. Halogens form compounds with each other called **interhalogen compounds.** Examples are BrCl, ClF_3, IF_5, IF_7, and ICl.

The noble gases or aerogens, group 0 These gaseous elements were once thought to participate in no chemical reactions; they were therefore called "inert." In recent years "inert" gas chemistry (especially that of xenon) has been studied. (This subject is treated in Ch. 20.)

The transition and inner transition elements These are all metals, ranging widely in physical properties. The highest melting, most brittle, hardest, and densest metals are found among these elements. (The general subject of metallic character will be treated in Chapter 23.)

3.7 □ ATOMIC STRUCTURE AND PERIODIC PROPERTIES OF ATOMS

Size of atoms and ions By virtue of the indefinite position of an electron about a nucleus, the actual radius of an atom or an ion is not a definite, physically measurable quantity. Nevertheless, electron and x-ray diffraction experiments yield definite information about internuclear distances, called "interatomic" distances in molecules and "interionic" distances in ionic solids. Interatomic distances are also known as **bond lengths,** or **bond distances.** If enough assumptions are made, a radius can be assigned to each atom or ion, whose addition for a pair of atoms or ions would equal the measured internuclear distance.

Following the monumental work of Sir William H. Bragg and his son, Sir Lawrence Bragg (Ch. 12), it became evident through the work of Maurice Huggins and Linus Pauling that the radius of an atom or an ion is a function of its environment, that is, the electrical nature of the neighboring atoms in molecules and the geometric arrangement of the ions in crystals. An extreme example is the hydrogen atom, which must be assigned one value when it combines with itself, a second value when it combines with elements of the second period, a third value when it combines with elements of the third period, and so on. Most ionic radii, however, may be treated as fairly constant. It is, therefore, advantageous for many purposes to assign a radius to each atom, called the **"covalent radius,"** and to each ion, called the **"crystal (ionic) radius."**

The variation in the size of atoms and ions within a period and a group is illustrated for the representative elements in Table 3.7.

Within a given period, the atomic radius decreases with increasing atomic number owing to the effect of increasing the nuclear charge without introducing another shell (principal quantum level). The irregularity at nitrogen may be associated with the stability of the half-filled p orbitals (p. 29). Within a given family, however, the atomic radius increases with atomic number because of the addition of another principal quantum level. Several notable exceptions occur in which the atomic radius within a family increases only very slightly or actually decreases with increasing atomic number; for example, $_{13}$Al 1.25 Å and $_{31}$Ga 1.25 Å; $_{46}$Pd 1.28 Å and $_{78}$Pt 1.29 Å; $_{40}$Zr 1.45 Å and $_{72}$Hf 1.44 Å. This results from the presence of the intervening transition elements and, in particular, the inner transition elements, in which the added electrons generally enter an inner principal quantum level without alteration in the outer principal quantum level. We would then expect the added electrons virtually to nullify the electrostatic attraction effect of the increased nuclear charge, maintaining thereby an approximately constant atomic size for each transition series. Actually, however, small but significant decreases occur in each of these series.* Hence, the size increase characteristic of families generally does not appear after these transition series.

Variations in the crystal radius are also regular. It increases downward in a group. However, it decreases across a period for **isoelectronic ions,** ions possessing the same electron configuration, illustrated by Li^+ $1s^2$, Be^{2+} $1s^2$, and by N^{3-} $1s^22s^22p^6$, O^{2-} $1s^22s^22p^6$, F^- $1s^22s^22p^6$, Na^+ $1s^22s^22p^6$, Mg^{2+} $1s^22s^22p^6$, Al^{3+} $1s^22s^22p^6$.

* This shrinkage, for the inner transition series, is known as the **lanthanide contraction.**

TABLE 3.7 ■ COVALENT* AND CRYSTAL (IONIC) RADII IN ANGSTROM UNITS

Period	I A	II A	III A	IV A	V A	VI A	VII A	
1	H {0.28, 0.38}							
	H^- 2.08							
2	Li 1.33	Be 0.90	B 0.80	C 0.77	N 0.73	O 0.74	F 0.71	
				C^{4+} 0.16				
	Li^+ 0.60	Be^{2+} 0.31	B^{3+} 0.23	C^{4-} 2.60	N^{3-} 1.71	O^{2-} 1.40	F^- 1.36	
3	Na 1.54	Mg 1.36	Al 1.25	Si 1.15	P 1.1	S 1.02	Cl 0.99	
	Na^+ 0.95	Mg^{2+} 0.65	Al^{3+} 0.50	Si^{4+} 0.42	P^{3-} 2.12	S^{2-} 1.84	Cl^- 1.81	
4	K 1.96	Ca 1.74	Ga 1.26	Ge 1.22	As 1.19	Se 1.16	Br 1.14	
	K^+ 1.33	Ca^{2+} 0.99	Ga^{3+} 0.62	Ge^{4+} 0.53	As^{3-} 2.22	Se^{2-} 1.98	Br^- 1.95	
5	Rb 2.16	Sr 1.92	In 1.44	Sn 1.41	Sb 1.38	Te 1.35	I 1.33	
	Rb^+ 1.48	Sr^{2+} 1.13	In^{3+} 0.81	Sn^{4+} 0.71	Sb^{3-} 2.45	Te^{2-} 2.21	I^- 2.16	
6	Cs 2.35	Ba 1.98	Tl 1.48	Pb 1.47	Bi 1.46	Po	At	
	Cs^+ 1.69	Ba^{2+} 1.35			Bi^{3+} 0.96			
7	Fr	Ra						

Period	III B	IV B	V B	VI B	VII B	VIII	VIII	VIII	I B	II B
5	Sc 1.44	Ti 1.36	V 1.36	Cr	Mn	Fe	Co	Ni	Cu 1.38	Zn 1.31
	Sc^{3+} 0.81								Cu^+ 0.96	Zn^{2+} 0.74
6	Y 1.62	Zr 1.48	Nb	Mo	Tc	Ru	Rh	Pd	Ag 1.53	Cd 1.48
	Y^{3+} 0.93								Ag^+ 1.26	Cd^{2+} 0.97
7	La 1.69	Hf	Ta	W	Re	Os	Ir	Pt	Au 1.50	Hg 1.49
									Au^+ 1.37	Hg^{2+} 1.10
	Ac									

* Covalent radii (taken mostly from *Table of Interatomic Distances*, London Chem. Soc., Special Publ., 1958) are applicable only to single-bonded atoms in mainly covalent molecules. Crystal (ionic) radii are from the publications of Linus Pauling.

Although wave mechanics negates the concept of a definite size, we can think of the radius as roughly the distance from the nucleus corresponding to the maximum density of electronic charge. This distance is related to the principal quantum number of the orbitals with the highest n value—3 for Na to Cl, 6 for Cs, 2 for N^{3-}, Na^+, and Al^{3+}, 3 for Cl^-, 5 for Cs^+—and is inversely proportional to the nuclear charge. The electrons occupying the orbitals with smaller values of n, however, partially neutralize the nuclear field.* This "screening effect" reduces the nuclear charge to a so-called "effective nuclear charge." Thus, across a period, the maximum n value remains constant but the effective nuclear charge increases; the electrons are more strongly attracted and the radius decreases. In the transition series, the addition of the electrons to the $3d$ orbitals and to the $4f$ orbitals almost cancels the increase in nuclear charge, resulting in a small decrease in radius. These relationships are shown in the following sequences.

Element	$_{11}$Na	$_{12}$Mg	$_{13}$Al	$_{14}$Si	$_{15}$P	$_{16}$S	$_{17}$Cl
Highest principal energy level, n	3	3	3	3	3	3	3
How many electrons in lower n levels?	10	10	10	10	10	10	10
"Effective nuclear charge"	+1	+2	+3	+4	+5	+6	+7
Covalent radii	Marked decrease → (due to increasing "nuclear charge," constant n)						

Element	$_{21}$Sc	$_{22}$Ti	$_{23}$V	$_{24}$Cr	$_{25}$Mn	$_{26}$Fe	$_{27}$Co	$_{28}$Ni	$_{29}$Cu
Highest n	4	4	4	4	4	4	4	4	4
How many electrons in lower n's?	19	20	21	23	23	24	25	26	28
"Effective nuclear charge"	+2	+2	+2	+1	+2	+2	+2	+2	+1
Covalent radii	Not much change → (due to practically constant "nuclear charge," constant n)								

Down a group, the maximum value of n and the nuclear charge increase; however, in going, for example, from $_{3}$Li to $_{55}$Cs, the screening effect largely cancels the increase in charge and the radius increases.

Ionization energy The minimum energy required to remove an electron from a gaseous atom in its ground state is called the first **ionization energy** or the first **ionization potential;** the second ionization energy refers to the removal of a second electron from the positive ion.

Ionization process	*Ionization energy*
$He(g) - 1e^- \longrightarrow He^+(g)$	567 kcal/mole† required
$He^+(g) - 1e^- \longrightarrow He^{2+}(g)$	1254 kcal/mole required

in which e^- represents an electron.

* The inner electrons do not prevent the nucleus from attracting the outer electrons, but merely add their own repulsive effect.

† A mole of an element is the quantity in 6.02×10^{23} atoms. See Ch. 4.

The ionization energy (potential) can be measured in a discharge tube containing the gas or vapor under low pressure. Initially, the current flow through the tube is practically zero; as the voltage between the cathode and anode is increased, a well-defined voltage is reached at which the gas is ionized; at this voltage, the electrons have acquired sufficient energy for impact of an electron upon an atom to produce ionization. This event is signaled by a sudden very large increase in the current flow through the tube. This well-defined value of the voltage is known as the ionization potential or the ionization energy, commonly expressed in units of volts or electron volts, convertible to ergs per atom. One electron volt is equivalent to 3.83×10^{-20} cal/atom, or 23.1 kcal/mole. With hydrogen atoms, for example, a very large increase in current flow occurs when the voltage is 13.60 V; the ionization energy of the hydrogen atom is then said to be 13.60 electron volts.

Among the representative elements, the first ionization energy increases with the completion of an s orbital, decreases upon the addition of an electron to the p orbital, and then increases across the period with an irregularity at p^3:

H							He	
$1s$							$1s^2$	
314							567	kcal/mole

Li	Be	B	C	N	O	F	Ne	
$2s^1$	$2s^2$	$2s^22p^1$	$2s^22p^2$	$2s^22p^3$	$2s^22p^4$	$2s^22p^5$	$2s^22p^6$	
124	215	191	260	335	314	402	497	kcal/mole

The high ionization energy of nitrogen is most likely due to the stability of the half-filled p orbitals. Within a group, the ionization energy decreases downward, with some exceptions:

$_{3}$Li	124 kcal/mole	$_{5}$B	191 kcal/mole
$_{11}$Na	118 kcal/mole	$_{13}$Al	138 kcal/mole
$_{19}$K	100 kcal/mole	$_{31}$Ga	138 kcal/mole
$_{37}$Rb	96.3 kcal/mole	$_{49}$In	133 kcal/mole
$_{55}$Cs	89.7 kcal/mole	$_{81}$Tl	141 kcal/mole

This neat pattern, however, is somewhat upset by the transition elements. We shall therefore content ourselves with the broad generalization that the first ionization energy increases across a period and decreases down a group.

Among the representative elements, metallic character—malleability, good conduction of heat and electricity, tendency to produce positive ions—decreases rapidly across a period, excepting period 1, and increases downward within a group. This gradation of metallic properties is related to the trends in ionization energy; *the smaller the ionization energy, the greater the metallic character.*

The properties of the elements are related not only to ionization energies, but also to excitation energies—the energy needed to promote an electron to the next available orbital. For example,

Be, $1s^22s^2$	He, $1s^2$
Excitation possible within occupied electron shell $\rightarrow 1s^22s^12p^1$. Therefore Be is reactive.	Excitation requires promotion to new electron shell $\rightarrow 1s^12s^1$. Requires $\sim 10^4$ times as much energy as for Be promotion. Therefore He is inert.

The aufbau rules (p. 29) do not necessarily apply to ions. For example, the configuration ${}_{21}$Sc $KL\ 3s^23p^63d^14s^2$ is correct; but the assumption that Sc^+, being isoelectronic with Ca, is $KL\ 3s^23p^63d^04s^2$ predicts that Sc^+ is not paramagnetic. But it is. Its magnetic moment corresponds to two unpaired electrons. This means that the ionization of Sc yields Sc^+ $KL\ 3s^23p^63d^14s^1$; thus for the Sc^+ ion, the $3d^14s^1$ state is more stable (lower energy) than the $3d^24s^0$ or $3d^04s^2$ state. The configuration of Sc^{2+} is $KL\ 3s^23p^63d^1$. Thus, for the ions of the first transition elements, the $3d$ sublevel is usually more stable than the $4s$. For example, the ionization of ${}_{27}$Co, $KL\ 3s^23p^63d^74s^2$, yields ${}_{27}Co^+$ with a $3d^84s^0$ configuration. On the other hand, the formation of ${}_{39}Y^+$ KLM $4s^24p^64d^05s^2$ shows that the $5s$ sublevel is more stable than the $4d$ for this ion. This means that, for any atom except hydrogen, the energy of one atomic orbital is actually dependent in a complicated manner on the number of electrons in all the other atomic orbitals. The forces involved, nucleus-electron attraction and repulsion between electrons, cannot be described or evaluated in a simple manner.

Electron affinity Most neutral atoms with incomplete orbitals have an attraction for electrons. The energy change accompanying the addition of an electron to a gaseous atom in its ground state is called the **electron affinity,** exemplified by the equations

$$Cl(g) + 1e^- \longrightarrow Cl^-(g) \qquad 88.2 \text{ kcal/mole released}$$
$$O(g) + 1e^- \longrightarrow O^-(g) \qquad 53.8 \text{ kcal/mole released}$$
$$O^-(g) + 1e^- \longrightarrow O^{2-}(g) \qquad 210 \text{ kcal/mole required}$$

We should expect the attraction to be greatest for the smallest atom within a given family and a given period, because the nucleus is then closest to the electrons. The electron affinity is difficult to measure, and meager available data indicate that it is unwise to generalize on the expected relationship between size and electron affinity of atoms.

PROBLEMS

1. Valence Assuming constant valence, write the missing formula in each set: (a) Rubidium oxide, Rb_2O; rubidium oxalate, $Rb_2C_2O_4$; strontium hydride,

SrH_2; strontium oxalate. (b) Osmium oxide, OsO_2; hydrogen telluride, H_2Te; osmium telluride. (c) Neodymium oxide, Nd_2O_3; ammonia, NH_3; neodymium nitride.

2. Nomenclature Complete the following table.

Name	Formula
potassium nitrate	KNO_3
potassium nitrite	———
sodium selenite	Na_2SeO_3
———	Na_2SeO_4
calcium hypochlorite	———
aluminum perchlorate	———

3. Mendeleev table Write the type formulas for the hydrides of elements of Groups I, II, and III, and the chlorides of elements of Groups IV, V, VI, and VII. Criticize the use of such type formulas in predicting the compositions of real compounds.

4. Modern periodic table If the atomic numbers in yet undiscovered elements were to follow a sequence analogous to the structure of the existing table, what atomic numbers would appear in (a) group VB, period 7; (b) group VIA, period 7? Predict the electron configurations of these elements.

5. Chemical properties Complete and balance the following equations:

(a) $Cs(c) + H_2O(l) \longrightarrow$

(b) $Rb(c) + Cl_2(g) \longrightarrow$

(c) $Sr(c) + F_2(g) \longrightarrow$

(d) $Si_2H_6(g) + O_2(g) \longrightarrow$

(e) $N_2(g) + Ca(c) \longrightarrow$

(f) $N_2H_4(g) + O_2(g) \longrightarrow NO_2(g) +$

(g) $PH_3(g) + O_2(g) \longrightarrow$

(h) $Bi(c) + F_2(g) \longrightarrow$

(i) $S_2(g) + Mg(g) \longrightarrow$

(j) $Al(c) + Br_2(l) \longrightarrow$

(k) $CH_4(g) + F_2(g) \longrightarrow$

6. Size, ionization energy, and electron affinity For each of the following pairs, state which one has the larger radius, which the greater ionization energy, and which the greater electron affinity. (When the electron affinity is high, the reaction $X + e^- \rightarrow X^-$ releases energy.)

(a) F,F^-; (b) Cl,Na; (c) O,Se; (d) Li,K; (e) Rb^+,Rb; (f) Br,I

7. Size From the radii given in Table 3.7 (p. 56), calculate (a) the interionic distance in a sodium chloride crystal; (b) the interatomic distance in a diamond; (c) the interatomic distance in a quartz (SiO_2) crystal.

8. Promotion energy (a) Pick the promotion requiring the smaller amount of energy:

(i) Mg $KL\ 3s^2 \longrightarrow$ Mg $KL\ 3s^1 3p^1$
(ii) Ar $KL\ 3s^2 3p^6 \longrightarrow$ Ar $KL\ 3s^2 3p^5 4s^1$

(b) Compare, qualitatively, the energy needed to promote an electron to the next available orbital for the noble gases with that needed for the other elements.

9. Valence and formula (a) What is the valence of the element combined with oxygen or hydrogen in each of the following compounds: Au_2O_3, RuO_4, CaH_2, UO_3, Cs_2O, N_2O_5, P_4O_{10}, SiH_4? (b) Does the definition of valence given on p. 41 imply that the valence must be a whole number? Are whole number valences applicable to the elements in the compounds Fe_3O_4 (magnetite), HN_3 (hydrazoic acid), Cu_4O (copper suboxide), and C_3O_2 (carbon suboxide)?

ADDITIONAL PROBLEMS

10. Medeleev table Advance arguments that might have been appropriate in 1871 for the inclusion of Cu, Ag, and Au in group I; do the same for inclusion of these metals in group VIII. Which set of arguments do you favor?

11. Ionization energy The ionization of molecular oxygen, $O_2 \rightarrow O_2^+ +$ electron, requires 282 kcal/mole. Oxygen combines with platinum hexafluoride vapor, PtF_6, to form the compound dioxygenyl hexafluoroplatinate(V), $O_2^+(PtF_6)^-$. The ionization energy of the "inert" gas xenon, $Xe \rightarrow Xe^+ +$ electron, is 280 kcal/mole. Would you undertake the preparation of the corresponding compound, $Xe^+(PtF_6)^-$? Explain your answer.

12. Ionization energy Plot the following graphs, with the first ionization energy along the y axis: (a) Ionization energy versus atomic number, for the elements of period 2, group IA, and of the transition series Sc to Cu (all of these on one graph). (b) Ionization energy versus covalent radii for these elements (all on another graph). Ionization energies of the transition series are given below; others appear on p. 58.

$_{21}Sc$	$_{22}Ti$	$_{23}V$	$_{24}Cr$	$_{25}Mn$	$_{26}Fe$	$_{27}Co$	$_{28}Ni$	$_{29}Cu$	
154	157	155	156	171	181	181	176	178	kcal/mole

What conclusions would you draw from these graphs?

13. Radii of isoelectronic particles S^{2-}, Cl^-, Ar, K^+, Ca^{2+}, and Sc^{3+} are all isoelectronic. Using the data in Table 3.7 for the ions, plot radius (y axis) versus atomic number. By graphic interpolation, estimate the atomic radius for argon.

14. Radii Account for the following data: the decrease in the radius for the seven elements Li to F is 0.63 Å, while the decrease for the 15 elements La to Lu is only 0.13 Å.

15. Ionic radii Account for the large difference between the ionic radii of Sb and Bi as shown in Table 3.7 (p. 56).

16. Ionic radii (a) Plot the assigned covalent radii (y axis) versus atomic number for the elements C through Si, as given in Table 3.7. (b) On the same graph, plot the corresponding ionic radii. (c) Account for the shape of plot in (a). (d) Explain the relation in size between each ion and its parent atom.

4 ATOMIC AND MOLECULAR WEIGHTS. THE MOLE

4.1 THE LAW OF CONSERVATION OF MATTER

Measurements show that, within the experimental error, the mass (weight)* of substances in a sealed container does not change during a chemical change. Thus, gas containing oxygen, O_2, and hydrogen, H_2, is sealed in a tube so that no matter can enter or escape. The tube and contents are weighed. Passage of an electric spark across the tube causes hydrogen and oxygen gases to disappear, and liquid water, H_2O, and heat are produced. After the original temperature is restored, the tube and contents are reweighed. The observed weights, before and after the chemical change, are identical within the precision of the balance. Such observations are summarized in the **law of conservation of matter:** *the mass of a chemically reacting system remains constant.* This law is consistent with the data obtained with the most precise balances available. If matter is created or destroyed, the quantity is less than can be detected with the best available balance.

4.2 THE LAW OF DEFINITE PROPORTIONS

Analyses of compounds show that when elements form a given compound, they always combine in the same ratio by weight. For example, independently of the source or method of formation, silicon dioxide, SiO_2, contains 46.7% by weight of silicon and 53.3% of oxygen. This knowledge is summarized in the **law of definite proportions:** *the weight composition of a given compound is constant, regardless of its source or method of preparation.*

* "Weight" and "mass" will be used synonymously (considered in Appendix 1).

4.3 □ THE ATOMIC THEORY

The weight relationships of substances participating in chemical reactions are clearly explained in terms of the atomic theory. Although John Dalton (1803) is generally recognized as the founder of the theory, he was anticipated by other scientists, particularly William Higgins (1789). The novel and central point of Dalton's activities was the attempt to determine the relative weights of atoms. This goal focused attention upon his theory, and initiated a new field of human endeavor that ultimately made chemistry a systematized body of knowledge.

The assumptions of the atomic theory are

(1) *The elements are composed of indivisible particles called atoms.*

(2) *All the atoms of a given element possess identical properties, for example, mass.*

(3) *The atoms of different elements differ in properties.*

(4) *These atoms are the units of chemical changes; chemical changes merely involve the combination or the rearrangement of atoms; atoms are not destroyed, created, or changed.*

(5) *When atoms combine, they combine in fixed ratios of whole numbers forming particles known as molecules.*

This theory differed fundamentally from previous theories of matter in that it endowed atoms with definite properties, particularly with definite masses, and limited the number of different kinds of atoms to the number of elements known*

The theory offers an acceptable explanation of the laws of chemical change.

First, the conservation of matter: Since chemical reactions proceed by the combination or rearrangement (not the destruction or creation) of atoms of fixed mass, the total mass must be conserved.

Second, the law of definite proportions: since the mass of the atoms of any given element is constant and since the atoms, by assumption, combine in fixed whole number ratios for a given combination, the weight ratio of the elements in a given compound must be constant. For example, 20×10^{10} C atoms combine with 20×10^{10} O atoms forming 20×10^{10} CO molecules. Each CO molecule contains 1 C atom and 1 O atom. Since each molecule contains the same number of C and O atoms and since the masses of these atoms are also constant, the mass ratio of carbon to oxygen must be constant in any given quantity of carbon monoxide.

* Present knowledge recognizes that the atoms of a given element do not possess uniform masses. Silver, for example, occurs as two isotopes (p. 12), but as found in nature, these silver atoms are always mixed in the same number ratio so that the average mass of the silver atoms is always the same.

4.4 □ THE LAW OF COMBINING VOLUMES; THE AVOGADRO HYPOTHESIS

As a result of his experimental studies of the volumes of gases involved in chemical changes, Joseph Gay-Lussac concluded (1808) that *the volumes of gases consumed and produced in a chemical reaction, measured at the same temperature and pressure, are in ratios of small whole numbers*. For example, at the same temperature and pressure,

10.6 ml hydrogen gas combines with 10.6 ml chlorine gas
to produce 21.2 ml hydrogen chloride gas (1:1:2).
20.4 ml hydrogen gas combines with 10.2 ml oxygen gas
to produce 20.4 ml water vapor (2:1:2).
12 ml hydrogen gas combines with 4.0 ml nitrogen gas
to produce 8.0 ml ammonia gas (3:1:2).
6.3 ml water vapor combines with solid carbon
to produce 6.3 ml hydrogen and 6.3 ml carbon monoxide gas (1:1:1).

In 1811 Amadeo Avogadro, whose theoretical work was ignored for half a century, showed that these results may be explained by the assumption that molecules of elements are composed of like atoms, molecules of compounds are composed of unlike atoms, and, further, that *equal volumes of all gases under the same conditions of temperature and pressure contain the same number of molecules* (p. 179). Thus, let N be the number of molecules in a given volume of gas; then, according to the Avogadro hypothesis, the experimental observation

$$\begin{matrix}\text{1 volume} \\ \text{of hydrogen}\end{matrix} + \begin{matrix}\text{1 volume} \\ \text{of chlorine}\end{matrix} \longrightarrow \begin{matrix}\text{2 volumes of} \\ \text{hydrogen chloride}\end{matrix}$$

means that

$$\begin{matrix}N\text{ molecules} \\ \text{of hydrogen}\end{matrix} + \begin{matrix}N\text{ molecules} \\ \text{of chlorine}\end{matrix} \longrightarrow \begin{matrix}2N\text{ molecules of} \\ \text{hydrogen chloride}\end{matrix}$$

and dividing by N yields

$$\begin{matrix}\text{1 molecule} \\ \text{of hydrogen}\end{matrix} + \begin{matrix}\text{1 molecule} \\ \text{of chlorine}\end{matrix} \longrightarrow \begin{matrix}\text{2 molecules of} \\ \text{hydrogen chloride}\end{matrix}$$

Since a molecule of hydrogen chloride must contain at least one atom of hydrogen and one atom of chlorine, the symbol HCl may be used to represent a hydrogen chloride molecule. But there are 2 hydrogen chloride molecules for every 1 hydrogen molecule and 1 chlorine molecule. Whence, to satisfy the law of conservation of matter, 1 hydrogen molecule must contain at least 2 hydrogen atoms and 1 chlorine molecule must contain at least 2 chlorine atoms, or

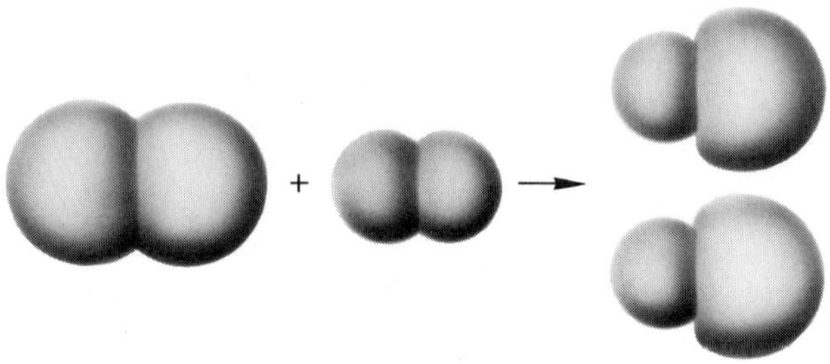

$$\begin{array}{lllll} H_2 & + & Cl_2 & \rightarrow & 2HCl \\ \text{1 molecule} & + & \text{1 molecule} & \rightarrow & \text{2 molecules} \\ \text{1 volume} & + & \text{1 volume} & \rightarrow & \text{2 volumes} \end{array}$$

This explanation, however, does not prove that hydrogen and chlorine molecules in the gaseous state are composed of 2 atoms (**diatomic**). Nor does it prove that hydrogen chloride molecules are composed of 1 atom of each constituent element. Thus, if we assume H_4, Cl_4, and H_2Cl_2 to be the molecular species, we can write

$$\begin{array}{lllll} H_4 & + & Cl_4 & \longrightarrow & 2H_2Cl_2 \\ \text{1 molecule} & + & \text{1 molecule} & \longrightarrow & \text{2 molecules} \\ \text{1 volume} & + & \text{1 volume} & \longrightarrow & \text{2 volumes} \end{array}$$

which also explains the volume relations experimentally observed, 1:1:2. We have, however, proved that the gaseous molecules of hydrogen and chlorine contain an even number of atoms, like H_2 or H_4, Cl_2 or Cl_4, and not an odd number, like H or H_3, Cl or Cl_3.

4.5 □ APPROXIMATE MOLECULAR WEIGHTS OF GASES. THE MOLE

We cannot weigh a single molecule. However, if molecules really exist, it should be possible to determine their *relative* weights by weighing large but identical number of molecules. Let us illustrate this with an analogy. Imagine that you are in a room with a large supply of coins, an automatic coin-changing machine, and a bathroom scale sensitive to ± 0.5 lb. You now state that you can determine within $\pm 1\%$ the value of the ratio $\frac{\text{wt. of 1 nickel}}{\text{wt. of 1 dime}}$, without counting any coins. Can you do it? Yes. Take any large number, N, of quarters (you need not count them), and insert them all in the change-maker. For each quarter, 2 dimes and a nickel will emerge. Then

$$N \text{ quarters} \longrightarrow 2N \text{ dimes} + N \text{ nickels}$$

Weigh all the dimes thus obtained, then all the nickels. If each set of coins weighs at least 50 lb, the error will not exceed 1%. Now if you divide *half* of your first result by your second, you have

$$\frac{\frac{1}{2}(\text{wt. of } 2N \text{ dimes})}{\text{wt. of } N \text{ nickels}} = \frac{\text{wt. of } N \text{ dimes}}{\text{wt. of } N \text{ nickels}} = \frac{\text{wt. of 1 dime}}{\text{wt. of 1 nickel}}$$

You have accomplished the task under the conditions specified.

Similarly, the Avogadro hypothesis teaches the chemist how to select equal numbers of molecules of gases so as to determine the ratios of the weights of *individual* molecules. For example,

$$\frac{\text{wt. of 1 liter of chlorine, 20°C, 1.2 atm}}{\text{wt. of 1 liter of hydrogen, 20°C, 1.2 atm}} = \frac{\text{wt. of } N \text{ molecules of chlorine}}{\text{wt. of } N \text{ molecules of hydrogen}} = \frac{\text{wt. of 1 molecule of chlorine}}{\text{wt. of 1 molecule of hydrogen}}$$

Thus when a chemist says, "The molecular weight of water is 18, the molecular weight of carbon dioxide is 44, and the molecular weight of sulfur dioxide is 64," he means there are as many molecules in 18 g of water as in 44 g of carbon dioxide, or in 64 g of sulfur dioxide. Also, since these masses contain the same number of molecules, 18, 44, and 64 may be used to represent the *relative* weights of one molecule of each of these substances.

It is convenient to refer each molecular weight to a common standard, rather than always to write ratios.

By common consent,* we take 31.9988 g of oxygen, using air as the source of oxygen. This quantity contains a definite number of molecules; we call this number the **Avogadro number** $\mathcal{N}$. Next, we determine the volume occupied by this quantity of oxygen at standard conditions (sc), 0°C and 1 atmosphere pressure.

EXAMPLE 1 The density of oxygen, determined by Gregory Baxter and Howard Starkweather in 1926, is 1.42765 g/liter at sc.* Calculate the volume occupied by 31.9988 g of oxygen at sc.

* By a majority vote of chemists taken in 1905 through the International Union of Pure and Applied Chemistry, 32.0000 g of oxygen was chosen to avoid a number smaller than 1 for the atomic weight of hydrogen. In 1961, this was changed to 31.9988 g with the adoption of the atomic weight scale based on the isotope of carbon, ^{12}C = exactly 12 (further treated on p. 75).

* This value is corrected for nonideality. It is the density that O_2 would have if it were an ideal gas. See Chapter 11.

ANSWER

$$\frac{31.9988\ \cancel{g}}{1.42765\ \dfrac{\cancel{g}}{\text{liter (sc)}}} = 22.4136 \text{ liters (sc)}$$

It follows, then, from the Avogadro hypothesis that 22.4 liters, sc, of any other gas also contains $\mathcal{N}$ molecules. Hence, the weight in grams of 22.4 liters at sc of any gas will be its molecular weight because it contains $\mathcal{N}$ molecules. The quantity containing $\mathcal{N}$ molecules, or in general $\mathcal{N}$ particles regardless of their nature, is known as a **mole.** Thus a mole of atoms contains $\mathcal{N}$ atoms, a mole of molecules contains $\mathcal{N}$ molecules, and a mole of protons contains $\mathcal{N}$ protons. The value of $\mathcal{N}$, experimentally determined (p. 182), is $(6.0225 \pm 0.0003) \times 10^{23}$ particles. Then, the quantity of matter in a mole may be expressed as 6.02×10^{23} particles per mole, 22.4 liters (sc) per mole of gas, and 22.4×10^3 ml (sc) per mole of gas. The dimensional units of molecular weight are grams per mole. Thus,

$$\text{moles} = \frac{\text{mass of substance } (\cancel{g})}{\text{molecular weight } (\cancel{g}/\text{mole})}$$

The volume (sc) occupied by one mole of a gas is known as the molar volume or the gram molecular volume.

EXAMPLE 2 Hydrocarbons are compounds composed of hydrogen and carbon. 0.179 g of the hydrocarbon methane occupies 250 ml as a vapor at sc. Calculate the molecular weight of methane.

ANSWER Since 250 ml weighs 0.179 g, the weight of 1.00 ml (which is also the density in grams per milliliter) is given by

$$\frac{0.179 \text{ g}}{250 \text{ ml (sc)}}$$

and therefore the weight of 22.4×10^3 ml is

$$\frac{0.179 \text{ g}}{250\ \cancel{\text{ml (sc)}}} \times 22.4 \times 10^3\ \frac{\cancel{\text{ml (sc)}}}{\text{mole}} = 16.0\ \frac{\text{g}}{\text{mole}}$$

The previous example shows that the molecular weight of a gas is directly related to its standard density:

$$\text{Molecular weight of a gas} = \text{density (g/ml at sc)} \times 22.4 \times 10^3 \text{ ml/mole}$$

or

$$= \text{density (g/liter at sc)} \times 22.4 \text{ liters/mole}$$

EXAMPLE 3 The molecular weight of the hydrocarbon butane is 58.1. Calculate the number of moles, the number of molecules, and the volume in liters (sc) corresponding to 12.0 g of butane.

ANSWER One mole weighs 58.1 g; hence the number of moles is given by

$$\frac{12.0\ \cancel{g}}{58.1\ \dfrac{\cancel{g}}{\text{mole}}} = 0.207 \text{ mole}$$

and since there are 6.02×10^{23} molecules in one mole,

$$0.207\ \cancel{\text{mole}} \times 6.02 \times 10^{23}\ \frac{\text{molecules}}{\cancel{\text{mole}}} = 1.25 \times 10^{23} \text{ molecules}$$

and since one mole occupies 22.4 liters (sc),

$$0.207\ \cancel{\text{mole}} \times 22.4\ \frac{\text{liters (sc)}}{\cancel{\text{mole}}} = 4.64 \text{ liters (sc)}$$

4.6 □ ATOMIC WEIGHTS FROM MOLECULAR WEIGHTS; THE CANNIZZARO METHOD

The first set of atomic weights which had the confidence of a large number of scientists was deduced from the molecular weights of substances by Stanislao Cannizzaro in 1858. The atomic weight of an element is not the weight of one atom; it is the weight of $\mathcal{N}$ or 6.0225×10^{23} atoms of the element. Thus when a chemist says, "The atomic weight of mercury, Hg, is 200.6, the atomic weight of oxygen, O, is 16.00, and the atomic weight of lead, Pb, is 207.2," he means that 6.02×10^{23} mercury atoms weigh 200.6 g, the same number of oxygen atoms weigh 16.00 g, and the same number of lead atoms weigh 207.2 g. The problem, then, is, "How do we know when we are weighing $\mathcal{N}$ atoms?"

Molecules are composed of *whole numbers of atoms; the molecular weight then must be the sum of the atomic weights of all the atoms in the molecule.* Also, it is reasonable to suppose that at least one compound of each element contains only one atom of the element per molecule. Thus, if hydrogen chloride molecules are composed of 1 hydrogen atom and 1 chlorine atom, then the molecular weight of hydrogen chloride is the sum of the atomic weight of hydrogen and the atomic weight of chlorine. If molecules of pentane are composed of 5 atoms of carbon and 12 atoms of hydrogen, then its molecular weight is the sum of 5 × atomic weight of carbon and 12 × atomic weight of hydrogen. If carbon monoxide molecules are composed of 1 carbon atom and 1 oxygen atom, then its molecular weight is the sum of the atomic weight of carbon and the atomic weight of oxygen. We can therefore conclude that *the weight of an element found in one mole of any of its compounds is the atomic weight or a whole number multiple of the atomic weight of the element.*

EXAMPLE 4 Find the weight in grams of hydrogen and chlorine in one mole of hydrogen chloride. The experimentally determined molecular weight is 36.5 g/mole and the weight composition is 2.74% hydrogen and, therefore, 97.26% chlorine.

ANSWER The weight percent means that in 100 g of the compound there are 2.74 g of hydrogen and 97.26 g of chlorine. Therefore in 36.5 g of the compound

$$36.5 \frac{\cancel{\text{g compd}}}{\text{mole}} \times \frac{2.74 \text{ g hydrogen}}{100 \cancel{\text{ g compd}}} = 1.00 \frac{\text{g hydrogen}}{\text{mole}}$$

$$36.5 \frac{\cancel{\text{g compd}}}{\text{mole}} \times \frac{97.26 \text{ g chlorine}}{100 \cancel{\text{ g compd}}} = 35.5 \frac{\text{g chlorine}}{\text{mole}}$$

there are 1.00 g hydrogen and 35.5 g chlorine.

We can conclude that 1.00 is either the atomic weight or a whole number multiple of the atomic weight of hydrogen. Similarly, 35.5 is either the atomic weight or an integral multiple of the atomic weight of chlorine.

EXAMPLE 5 The molecular weight of pentane is 72.0 g per mole and its weight composition is 16.7% hydrogen and 83.3% carbon. What statement can you make regarding the atomic weights of hydrogen and carbon?

ANSWER First, calculate the weight of each element in one mole of the compound,

$$72.0 \frac{\cancel{\text{g compd}}}{\text{mole}} \times \frac{16.7 \text{ g hydrogen}}{100 \cancel{\text{ g compd}}} = 12.0 \frac{\text{g hydrogen}}{\text{mole}}$$

$$72.0 \frac{\cancel{\text{g compd}}}{\text{mole}} \times \frac{83.3 \text{ g carbon}}{100 \cancel{\text{ g compd}}} = 60.0 \frac{\text{g carbon}}{\text{mole}}$$

We can therefore state that 12.0 is either the atomic weight or an integral multiple of the atomic weight of hydrogen; similarly, 60.0 is either the atomic weight or an integral multiple of the atomic weight of carbon.

We are now at an important juncture. The calculations with hydrogen chloride lead us to the conclusion that 1.0 may be the atomic weight or a multiple of the atomic weight of hydrogen, whereas our calculations with pentane lead us to the value of 12.0 as the atomic weight or a multiple of the atomic weight of hydrogen. Since 12.0 is larger than 1.0, 12.0 cannot be the atomic weight of hydrogen. But 1.0 is a multiple of $\frac{1}{4}$, $\frac{1}{2}$, or some other such fraction; how can we decide whether the atomic weight of hydrogen is $\frac{1}{4}$, $\frac{1}{3}$, $\frac{1}{2}$, or 1.0? Let us continue our investigation, as in Examples 4 and 5, determining the weight of hydrogen in 1 mole of many more compounds of hydrogen. If $\frac{1}{4}$, $\frac{1}{3}$, or $\frac{1}{2}$ is the atomic weight of hydrogen, then we should discover $\frac{1}{4}$ g, $\frac{1}{3}$ g, or $\frac{1}{2}$ g in 1 mole of at least one hydrogen compound. But, although hundreds of thousands of hydrogen compounds have been studied, the smallest weight ever found is 1.0 g of hydrogen per mole of compound and the weight of hydrogen in a mole of every compound is a multiple of 1.0 g. Hence, 1.0 is almost certainly the atomic weight of hydrogen. Generalizing, *the atomic weight of an element is the smallest weight of the element that can ever be found in one mole of its compounds.*

EXAMPLE 6 Determine the probable atomic weight of carbon from the following data:

Compound	Molecular weight (g/mole)	Weight % of carbon	Grams of carbon (per mole of compound)
Pentane	72.0	83.3	60.0
Carbon monoxide	28.0	42.9	12.0
Benzene	78.0	92.3	72.0
Chloroform	119.5	10.0	12.0

ANSWER 12.0 g is the smallest weight of carbon found in the gram molecular weight of its compounds and all the weights of carbon per mole are divisible by 12.0; therefore, 12.0 is probably the atomic weight of carbon.

The precision of atomic weights so obtained may be quite good. Nevertheless, we can still ask, "How certain can we be that 1.0 is the atomic weight of hydrogen, and 12.0 the atomic weight of carbon?" What, for example, is the probability of discovering a compound that contains 6.0 grams of carbon per mole? Because the number of compounds studied is very large, the probability of such an event is vanishingly small. Furthermore, these atomic weights have been verified by an independent method (p. 75). Such verification increases our confidence in the existence of atoms and molecules.

A **gram-atom** is defined as a mole of atoms (one Avogadro number of atoms). Equivalently, it is that quantity whose weight in grams is the atomic weight of the element. The dimensional units of an atomic weight are grams per mole or grams per gram-atom (g/g-atom).

4.7 □ MOLECULAR FORMULAS

A chemical symbol such as C represents not only the element C but also a definite quantity of carbon, one g-atom or 12.0 g. A molecular formula such as C_3H_8, propane, gives the actual composition in number of atoms per molecule or in number of gram-atoms per mole of compound; it also represents a definite quantity of the substance, 1 mole or 44 g ($3 \times 12.0 + 8 \times 1.0$). A molecular formula converts the weight composition into gram-atoms of each element per mole of compound. Hence, to calculate such a formula, the molecular weight, the weight composition of the substance, and the atomic weights of the constituent elements must be known. For example, in Example 5 (p. 70) from the molecular weight and weight composition, we calculated that pentane contains 12.0 g hydrogen per mole and 60.0 g carbon per mole. Since the atomic weight of hydrogen is 1.0 g/g-atom and that of carbon is 12.0 g/g-atom, the number of gram-atoms per mole is given by

$$\frac{12.0\ \frac{\cancel{\text{g hydrogen}}}{\text{mole}}}{1.0\ \frac{\cancel{\text{g hydrogen}}}{\text{g-atom}}} = 12.0\ \frac{\text{g-atoms}}{\text{mole}} = 12.0\ \frac{\text{atoms}}{\text{molecule}}$$

$$\frac{60.0\ \frac{\cancel{\text{g carbon}}}{\text{mole}}}{12.0\ \frac{\cancel{\text{g carbon}}}{\text{g-atom}}} = 5.00\ \frac{\text{g-atoms}}{\text{mole}} = 5.00\ \frac{\text{atoms}}{\text{molecule}}$$

The molecular formula of pentane is therefore C_5H_{12}.

EXAMPLE 7 Hexachlorocyclohexane, synthesized from benzene and chlorine and used as an insecticide, has a molecular weight of 291. Analysis shows 24.7% carbon, 2.06% hydrogen, and 73.2% chlorine. Using 12.0, 1.0, and 35.5 as the atomic weights, respectively, of these elements, calculate the molecular formula of hexachlorocyclohexane.

ANSWER Since a molecular formula represents the number of gram-atoms of the elements per mole of compound, first calculate the weight of each element in one mole:

$$291\ \frac{\cancel{\text{g compd}}}{\text{mole}} \times \frac{24.7\ \text{g carbon}}{100\ \cancel{\text{g compd}}} = 71.9\ \frac{\text{g carbon}}{\text{mole}}$$

$$291\ \frac{\cancel{\text{g compd}}}{\text{mole}} \times \frac{2.06\ \text{g hydrogen}}{100\ \cancel{\text{g compd}}} = 6.00\ \frac{\text{g hydrogen}}{\text{mole}}$$

$$291\ \frac{\cancel{\text{g compd}}}{\text{mole}} \times \frac{73.2\ \text{g chlorine}}{100\ \cancel{\text{g compd}}} = 213\ \frac{\text{g chlorine}}{\text{mole}}$$

Then convert these weights to the corresponding number of gram-atoms:

$$\frac{71.9\ \frac{\cancel{\text{g carbon}}}{\text{mole}}}{12.0\ \frac{\cancel{\text{g carbon}}}{\text{g-atom}}} = 5.99\ \frac{\text{g-atoms}}{\text{mole}} = 6.00^{*}\ \frac{\text{g-atoms}}{\text{mole}} = 6.00\ \frac{\text{atoms}}{\text{molecule}}$$

* The difference between 5.99 and 6.00 is within the experimental error.

$$\frac{6.00 \frac{\cancel{\text{g hydrogen}}}{\text{mole}}}{1.0 \frac{\cancel{\text{g hydrogen}}}{\text{g-atom}}} = 6.00 \frac{\text{g-atoms}}{\text{mole}} = 6.00 \frac{\text{atoms}}{\text{molecule}}$$

$$\frac{213 \frac{\cancel{\text{g chlorine}}}{\text{mole}}}{35.5 \frac{\cancel{\text{g chlorine}}}{\text{g-atom}}} = 6.00 \frac{\text{g-atoms}}{\text{mole}} = 6.00 \frac{\text{atoms}}{\text{molecule}}$$

The molecular formula is, therefore, $C_6H_6Cl_6$.

4.8 □ EMPIRICAL FORMULAS; IONIC SOLIDS

It is evident that if the molecular weight of a substance is unknown, then its molecular formula cannot be calculated. For such substances, it is, however, possible to calculate the **empirical formula**—*the simplest integral ratio in which the atoms combine*. For example, the empirical formula corresponding to $C_6H_6Cl_6$ is CHCl. The weight composition expresses the relative weights of the constituent elements in a particular compound. If we divide each relative weight by the atomic weight of the corresponding element, the relative number of gram-atoms of each constitutent element is obtained. See the back cover for a table of atomic weights.

EXAMPLE 8 Potassium osmate, a pink solid synthesized in 1960, contains 21.3% potassium, 51.6% osmium, 26.1% oxygen, and 1.10% hydrogen by weight. What is its simplest formula?

ANSWER Dividing the relative weight of each element by its atomic weight gives the relative number of gram-atoms of the elements,

$$\text{for potassium: } \frac{21.3\ \cancel{\text{g}}}{39.1 \frac{\cancel{\text{g}}}{\text{g-atom}}} = 0.545 \text{ g-atom K}$$

$$\text{for osmium: } \frac{51.6\ \cancel{\text{g}}}{190 \frac{\cancel{\text{g}}}{\text{g-atom}}} = 0.272 \text{ g-atom Os}$$

$$\text{for oxygen: } \frac{26.1\ \cancel{\text{g}}}{16.0 \frac{\cancel{\text{g}}}{\text{g-atom}}} = 1.63 \text{ g-atoms O}$$

$$\text{for hydrogen: } \frac{1.10\ \cancel{\text{g}}}{1.01 \frac{\cancel{\text{g}}}{\text{g-atom}}} = 1.09 \text{ g-atoms H}$$

The ratios of the numbers of gram-atoms of the elements are thus

$$0.545 \text{ K} : 0.272 \text{ Os} : 1.63 \text{ O} : 1.09 \text{ H}$$

To find the relative whole number ratios, divide by the smallest number, 0.272; thus

$$\frac{0.545}{0.272}\text{ K} : \frac{0.272}{0.272}\text{ Os} : \frac{1.63}{0.272}\text{ O} : \frac{1.09}{0.272}\text{ H}$$

yielding

$$2 \text{ K} : 1 \text{ Os} : 6 \text{ O} : 4 \text{ H}$$

For each Os atom, there are 2 K atoms, 6 O atoms, and 4 H atoms; the simplest formula is therefore $K_2OsO_6H_4$.

The sum of the atomic weights of all the atoms indicated by an empirical formula is called the empirical formula weight; the molecular weight is, of course, the formula weight multiplied by a whole number. By general usage, one mole of a substance means one gram molecular weight or one formula weight expressed in grams; the dimensional units of a gram formula weight are then grams per mole.

EXAMPLE 9 What is the formula weight of potassium osmate, $K_2OsO_6H_4$?

ANSWER The formula weight = 2(atomic weight of potassium) + 1(atomic weight of osmium) + 6(atomic weight of oxygen) + 4(atomic weight of hydrogen) = 368. The formula weight is 368 or 368 g/mole.

Formulas derived from molecular weights of substances in the gaseous state are valid only for the gaseous state. We are not justified in assuming that the ultimate particles of liquids and solids must be identical with the molecules of the corresponding gaseous state. For example, the molecular weight of sodium chloride vapor is about 59, determined at 1970°C and atmospheric pressure by Walther Nernst in 1904. This indicates that the vapor consists essentially of NaCl molecules; NaCl is thus the molecular formula for sodium chloride in the vapor state at about 2000°C. But *solid sodium chloride* is a typical ionic compound. The solid is not composed of NaCl or $(NaCl)_2$ or $(NaCl)_x$ molecules, nor is it composed of sodium atoms and chlorine atoms. Rather, it is composed of positively charged sodium ions (Na^+) and negatively charged chloride ions (Cl^-). It is therefore manifestly impossible to write a molecular formula for an ionic solid; an empirical formula* is, however, acceptable. Nevertheless, we do refer to the gram formula weight of an ionic substance as one mole, so that the dimensional units of gram formula weight of an ionic substance are grams per mole.

* Ionic solids such as sodium chloride and aluminum fluoride might be more clearly represented by ionic formulas, Na^+Cl^-, $Al^{3+}(F^-)_3$, but common practice dictates the use of empirical formulas, NaCl and AlF_3.

4.9 □ MASS SPECTROSCOPY; ISOTOPES; ACCURATE ATOMIC WEIGHTS

The modern scale of atomic weights (inside back cover) is based on the standard value for the isotope of carbon, $^{12}C = 12$ (exactly). The atomic weight of natural carbon, which is a mixture of isotopes, has been found to be 12.01115 ± 0.00005.

Very precise atomic weights can be obtained by the method of **mass spectroscopy.** The early pioneers in this field were Sir J. J. Thomson (about 1910), Francis Aston, Arthur Dempster, Kenneth Bainbridge, and Alfred Nier. The principle of mass spectroscopy is that when a charged particle enters a magnetic field perpendicularly, the particle moves in a circular path, and, for given charge, velocity, and magnetic field strength, the radius of the path is directly proportional to the mass of the particle (Fig. 4.1). In practice, an apparatus is set up so that a beam of particles of unit charge and uniform velocities enters a magnetic field. The heavier particles are deflected less (greater radius) and the lighter ones are deflected more (smaller radius). The result is a separation of the particles according to their masses; this separation, detected by a photographic plate or other device, is a **mass spectrum.**

Fig. 4.1 Schematic representation of the Bainbridge mass spectrograph. The apparatus is vacuum-sealed; gas pressure is about 10^{-6} torr. The velocity selector consists of a magnetic field perpendicular to the plane of the paper and an electric field in the plane of the paper. Then only those ions pass through the selector which have the same velocity, fixed by the ratio of the intensity of the electric field to the strength of the magnetic field. The radius, r, is measured in centimeters.

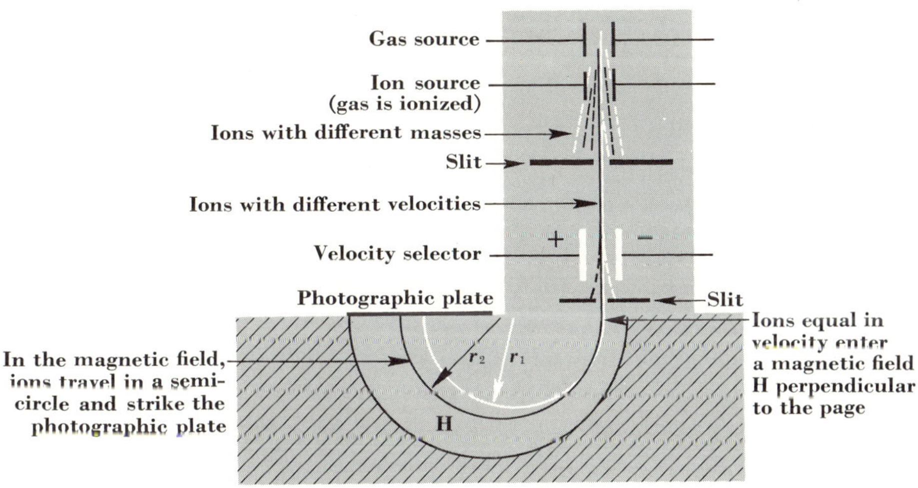

EXAMPLE 10 Atoms of the element nitrogen, with identical positive charges and velocities, produce two circular paths in a magnetic field; the radius ratio is 1.0712. What are the relative weights of the atoms of the element nitrogen?

ANSWER Since the masses of the ions are directly proportional to the radii of the circular paths, the relative masses (weights) of the ions are in the ratio of 1.0712, or one atom of nitrogen is 1.0712 times heavier than the other atom.

The atoms of the same element having different weights are called **isotopes** (p. 12); nitrogen thus consists of two isotopes. Oxygen contains three isotopes with relative weights of 1.0000:1.06277:1.12531. To compare the weights of the atoms of the elements, a standard must be established; the presently accepted standard is the isotope of carbon ^{12}C = exactly 12. 12.0000 g of ^{12}C contains as many atoms as there are atoms of oxygen in 15.9994 grams of oxygen obtained from air.

From the radii described by ^{12}C and ^{16}O, the atomic weight of the isotope ^{16}O is 15.9949. The atomic weights of the other isotopes of oxygen are thus 15.9949×1.06277 and 15.9949×1.12531, or 16.9991 and 17.9992, abbreviated as ^{17}O and ^{18}O. The atomic weights of the isotopes of nitrogen are 14.003 and 14.003×1.0712, or 14.003 and 15.000, abbreviated as ^{14}N and ^{15}N. In these symbols, the superscript, the nearest whole number to the atomic weight of an isotope, is called the **mass number.**

The chemical methods yield an average value of the atomic weights of the isotopes of a particular element. But to convert the atomic weights of the isotopes of an element to an atomic weight of the element, the relative number of atoms of each isotope present must also be known. In mass spectroscopy, the relative abundance of isotopes is obtained from the relative darkness of the lines developed on the photographic plate or from the relative currents produced by the ions of the isotopes.

EXAMPLE 11 Calculate the atomic weight of oxygen found in air and of naturally occurring nitrogen from the previous and the following data. Atmospheric oxygen contains 99.7587% ^{16}O, 0.0374% ^{17}O, and 0.2039% ^{18}O. Natural nitrogen contains 99.625% ^{14}N and 0.375% ^{15}N.

ANSWER To obtain the atomic weight of an element from mass spectroscopic data, calculate the average atomic weight from the atomic weight and the relative abundance of each isotope. Thus, the atomic weight of atmospheric oxygen is

$$(15.9949 \times 0.997587) + (16.9991 \times 0.000374) + (17.9992 \times 0.002039) = 15.9994$$

and the atomic weight of nitrogen is

$$(14.003 \times 0.99625) + (15.000 \times 0.00375) = 14.007$$

The International Commission on Atomic Weights was established in 1900 by the International Union of Pure and Applied Chemistry and given the duty of issuing periodically a table of atomic weights after consideration of all papers dealing with the subject. The values chosen by the Commission (see back cover) become the **accepted atomic weights.**

PROBLEMS

1. Definite proportions and conservation A bar of metal is weighed four times on the same balance. Results: 5.012 g, 5.010 g, 5.012 g, 5.011g. With this balance, the following data were obtained on heating silver oxide, the reaction being

$$\text{silver oxide } (c) \xrightarrow{\text{(heated)}} \text{silver } (c) + \text{oxygen } (g)$$

Experiment	Weight of silver oxide heated (*g*)	Weight of silver remaining (*g*)
1	2.157	1.998
2	4.315	3.996
3	6.471	5.994

(a) Calculate the weight of oxygen liberated in each experiment. (b) Calculate the ratio of the weight of silver to the weight of oxygen from Experiment 1. (c) Does this series of experiments support the law of definite proportions? Explain, using the data to illustrate. (d) What additional measurement would be needed to test the law of conservation of matter?

2. Dalton's theory In terms of the Dalton atomic theory, name the two properties of atoms that must be fixed to explain the law of definite proportions.

3. Atomic weights If 6×10^{23} atoms of X weigh 12 g and 12×10^{23} atoms of Y weigh 8 g, what is the ratio of the weight of one atom of X to the weight of one atom of Y?

4. Cannizzaro method

Gaseous substance	Molecular weight	Wt. % of element		
		Nitrogen	Hydrogen	Oxygen
Ammonia	17 g/mole	82.5	17.5	
Nitrous oxide	44	63.7		36.3
X	75	93.3	6.6	
Nitrogen	28	100		

(a) What is the atomic weight of nitrogen? Show your calculations. (b) What is the molecular formula of nitrogen? Show how you arrive at your answer. (c) If the atomic weight of hydrogen is 1.0, what is the molecular formula of compound X?

5. Mole (a) One gross of nickels weighs 1.6 lb. Calculate the weight and number of nickels in 0.25 gross. (b) One mole of silicon tetrabromide weighs 348 g. Calculate the weight and number of silicon tetrabromide molecules in 0.250 mole.

6. Empirical and molecular formulas (a) What is the empirical formula of a substance that is 50% sulfur and 50% oxygen by weight? (b) The density (sc) of the substance as a gas is 2.9 g/liter. What is its molecular formula?

7. Combining volumes; Avogadro hypothesis (a) One liter of N_2O_3 gas is oxidized to NO_2 gas. If the temperature and pressure remain constant, what is the volume of the NO_2 produced? (b) One volume of an oxide of carbon, C_xO_y, reacts with 2 volumes of O_2 to give 3 volumes of CO_2 (all gases measured at the same temperature and pressure). What is the formula of the unknown oxide?

8. Molecular weight (a) The weight of 5.60 liters of carbon monoxide gas at standard conditions (sc) is 7.00 g. Calculate the molecular weight of carbon monoxide. (b) A gaseous hydrocarbon whose empirical formula is CH_2 has the same density as carbon monoxide. What is the molecular formula of the hydrocarbon? (c) The molecular weight of nitrogen is 28 g/mole. Calculate the weight of 5.60 liters (sc) of nitrogen. (d) Calculate the number of moles in (*i*) 280 g of nitrogen; (*ii*) 280 g of the hydrocarbon; (*iii*) 280 liters of carbon monoxide.

9. Mass spectroscopy An element A is composed of two isotopes of masses 4.20×10^{-23} g/atom and 4.27×10^{-23} g/atom, respectively. The isotopic abundance is, respectively, 3 atoms to 4 atoms. Calculate the atomic weight of this element.

ADDITIONAL PROBLEMS

10. Conservation J. J. Manley performed the following experiment in 1913: A small flask containing sodium sulfate, Na_2SO_4, solution was placed in a large jar containing barium chloride, $BaCl_2$, solution; the jar was then shaped into a flask and sealed, thereby producing a closed system containing two unmixed solutions. The system was then weighed; when the system was tilted, the two solutions mixed and a chemical change occurred, producing insoluble barium sulfate, $BaSO_4$, and sodium chloride, NaCl. After the original temperature was restored, the system was reweighed. The data obtained may be reproduced as

$$\text{weight of system before reaction} = 109.000000 \text{ g}$$
$$\text{weight of system after reaction} = 109.000001 \text{ g}$$

The precision of the balance is 6 parts in 100,000,000. What conclusion can you draw regarding the creation of matter in this experiment? Justify your answer.

11. Conservation An experiment is set up as follows. Hydrogen gas from a cylinder is streamed through a bed of hot copper oxide. All the hydrogen that leaves the cylinder reacts with the copper oxide. One product of the reaction, water vapor, is swept into a tube of drying agent where it is absorbed and retained. The other product, copper, remains in the bed with the oxide. (a) Write the correct sentence by choosing the proper words or symbols from the parentheses in the following statement: The (gain, loss) of weight of the hydrogen cylinder (+, −, =) the (gain, loss) of weight of the copper-copper oxide bed (+, −, =) the (gain, loss) of weight of the drying tube. (b) The following data were obtained in one experiment:

Initial weight of copper oxide bed	61.31 g
Final weight of copper-copper oxide bed	27.84 g
Initial weight of drying tube	87.02 g
Final weight of drying tube	124.67 g

Calculate (*i*) the weight of oxygen used to produce the water, (*ii*) the weight of the hydrogen used to produce the water; (*iii*) the weight of hydrogen that combines with 16.0 g of oxygen; (*iv*) the composition of water by weight.

12. Definite proportions Are the following data on silver bromide, taken from Baxter (1906), in accord with the law of definite proportions?

Initial weight of silver	Weight of silver bromide formed (from weight of silver given in column 1)
4.80711 g	8.36827 g
6.23696 g	10.85722 g
8.13612 g	14.16334 g

Justify your answer.

13. Definite proportions and isotopes The ratio of oxygen to hydrogen is given below for each of several samples of water. In each case, the sample was purified first, so that the data refers to the composition of the pure water.

Source of water	Wt. oxygen/wt. hydrogen (±0.0002)
Atlantic Ocean	7.9364
Emmons Glacier, Mt. Rainier	7.9365
Kitchen faucet, Cleveland, Ohio	7.9363
Water labeled "bideuterium oxide" purchased from a manufacturer of isotopic chemicals	3.9710
Water formed by burning hydrogen that was produced by the electrolysis of water	7.9378

Is the law of definite proportions supported or contradicted, or both, by these data? Explain your answer. What type of process occurs in distillation or electrolysis that could change the atomic weight of an element?

14. Atomic weights There are two atoms of element X and three atoms of element Y per molecule of a compound that contains 20 g of X to every 240 g of Y. What are the relative weights of the atoms of X and Y?

15. Atomic weights 6.02×10^{23} atoms of carbon weigh 12.0 g. A magnesium atom is 2.02 times as heavy as a carbon atom. Calculate the weight of 6.02×10^{10} magnesium atoms. How many atoms are there in 12.0 g of magnesium?

16. Atomic weights The atomic weight scale is based on $^{12}C = 12$ (exactly). The atomic weight of natural carbon is 12.01115. If the standard were changed to natural carbon = 12 (exactly), state whether each of the following would be increased, decreased, or unchanged: (a) the weight of an atom of carbon in grams; (b) the atomic weight of ^{12}C; (c) the atomic weight of natural carbon; (d) the atomic weight of oxygen.

17. Atomic theory Complete the following table.

	Formula	Composition	Atomic weight ratio
(a)	AB	50% A	at. wt. A/at wt. B = ?
(b)	C_2D	10 g C to 20 g D	at. wt. C/at. wt. D = ?
(c)	$E_?F_?$	50% E	at. wt. E/at. wt. F = 1.5
(d)	GH_3	?	at. wt. G/at. wt. H = 0.9
(e)	IJ_4	1 g I to 3 g J	at. wt. I/at. wt. J = ?
(f)	$K_?L_?$	3 g K to 8 g L	at. wt. K/at. wt. L = 0.5

18. Avogadro hypothesis Consider the following set of statements: (*i*) The Avogadro hypothesis is true. (*ii*) One liter of oxygen gas + solid carbon → 2 liters of carbon monoxide gas (constant temperature, pressure). (*iii*) An oxygen molecule consists of 1 atom of oxygen, O. (a) Is it possible for all three statements to be true? (b) Can any set of two statements be true if the third is false? (c) Statement (*ii*) is an experimental fact. Show that if (*i*) is true, then (*iii*) must be false.

19. Mole Express in your own words and illustrate the meaning of the following definition adopted by the International Union of Pure and Applied Chemistry: "The mole is defined as the amount of substance which contains the same number of molecules (or ions, or atoms, or electrons, as the case may be) as there are atoms in exactly 12 grammes of the pure carbon nuclide ^{12}C."

20. Empirical formula When a mixture of fluorine and xenon in a nickel container is heated to 400°C and then cooled rapidly to room temperature, a number of compounds are obtained. The composition of one of these compounds expressed in milliliters (sc) is

Experiment number	Xe (ml)	F_2 (ml)
(1)	2.248	4.558
(2)	2.745	5.453

Calculate the empirical formula of the compound.

21. Molecular weight The molecular weights of carbon dioxide, CO_2, and propane, C_3H_8, are both 44 g/mole. (a) What is the weight of 22.4 liters (sc) of CO_2? Of C_3H_8? (b) What is the weight of 1.00 liter (sc) of CO_2? Of C_3H_8? (c) What is the sum of the weights of 21.4 liters CO_2 + 1.00 liter C_3H_8 (both sc)? Of x liters CO_2 + $(22.4 - x)$ liters C_3H_8 (both sc)? (d) Does the density of a mixture of CO_2 and C_3H_8 depend on the composition? Explain. (e) Is it possible to determine the "molecular weight" of a mixture of CO_2 and C_3H_8 of unknown composition? Of a mixture of unknown gases? Explain.

22. Molecular weight (a) The density of air at sc is 1.29 g/liter. Calculate the "molecular weight" of air. (b) How many molecules are there in 22.4 liters (sc) of air?

23. Molecular formula The molecular weight of "sulfur monochloride" is 135 g/mole. The substance contains 47.4% sulfur by weight. What is its molecular formula?

24. Molecular formula (a) The molecular weight of nicotine, a colorless oil, is 162.1 and it contains 74.0% carbon, 8.7% hydrogen, and nitrogen. Using three significant figures for the atomic weights, calculate the molecular formula of nicotine. (b) If a scientist synthesized a nitrogen compound containing 7.0 g of nitrogen per mole, would this affect the formula for nicotine? If your answer is yes, explain your answer and give the "new" formula for nicotine.

25. Cannizzaro method Given the experimental data:

Substance	Molecular weight (g/mole)	Weight % composition		
		Sulfur	Oxygen	Fluorine
Oxygen	32.0	—	100.0	—
Fluorine fluorosulfonate	118.1	27.2	40.6	32.2
Thionyl tetrafluoride	124.1	25.9	12.9	61.2
Pentafluorosulfur hypofluorite	162.1	19.8	9.88	70.4
Sulfur monofluoride	102.2	62.9	—	37.2
Hydrogen fluoride	20.1	—	—	94.5

Calculate the atomic weight of sulfur, oxygen, and fluorine and the molecular formulas of each of the substances. What data would you have to discover to compel a change in the atomic weight of sulfur or oxygen? Would you undertake such a research project?

26. Nuclear composition from atomic weights What is the nuclear composition of the isotope of calcium with each of the following atomic (isotopic) weights? 39.963, 41.959, 42.959, 43.955, 45.954, 47.952.

27. Mass spectroscopy The isotopic abundance ratio of natural chlorine, redetermined recently at the U.S. National Bureau of Standards, is

$$^{35}Cl/^{37}Cl = 3.1272 \pm 0.0080$$

On the scale $^{12}C = 12$, the isotopic weights are $^{35}Cl = 34.9689$ and $^{37}Cl = 36.9659$. Calculate the atomic weight of natural chlorine.

ANSWERS

3. 3.

4. N_5H_5.

7. (a) 2 liters; (b) C_3O_2, carbon suboxide.

9. 25.5 g/g-atom.

11. (b) (*i*) 33.47 g; (*ii*) 4.18 g; (*iii*) 2.00 g; (*iv*) 88.9% oxygen.

17. (c) E_2F_3; (d) 23% G.

22. (a) 28.9 g/mole; (b) 6.02×10^{23} molecules.

23. S_2Cl_2.

5 STOICHIOMETRY—THE ARITHMETIC OF CHEMISTRY

5.1 CHEMICAL EQUATIONS

Chemical equations tell us what substances react and what substances are produced. The products of a reaction are determined by experimentation. Thus propane, a typical hydrocarbon, burns in the presence of oxygen to form carbon dioxide and water. First, since the formulas for these substances are known, we may write

$$\underbrace{C_3H_8(g) + O_2(g)}_{\textit{reactants}} \longrightarrow \underbrace{CO_2(g) + H_2O(g)}_{\textit{products}}$$

Second, since atoms are not created or destroyed but merely rearranged in chemical changes, we **balance the equation** by making the number of atoms of each element in the reactants the same as that appearing in the products. This is accomplished by placing the required number before each formula.* Thus, the 8 atoms of hydrogen in the C_3H_8 must form 4 molecules of H_2O, while the 3 atoms of carbon must form 3 molecules of CO_2; but the 3 molecules of CO_2 and the 4 molecules of H_2O require 10 atoms or 5 molecules of oxygen; hence

$$C_3H_8(g) + 5O_2(g) \longrightarrow 3CO_2(g) + 4H_2O(g)$$

The number before each formula is known as its **coefficient.** The coefficient is a multiplier for the entire formula, never for only a part of it. Thus, $3CaCl_2(H_2O)_6$ includes 3 Ca atoms, 36 H atoms, and 18 O atoms. *The subscripts in the formulas must not be altered.* This would alter the nature of the substances involved, thereby violating the experimental observation. Also, the balanced equation is a statement of the stoi-

* A systematic method is considered later for more complicated reactions.

chiometric relations between the reactants and products involved in a chemical change; it does not, however, show the actual processes (Chapter 19) by which reactants are converted to products.

5.2 ▫ QUANTITATIVE INFORMATION FROM CHEMICAL EQUATIONS

The subject referred to as stoichiometry, in a generalized sense, deals with the weight and volume relationships determined by formulas and chemical equations. The balanced equation, for example,

$$C_3H_8(g) + 5O_2(g) \longrightarrow 3CO_2(g) + 4H_2O(g)$$

summarizes the chemical reaction and also gives the relative amounts of reactants and products involved in the chemical change. Thus, for every 1 molecule of propane, 5 oxygen molecules react, producing 3 carbon dioxide molecules and 4 water molecules; or, in more practical terms, for every 6.02×10^{23} molecules of propane consumed, $5(6.02 \times 10^{23})$ molecules of oxygen react, producing $3(6.02 \times 10^{23})$ molecules of carbon dioxide and $4(6.02 \times 10^{23})$ molecules of water, abbreviated as

$$1 \text{ molecule } C_3H_8 + 5 \text{ molecules } O_2 \longrightarrow 3 \text{ molecules } CO_2 + 4 \text{ molecules } H_2O$$

or

$$6.02 \times 10^{23} \text{ molecules } C_3H_8 + 5(6.02 \times 10^{23}) \text{ molecules } O_2 \longrightarrow 3(6.02 \times 10^{23}) \text{ molecules } CO_2 + 4(6.02 \times 10^{23}) \text{ molecules } H_2O$$

or

$$1 \text{ mole } C_3H_8 + 5 \text{ moles } O_2 \longrightarrow 3 \text{ moles } CO_2 + 4 \text{ moles } H_2O$$

or

$$22.4 \text{ liters (sc) } C_3H_8 + 5(22.4) \text{ liters (sc) } O_2 \longrightarrow 3(22.4) \text{ liters (sc) } CO_2 + 4(22.4) \text{ liters (sc) } H_2O$$

or

$$44.1 \text{ g } C_3H_8 + 5(32.0) \text{ g } O_2 \longrightarrow 3(44.0) \text{ g } CO_2 + 4(18.0) \text{ g } H_2O$$

or

$$1 \text{ mole } C_3H_8 + 5(6.02 \times 10^{23}) \text{ molecules } O_2 \longrightarrow 3(22.4) \text{ liters (sc) } CO_2 + 4(18.0) \text{ g } H_2O$$

Note that other units of weight, like kilograms, may also be used. Similarly, from the coefficients in the balanced equation.

$$3H_2(g) + Fe_2O_3(c) \longrightarrow 2Fe(c) + 3H_2O(g)$$

$$3 \text{ moles } H_2 + 1 \text{ mole } Fe_2O_3 \longrightarrow 2 \text{ moles Fe} + 3 \text{ moles } H_2O$$

or

$$3(22.4) \text{ liters (sc) } H_2 + 160 \text{ g } Fe_2O_3 \longrightarrow 2(55.9) \text{ g Fe} + 3(22.4) \text{ liters (sc) } H_2O$$

or

$$3(2.02) \text{ g } H_2 + 160 \text{ g } Fe_2O_3 \longrightarrow 2(55.9) \text{ g Fe} + 3(18.0) \text{ g } H_2O$$

If a volume of gas is measured at other than standard conditions, we must correct the volume to the value it would have at standard conditions. These corrections are explained in Chapter 11 on gases.

Note that each *quantity includes a unit;* refer to Appendix 1 for a review of dimensional numbers. The rigorous use of dimensional units is particularly important in chemical stoichiometry.

Chemical equations thus make it possible to calculate quantities of materials required to produce a definite quantity of a desired product.

EXAMPLE 1 A process for manufacturing the dry cleaning fluid tetrachloroethylene, C_2Cl_4, from acetylene, C_2H_2, may be summarized as

$$C_2H_2(g) + 3Cl_2(g) + Ca(OH)_2(c) \longrightarrow C_2Cl_4(l) + CaCl_2(c) + 2H_2O(l)$$

(a) What weight in grams of C_2Cl_4 is produced from 50.0 g of Cl_2?

ANSWER From the balanced equation

3 moles Cl_2 yields 1 mole C_2Cl_4

or, since the molecular weight of Cl_2 is 71.0 and that of C_2Cl_4 is 166 ($2 \times 12.0 + 4 \times 35.5 = 166$),

3(71.0) g Cl_2 yields 166 g C_2Cl_4

Hence, the number of grams of C_2Cl_4 obtained from 1 g of chlorine is

$$\frac{166 \text{ g } C_2Cl_4}{3(71.0) \text{ g } Cl_2}$$

Then the number of grams of C_2Cl_4 obtained from 50.0 g of Cl_2 is

$$50.0 \text{ g } \cancel{Cl_2} \times \frac{166 \text{ g } C_2Cl_4}{3(71.0) \text{ g } \cancel{Cl_2}} = 39.0 \text{ g } C_2Cl_4$$

(b) What volume of Cl_2 in liters (sc) is required to produce 100 g of C_2Cl_4?

ANSWER From the balanced equation

1 mole of C_2Cl_4 requires 3 moles Cl_2

or

166 g of C_2Cl_4 requires 3(22.4) liters (sc) Cl_2

Hence, the number of liters (sc) of Cl_2 required for 1 gram of C_2Cl_4 is

$$\frac{3(22.4) \text{ liters (sc) } Cl_2}{166 \text{ g } C_2Cl_4}$$

Then the number of liters of Cl_2 (sc) required for 100 g of C_2Cl_4 is

$$100 \text{ g } \cancel{C_2Cl_4} \times \frac{3(22.4) \text{ liters (sc) } Cl_2}{166 \text{ g } \cancel{C_2Cl_4}} = 40.5 \text{ liters (sc) } Cl_2$$

(c) What weight in grams of chlorine will react with 75.0 liters of C_2H_2, measured at sc?

ANSWER From the balanced equation

3 moles Cl_2 reacts with 1 mole C_2H_2

or

3(71.0) g Cl_2 reacts with 22.4 liters (sc) C_2H_2

Hence, the number of grams of Cl_2 that reacts with 1 liter (sc) of C_2H_2 is

$$\frac{3(71.0)\text{ g Cl}_2}{22.4\text{ liters (sc) C}_2\text{H}_2}$$

Then the number of grams of Cl_2 that reacts with 75.0 liters (sc) of C_2H_2 is

$$75.0\text{ liters (sc) C}_2\text{H}_2 \times \frac{3(71.0)\text{ g Cl}_2}{22.4\text{ liters (sc) C}_2\text{H}_2} = 713\text{ g Cl}_2$$

Notice that the *choice of units** is determined by the statement of the problem; also notice that the use of units verifies the method used for the solution of these problems, because the correct unit is obtained for the answer. If we inadvertently had inverted the ratio in the last step of the previous problem

$$75.0\text{ liters (sc) C}_2\text{H}_2 \times \frac{22.4\text{ liters (sc) C}_2\text{H}_2}{3(71.0)\text{ g Cl}_2} = 7.88\,\frac{[\text{liters (sc)}]^2\text{ C}_2\text{H}_2}{\text{g Cl}_2}$$

the error would be obvious since

$$\frac{[\text{liters (sc)}]^2\text{ C}_2\text{H}_2}{\text{g Cl}_2}$$

is irrelevant to the desired answer, grams Cl_2. Note finally that the arithmetical procedure is similar to that used in everyday life; a typical problem is, "What is the cost in dollars of 12.2 gross of nails selling at 11.2 cents per hundred?" From the price, the number of cents for one nail is

$$\frac{11.2\text{ cents}}{100\text{ nails}}$$

Then the number of cents required for 12.2 gross of nails is

$$12.2\text{ gross} \times \frac{144\text{ nails}}{\text{gross}} \times \frac{11.2\text{ cents}}{100\text{ nails}}$$

and, converting to dollars,

$$12.2\,\cancel{\text{gross}} \times \frac{144\,\cancel{\text{nails}}}{\cancel{\text{gross}}} \times \frac{11.2\,\cancel{\text{cents}}}{100\,\cancel{\text{nails}}} \times \frac{1\text{ dollar}}{100\,\cancel{\text{cents}}} = 1.97\text{ dollars}$$

* See Appendix 1 for further discussion of measurement units.

Terms such as

$$\frac{100 \text{ cents}}{1 \text{ dollar}} \quad \text{or} \quad \frac{12 \text{ in.}}{1 \text{ ft}} \quad \text{or} \quad \frac{91.4 \text{ cms}}{1 \text{ yard}}$$

are known as **conversion factors** because they are used to convert a given unit to another. But the terms used in Example 1,

$$\frac{166 \text{ g } C_2Cl_4}{3(71) \text{ g } Cl_2}, \quad \frac{3(22.4) \text{ liters (sc) } Cl_2}{166 \text{ g } C_2Cl_4}, \quad \frac{3(71.0) \text{ g } Cl_2}{22.4 \text{ liters (sc) } C_2H_2}$$

are also "conversion factors." The first term, for example, "converted" the given quantity of Cl_2 to the quantity of C_2Cl_4 that it produces in this particular reaction. This offers an alternative method for solving problems involving quantity relations in chemical equations.

In analytical chemistry, "conversion factors" involving gram per gram relations are called **"gravimetric factors."**

EXAMPLE 2 The calcium content of a substance may be determined by converting all the calcium to $CaCO_3$, which is then weighed. What gravimetric factor can the analyst use to convert the weight of $CaCO_3$ to the weight of Ca in the original sample? How is the factor used?

ANSWER The gravimetric factor is

$$\frac{\text{at. wt. Ca}}{\text{mol. wt. } CaCO_3} = \frac{40.08 \text{ g Ca}}{100.09 \text{ g } CaCO_3} = 0.4004 \frac{\text{g Ca}}{\text{g } CaCO_3}$$

Then, weight of Ca (g) = weight of $CaCO_3$ (g) × gravimetric factor

$$= \text{weight of } CaCO_3 \text{ (g)} \times 0.4004 \frac{\text{g Ca}}{\text{g } CaCO_3}$$

EXAMPLE 3 When iodic acid, HIO_3, is mixed with an excess of iodine, I_2, in sulfuric acid, a solution is formed having the brown-red color of the ion I_3^+ formed by the reaction

$$HIO_3 + 7I_2 + 8H_2SO_4 \longrightarrow 5I_3^+ + 3H_3O^+ + 8HSO_4^-$$

How many I_3^+ ions are produced when 1.5 g of HIO_3 so reacts?

ANSWER Essentially, we wish to "convert" (by calculation) 1.5 g HIO_3 to a number of ions of I_3^+. From the given chemical equation, the relation between these two quantities is

$$\frac{5 \text{ moles } I_3^+}{1 \text{ mole } HIO_3} \quad \text{or} \quad \frac{5(6.0 \times 10^{23}) \text{ ions } I_3^+}{176 \text{ g } HIO_3}$$

Then, following the same arithmetical procedure by which feet are converted to inches, we obtain

$$1.5 \text{ g } HIO_3 \times \frac{5(6.0 \times 10^{23}) \text{ ions } I_3^+}{176 \text{ g } HIO_3} = 2.6 \times 10^{22} \text{ ions } I_3^+$$

EXAMPLE 4 2.00 g of lead nitrate, $Pb(NO_3)_2$, is added to a solution containing 1.00 g of potassium iodide, KI. Insoluble lead iodide forms,

$$Pb(NO_3)_2(aq) + 2KI(aq) \longrightarrow PbI_2(c) + 2KNO_3(aq)$$

How many grams of PbI_2 are formed? Which reactant is present in excess and by how many grams?

ANSWER Let us first determine which reactant is present in excess by calculating the amount of one reactant required to react with the given amount of the second reactant. This calculation, *regardless of the choice* we make for the first reactant, decides which reactant is in excess. Thus, suppose we decide to calculate how many grams of $Pb(NO_3)_2$, molecular weight 331, are needed for the 1.00 g of KI, molecular weight 166,

$$1.00\ \text{g KI} \times \frac{331\ \text{g Pb(NO}_3)_2}{2 \times 166\ \text{g KI}} = 0.997\ \text{g Pb(NO}_3)_2 \tag{1}$$

However, 2.00 g of $Pb(NO_3)_2$ is given; therefore, $Pb(NO_3)_2$ is in excess by 1.00 g.

If we had decided to calculate how many grams of KI are needed for the 2.00 g of $Pb(NO_3)_2$,

$$2.00\ \text{g Pb(NO}_3)_2 \times \frac{2 \times 166\ \text{g KI}}{331\ \text{g Pb(NO}_3)_2} = 2.01\ \text{g KI}$$

we would have concluded that 1.00 g of KI is insufficient to react with 2.00 g of $Pb(NO_3)_2$. Then, to determine the quantity of $Pb(NO_3)_2$ in excess, we would have to proceed as above (Equation 1).

These calculations also show that the quantity of PbI_2, molecular weight 461, formed is determined by the 1.00 g of KI [or 0.997 g of $Pb(NO_3)_2$] consumed. Hence

$$1.00\ \text{g KI} \times \frac{461\ \text{g PbI}_2}{2 \times 166\ \text{g KI}} = 1.39\ \text{g PbI}_2$$

or

$$0.997\ \text{g Pb(NO}_3)_2 \times \frac{461\ \text{g PbI}_2}{331\ \text{g Pb(NO}_3)_2} = 1.39\ \text{g PbI}_2$$

Equations containing fractional coefficients, such as

$$H_2(g) + \tfrac{1}{2}O_2(g) \longrightarrow H_2O(l)$$

are frequently used; this balanced equation is read as follows: 1 mole of H_2 combines with $\frac{1}{2}$ mole of O_2 to form 1 mole of H_2O (not as 1 molecule of H_2 combining with $\frac{1}{2}$ molecule of O_2).

5.3 □ CONVERSION AND SELECTIVITY (YIELD)

Reactants often yield quantities of products that are *less* than those calculated from the balanced chemical equation. Reasons for such deficiencies may be that (a) some of the original material (called "feed stock" or "feed") fails to undergo reaction, and such unreacted feed may be partially or completely recovered; (b) some of the feed reacts to give unexpected or undesired products; (c) some of the expected products continue to react to form other products or revert to the original feed; (d) not all the product is recovered.

The percentage of feed that is not recovered, and therefore presumably is converted to something else, is called the **conversion.** The percentage of desired product obtained, based on the theoretically possible amount from the converted feed, is called the **selectivity,** or sometimes the **percent yield based on unrecovered feed.**

$$\text{Conversion} = \left(\frac{\text{total feed} - \text{recovered feed}}{\text{total feed}}\right) \times 100$$

$$= \left(\frac{\text{unrecovered feed}}{\text{total feed}}\right) \times 100$$

$$\text{Selectivity (\% yield)} = \frac{\text{weight of desired product obtained}}{\text{theoretical weight obtainable from unrecovered feed}}(100)$$

EXAMPLE 5 Ethylene oxide, C_2H_4O, is manufactured by the oxidation of ethylene in air,

$$C_2H_4 + \tfrac{1}{2}O_2 \longrightarrow C_2H_4O$$

Undesirable events are (a) the failure of some ethylene to react; (b) the more complete oxidation of some ethylene to formaldehyde (H_2CO), CO, and CO_2; (c) the decomposition of some ethylene to carbon (smoke) and other products. If 60 kg of C_2H_4O is obtained from a feed of 50 kg of C_2H_4, with the recovery of 8 kg of C_2H_4, what is the conversion and the selectivity?

ANSWER The conversion is

$$\frac{(50\text{ kg} - 8\text{ kg})}{50\text{ kg}} \times 100 = 84\%$$

The amount of unrecovered feed, which is the quantity we assume reacted in some way, is (50 − 8) kg, or 42 kg. The theoretical weight of C_2H_4O that could be obtained from 42 kg of C_2H_4 is

$$42\text{ kg } C_2H_4 \times \frac{44\text{ kg } C_2H_4O}{28\text{ kg } C_2H_4} = 66\text{ kg } C_2H_4O$$

The actual weight obtained, however, is 60 kg C_2H_4O. The selectivity or percent yield is therefore

$$\frac{60\ \text{kg}\ C_2H_4O}{66\ \text{kg}\ C_2H_4O}(100) = 91\%$$

PROBLEMS*

1. Chemical equations Each of the following reactions is the basis of a process used in the industry shown in parentheses. Write the balanced equation for each reaction:

(a) (Fertilizer) $CO_2(g) + NH_3(g) \longrightarrow OC(NH_2)_2(c) + H_2O(g)$

(b) (Fertilizer) $P_4O_{10}(c) + H_2O(l) \longrightarrow H_3PO_4(aq)$

(c) (Solvent) $S_2Cl_2(l) + C(amorph) \longrightarrow CCl_4(l) + S(c)$

(d) (Soap) $(C_{15}H_{31}COO)_3C_3H_5(c) + NaOH(aq) \longrightarrow C_{15}H_{31}COONa(c) + C_3H_5(OH)_3(l)$

(e) (Welding) $Al(c) + Fe_2O_3(c) \longrightarrow Al_2O_3(c) + Fe(c)$

2. Chemical equations

$$CH_5N(g) + xO_2(g) \longrightarrow CO_2(g) + 2\tfrac{1}{2}H_2O(l) + NO_2(g)$$

(a) Calculate the value of x needed to balance the equation above if no other coefficients are changed. (b) Can a fractional coefficient refer to a number of molecules? A number of moles? Can stoichiometric calculations that use fractional coefficients yield correct answers?

3. Conversion factors Write the conversion factors you would use in solving each of the following problems:
(a) How many square feet in 15 square yards?
(b) How many milligrams in one kilogram?
(c) How many grams of Fe_2O_3 may be prepared by the reaction $4FeS_2 + 11O_2 \longrightarrow 2Fe_2O_3 + 8SO_2$ (*i*) if 4 moles of iron sulfide react? (*ii*) if 1.1 g of sulfur dioxide is produced? (*iii*) if 1.12 liter (sc) of oxygen gas reacts?

4. Chemical arithmetic On heating, sodium hypochlorite decomposes, $3NaClO(c) \longrightarrow 2NaCl(c) + NaClO_3(c)$. (a) How many grams of sodium hypochlorite will yield 60.0 g of NaCl? (b) What weight of sodium chlorate will be formed at the same time?

5. Chemical arithmetic How many liters (sc) of nitrogen may be obtained by the complete decomposition of 6.40 g ammonium nitrite, according to the equation

$$NH_4NO_2(c) \longrightarrow N_2(g) + 2H_2O(l)?$$

6. Chemical arithmetic (a) How many grams of magnesium iodide, MgI_2, can be prepared by heating 5.00 g Mg with 5.00 g I_2? (b) Which reactant is left over? (c) How much of the reactant named in (b) is left over?

7. Chemical arithmetic Equal volumes (sc) of $H_2(g)$ and $O_2(g)$ are mixed and

* See Problem 30 (p. 194) for gas volume corrections in stoichiometry.

ignited. The products are then adjusted to their original temperature and pressure. (a) Write the equation for the reaction. (b) What percentage by volume of each gas is consumed? By weight?

8. Chemical arithmetic and selectivity (percent yield) The plastics industry produces phthalic anhydride by the controlled oxidation of naphthalene,

$$2C_{10}H_8(c) + 9O_2(g) \longrightarrow 2C_8H_4O_3(c) + 4CO_2(g) + 4H_2O(l)$$

(a) What volume (sc) of oxygen is theoretically needed to oxidize 320 g of naphthalene? (b) What weight of CO_2 would be produced at the same time? (c) How many moles of phthalic anhydride are produced from one mole of naphthalene, assuming the reaction is complete? (d) In a commercial process a selectivity (percent yield) of 85% of phthalic anhydride is obtained. What weight of naphthalene is needed to yield 1000 kg of the anhydride?

9. Conversion and selectivity (percent yield) In the laboratory, oxygen may be prepared by the thermal decomposition of solid potassium chlorate. During the decomposition of potassium chlorate, $2KClO_3(c) \longrightarrow 2KCl(c) + 3O_2$, a number of side reactions occur; for example, oxides of chlorine are also produced. In one experiment, a sample of 20.0 g of $KClO_3$ was heated, and yielded 7.2 g of oxygen, with recovery of 0.6 g of the starting material. Calculate (a) the conversion and (b) the selectivity (percent yield) of the reaction.

ADDITIONAL PROBLEMS

10. Balancing chemical equations (continuation of problem 1)

(a) (Plastics) $C_6H_6(l) + O_2(g) \longrightarrow C_4H_2O_3(c) + CO_2(g) + H_2O(l)$

(b) (Bleach) $NaOH(aq) + Cl_2(g) \longrightarrow NaOCl(aq) + NaCl(aq) + H_2O(l)$

(c) (Welding) $C_2H_2(g) + O_2(g) \longrightarrow CO_2(g) + H_2O(l)$

(d) (Metallurgy) $Fe(CrO_2)_2(c) + C(amorph) \longrightarrow Fe(c) + Cr(c) + CO(g)$

(e) (Metallurgy) $Fe_2O_3(c) + CO(g) \longrightarrow Fe(c) + CO_2(g)$

11. Conversion factors Write the conversion factor you would use in solving each of the following problems: (a) How many liters are in 30 ml? (b) How many molecules of hydrogen fluoride are obtained from 6×10^{-3} mole of XeF_4 in the reaction $XeF_4(g) + 2H_2(g) \longrightarrow Xe(g) + 4HF(g)$? (c) How many grams of bromine are in a sample that yields 0.4203 g of silver bromide when all the bromine in the sample reacts as follows?

$$2Ag(c) + Br_2(l) \longrightarrow 2AgBr(c)$$

12. Chemical arithmetic 10.0 g of silicon dust, Si, is exploded with 100.0 g of oxygen, O_2, forming silicon dioxide, SiO_2. (a) How many grams of SiO_2 are formed and (b) how many grams of O_2 remain uncombined?

13. Chemical arithmetic 4.0 g of aluminum wire is ignited in 4.0 g pure oxygen, producing aluminum oxide,

$$4Al(c) + 3O_2(g) \longrightarrow 2Al_2O_3(c)$$

(a) Which reactant is present in excess? (b) What is the weight, in grams, of the excess? (c) How many grams of Al_2O_3 are formed?

14. Chemical arithmetic (a) How many grams of oxygen are obtained from the complete decomposition of 11.0 g of silver oxide? The equation is

$$2Ag_2O(c) \longrightarrow 4Ag(c) + O_2(g)$$

(b) A sample of silver oxide was heated until all the oxygen was driven off. It was found that the loss of weight of the sample was 3.12 g. Calculate the weight of the silver produced.

15. Chemical arithmetic The production of **sponge iron** from iron oxides and methane gas, which requires no blast furnace, may be represented as

$$4Fe_2O_3(c) + 3CH_4(g) \longrightarrow 8Fe(c) + 3CO_2(g) + 6H_2O(g)$$

(a) How many liters (sc) of methane would be required to produce a kilogram of iron? (b) How many liters of gas (CO_2 and H_2O), measured at sc, would be produced from every liter (sc) of methane that reacts? (c) How many grams of iron oxide are needed to produce a kilogram of iron?

16. Chemical arithmetic When diborane, B_2H_6, is passed at low pressure through a hot tube, the high-energy fuel pentaborane, B_5H_9, is obtained:

$$5B_2H_6(g) \longrightarrow 2B_5H_9(g) + 6H_2(g)$$

(a) What weight in grams of B_2H_6 is needed to produce 25.0 g B_5H_9 (assume complete conversion)? (b) What is the weight in grams of the hydrogen produced by 5.00 liters (sc) of B_2H_6? (c) What volume in liters (sc) of B_2H_6 is needed to produce 10 g of B_5H_9?

17. Chemical arithmetic When ammonium chloride, NH_4Cl, and calcium oxide, CaO, are heated together they react to form ammonia, NH_3, calcium chloride, $CaCl_2$, and water. (a) Write the balanced equation for this reaction. (b) If equal weights of reactants are used, which one is present in excess? (c) If 1.024 g of NH_4Cl is used with an excess of the other reactant, what volume of NH_3 (sc) is produced?

18. Chemical arithmetic (a) An electric current passed through a solution of ammonium hydrogen sulfate produces the reaction,

$$2NH_4HSO_4(aq) \longrightarrow H_2(g) + (NH_4)_2S_2O_8(aq)$$

Calculate how many grams of NH_4HSO_4 would be needed to produce 1.00 liter (sc) of hydrogen gas. How many grams of $(NH_4)_2S_2O_8$, ammonium peroxydisulfate, are produced? (b) When steam is passed through the dissolved product of part (a), the reaction

$$(NH_4)_2S_2O_8(aq) + 2H_2O(l) \longrightarrow 2NH_4HSO_4(aq) + H_2O_2(aq)$$

occurs. How many grams of hydrogen peroxide, H_2O_2, would be produced from the quantity of $(NH_4)_2S_2O_8$ formed in part (a)? (c) Write the equation for the net reaction that occurs (the sum of the two reactions).

19. Chemical arithmetic In a lead storage battery, the overall reaction is

$$Pb(c) + PbO_2(c) + 2H_2SO_4(l) \longrightarrow 2PbSO_4(c) + 2H_2O(l)$$

(a) What quantity of lead sulfate could be produced from 135 g of sulfuric acid, assuming 100% conversion? Assuming 25% conversion? (b) What quantities of lead and of lead dioxide would be required to react with this amount of sulfuric acid, assuming 100% conversion?

20. Chemical arithmetic Isopropanol or rubbing alcohol is combustible:

$$2C_3H_7OH(l) + 9O_2(g) \longrightarrow 6CO_2(g) + 8H_2O(l)$$

(a) What volume of oxygen (sc) would be required to burn completely 1.00 liter of isopropanol (density 0.79 g/ml)? (b) How many grams of CO_2 would be produced in (a)?

21. Chemical arithmetic Phosphorus can be prepared in an electric furnace according to the equation

$$2Ca_3(PO_4)_2(c) + 6SiO_2(c) + 10C(amorph) \longrightarrow 6CaSiO_3(c) + 10CO(g) + P_4(g)$$

(a) How many tons of coke would be consumed in preparing 5 tons of phosphorus? (b) How many liters of carbon monoxide (sc) would be liberated per kilogram of phosphorus produced?

22. Chemical arithmetic Lithium oxide, Li_2O, has a very high capacity, on a weight basis, for absorbing carbon dioxide. One mole of Li_2O combines chemically with one mole of CO_2. (a) Write the equation for this combination. (b) Calculate the weight ratio of CO_2 to Li_2O in this combination. (c) Calculate the weight of Li_2O needed to purify the air exhaled by an astronaut during a 10 hour flight, assuming a production of 0.35 liter (sc) of CO_2 per minute. (d) Calcium oxide (lime), CaO, also combines with CO_2 in a 1:1 molar ratio. Write the equation for this combination. (e) What is the ratio of the capacity of Li_2O for absorbing CO_2 to the capacity of CaO for absorbing CO_2? How is this ratio related to the molecular weights of the two absorbents? (Capacities refer to weight basis.)

23. Balancing equations and chemical arithemtic The following is a word description of the operation of an industrial process that converts waste hydrogen sulfide gas to sulfur: "Impure hydrogen sulfide (H_2S) is used as a feed stream to the sulfur processing plant. Enough air is then added to the stream to burn one third of the hydrogen sulfide to sulfur dioxide (SO_2) and water vapor. The resulting stream, consisting of the SO_2 and the unburned H_2S feedstock, is fed to a catalyst where it is converted to sulfur vapor (S_2) and H_2O." (a) Write the balanced equation for each of the reactions cited. (b) Add the two equations to give an overall equation for the process. (c) Explain why it is advantageous to burn just one third of the H_2S. (d) Calculate the weight of sulfur that can theoretically be produced from 1.00 kg of H_2S. (e) What volume of oxygen (sc) is required for this conversion?

24. Selectivity Byproduct SO_2 is converted to carbon disulfide with coke according to the reaction

$$5C(amorph) + 2SO_2(g) \longrightarrow CS_2(l) + 4CO(g)$$

(a) What weight of CS_2 can be produced theoretically from 100 liters (sc) of waste SO_2? (b) If 100 g of CS_2 is produced from 100 liters (sc) of SO_2, what is the selectivity (percent yield)?

ANSWERS

3. (c) (*i*) 79.8 g Fe_2O_3/mole FeS_2; (*ii*) 0.623 g Fe_2O_3/g SO_2; (*iii*) 1.30 g Fe_2O_3/liter (sc) O_2.

5. 2.24 liters.

8. (a) 252 liters; (b) 220 g; (c) 1 mole; (d) 1.02×10^3 g.

12. (a) 21.4 g; (b) 88.6 g.

14. (a) 0.760 g; (b) 4.21 g.

16. (a) 27.4 g; (b) 5.40 g; (c) 8.87 liters.

18. (a) 10.3 g; (b) 1.54 g.

20. (a) 1.3×10^3 liters; (b) 1.7×10^3 g.

23. (d) 0.94 kg; (e) 329 liters.

24. (a) 170 g; (b) 59%.

6 THE FIRST LAW OF THERMODYNAMICS; THERMOCHEMISTRY

6.1 CONSERVATION OF ENERGY

Chemical changes are nearly always accompanied by energy changes. The combustion of fossil fuels, such as coal, petroleum products, and natural gas, now constitutes man's major energy source. Chemical reactions evolving heat to the surroundings are called **exothermic;** if heat is absorbed from the surroundings the reaction is **endothermic.** It is assumed that the chemical system itself is restored to the initial temperature, so that only the surroundings are warmed or cooled.

In our investigation of processes that involve energy changes we assume the **law of conservation of energy,** expressed by any of the following statements: (a) The energy of the universe is constant. (b) In any process, energy is neither created nor destroyed. (c) The energy of an undisturbed portion of matter (called a **system**) in a definite condition or state (of pressure volume, temperature, and so on) is fixed and independent of the method of preparation. This law is not derived from any previous assumptions; it summarizes our experience with material processes—for example, the fact that no one has ever invented a "perpetual motion" machine that can provide work without consuming fuel.

6.2 THE FIRST LAW OF THERMODYNAMICS

When a system undergoes a change of state, its energy also changes. For example, the energy of 1.72643 g of natural nitrogen gas at 425.270°K and 2.54361 atm is a fixed value, E_1. If the temperature of the nitrogen is raised to 526.371°K and the pressure is increased to

4.53205 atm, the energy of the system in the second state reaches a new fixed value, E_2, independent of the method of preparation. Then the difference in the energies of these two states, $E_2 - E_1$, must also have a definite value regardless of the method of arriving at these two states: $E_2 - E_1 = \Delta E$, regardless of how state 1 is changed to state 2. The energy of a system can be changed by two means—the transfer of heat (**"thermo"**) and the performance of work (**"dynamics"**). The conventional designations are

Change in energy of system	Performance of work	Transfer of heat
Increase in energy, $+\Delta E$	Work done on the system, $-w$	Heat added to (absorbed by) the system, $+q$ (**endothermic**)
Decrease in energy, $-\Delta E$	Work done by the system, $+w$	Heat removed from the system, $-q$ (**exothermic**)

Heat added to a system (q) appears as an increase in the internal energy of the system ($+\Delta E$) plus any work done by the system ($+w$), or

$$q = \Delta E + w$$

This expression of energy conservation is the **first law of thermodynamics.** It is also commonly written in the form

$$\Delta E = q - w$$

When the final and initial states of the system are fixed, ΔE is fixed, but the heat gained or lost by the system and the work done on or by the system are not fixed. For example, let us assign arbitrarily an energy value of 90 cal to a particular quantity of a gas in an initial state, and a value of 100 cal in a final state; then

$$\Delta E = E_2 - E_1 = 100 \text{ cal} - 90 \text{ cal} = 10 \text{ cal}$$

This change may be effected by any of an infinite number of methods (paths). For example,

Path I Add 10 cal of heat to the gas, arranging the experiment so that no work is done by or on the gas; then

$$\Delta E = q - w = 10 \text{ cal} - 0 = 10 \text{ cal}$$

Path II Do 10 cal of work on the gas without permitting the evolution or absorption of heat; then

$$\Delta E = 0 \text{ cal} + 10 \text{ cal} = 10 \text{ cal}$$

Path III Add 35 cal of heat and let the system perform 25 cal of work; then

$$\Delta E = 35 \text{ cal} - 25 \text{ cal} = 10 \text{ cal}$$

Path IV Do 40 cal of work on the gas and let the gas evolve 30 cal of heat; then

$$\Delta E = -30 \text{ cal} + 40 \text{ cal} = 10 \text{ cal}$$

The student can imagine many other schemes by which this gas may pass from the initial to the final state. In words, we say ΔE *is independent of the path between the initial and the final state but q and w are path-dependent.*

The work done by a gas in expanding *against* a constant pressure P is the pressure times the change in volume, ΔV or $w = P\,\Delta V$ (see Appendix 1).

The first law, when the only work is $P\,\Delta V$, may then be written

$$\Delta E = q - P\,\Delta V$$

Let us now define

q_v = heat transferred to or from system at constant volume;
q_p = heat transferred to or from system at constant pressure.

For chemical reactions that occur in a container of *fixed volume*, $P\,\Delta V = 0$ and

$$\Delta E = q_v \tag{1}$$

However, chemical reactions occur more frequently in open vessels at constant barometric pressure; then

$$\Delta E = q_p - P\,\Delta V$$

In words, the change in energy is equal to the heat absorbed by the system at constant pressure, q_p, minus the work done at constant pressure. Rearranging,

$$q_p = \Delta E + P\,\Delta V \tag{2}$$

Since it is common practice to conduct reactions at constant pressure, we define a new quantity, called the **enthalpy,*** H, such that the enthalpy change, ΔH, is the heat evolved or absorbed at constant tem-

* The term "enthalpy" is derived from the Greek word *enthalpo*, meaning "warming up." The more descriptive but misleading term, "heat content," is sometimes used synonymously with enthalpy.

perature and *constant pressure*. Then

(a) when there is a change of volume

$$\Delta E + P\,\Delta V = q_p = \Delta H \tag{3}$$

(b) when there is no change of volume, $P\,\Delta V = 0$, and

$$\Delta E = q_v = \Delta H$$

Thus, for chemical reactions *involving only solids and liquids*, which undergo relatively small expansions and contractions, ΔV is negligible and q_v is practically equal to q_p. However, if *gases are involved* in the chemical reaction, appreciable volume changes may occur and the heat absorbed or evolved at constant pressure may differ appreciably from the heat absorbed or evolved at constant volume. Thermochemical data are usually expressed in terms of the enthalpy change, ΔH, representing heat absorbed or evolved at constant pressure.

EXAMPLE 1 For the decomposition of nickel oxide

$$NiO(c) \longrightarrow Ni(c) + \tfrac{1}{2}O_2(g)$$

$$\Delta E = +57.0\,\frac{\text{kcal}}{\text{mole}} \qquad \text{and} \qquad \Delta H = +57.3\,\frac{\text{kcal}}{\text{mole}}$$

(a) Is there a volume change, and if so, in what direction, in a rigid closed container? In an open container at constant pressure? (b) In each case, how much work is done on or by the system? (c) Under which condition, constant volume or constant pressure, is the reaction more endothermic?

ANSWER (a) There is no volume change, of course, in a rigid closed container. At constant pressure, however, the volume increases by the volume of $\frac{1}{2}$ mole of gas per mole of oxide. The volumes of the solids are negligible.

(b) No work is done at constant volume ($P\,\Delta V = 0$). At constant pressure (Equation 3), the work done is

$$P\,\Delta V = \Delta H - \Delta E = +57.3 - 57.0 = +0.3\,\frac{\text{kcal}}{\text{mole}}$$

The positive sign indicates that work is done by the system on the surroundings.

(c)

$$\Delta H = q_p = +57.3\,\frac{\text{kcal}}{\text{mole}} \qquad \text{and} \qquad \Delta E = q_v = +57.0\,\frac{\text{kcal}}{\text{mole}}$$

q_p has the more positive value and therefore the reaction is more endothermic at constant pressure.

6.3 □ THERMOCHEMISTRY

The amount of heat removed in an exothermic reaction or added in an endothermic reaction is that necessary to restore the reaction mixture to its original temperature. Heat measurements are made by mixing known amounts of reactants in a calorimeter (Fig. 6.1). The heat evolved by the reaction is equal to the heat absorbed by a known quantity of water, the metal bucket, the metal reaction chamber containing known quantities of reactants, the stirrer, and the thermometer. The **water equivalent** of a calorimeter is the quantity of water required to absorb the same quantity of heat absorbed by the calorimeter and its component parts.

EXAMPLE 2 The combustion of 0.100 g of liquid benzene, C_6H_6, in a calorimeter whose "water equivalent," including that of its liquid bath, is 383.6 g of water produced a temperature rise of 2.609°C. Calculate the molar **heat of combustion** of C_6H_6 (the amount of heat involved in the combustion of one mole of C_6H_6). The molecular weight of C_6H_6 is 78.1. The specific heat of water is 0.9983 cal per gram per degree. There are 10^3 cal in a kilocalorie (kcal).

ANSWER The quantity of heat emitted is

$$0.9983 \frac{\text{cal}}{\cancel{\text{g}}\,\cancel{\text{deg}}} \times 383.6\,\cancel{\text{g}} \times 2.609\,\cancel{\text{deg}} = 999 \text{ cal}$$

Since 999 cal are emitted during the combustion of 0.100 g of C_6H_6, the heat emitted by 1 g of C_6H_6 is

$$\frac{999 \text{ cal}}{0.100 \text{ g}}$$

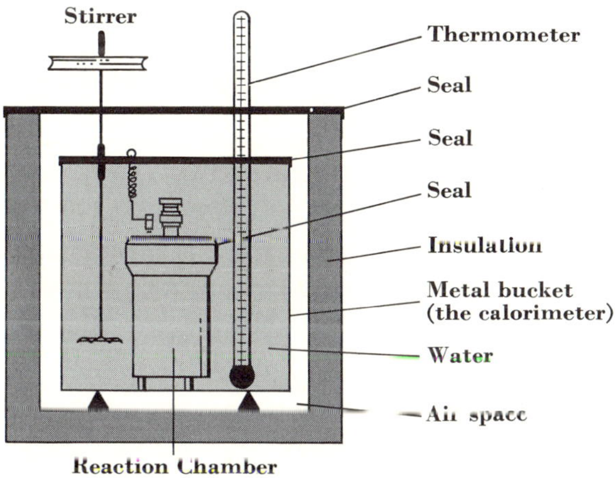

Fig. 6.1 Schematic representation of a calorimeter. Air space and insulation minimize heat exchange with the outside. The stirrer maintains uniform temperature.

Then the heat emitted by 1 mole of C_6H_6 is

$$\frac{999\ \cancel{\text{cal}}}{0.100\ \cancel{\text{g}}} \times 78.1\ \frac{\cancel{\text{g}}}{\text{mole}} \times \frac{1\ \text{kcal}}{10^3\ \cancel{\text{cal}}} = 780\ \frac{\text{kcal}}{\text{mole}}$$

Therefore $q_v = -780$ kcal/mole.

Since calorimetric measurements are generally made in a container of fixed volume, *the chemical reaction occurs at constant volume.* From Equation 1, the value obtained in such an experiment, q_v, is equal to the ΔE of the reaction. Thus the thermochemical equation for the combustion of liquid C_6H_6,

$$C_6H_6(l) + 7\tfrac{1}{2}O_2(g) \longrightarrow 3H_2O(l) + 6CO_2(g) \qquad \Delta E = -780\ \text{kcal}$$

(exothermic reaction)

means that when 1 mole of liquid benzene reacts with $7\frac{1}{2}$ moles of oxygen, forming 3 moles of liquid water and 6 moles of carbon dioxide at *constant volume*, and the temperature of the products is *restored to the original temperature* of the reactants, 780 kcal is evolved. The energy of the 3 moles of $H_2O(l)$ and 6 moles of $CO_2(g)$ is therefore 780 kcal less than the energy of 1 mole of $C_6H_6(l)$ and $7\frac{1}{2}$ moles of $O_2(g)$.

Since the energy of a substance is a function of its physical state and temperature, the ΔE of a reaction is also a function of these variables. Thus the state of each reactant and product must be clearly indicated. The data given in this text are for substances in their stable states at 25.0°C and 1 atm pressure; graphite is the stable state for carbon.

When the reaction is carried out at constant pressure, the volume decreases ($7\frac{1}{2}$ moles of gaseous O_2 are converted to 6 moles of gaseous CO_2), some work is thus done on the system by the surroundings, and more heat is therefore evolved. The thermochemical equation is

$$C_6H_6(l) + 7\tfrac{1}{2}O_2(g) \longrightarrow 3H_2O(l) + 6CO_2(g) \qquad \Delta H = -781\ \text{kcal}$$

This means that when 1 mole of liquid benzene reacts with $7\frac{1}{2}$ moles of oxygen forming 3 moles of liquid water and 6 moles of carbon dioxide at *constant pressure*, and the temperature of the products is restored to the original temperature of the reactants, 781 kcal is evolved; therefore the enthalpy of the 3 moles of $H_2O(l)$ and 6 moles of $CO_2(g)$ is 781 kcal less than the enthalpy of 1 mole of $C_6H_6(l)$ and $7\frac{1}{2}$ moles of $O_2(g)$. When ΔH is +, heat is absorbed and the reaction is endothermic. For example, the thermochemical equation for the formation of hydrogen iodide is

$$\tfrac{1}{2}H_2(g) + \tfrac{1}{2}I_2(c) \longrightarrow HI(g) \qquad \Delta H = +6.20\ \text{kcal}$$

This means that when $\frac{1}{2}$ mole of hydrogen reacts with $\frac{1}{2}$ mole of solid iodine forming 1 mole of gaseous hydrogen iodide at constant temperature and *constant pressure*, 6.20 kcal is absorbed from the surroundings; the enthalpy of the products is therefore larger than the enthalpy of reactants.

6.4 □ HESS'S LAW

In accord with the law of conservation of energy, thermochemical equations possess two important properties:

(a) They may be reversed, in which case the sign of ΔE or ΔH is changed. For example,

$$H_2(g) + \tfrac{1}{2}O_2(g) \longrightarrow H_2O(l) \qquad \Delta H = -68.3 \text{ kcal}$$
$$H_2O(l) \longrightarrow H_2(g) + \tfrac{1}{2}O_2(g) \qquad \Delta H = +68.3 \text{ kcal}$$

means that the heat evolved in the formation of 1 mole of liquid water is equal to the heat required to decompose 1 mole of liquid water.

(b) They may be added and subtracted as ordinary algebraic equations. This generalization is known as **Hess's law.** For example, the addition of the two thermochemical equations

$$C_2H_4(g) + H_2(g) + \cancel{3\tfrac{1}{2}O_2}(g) \longrightarrow \cancel{2CO_2}(g) + \cancel{3H_2O}(l) \qquad \Delta H = -405.5 \text{ kcal}$$
$$\cancel{2CO_2}(g) + \cancel{3H_2O}(l) \longrightarrow C_2H_6(g) + \cancel{3\tfrac{1}{2}O_2}(g) \qquad \Delta H = +372.8 \text{ kcal}$$

predicts that the heat of hydrogenation of ethylene, C_2H_4, to ethane, C_2H_6, is

$$C_2H_4(g) + H_2(g) \longrightarrow C_2H_6(g) \qquad \Delta H = -405.5 + 372.8 = -32.7 \text{ kcal}$$

When a formula appears on the right of one equation and on the left of another, the substance will not appear in the final equation if it is produced and consumed in equal quantities.

The experimentally determined heat of reaction is

$$C_2H_4(g) + H_2(g) \longrightarrow C_2H_6(g) \qquad \Delta H = -32.6 \pm 0.1 \text{ kcal}$$

Hess's law teaches us that the heat of a reaction, ΔE or ΔH, depends only on the nature of the final products and initial reactants, and is independent of the path (how the chemical change is carried out, in one or several steps).

Since the enthalpy of a substance is a function of its physical state and temperature, the ΔH of a reaction is also a function of these variables. For example, the enthalpy of $H_2O(g)$ is greater than that of $H_2O(l)$ by 10.5 kcal/mole,

$$H_2O(l) \longrightarrow H_2O(g) \qquad \Delta H = +10.5 \text{ kcal}$$

EXAMPLE 3 Use the thermochemical equations

$$\tfrac{1}{2}H_2(g) + \tfrac{1}{2}I_2(c) \longrightarrow HI(g) \qquad \Delta H = +6.20 \text{ kcal} \qquad (4)$$

$$\tfrac{1}{2}H_2(g) \longrightarrow H(g) \qquad \Delta H = +52.1 \text{ kcal} \tag{5}$$

$$\tfrac{1}{2}I_2(g) \longrightarrow I(g) \qquad \Delta H = +18.1 \text{ kcal} \tag{6}$$

$$I_2(c) \longrightarrow I_2(g) \qquad \Delta H = +14.9 \text{ kcal} \tag{7}$$

to calculate ΔH for the formation of gaseous hydrogen iodide, HI, from gaseous atomic hydrogen and gaseous atomic iodine,

$$H(g) + I(g) \longrightarrow HI(g)$$

ANSWER Basically, the problem is to manipulate the given chemical equations as we would algebraic equations so as to eliminate those substances not appearing in the final desired equation. Thus, transpose Equations 5, 6, and 7 and also multiply Equation 7 by $\frac{1}{2}$, yielding

$$\tfrac{1}{2}H_2(g) + \tfrac{1}{2}I_2(c) \longrightarrow HI(g) \qquad \Delta H = +6.20 \text{ kcal} \tag{4}$$

$$H(g) \longrightarrow \tfrac{1}{2}H_2(g) \qquad \Delta H = -52.1 \text{ kcal} \tag{8}$$

$$I(g) \longrightarrow \tfrac{1}{2}I_2(g) \qquad \Delta H = -18.1 \text{ kcal} \tag{9}$$

$$\tfrac{1}{2}I_2(g) \longrightarrow \tfrac{1}{2}I_2(c) \qquad \Delta H = -7.5 \text{ kcal} \tag{10}$$

which on addition yield

$$H(g) + I(g) \longrightarrow HI(g)$$

$$\Delta H = +6.20 - 52.1 - 18.1 - 7.5 = -71.5 \text{ kcal}$$

A convenient thermochemical reference value is the **heat of formation** of a substance, ΔH_f, which is defined as the increase in enthalpy when one mole of the substance is formed from the elements. It is assumed that the elements are initially in their stable states at ordinary temperature (25°C) and 1 atm pressure.

EXAMPLE 4 Given the following heats of formation (ΔH_f, kcal/mole): acetylene, C_2H_2, +54.2; CO_2, −94.1; and H_2O, −68.3. Calculate the heat of combustion of acetylene.

ANSWER The thermochemical equations that express the heats of formation are

$$2C(graphite) + H_2(g) \longrightarrow C_2H_2(g) \qquad \Delta H = +54.2 \text{ kcal}$$
$$C(graphite) + O_2(g) \longrightarrow CO_2(g) \qquad \Delta H = -94.1 \text{ kcal}$$
$$H_2(g) + \tfrac{1}{2}O_2(g) \longrightarrow H_2O(l) \qquad \Delta H = -68.3 \text{ kcal}$$

To obtain the relationship for the heat of combustion, transpose the first equation, double the second, and then add all three. The sum is

$$C_2H_2(g) + 2\tfrac{1}{2}O_2(g) \longrightarrow 2CO_2(g) + H_2O(l)$$
$$\Delta H = -54.2 - 2(94.1) - 68.3 = -310.7 \text{ kcal/mole}$$

6.5 □ BOND DISSOCIATION ENERGY

The formation of 1 mole of HI(g) from gaseous atoms evolves 71.5 kcal while the dissociation of 1 mole of HI(g) into gaseous atoms requires

71.5 kcal. The quantity of energy required to decompose a gaseous molecule into two individual fragments or to break one bond in a molecule is called the **bond dissociation energy,*** commonly expressed as ΔH in kilocalories per mole. Bond dissociation energies for a number of bonds are given in Table 6.1.

TABLE 6.1 ■ BOND DISSOCIATION ENERGIES[a] FOR THE REACTION IN WHICH ONE BOND PER MOLECULE IS BROKEN: $R_1R_2(g) \longrightarrow R_1(g) + R_2(g)$

HF	136	H_2	104.2	N_2	225.9
HCl	103	F_2	38	$NF_3 \longrightarrow NF_2 + F$	74.0
HBr	87.4	Cl_2	57.9	$NF_2 \longrightarrow NF + F$	62.6
HI	71.5	Br_2	46.1	$NF \longrightarrow N + F$	62.6

[a] ΔH in kilocalories per mole.

The heats of dissociation illustrated in Table 6.1 are obtained without assumptions regarding the nature of the forces holding atoms together in molecules.

From the data in Table 6.1, Hess's law predicts that the decomposition of nitrogen trifluoride, NF_3, into its constituent atoms, N + 3F, will require 199.2 kcal/mole:

$$
\begin{array}{llr}
NF_3(g) \longrightarrow NF_2(g) + F(g) & \Delta H = & +74.0 \\
NF_2(g) \longrightarrow NF(g) + F(g) & \Delta H = & +62.6 \\
NF(g) \longrightarrow N(g) + F(g) & \Delta H = & +62.6 \\
\hline
NF_3(g) \longrightarrow N(g) + 3F(g) & \Delta H = & +199.2
\end{array}
$$

Now let us imagine that we did not have the ΔH values for the successive dissociations of each of the three F atoms from NF_3, but only the value for complete dissociation of NF_3, +199.2 kcal/mole. We could assume that equal energies are required to remove the fluorine atoms successively from the nitrogen atom; then we may assign to each N—F bond an average energy of (199.2)/3 or 66.4 kcal/mole, called the **average bond energy** of the N—F bond. Since no experimental data for the stepwise dissociation of most polyatomic molecules exist, the assignment of average bond energies to chemical bonds is common practice.

EXAMPLE 5 Calculate the average bond energy for the C—H bond from the following experimental data:

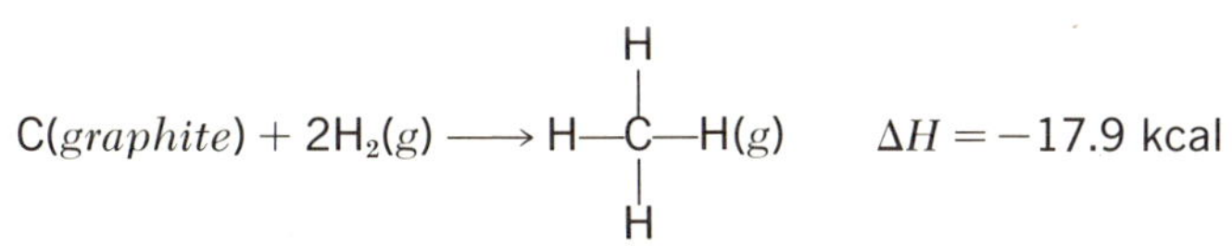

methane

* Sometimes also called **bond strength.** Though referred to as "energy," it is usually the enthalpy change that is listed.

$$\mathrm{C}(g) \longrightarrow \mathrm{C}(graphite) \qquad \Delta H = -170.9 \text{ kcal}$$
$$2\mathrm{H}(g) \longrightarrow \mathrm{H_2}(g) \qquad \Delta H = -104.2 \text{ kcal}$$

ANSWER First, multiplying the third equation by 2 and adding the three equations yields

$$\mathrm{C}(g) + 4\mathrm{H}(g) \longrightarrow \mathrm{CH_4}(g) \qquad \Delta H = -397.2 \text{ kcal}$$

Notice that the only reaction now occurring is the formation of four C—H bonds; *assuming* they are identical, it follows that in forming one bond

$$\mathrm{C}(g) + \mathrm{H}(g) \longrightarrow \mathrm{C{-}H}(g) \qquad \Delta H = \frac{-397.2}{4} = -99.3 \frac{\text{kcal}}{\text{mole}}$$

or breaking one bond

$$\mathrm{C{-}H}(g) \longrightarrow \mathrm{C}(g) + \mathrm{H}(g) \qquad \Delta H = +99.3 \frac{\text{kcal}}{\text{mole}}$$

The average bond energy for the C—H bond is therefore 99.3 kcal/mole.

TABLE 6.2 AVERAGE BOND ENERGY VALUES (ΔH, KCAL/MOLE) FOR THE PROCESS: BOND AB IN A GASEOUS MOLECULE $\longrightarrow$ A(g) + B(g)

C—H	99.3	C—F	116	H—F	136
		C—Cl	78	H—Cl	103
		C—Br	68	H—Br	87.4
		O—H	111	H—I	71.5

Average bond energies are usually called simply **"bond energies"**. Table 6.2 lists a number of bond energy values.

The term "bond energy" may be confusing because it connotes a bond full of vim and vigor, ready to explode. Such a notion is entirely misleading. "Bond energy" is rather the energy that the bonded molecule does *not* have, and will not get until some outside source of energy comes along and breaks the bond. Thus a high bond energy means the bond is stable. An explosive compound like TNT or nitroglycerin can do its work because the reaction converts bonds of lower bond energy to bonds of higher bond energy:

	TNT		**Explosion products**			
	$\mathrm{C_7H_5(NO_2)_3}$	$\xrightarrow{bang!}$	$2\frac{1}{2}\mathrm{H_2O}$ +	$3\frac{1}{2}\mathrm{CO}$ +	$1\frac{1}{2}\mathrm{N_2}$	+ $3\frac{1}{2}\mathrm{C}$
Bond energies, kcal/mole	H—C, 99		H—O, 111	C≡O, 256	N≡N, 226	
	C—C, 83					
	C—N, 73		*higher bond energies, more stable*			
	N—O, 53					
	lower bond energies, less stable					

6.6 □ THE INTERCONVERTIBILITY OF MATTER AND ENERGY

In any type of transformation, the conversion of one form of matter to another involves the conversion of matter to energy if the reaction is exothermic; conversely, in an endothermic reaction, the heat energy absorbed is converted into matter. This means matter and heat energy are not conserved separately; rather, the principle of conservation of energy is broadened to include matter as another form of energy. Thus, the quantity of energy, expressed in grams, liberated or absorbed is *exactly equal* to the quantity of matter, expressed in grams, destroyed or created.

The interconvertibility of matter and energy is described by the equation $\Delta E = c^2 \, \Delta m$, predicted by Albert Einstein in 1905; ΔE is the energy in ergs when Δm, the quantity of matter converted, is in grams; the constant c, the speed of light, is 3.00×10^{10} cm/sec. An erg has the dimensions of g cm^2/sec^2, and there are 4.18×10^7 ergs in one calorie.

EXAMPLE 6 In an explosion of gasoline-air mixture 4.9×10^6 cal are evolved,

$$\text{reactants} \longrightarrow \text{products} \qquad \Delta E = -4.9 \times 10^6 \text{ cal}$$

Calculate the decrease in mass (weight) in grams accompanying this reaction.

ANSWER The mass (weight) of the products will be less by an amount calculated from the Einstein equation ($\Delta E = c^2 \, \Delta m$),

$$4.9 \times 10^6 \, \cancel{\text{cal}} \times 4.2 \times 10^7 \, \frac{\text{ergs}}{\cancel{\text{cal}}} = \Delta m (3.0 \times 10^{10})^2 \, \frac{\text{cm}^2}{\text{sec}^2}$$

from which

$$\Delta m = 2.3 \times 10^{-7} \text{ ergs} \, \frac{\text{sec}^2}{\text{cm}^2}$$

and since an erg has the dimensions of g $\dfrac{\text{cm}^2}{\text{sec}^2}$

$$\Delta m = 2.3 \times 10^{-7} \, \frac{\text{g} \, \cancel{\text{cm}^2}}{\cancel{\text{sec}^2}} \, \frac{\cancel{\text{sec}^2}}{\cancel{\text{cm}^2}} = 2.3 \times 10^{-7} \text{ g}$$

Thus 2.3×10^{-7} g is the quantity of matter liberated as heat to the surroundings.

This quantity is so small that unless nuclear reactions (Chapter 24) are involved, we can say that the total mass of matter of a chemically reacting system is *practically* constant.

PROBLEMS

1. Definitions (a) State the first law of thermodynamics in a complete sentence, without using symbols or abbreviations. (b) Write your favorite statement of the law of conservation of energy. In what way do you prefer this to other statements of the law? Do you consider all the statements of this law to be entirely equivalent? (c) State Hess's law in a complete sentence.

2. Conservation of energy (a) Potassium chlorate ($KClO_3$) is stable at room temperature. Above its melting point (356°C) it begins to decompose, with evolution of oxygen; the higher the temperature, the more rapid the decomposition. Can you deduce from these statements whether the decomposition is exothermic or endothermic? (b) A quantity of $KClO_3$ is warmed to 385°C and then is placed in a closed insulated container. As the material decomposes, the temperature of the system rises. Can you deduce whether the decomposition is exothermic or endothermic? If so, which is it?

SAFETY NOTE: THIS PROBLEM IS PRESENTED FOR THEORETICAL CONSIDERATION ONLY, NOT FOR EXPERIMENT. CHLORATES ARE HAZARDOUS CHEMICALS, AND FREQUENTLY EXPLODE IN THE PRESENCE OF IMPURITIES. IN GENERAL, THE STUDENT MUST NOT HEAT SUBSTANCES IN CLOSED CONTAINERS.

3. The first law of thermodynamics The heat of combustion of acetic acid

$$C_2H_6O_2(l) + 2\tfrac{1}{2}O_2(g) \longrightarrow 2CO_2(g) + 3H_2O(l)$$

is 209 kcal given out per mole of acetic acid at atmospheric pressure. Under these conditions, the amount of work involved is 0.3 kcal per mole of acetic acid. (a) What is the value of ΔH for the reaction? (b) Is there a volume change, and, if so, in what direction, at atmospheric pressure? (c) If the reaction had been carried out in a rigid closed container, would there have been a change of volume? (d) Pick the correct words from the parentheses: At constant pressure, the reaction is (exo-, endo-) thermic and (some, no) work is done (on, by) the system. At constant volume the reaction is (exo-, endo-) thermic and (some, no) work is done (on, by) the system. The reaction is (more, less) (exo-, endo-) thermic at constant (pressure, volume) than at constant (pressure, volume). (e) What is the value of ΔE for the reaction?

4. Thermochemistry Given the following thermochemical equations:

$$C(graphite) + O_2(g) \longrightarrow CO_2(g) \qquad \Delta H = -94.1\ \frac{\text{kcal}}{\text{mole}}$$

$$C(graphite) \longrightarrow C(diamond) \qquad \Delta H = +0.5\ \frac{\text{kcal}}{\text{mole}}$$

(a) Calculate ΔH for the burning of diamond to CO_2. (b) Calculate the quantity of graphite that must be burned to evolve 1000 kcal of heat. (c) Which will evolve more heat in burning, 1 lb of graphite or 1 lb of diamond? (d) At atmospheric pressure, which substance is in a condition of lower energy, graphite or diamond?

5. Thermochemistry Given the thermochemical equation

$$CaCO_3(c) \longrightarrow CaO(c) + CO_2(g) \qquad \Delta H = +43\ \text{kcal/mole}$$

and the heats of formation, ΔH_f, for $CO_2(g)$, −94 kcal/mole, and for $CaCO_3(c)$, −289 kcal/mole, calculate the heat of formation of CaO.

6. Bond energies (a) How much energy (ΔH) in kilocalories per mole and in calories per molecule is needed to split gaseous HBr into its constituent gaseous atoms? (b) Calculate the energy (ΔH) in kilocalories per mole required to decompose gaseous methyl chloride,

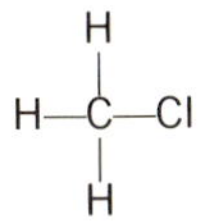

into isolated atoms.

7. $E = mc^2$ (a) How much energy in kilocalories can be produced in a nuclear reactor in which 1.0 g of mass is converted to energy? (b) What weight of carbon would have to be burned to CO_2 to yield this quantity of heat? (ΔH for $C + O_2 \longrightarrow CO_2$ is -94 kcal/mole.)

ADDITIONAL PROBLEMS

8. Conservation of energy A friend asks you to invest your savings in the following proposition, which he claims is a cheap method for electrical heating of homes. Hydrogen and oxygen are produced by the electrolysis of water. The gases are diluted in argon below the concentrations at which they will explode and are passed through a catalyst where they recombine to water. The entire system is self-contained, the argon is recirculated, and the water is returned to the cell. Heat is produced in the catalytic bed, which thus serves as a room radiator. Your friend says the catalyst is so effective that you can produce heat with very little cost of electricity. What would you think of such an investment opportunity? Would the truth of your friend's claim depend on the effectiveness of his catalyst? Explain.

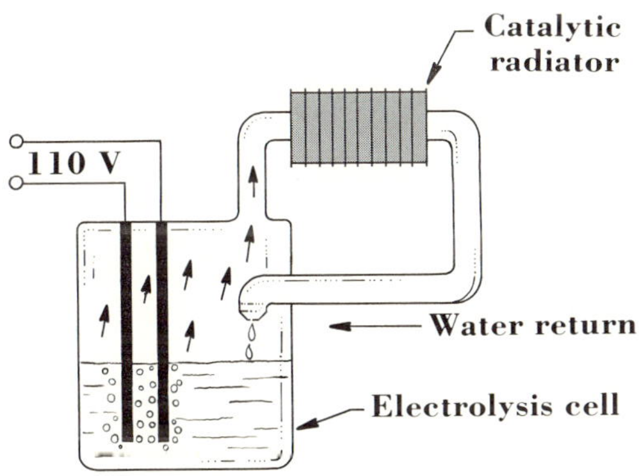

9. The first law of thermodynamics The enthalpy change for the combustion of graphite

$$C(graphite) + O_2(g) \longrightarrow CO_2(g)$$

is $\Delta H = -94.1$ kcal/mole. What is the value of ΔE for the reaction? Explain.

10. Thermochemistry Given the following heats of formation (ΔH_f) in kilocalories per mole: NO, +21.60; NH_3, −11.04; H_2O, −57.80. Calculate ΔH for the reaction.

$$4NH_3(g) + 5O_2(g) \longrightarrow 4NO(g) + 6H_2O(g)$$

11. Thermochemistry Given the following thermochemical equations:

$$C(graphite) + O_2(g) \longrightarrow CO_2(g) \qquad \Delta H = -94.1 \text{ kcal}$$

$$CO(g) + \tfrac{1}{2}O_2(g) \longrightarrow CO_2(g) \qquad \Delta H = -67.7 \text{ kcal}$$

(a) Calculate ΔH for the oxidation of carbon to carbon monoxide. (b) Calculate the amount of heat that can be produced by the oxidation of 1.00 kg of graphite to CO_2. (c) Show how, if Hess's law were violated, it would be possible to create energy from the progress or reversal of one or more of the reactions described above.

12. Thermochemistry (a) An exothermic reaction is allowed to occur very rapidly in the air. Will the temperature of the surrounding air increase or decrease? (b) An endothermic reaction is allowed to occur very rapidly in the air. Will the temperature of the surrounding air increase or decrease? (c) A person pours a liquid into the palm of his hand. As the liquid evaporates, his hand feels cold. Is the evaporation of this liquid an exothermic or an endothermic process?

13. Thermochemistry (a) Do the products of an exothermic reaction contain more or less energy than the reactants at the same temperature? Explain your answer. (b) For each of the following reactions, explain whether the heat evolved at constant pressure is smaller than, larger than, or the same as the heat evolved at constant volume:

$$C(graphite) + \tfrac{1}{2}O_2(g) \longrightarrow CO(g)$$

$$NO(g) + \tfrac{1}{2}O_2(g) \longrightarrow NO_2(g)$$

$$4HCl(g) + O_2(g) \longrightarrow 2H_2O(g) + 2Cl_2(g)$$

$$\tfrac{1}{2}H_2(g) + \tfrac{1}{2}Cl_2(g) \longrightarrow HCl(g)$$

$$C_2H_2(g) + 1\tfrac{1}{2}O_2(g) \longrightarrow H_2O(l) + 2CO(g)$$

14. Explosives The bond energy of the O—O bond in peroxides, such as hydrogen peroxide, H—O—O—H, is 35 kcal/mole, and that of O—H is about 111 kcal/mole. The bond dissociation energy of molecular oxygen, O_2, is 119 kcal/mole. (a) Write the balanced equation for the decomposition of H_2O_2 to produce water and oxygen. (b) From an inspection of the bond energy data cited above, but without calculating ΔH for the reaction, predict whether the decomposition is exothermic or endothermic. (c) The peroxide linkage occurs in other compounds such as ethyl peroxide, peroxyacetic acid, peroxyacetyl nitrate, and many others. Would you predict that peroxides as a class might be explosives? What would be the basis for such a prediction? (SAFETY NOTE: REFER TO THE ANSWER TO THIS QUESTION, P. 109).

15. $E = mc^2$ The heat of formation of solid barium sulfate from its ions is given by the following equation:

$$Ba^{2+}(aq) + SO_4^{2-}(aq) \longrightarrow BaSO_4(c) \qquad \Delta H = -5 \frac{\text{kcal}}{\text{mole}}$$

Refer to Problem 10, p. 78, which deals with Manley's verification of conservation of mass in 1913. Assume that the reaction occurring in his experiment was that shown in the equation above, and that he produced 23 g (0.10 mole) of $BaSO_4$, with the consequent liberation of 0.5 kcal of heat. Calculate the decrease in mass that accompanied his reaction. To how many *more* places after the decimal point should Manley's balance have been sensitive for him to detect this decrease in mass? Do you think that balances of such precision, with a capacity of about 100 g, can be manufactured today?

16. Heat of formation Referring to Example 3, p. 101–102, explain why the heat of formation of HI is +6.20 kcal/mole, not −71.5 kcal/mole.

ANSWERS **5.** −152 kcal/mole.

6.(b) +376 kcal/mole.

10. −216.24 kcal.

14. Yes, peroxides are explosive.

7 ■ TYPES OF CHEMICAL BONDS

7.1 □ INTRODUCTION

Most molecules that are stable under terrestrial conditions are destroyed at temperatures above about 10^3 or 10^4°C, and other species of matter become predominant. Typical among these are N, O, CaH, AlCl, $(LiF)_3$, Na_2Cl^+, and $NaCl_2^-$. At still higher temperatures, more extensive ionization occurs until matter consists largely of individual nuclei which move about independently of their electrons. This condition is called the **plasma** state. At high pressures, 10^5–10^6 atm, most substances become metallic (Ch. 23); wood and chalk ($CaCO_3$), for example, become metallic conductors under pressures of about 10^5 atm.

These remarks emphasize that our study of chemical bonding is largely directed to phenomena that occur under conditions that are familiar to us on earth.

7.2 □ WHAT TYPE OF ATTRACTIVE FORCES HOLD ATOMS TOGETHER IN CHEMICAL BONDS?

We know of three types of attractive forces among macroscopic bodies —gravitational, magnetic, and electrostatic. Of these, only electrostatic forces are strong enough to account for observed bond energies. The simplest notion we might entertain, then, is that chemical bonds are the result of Coulombic forces of attraction between individual positive and negative ions. Such an idea was the basis for the **dualistic theory** of Jöns Jakob Berzelius, Humphry Davy, and others in the nineteenth century. Thus, sodium atoms were thought to be positively charged, or at least to acquire such a charge near other atoms. Oxygen atoms, on the other hand, were believed to be negative. A positive charge was considered to characterize a "basic" substance, a negative charge an "acidic" substance. The resulting attraction accounted for the formation of the compound Na_2O. But the positive charge on 2Na was thought to be greater than the negative charge on O, with the result that Na_2O was considered to have some residual positive charge, accounting for its basic character. Na_2O would then further combine with a compound that had a residual negative charge (acid character), like SO_3, to form a "neutral" compound, $Na_2O \cdot SO_3$ (now written Na_2SO_4). Being neither acid nor basic meant being neither plus nor minus, hence "neutral." This theory could not suggest what charge condition of carbon enables it to form bonds with "positive" hydrogen, in CH_4, as well as with "negative" chlorine, in CCl_4, or even with both in the same molecule, as in chloroform, $CHCl_3$. Despite such difficulties, the belief that electrostatic attraction accounts for the bonding in electrolytic substances like Na_2O and sodium chloride, NaCl, persisted.

Around the turn of the century the gases argon, krypton, xenon, neon, and helium were discovered. What was remarkable about these gases was their chemical inertness. It was also evident that they constituted a new periodic group; this was designated group 0. The relationship between the bonding of atoms and the electronic configuration of noble gases was proposed in 1916 independently by Walther Kossel and Gilbert N. Lewis, and extended in 1919 by Irving Langmuir. It was suggested that *atoms interact by changing the number of their electrons so as to acquire the electronic structure of* (*become isoelectronic with*) *a noble gas*. With the exception of helium, which has a $1s^2$ electronic configuration, each noble gas has eight electrons with an s^2p^6 distribution in its highest shell (p. 30). The need for eight electrons gives the name **octet rule** to this concept. There are, however, many exceptions to the octet rule—even compounds of the noble gases have been synthesized.

7.3 □ LEWIS SYMBOLS

In the following discussion, Lewis symbols are used to represent atoms and ions. In these symbols the usual letter designations represent the element and dots represent all the *valence* electrons. The term **valence electrons** defines all the electrons that are capable of being involved

LEWIS SYMBOLS
Periodic Group Number, *G*

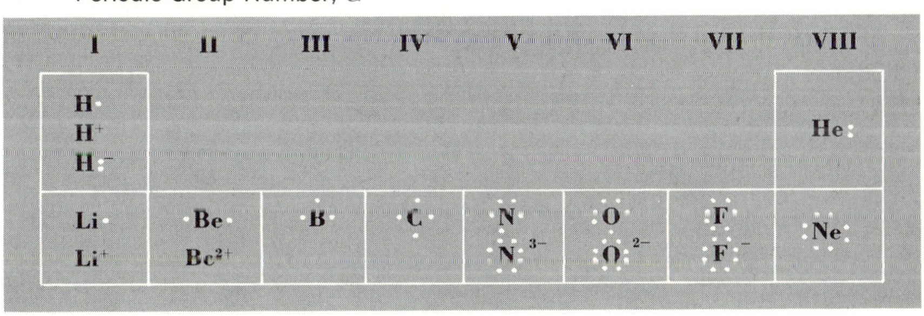

in bonding. For the representative elements, these are usually taken to be all the electrons in the highest s and p subshells. The number of valence electrons then is the same as the periodic group number of the element.

Difficulties arise when Lewis symbols are used for the transition elements, because bonding of the transition elements can involve electrons in more than one shell.

7.4 □ THE IONIC BOND

We pour a little liquid bromine, a reddish-brown, volatile, highly toxic (in fact, deadly) element, into a large flask. Next we cut a small piece of the metallic element potassium, clean off any surface impurities, and drop the metal into the bromine.

SAFETY NOTE: THIS IS A HAZARDOUS EXPERIMENT THAT DEMANDS EXPERIENCE, JUDGMENT, A FUME HOOD, A SAFETY SHIELD, THE USE OF SMALL QUANTITIES OF REACTANTS, AND PERSONAL PROTECTION LIKE GOGGLES AND GLOVES. THE STUDENT SHOULD NOT ATTEMPT IT.

The materials react on contact, producing a loud report and a flash of light; the reaction product, a white deposit of potassium bromide, flecks the inner walls of the flask. How shall we interpret this chemical change in accordance with our ideas of the octet rule and the electrostatic attractions of ions? We could say this:

> The reaction started with (a) potassium atoms that tend to lose one electron and become isoelectronic with argon ($K\cdot - e^- \longrightarrow K^+$) and (b) bromine atoms that tend to gain one electron and become isoelectronic with krypton ($:\ddot{B}r\cdot + e^- \longrightarrow :\ddot{B}r:^-$). These tendencies bring about a transfer of electrons to produce KBr.

One of the statements in the above paragraph is correct. The others are false or misleading. Let us examine them one by one.

We started with potassium atoms. Misleading. We started with the metal, in which potassium atoms are bonded together to produce a solid. This solid was destroyed in the reaction, so all the bonds must have been broken. This action is energetically equivalent to vaporizing the metal:

$$\textit{Heat of sublimation} \quad K(c) \longrightarrow K(g) \qquad \Delta H = +21.5 \text{ kcal/mole}$$

Potassium atoms tend to lose electrons. False. Work must be done to pull the electron away from the cation it leaves behind:

$$\textit{Ionization energy} \quad K(g) - e^- \longrightarrow K^+(g) \qquad \Delta H = +100.0 \text{ kcal/mole}$$

It is true that potassium loses electrons more easily than most atoms, but the ΔH is still positive, and, indeed, quite high compared with that of many other decompositions of strong bonds. Thus, it is almost as difficult to remove an electron from a potassium atom as it is to break and O—H bond in water (bond energy 111 kcal/mole).

We started with bromine atoms. Again false. The element in its usual state is Br_2, not Br. The liquid must be vaporized and the bonds in these diatomic molecules must all be broken. Then,

$\frac{1}{2}$ *Heat of vaporization*	$\frac{1}{2}Br_2(l) \longrightarrow \frac{1}{2}Br_2(g)$	$\Delta H = \frac{1}{2} \times 7.2$ kcal/mole
$\frac{1}{2}$ *Bond energy*	$\frac{1}{2}Br_2(g) \longrightarrow Br(g)$	$\Delta H = \frac{1}{2} \times 45.4$ kcal/mole
		Sum $= \frac{1}{2} \times 52.6$ kcal/mole
		$= 26.3$ kcal/mole

Bromine atoms tend to gain electrons. True, for a change. The ΔH is negative.

$$\textit{Electron affinity}\quad Br(g) + e^- \longrightarrow Br^-(g) \qquad \Delta H = -81.6 \text{ kcal/mole}$$

The sum of all these ΔH values is $+21.5 + 100.0 + 26.3 - 81.6 = +66.2$ kcal/mole. Then

$$K(c) + \tfrac{1}{2}Br_2(l) \longrightarrow K^+(g) + Br^-(g) \qquad \Delta H = +66.2 \text{ kcal/mole}$$

This means that the conversion of liquid bromine and solid potassium to individual Br^- ions and K^+ ions is energetically *unprofitable*. Thus, the last statement in the shady paragraph above is also false.

What, then, is the major source of the energy that makes the reaction go, with enough calories left over for the noise and the flash? It is the formation of the crystalline solid, potassium bromide, as a result of the electrostatic attraction of the individual ions, and their close arrangement in an ionic lattice* to produce the ionic solid. The **lattice energy** of an ionic solid is the energy released when one mole of crystal is produced from isolated ions. For KBr, the value is

$$\textit{Lattice energy}\quad K^+(g) + Br^-(g) \longrightarrow KBr(c) \qquad \Delta H = -159.9 \text{ kcal/mole}$$

The sum of all these ΔH values, including that involved in formation of the lattice, leaves enough energy for the fireworks (all units are kilocalories per mole): Sublimation of $K(c)$, $+21.5$; and ionization of $K(g)$, $+100.0$; and vaporization of $\frac{1}{2}Br_2(l)$, $+3.6$; and dissociation of $\frac{1}{2}Br_2(g)$, $+22.7$; and capture of an electron by $Br(g)$, -81.6; and formation of the $KBr(c)$ crystal lattice, -159.9; all added together give

$$\textit{Heat of formation}\quad K(c) + \tfrac{1}{2}Br_2(l) \longrightarrow KBr(c) \qquad \Delta H = -93.7 \text{ kcal/mole}$$

The sequence of thermochemical equations typified by those shown above is called the **Born-Haber cycle.**

* The solid state is discussed in Chapter 12. The KBr lattice is not 100% ionic because the transfer of electrons from K to Br is not complete. In general, all lattices of binary compounds retain some covalent (electron sharing) character.

This discussion has been concerned with the thermodynamics of the formation of the ionic bond in KBr. Remember that the ΔH value tells us only the difference in enthalpy between the initial states (solid potassium and liquid bromine) and the final state (crystalline KBr), and that this difference does not depend on the path, or sequence of steps, between the initial and final states. Our treatment was thus thermodynamic bookkeeping, and does not imply that the various equations represent distinct events. An elucidation of the real events is the concern of chemical kinetics (Chapter 19). We believe that the sequence of steps in the reaction of potassium with bromine is not the same as those indicated by the preceding thermochemical equations. Some portion of the reaction sequence occurs as a **concerted process,** which is one that happens all at once with no discernible separate steps.

The six ΔH values in this Born-Haber cycle are related to each other by Hess's law, and we therefore need determine only five of them and can calculate the value of the sixth. Heats of formation, heats of sublimation, ionization energies, and bond energies can may determined experimentally with good precision. Lattice energies may be determined indirectly from a knowledge of the ionic radii, the arrangements of ions in the crystal, and from the assumption of Coulomb's law (App. 1), or may be measured more directly by the vaporization and ionization of a crystal. Direct determination of electron affinities is difficult; the most satisfactory measurements have been made for the halogens. It is also possible to arrive at electron affinity values from calculated lattice energies via the Born-Haber treatment.

7.5 ▫ THE COVALENT BOND

The guests at a party in Paris one evening during the short reign of King Charles X (1824–1830) were irritated by unusually pungent fumes coming from the burning candles. The chemist Jean B. A. Dumas, investigating the problem, learned that the candles had been bleached with chlorine, and determined that the fumes were HCl. He concluded that the chlorine had become chemically bound to the carbon in the candle during the bleaching. Dumas' notion that chlorine could take the place of hydrogen in a chemical combination came to be called the **substitution theory.** But this concept presumed a type of chemical bonding that was inconsistent with the tenets of the dualistic theory, and Dumas' theory was therefore rejected with ridicule by the dualists. Thus began the long sequence of experiments and theories, still in progress, from which grew our present concept of the covalent bond.

We might start with a simpler substance than chlorinated candles, namely, hydrogen gas, H_2. In the combination reaction $2H \longrightarrow H_2$, if we assume that each H atom must become isoelectronic with helium, then each atom needs another electron. This requirement cannot be satisfied by electron transfer because the entire H_2 molecule has a total of only 2 electrons. If we assume, instead, that the H atoms *mutually share their electrons*, then we may assign the two electrons to *both* atoms. By this assignment, each H atom is considered to have 2 electrons, and thus to be isoelectronic with helium. The fact is that a strong bond is formed, and the resulting molecule, H_2, is very stable even though it is not ionic:

$$\mathrm{H\cdot} + \mathrm{H\cdot} \longrightarrow \mathrm{H:H} \qquad \Delta H = -104 \text{ kcal/mole}$$

Such a sharing of a pair of electrons is called a **covalent bond,** and is depicted by a dash, H—H, or a pair of dots, H : H. Once the covalent bond has formed, each of the two bonding electrons is attracted by two nuclei instead of one, and the bonded state is therefore more stable than the nonbonded state.

The number of covalent bonds formed by an atom is termed its **covalency.** Some of the values for the more common elements are as follows: hydrogen and the halogens, 1; oxygen and sulfur, 2; nitrogen and phosphorus, 3; carbon and silicon, 4. In each case cited, the covalency is equal to the number of electrons the atom needs to become isoelectronic with a noble gas. **Structural formulas** use a dash to indicate a covalent bond. Lewis formulas also show unbonded valence electrons. Some examples follow:

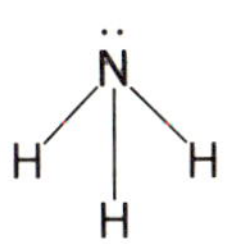

ammonia

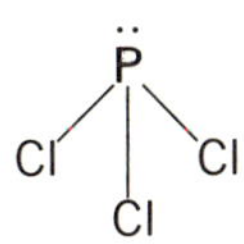

phosphorus trichloride

silicon tetrafluoride

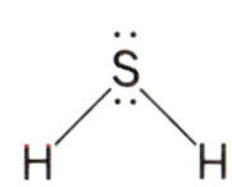

hydrogen sulfide

If we know the molecular formula of a substance and the covalencies of its elements, we can often write the structural or the Lewis (electronic) formula.

EXAMPLE 1 Write the structural formula for CH_4O.

ANSWER The only structure that satisfies the octet rule and the covalency requirements of C = 4, O = 2, and H = 1, is

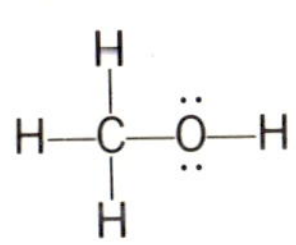

methyl alcohol

If all four H atoms were bonded to carbon, the O atom would remain without an octet, and if two H atoms were bonded to the O atom, the carbon atom would remain without an octet.

7.6 □ MULTIPLE BONDS

To satisfy their covalency requirements, two atoms must often share more than one pair of electrons. Such an arrangement is depicted as a **multiple bond.** The sharing of two pairs of electrons constitutes a **double bond;** the sharing of three pairs of electrons, a **triple bond.** Examples of multiply bonded molecules are phosgene, $COCl_2$, which contains a double bond; N_2 and HCN, each of which contains a triple bond; and CO_2, in which carbon has two double bonds:

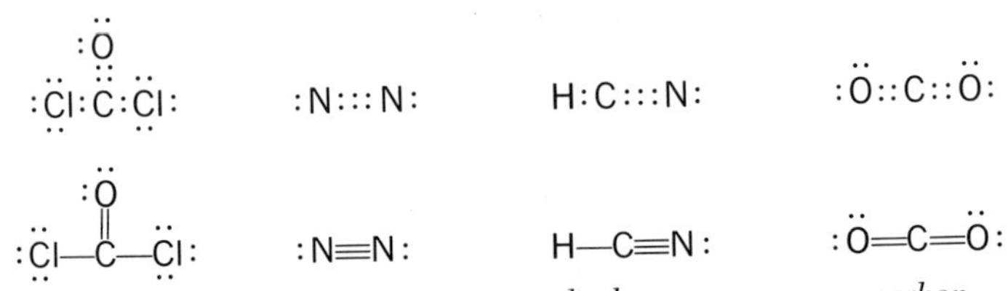

phosgene *nitrogen* *hydrogen cyanide* *carbon dioxide*

In general, within a given group in the periodic table, the ability to form multiple bonds diminishes with increasing size of the atom. With few exceptions, such as S, only atoms in the second period, for example, C, N, and O, can have a multiple bond and an octet of electrons. Examples are

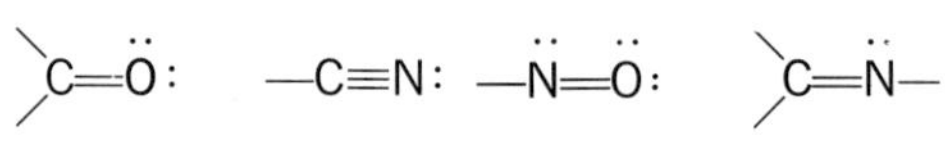

The number of compounds in which sulfur forms multiple bonds are few, and the bonds are much weaker than their oxygen analogs.

Let us consider what effects on the structure of molecules may result from the tendency or the disinclination of atoms to undergo multiple bonding. Compare the oxides of empirical formulas CO_2 and SiO_2. Both C and Si have a covalence of 4, but C forms multiple bonds in CO_2 and Si is singly bonded in SiO_2. For CO_2, we can write :Ö═C═Ö: corresponding to the molecular formula CO_2, which is identical with the empirical formula. The molecular weight of carbon dioxide can therefore be calculated from this formula: the value is 44 g/mole. Now consider SiO_2. In the absence of multiple bonds, the covalency of 4 for Si is satisfied by writing

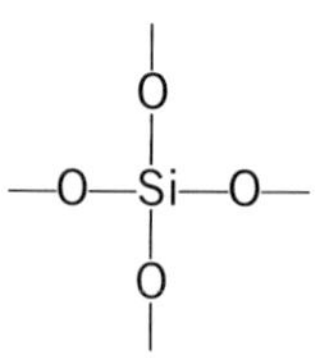

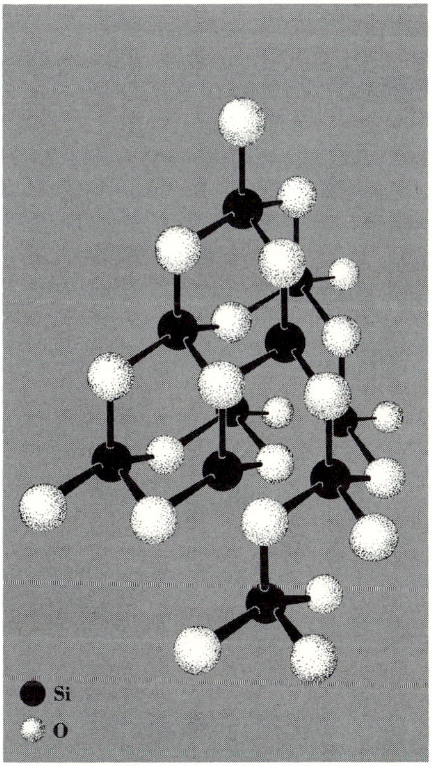

Fig. 7.1 Silicon dioxide.

But the covalence of 2 for O is satisfied only by writing

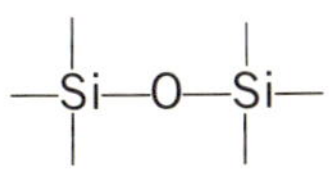

If we start to write a structure to satisfy both these requirements, we write indefinitely; there is no end to the "molecule." This circumstance has its counterpart in the fact that the real substance silicon dioxide, a component of sand and rocks, is actually a three-dimensional network of bonded atoms in space, to which the concept of "molecule" is inapplicable. "SiO_2" is therefore an empirical formula, not a molecular one. Figure 7.1 shows a model of the SiO_2 lattice.

7.7 □ PROPERTIES OF IONIC AND COVALENT COMPOUNDS

The following properties distinguish molecular covalent, macromolecular covalent (network), and ionic substances from each other.

(a) The **molecular covalent** type, exemplified by CO_2, I_2, P_4, and Cl_2, consists of separable molecules. The atoms in the molecules are held

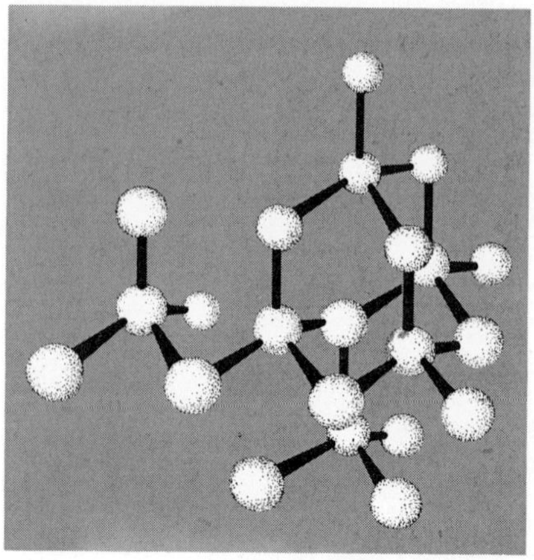

Fig. 7.2 Structure of diamond.

together by strong covalent bonds, but the forces between the molecules themselves are very weak. As a result, molecular covalent compounds have relatively low melting points and boiling points. It is rare to find a molecular compound that melts above 300°C or boils above 500°C.

(b) The **macromolecular** species is exemplified by silicon dioxide (Fig. 7.1) and diamond (Fig. 7.2). These substances are continuous, three-dimensional aggregates of covalently bonded atoms. The attractive forces between the atoms are very strong. The compounds are therefore invariably solids with very high melting and boiling points. Thus, silicon dioxide (quartz) melts at about 1710°C and boils at 2230°C, and diamond melts at 3500°C and boils at 4200°C.

(c) An **ionic** compound is an electrolyte when in the liquid state—either molten or in a solution. As an electrolyte, it conducts electricity by the motion of its ions, with accompanying chemical changes at the electrodes (See Sec. 15.3). Ionic compounds usually have relatively high melting points and boiling points, both above about 500°C. This property is a result of the large amount of heat energy that must be supplied to overcome the strong interionic attractive force. However, the melting and boiling points of ionic compounds are not so high as those of macromolecular covalent solids. For example, NaCl melts at 800°C and boils at 1413°C, and NaOH melts at 318°C and boils at 1390°C.

7.8 □ POLAR COVALENT BONDS; ELECTRONEGATIVITY

Equal sharing of a pair of electrons occurs in homodiatomic* molecules like H:H, and between identical atoms, with identical neighbors, as illustrated by the two C atoms in ethane, $H_3C:CH_3$. But if the two bonded

* Homodiatomic means "consisting of two like atoms."

TABLE 7.1 ▪ ELECTRONEGATIVITIES OF THE REPRESENTATIVE ELEMENTS

$_{1}$H 2.1							
$_{3}$Li 1.0	$_{4}$Be 1.5	$_{5}$B 2.0	$_{6}$C 2.5	$_{7}$N 3.0	$_{8}$O 3.5	$_{9}$F 4.0	
$_{11}$Na 0.9	$_{12}$Mg 1.2	$_{13}$Al 1.5	$_{14}$Si 1.8	$_{15}$P 2.1	$_{16}$S 2.5	$_{17}$Cl 3.0	
$_{19}$K 0.8	$_{20}$Ca 1.0	$_{31}$Ga 1.6	$_{32}$Ge 1.8	$_{33}$As 2.0	$_{34}$Se 2.4	$_{35}$Br 2.8	
$_{37}$Rb 0.8	$_{38}$Sr 1.0	$_{49}$In 1.7	$_{50}$Sn 1.7	$_{51}$Sb 1.9	$_{52}$Te 2.1	$_{53}$I 2.4	
$_{55}$Cs 0.7	$_{56}$Ba 0.9	$_{81}$Tl	$_{82}$Pb 1.7	$_{83}$Bi 1.9	$_{84}$Po	$_{85}$At	
$_{87}$Fr	$_{88}$Ra						

atoms are dissimilar, as in H:Cl, or are identical but in different molecular surroundings, as the two C atoms in H_3C:CCl_3, the sharing is unequal; one atom is likely to attract electrons more strongly than the other. The atom that attracts electrons more strongly develops some negative charge; the other atom develops some positive charge. These fractional charges are designated as δ^+ and δ^-, not to be confused with unit charges. For example, since a Cl atom is more electron-attracting than an H atom, hydrogen chloride is depicted as

$$\overset{\delta^+}{\mathrm{H}}\text{—}\overset{\delta^-}{\mathrm{Cl}}$$

Such covalent bonds are said to be **polar,** as distinguished from the Cl:Cl or H:H bonds, which are called **nonpolar.** The limiting case of unequal sharing of a pair of electrons exists in separate, independent ions in the gaseous state, which have integral charge units, like $Ca^{2+}(g)$ or $Na^+(g)$. In an ionic solid like NaCl(c), the separation of charge is not complete; the actual charges on these atoms are about +0.67 for Na and −0.67 for Cl. These values are high enough, however, for NaCl to assume ionic character. Thus, the nonpolar covalent bond and the ionic bond are the observed extremes for the distribution of a pair of electrons between two nuclei. Between these extremes are the many intermediate conditions of charge distribution.

The relative tendency of a bonded atom in a molecule to attract electrons is expressed by the term **electronegativity.** This word does not mean the actual content of negative charge, just the tendency to acquire it. Thus, F is highly electronegative; F^- is not. Electronegativity values are not measured directly; they are derived from other data and from various sets of assumptions on which there is not unanimous agreement among chemists. Some electronegativity tables are based on electron affinities and ionization energies. The values given in Table 7.1 are derived from bond energy data. Notice that *electronegativities increase from left to right in a period* and, with a number of exceptions, *from the bottom to the top of a group.* Notice also that fluorine and oxygen are the most electronegative of the elements. In most cases a difference 1.7 or more in electronegativities between two bonding atoms engenders an ionic bond, and a difference of less than 1.7 results in a polar covalent bond. When the difference is zero the bond is nonpolar. The hydrides of the metals in group IA, and of Ca, Sr, and Ba in group IIA are exceptions to these rules, because they are ionic despite the fact that the electronegativity difference is less than 1.7.

7.9 □ EXCEPTIONS TO THE OCTET RULE

Although the octet rule is useful, there are many exceptions:

(a) Atoms with less than an octet of electrons When an atom with fewer than four valence electrons shares them to form covalent bonds, it may have less than an octet of electrons. Boron, ·B: , the first member of group IIIA, and beryllium, Be:, the first member of group IIA, are typical examples. Boron and fluorine form the covalent compound boron trifluoride, BF_3 (b.p. = −101°C),

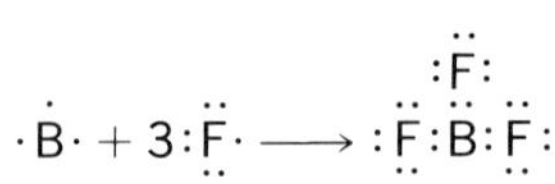

in which boron has only six electrons. However, the tendency to acquire an octet predisposes BF_3 to further reaction.

When beryllium reacts with fluorine, it forms an ionic compound, BeF_2; but with the less electronegative chlorine it forms a covalent compound, $BeCl_2$, in which Be has only four electrons.

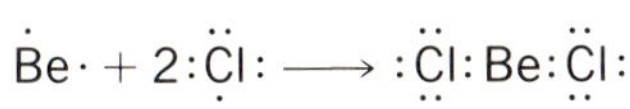

(b) Free radicals Nitric oxide, NO, is another example of a molecule possessing an atom with fewer than eight electrons. A possible electronic structure of the molecule

shows the oxygen atom surrounded by eight and the nitrogen atom surrounded by seven electrons. Molecules in which one or more electrons are unpaired are known as **free radicals.** Substances consisting of free radicals are characterized by two properties: they are paramagnetic and colored. Thus, nitrogen dioxide

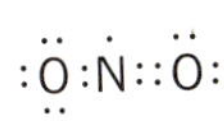

is a brown gas, and chlorine dioxide

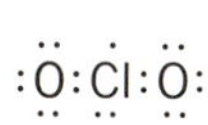

is a yellow gas; NO is colorless in the gaseous state, blue in the liquid state.

Although an electronic structure consistent with the octet rule can be written for O_2, experiment shows that O_2 has a permanent magnetic moment corresponding to two unpaired electrons. Consequently, molecular oxygen must be a free radical with an unpaired electron on each atom,

a possible structure for O_2

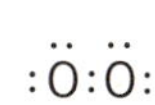

a free radical structure for O_2

We shall see (p. 136) that neither of these structures accurately represents the O_2 molecule.

(c) Atoms with more than eight outer electrons The electron octet is never exceeded by atoms in the second period of the periodic table; Li, Be, B, C, N, O, and F. It is possible, however, for atoms in the higher periods to be surrounded by more than eight valence electrons.

Stable molecules and ions such as SiF_6^{2-}, PCl_5, SF_6, and ICl_3 are known. The structural formulas are shown below.

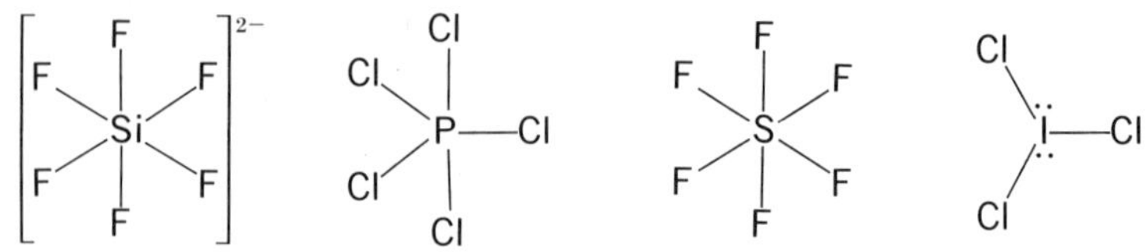

SiF_6^{2-} *hexafluorosilicate anion* | PCl_5 *phosphorus pentachloride* | SF_6 *sulfur hexafluoride* | ICl_3 *iodine trichloride*

(d) Ions of Transition Elements and of Certain Representative Elements The ions of many transition elements and heavier representative elements are not isoelectronic with noble gases, and hence do not necessarily achieve an octet of valence electrons. Examples include Fe^{2+} $3s^23p^63d^5$ and Pb^{2+} $6s^2$.

7.10 ▫ FORMAL CHARGE AND OXIDATION NUMBER

In this section we discuss some conventions of assigning charge values to atoms.

Formal charge Consider the following sequence:

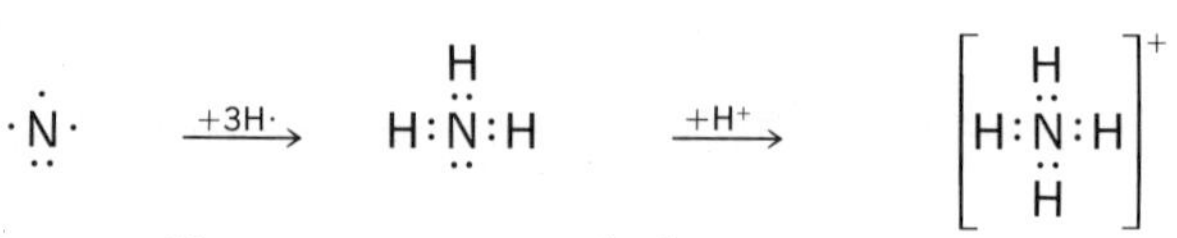

N atom *H atoms* *ammonia molecule* *hydrogen ion (a proton)* *ammonium ion*

The neutral N atom has five valence electrons; the neutral H atom has one. In the ammonia molecule, the N has an octet of valence electrons, but 6 of these are shared with H atoms. If we assume, by convention, that these 6 are shared equally, we assign 3 of them to the N for the purpose of calculating the formal charge on the nitrogen. Then the number of valence electrons assigned to N is

$$2 \text{ unshared electrons} + \tfrac{1}{2}(6 \text{ shared electrons}) = 5 \text{ electrons}$$

This is the same number of valence electrons as in the neutral N atom. We therefore say that the formal charge of N in NH_3 is zero.

The same calculation for N in NH_4^+ leads to

$$\begin{aligned}\text{number of assigned valence electrons} &= \tfrac{1}{2}(8 \text{ shared electrons})\\ &= 4 \text{ electrons}\end{aligned}$$

This is one *less* than the 5 valence electrons of the neutral N atom, and so we say the formal charge on N in NH_4^+ is +1. These conventions lead to the following definitions.

$$\text{formal charge} = \text{number of valence electrons in neutral atom} - \text{number of valence electrons assigned to atom on the basis of equal sharing in covalent bonds} \quad (1)$$

$$\text{number of assigned electrons} = \text{all unshared electrons} + \tfrac{1}{2}(\text{shared electrons}) \quad (2)$$

For the representative elements, the number of valence electrons is equal to the periodic group number, G. Each bond represents 2 shared electrons. Applying these values to Equation 1, we have

$$\text{formal charge} = G - \text{unshared electrons} - \text{covalent bonds} \quad (3)$$

EXAMPLE 1 Calculate the formal charges of atoms in nitric acid.

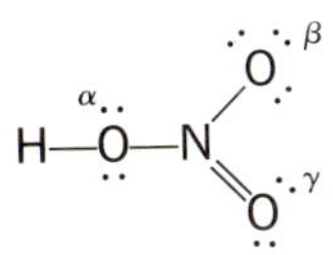

ANSWER

formal charge of H $= 1 - 0 - 1 = 0$
formal charge of αO $= 6 - 4 - 2 = 0$
formal charge of βO $= 6 - 6 - 1 = -1$
formal charge of γO $= 6 - 4 - 2 = 0$
formal charge of N $= 5 - 0 - 4 = +1$
Sum $= 0$

The sum of the formal charges equals zero for a molecule, or the ionic charge for an ion.

EXAMPLE 2 Which of the following structures are ions? Determine their charges.

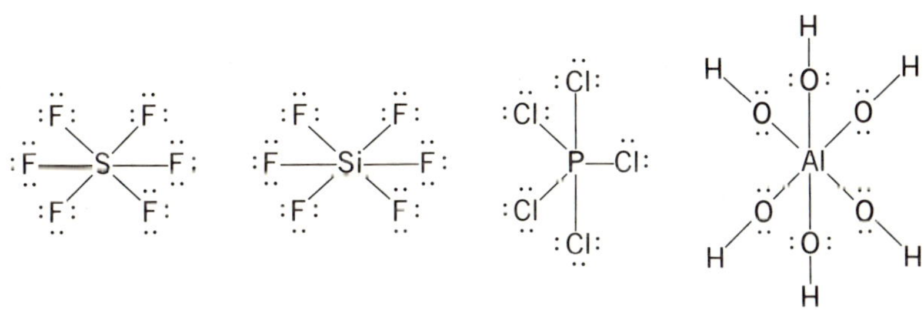

(*a*) (*b*) (*c*) (*d*)

ANSWER

(a) Formal charge of S $= 6 - 0 - 6 = 0$.
Formal charge of F $= 7 - 6 - 1 = 0$; SF_6 is a molecule.

(b) Formal charge of Si $= 4 - 0 - 6 = -2$.
Formal charge of F $= 7 - 6 - 1 = 0$;
SiF_6 is an ion; charge $= -2$.

(c) Formal charge of P $= 5 - 0 - 5 = 0$.
Formal charge of Cl $= 7 - 6 - 1 = 0$; PCl_5 is a molecule.

(d) Formal charge of Al $= 3 - 0 - 6 = -3$.
Formal charge of O $= 6 - 4 - 2 = 0$.
Formal charge of H $= 1 - 0 - 1 = 0$;
$Al(OH)_6$ is an ion; charge $= -3$.

Oxidation number Consider the combination

$$\mathrm{H}\cdot + \cdot\ddot{\underset{\cdot\cdot}{\mathrm{Cl}}}\!: \longrightarrow \mathrm{H}:\ddot{\underset{\cdot\cdot}{\mathrm{Cl}}}\!:$$

The assignment of formal charges (zero for each atom), which derives from the assumption that the bonding electrons are equally shared, *ignores the polar quality of the bond.* If we arbitrarily *assign both electrons to the more electronegative atom*, Cl, then we consider Cl to have eight electrons and hydrogen none, as though HCl consisted of separate ions. Charges assigned on such a basis are called *oxidation numbers*, and are said to represent the *oxidation states* of the element.

$$\text{oxidation number} = \left\{\begin{matrix}\text{number of}\\ \text{valence elec-}\\ \text{trons in neu-}\\ \text{tral atom}\end{matrix}\right. - \left\{\begin{matrix}\text{number of valence elec-}\\ \text{trons assigned to atom}\\ \text{on the basis that all the}\\ \text{bonded electrons be-}\\ \text{long to the more elec-}\\ \text{tronegative atom}\end{matrix}\right. \tag{4}$$

$$= G - \left\{\begin{matrix}\text{unshared}\\ \text{electrons}\end{matrix}\right. - \left\{\begin{matrix}\text{assigned bonded}\\ \text{electrons}\end{matrix}\right. \tag{5}$$

EXAMPLE 3 Calculate the oxidation numbers of the atoms in nitric acid.

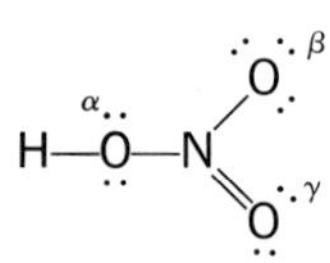

ANSWER Since O is more electronegative than N or H, we have

oxidation number of H $= 1 - 0 - 0 = +1$
oxidation number of αO $= 6 - 4 - 4 = -2$
oxidation number of βO $= 6 - 6 - 2 = -2$
oxidation number of γO $= 6 - 4 - 4 = -2$
oxidation number of N $= 5 - 0 - 0 = \underline{+5}$
Sum $= 0$

EXAMPLE 4 Calculate the oxidation number of each atom in the thiosulfate ion, $S_2O_3^{2-}$, whose electronic structure is

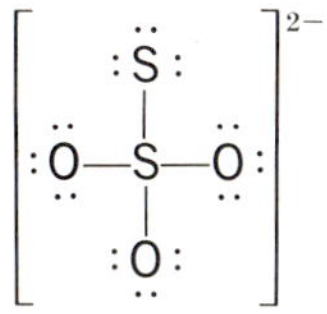

ANSWER Oxygen is more electronegative than sulfur, so all electrons shared by these atoms are allotted to oxygen. Since the S atoms have the same electronegativity, the electrons are presumed to be equally shared, and one electron is allotted to each S atom. Thus, we have

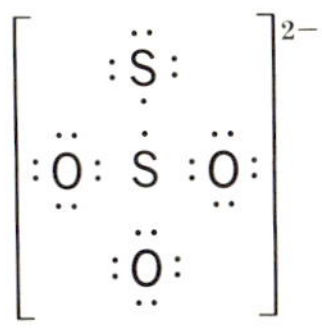

oxidation number of central S $= 6 - 1 - 0 = +5$
oxidation number of outer S $= 6 - 1 - 6 = -1$
oxidation number of each O $= 6 - 2 - 6 = -2$

Note that the sum of the oxidation numbers of all the atoms equals the charge on the ion, $3(-2) + 5 + (-1) = -2$.

This method of calculating oxidation numbers requires a prior knowledge of Lewis electronic formulas. Such formulas may be difficult to write for molecules and ions containing transition elements. Fortunately, operational rules, derived from the method just discussed, can be used to deduce oxidation numbers directly from molecular or empirical formulas.

Rule (1) The order of decreasing electronegativities of the more prominent electronegative atoms is

$$\text{F}(-1) > \text{O}(-2) > \text{Cl}(-1) > \text{N}(-3) > \text{Br}(-1) > \text{S}(-2) > \text{I}(-1) > \text{C}(-4) > \text{P}(-3) > \text{H}(-1)$$

Each value shown in parentheses is the oxidation number for the element when it is bonded to a *less electronegative atom.* The less electronegative atom is then assigned an oxidation number so that the total of the oxidation numbers of all atoms is equal to the charge on the molecule or ion. This rule is illustrated as follows (the number below the symbol for the atom representing the oxidation number):

H_2O	CaH_2	SiF_6^{2-}	H_2S	SO_2	OF_2	NH_3
+1 −2	+2 −1	+4 −1	+1 −2	+4 −2	+2 −1	−3 +1
NCl_3	NI_3	CH_4	CO_2	MnO_4^-	$Cr_2O_7^{2-}$	$S_2O_3^{2-}$
+3 −1	−3 +1	−4 +1	+4 −2	+7 −2	+6 −2	+2 −2

The detailed calculation of the oxidation number of the dichromate ion, $Cr_2O_7^{2-}$, is shown below. The charge on the ion is

$$(2 \times \text{oxidation number of Cr}) + (7 \times \text{oxidation number of O}) = -2$$

Let x = oxidation number of Cr. Since −2 is the oxidation number of O when bonded to a less electronegative atom, then

$$2x + 7(-2) = -2$$
$$x = +6$$

Similarly, for $S_2O_3^{2-}$, let x = oxidation number of S. Then

$$2x + 3(-2) = -2$$
$$x = +2$$

Note that the value obtained for the S atoms in thiosulfate ion, $S_2O_3^{2-}$, is the average of the values obtained for the individual S atoms by the method of allotting electrons, as done in Example 3 (p. 124).

Rule (2) Whenever an H atom is bonded to a more electronegative atom, it has an oxidation number of +1.

Rule (3) Whenever an atom is found in its elementary state, its oxidation number is zero, for example, Cl_2, N_2, and C.

Rule (4) In salts, each ion is treated separately, as shown for ammonium sulfate, $(NH_4)_2SO_4$,

NH_4^+	SO_4^{2-}
−3 +1	+6 −2

The total of the oxidation numbers of all atoms in the salt must equal zero. *The oxidation number for monatomic ions such as* K^+, Ca^{2+}, Al^{3+}, Cl^-, *and* S^{2-} *is the ionic charge.*

In summary, the assignment of electrons for calculating the oxidation number recognizes the ionic character of the bond, whereas the method for assigning formal charge recognizes the covalent character of the bond. Since most bonds are somewhere between purely covalent and purely ionic, neither assigned charge is real.

We might ask, "Which convention is closer to reality?" Following are data on charge condition and assignments on bromine in several compounds.

Compound	Actual charge condition* of Br	Formal charge of Br	Oxidation number of Br
HBr	−0.12	0	−1
Li^+Br^-	−0.61	−1	−1
CBr_4	−0.04	0	−1
$BiBr_3$	−0.05	0	−1

Obviously, the formal charge is a much more faithful approximation of the actual condition of bromine than is the oxidation number. Oxidation numbers are vestiges of the dualistic theory of the early nineteenth century, but we retain them because they are useful to us in systematizing data on reactions that involve significant changes in the charge conditions of atoms, reactions that we call **oxidations** and **reductions** (see Sec. 15.7).

7.11 ▫ PERIODICITY OF CHEMICAL BONDING

If a representative atom is to form a *monoatomic anion* (transition elements seldom form anions) that is isoelectronic with the next noble gas of higher atomic number, then the *anionic charge* must be the group number (G) minus 8, or ($G - 8$). Thus, sulfur (group VIA) would become S^{2-}, and nitrogen (group VA) would become N^{3-}. Certain periodic trends of anion formation are discernible:

(a) *The ability of an element to participate in ionic bonding by forming an anion increases as the group number increases.* Thus, there are more ionic fluorides ($G = 7$) than nitrides ($G = 5$). This trend is a consequence of the fact that fluorine must gain only one electron to form F^-, whereas nitrogen must gain three electrons to form N^{3-}. It becomes increasingly difficult to acquire successive electrons.

(b) *The ability of elements, within a given group, to form anions, increases with decreasing atomic weight.* Thus, there are more ionic fluorides than chlorides, bromides, or iodides.

If an atom is to form a *cation* that is isoelectronic with the noble gas of the nearest lower atomic number, then the cationic charge must be numerically equal to the group number (G). The most favorable circumstances for the existence of cations is in combination as anhydrous salts (free from water) with polyatomic anions, such as perchlorate, ClO_4^-; nitrate, NO_3^-; sulfate, SO_4^{2-}; carbonate, CO_3^{2-}; and phosphate, PO_4^{3-}. If an element does not form a saltlike compound even with these anions, then the element cannot sustain a cationic state in binary compounds, and it will *share* rather than *lose* electrons. The ability of an element to support a positive charge depends on the **charge density of**

* Calculated from electronegativity values.

the cation. The definition of the cationic charge density is

$$\text{cationic charge density} = \frac{\text{magnitude of positive charge}}{\text{volume of the cation*}}$$

Table 7.2 gives periodic trends in the charge densities of real and imaginary cations. The values given in this table are *relative* charge densities, the value for Fr^+, the ion of least charge density, being taken as 1.

Small values, such as those of group IA, predict stable cations; large values such as those for C^{4+}, H^+, and N^{5+} predict that the cations do not exist. Unfortunately, an unequivocal relationship cannot be drawn between the magnitude of the cationic charge density of an element and the element's tendency to form ionic salts. We have already seen (p. 113) that the crystal lattice energy is an important factor in determining the heat of formation of a crystalline salt. Thus, we find that although aluminum has the fairly high cationic charge density of 130, it nevertheless forms a stable sulfate salt. The formation of aluminum sulfate is accounted for mainly because of the large crystal lattice energy arising from the strong electrostatic attraction between the Al^{3+} and SO_4^{2-} ions. Some approximate but useful generalizations, nevertheless, can be made:

(a) Elements can form ionic compounds in those cationic states with charge densities of about 50 or less.

(b) Elements with cationic charge densities ranging from 50 to about 100 form ionic bonds in some compounds and covalent bonds in others.

(c) Elements with values above 100 form mainly, but not exclusively, covalent bonds or negative ions.

Certain definite trends are discernible from this table. These trends are helpful in organizing and generalizing the known chemistry of many elements:

(a) As we proceed down a group, the cationic charge density decreases and hence the likelihood of finding a cationic state increases. Thus, there are more ionic compounds of thallium (Tl) than of aluminum (Al).

(b) Among the representative elements, the drop in cationic charge density is most pronounced on proceeding from the second to the third period. For this reason the chemistry of the elements in the second period is atypical of their group. Thus, we find that although all the alkali metals (group IA) form cations, a tendency toward covalent bonding is sometimes observed for lithium. For example, when combined with carbon, as in ethyllithium, CH_3CH_2Li, lithium forms bonds with some covalent character. Unlike the other members of their respective groups, beryllium (group IIA) rarely forms Be^{2+}, and boron (group IIIA) never forms B^{3+}.

* Volumes of those cations that are imaginary are derived from quantum mechanical calculations.

TABLE 7.2 RELATIVE REAL AND HYPOTHETICAL CATIONIC CHARGE DENSITIES OF THE ELEMENTS

Energy levels	Periods	*I A*	*II A*	*III A*	*IV A*	*V A*	*VI A*	*VII A*	*0*
1*s*	1	**H** 1; 5.4×10^{15}							**He** 2
2*s*, 2*p*	2	**Li** 3; 25	**Be** 4; 370	**B** 5; 2000	**C** 6; 6500	**N** 7; 20,000	**O** 8; 45,000	**F** 9; 110,000	**Ne** 10
3*s*, 3*p*	3	**Na** 11; 6.4	**Mg** 12; 40	**Al** 13; 130	**Si** 14; 320	**P** 15; 690, 179.5	**S** 16; 1,300, 413.8	**Cl** 17; 22,000, 460	**Ar** 18
4*s*, 4*p*	4	**K** 19; 2.3	**Ca** 20; 11	**Ga** 31; 69, 3.8	**Ge** 32; 150	**As** 33; 260, 93.6	**Se** 34; 400	**Br** 35; 640	**Kr** 36
5*s*, 5*p*	5	**Rb** 37; 1.7	**Sr** 38; 7.6	**In** 49; 31, 4.8	**Sn** 50; 61, 8.1	**Sb** 51; 110, 22	**Te** 52; 190, 31	**I** 53; 310, 33	**Xe** 54
6*s*, 6*p*	6	**Cs** 55; 1.1	**Ba** 56; 4.4	**Tl** 81; 19, 3.6	**Pb** 82; 37, 6.3	**Bi** 83; 67, 18.5	**Po** 84	**At** 85	**Rn** 86
7*s*	7	**Fr** 87; 1.0	**Ra** 88						

Energy levels	Periods	*III B*	*IV B*	*V B*	*VI B*	*VII B*	*VIII*	*VIII*	*VIII*	*I B*	*II B*
3*d*	4	**Sc** 21; 31	**Ti** 22; 69, 37, 15	**V** 23; 130, 100, 40, 16	**Cr** 24; 230, 120, 50, 13	**Mn** 25; 390, 140, 57, 14	**Fe** 26; 62, 25	**Co** 27; 65, 27	**Ni** 28; 69, 29	**Cu** 29; 62, 29	**Zn** 30; 27
4*d*	5	**Y** 39; 20	**Zr** 40; 43	**Nb** 41; 79, 66	**Mo** 42; 140, 69	**Tc** 43	**Ru** 44; 79, 21	**Rh** 45; 50, 21	**Pd** 46; 17	**Ag** 47; 14, 2.7	**Cd** 48; 11
5*d*	6	**La** 57; 9	**Hf** 72; 41	**Ta** 73; 76	**W** 74; 120, 70	**Re** 75; 300	**Os** 76; 73, 16	**Ir** 77; 76, 14	**Pt** 78; 6	**Au** 79; 2.1	**Hg** 80; 8.2
6*d*	7	**Ac** 89; 10									

Key:

	Cationic charge
	+6
61	+4
8.1	+2
Sn	

Charge density of +2 cation of +4 cation

Relative to $Fr^{+}=1$

(c) For any element, the lower the charge state, the lower the cationic charge density and the more likely it is that the cation exists. Therefore, metals are more likely to form ionic compounds in their lower valence states than in their higher ones. (Compare the ionic species $PbCl_2$, m.p. 501°C, with the covalent species $PbCl_4$, m.p. −15°C, in terms of cationic charge densities.)

PROBLEMS

1. Structural formulas (single bonds only) Write Lewis formulas, consistent with the octet rule and covalency requirements, for the following: (a) CH_4S; (b) CCl_2F_2; (c) C_2H_6S (two structural possibilities); (d) NI_3; (e) CH_5N.

2. Structural formulas (with multiple bonds) Write Lewis formulas, consistent with the octet rule and covalency requirements, for (a) C_2H_4; (b) $SOCl_2$; (c) C_2H_2; (d) NOBr; (e) H_2CO.

3. Born-Haber cycle Calculate the lattice energy of NaCl from the following data:

		ΔH, kcal/mole
Heat of sublimation	$Na(c) \longrightarrow Na(g)$	+26.0
Ionization energy	$Na(g) \longrightarrow Na^+(g) + e^-$	+118.0
Bond energy	$Cl_2(g) \longrightarrow 2Cl(g)$	+57.2
Electron affinity	$Cl(g) + e^- \longrightarrow Cl^-(g)$	−88.2
Heat of formation	$Na(c) + \frac{1}{2}Cl_2(g) \longrightarrow NaCl(c)$	−98.2

4. Properties of ionic and covalent compounds State whether each of the following substances has a molecular covalent, a macromolecular (network) covalent, or an ionic structure. Give your reason for each answer. (a) HCl, m.p. −112°C, b.p. −84°C; the liquid is a poor conductor of electricity; (b) SiC, covalency of each atom is 4, no multiple bonds; (c) $BaCl_2$, m.p. 963°C, b.p. 1560°C; the molten substance conducts electricity; (d) Sodium chlorate, $NaClO_3$, m.p. 250°C; yields gaseous products at surfaces of electrodes when an electric current is passed through its aqueous solution.

5. Lewis formulas; electronegativity Write the Lewis formula for the binary compound formed from each pair of listed elements. Base your choice between ionic and covalent bonding on the electronegativities of the elements (Table 7.1, p. 119). Are the physical properties of each compound (refer to a chemical handbook) consistent with your choice of bond type? (a) K, S; (b) H, Se; (c) Na, N; (d) Sr, F; (e) Be, I; (f) Al, O.

6. Electronegativities Indicate by δ^+ or δ^- signs the fractional charge, if any, on the atoms in each of the following bonds (see Table 7.1, p. 119): (a) B—Cl; (b) C—Cl; (c) N—Cl; (d) O—Cl; (e) F—Cl.

7. Free radicals Chlorine forms the oxides Cl_2O, ClO_2, ClO_4, Cl_2O_8, and Cl_2O_7. Which ones would you expect to have free radical character? Explain.

8. Oxidation number Evaluate the oxidation number of the italicized atom in (a) $\mathit{C}O_2$; (b) $\mathit{C}OCl_2$; (c) $K\mathit{Cl}$; (d) $K\mathit{Cl}O$; (e) $K\mathit{Cl}O_2$; (f) $K\mathit{Cl}O_3$; (g) $K\mathit{Cl}O_4$; (h) $H_2\mathit{C}{=}O$; (i) $(\mathit{Ti}O)^{2+}(SO_4)^{2-}$.

9. Formal charge and oxidation number For the following compounds (a) write the structural formulas (note that in each of the sulfur compounds, S is the central atom); (b) locate all formal charges that are not zero; (c) determine the oxidation number of each atom: Sulfurous acid, $(HO)_2SO$; sulfuric acid, $(HO)_2SO_2$; sulfamic acid, H_2NSO_2OH; periodic acid, HIO_4.

ADDITIONAL PROBLEMS

10. Structural formulas (single bonds only) (continuation of Problem 1) Write Lewis formulas for (a) PH_3; (b) $C_2H_6S_2$ (two structural possibilities); (c) H_4N_2; (d) H_3NO; (e) S_2Cl_2.

11. Structural formulas (with multiple bonds) (continuation of Problem 2) Write Lewis formulas for (a) C_3H_6; (b) NOCl; (c) N_2F_2; (d) H_2C_2O; (e) H_2CO_2.

12. Born-Haber cycle Calculate the heat of formation of magnesium oxide, $Mg(c) + \frac{1}{2}O_2(g) \longrightarrow MgO(c)$, from the following data:

		ΔH, kcal/mole
Electron affinity	$O(g) + 2e^- \longrightarrow O^{2-}(g)$	+156
Bond dissociation energy	$O_2(g) \longrightarrow 2O(g)$	+119.1
Heat of sublimation	$Mg(c) \longrightarrow Mg(g)$	+ 36.5
Ionization energy	$Mg(g) - 2e^- \longrightarrow Mg^{2+}(g)$	+523
Lattice energy	$Mg^{2+}(g) + O^{2-}(g) \longrightarrow MgO(c)$	−919

13. Properties of ionic and covalent compounds (continuation of Problem 4) Predict the bonding of: (a) acetic acid, CH_3CO_2H, m.p. 17°C, b.p. 118°C, poor conductor of electricity; (b) beryllium nitride, Be_3N_2, m.p. 2200°; (c) CS_2 (sulfur has covalency of 2 and is doubly bonded to carbon).

14. Lewis formulas; electronegativity (continuation of Problem 5) Write the Lewis formula for the compound formed from (a) H, Ba; (b) N, Cl; (c) Bi, F; (d) Si, Cl; (e) Li, Sb.

15. Electronegativity (a) List three plausible covalent bonds between dissimilar atoms which would show no polarity. (b) List three ionic bonds which you think exhibit extreme ionic character. (c) List three covalent bonds you think are highly polar.

16. Electronegativity Indicate by δ^+ or δ^- signs the fractional charge, if any, on each of the following atoms when bonded with oxygen (see Table 7.1, p. 119): (a) Be; (b) B; (c) C; (d) N; (e) F.

17. Free radicals For each of the following, write a Lewis formula and indicate whether or not it is a free radical: (a) O^{2-}; (b) OH^-; (c) H_2O_2; (d) HO_2^-; (e) O_2^{2-}; (f) O_2^-.

18. Free radicals Tetramethyl lead, $(CH_3)_4Pb$, is a liquid, b.p. 110°C. When heated, it decomposes, leaving lead and a very reactive substance that rapidly becomes converted to ethane, $H_3C:CH_3$. (a) Write the structural formula for $(CH_3)_4Pb$. (b) Write the Lewis formula for the reactive substance. (c) Write the equation for the conversion of the reactive substance to ethane.

19. Oxidation number (continuation of Problem 8) Write the oxidation number of the italicized atom: (a) $P\mathrm{Cl}_5$; (b) $\mathrm{K}Mn\mathrm{O}_4$; (c) $N\mathrm{H}_3$; (d) $N\mathrm{H}_4\mathrm{Cl}$; (e) $N_2\mathrm{H}_4$; (f) N_2; (g) $\mathrm{NH}_4(V\mathrm{O}_3)$; (h) $\mathrm{CH}_3\overset{\mathrm{OH}}{\overset{|}{C}}{=}\mathrm{O}$.

20. Formal charge and oxidation number (continuation of Problem 9) Write the formulas, giving formal charges and oxidation numbers, of phosphorous acid, $HPO(OH)_2$; phosphoric acid, $(HO)_3PO$; NH_3BF_3; bromic acid, $HBrO_3$.

21. Charge densities Aluminum chloride dissolves in water with violent evolution of heat. Explain in terms of the calculated charge density of Al^{3+}.

22. Bond type (a) Select A and B from among the elements of atomic number 1–9. Write electronic structures and indicate the type of bond formed in each compound (molecules with multiple bonds):

(*i*) 3 compounds of formula AB_3 or A_3B;
(*ii*) 5 compounds of formula AB_2 or A_2B;
(*iii*) 4 compounds of formula AB;
(*iv*) 2 compounds of formula AB_4 or A_4B;
(*v*) 2 compounds of formula A_2B_3 or A_3B_2.

(b) Indicate which compounds do not fit the octet rule. (c) Which compounds are likely to have a macromolecular covalent structure?

23. Formal charge and oxidation number (a) Write a Lewis structure for hydrazoic acid, HN_3, whose skeleton is H—N—N—N. Make certain the structure obeys the octet rule. (b) Locate all formal charges. (c) Calculate the oxidation number for each atom. (d) Account for the fact that, for the middle N atom, the oxidation number and formal charge are identical.

24. Mixed bonding types Write Lewis formulas for the following ionic compounds, consistent with the octet rule: (a) calcium carbide, CaC_2; (b) sodium peroxide, Na_2O_2; (c) magnesium cyanide, $Mg(CN)_2$; (d) sodium formate, HCOONa.

25. Charge density Calculate the charge density (coulombs per cubic centimeter) of (a) Ca^{2+} (ionic radius 0.99 Å) and (b) Li^+ (ionic radius 0.60 Å).

26. Periodicity of chemical bonding Account for each of the following facts. (a) There are fewer ionic nitrides than oxides. (b) There are fewer ionic tellurides than sulfides. (c) There are fewer ionic compounds of germanium than of lead. (d) SbF_3 is ionic; SbF_5 is covalent.

27. Periodicity of chemical bonding The **"diagonal rule"** of the periodic table states that the elements of the second period resemble the elements "diagonally to the right" (that is, the elements of the next higher group in the third period) more than they do the elements of their own group. Rationalize this generalization in terms of cationic charge densities, especially for the pairs Li, Mg and Be, Al.

8 THE COVALENT BOND

8.1 THE MOLECULAR ORBITAL

The force of attraction between bonded atoms is electrostatic. We find this easy to understand for ionic substances because we imagine that the ions are charges existing at points in space, separated from each other by the interatomic distance. Thus, we say that the Na^+ ion attracts the Cl^- ion, just as a proton attracts an electron. We have seen that the real situation in ionic bonding is not so simple, but the qualitative picture is nonetheless satisfying.

The lack of polarity in a molecule like H_2 denies us the conceptual model of an attractive force exerted by individual atoms as though they were points in space bearing some electrical charge. Instead, we must consider attractions and repulsions among the constituent nuclei and electrons of the entire molecule. Thus, in H_2, each of the two nuclei attracts each of the two electrons; the two nuclei repel each other; and the two electrons repel each other. In aggregate, the attractive forces exceed the repulsive forces. Figure 8.1 shows the distribution of electron density in an H_2 molecule. Notice the following: (a) The charge density, or probability that the electron is found in a given location, can be depicted in space; this representation is a **molecular orbital.** Each electron is thus considered to be a part of the entire molecule,

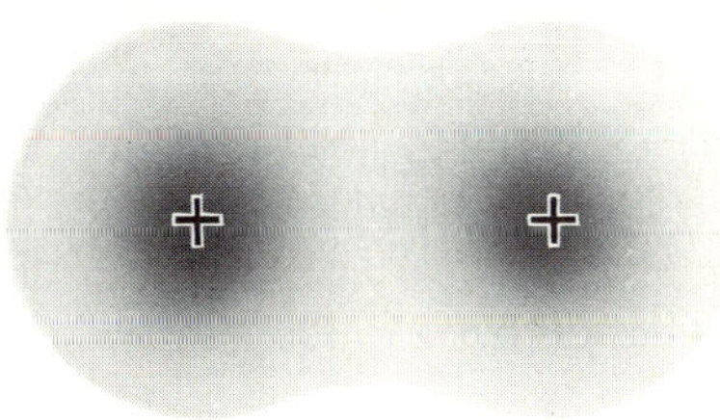

Fig. 8.1 Charge distribution in an H_2 molecule.

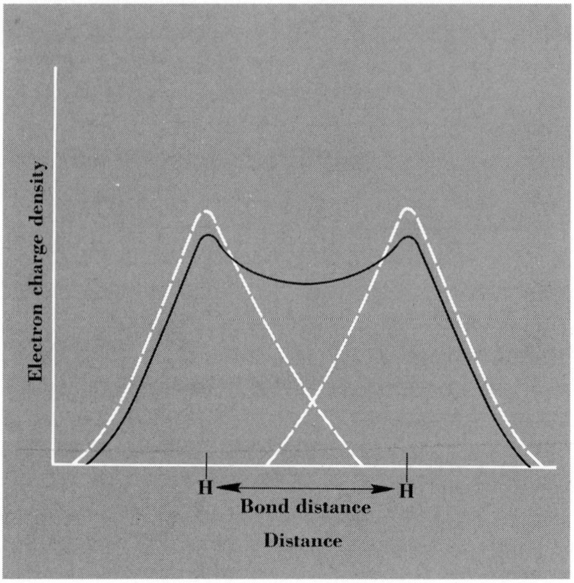

Fig. 8.2 Plots of electron charge density for (a) individual atomic orbitals (dotted line), and (b) bonding molecular orbital (solid line).

rather than of either individual atom. (b) In H_2, the molecular orbital is symmetrical with respect to the two nuclei; as a result, the molecule is nonpolar. (c) The strength of the bond results from the fact that *there is a greater electron density between the nuclei than there would be if the two atomic orbitals existed independently*. This is a crucial point that merits further illustration. Figure 8.2 shows the charge densities of the individual atomic orbitals which would exist if there were no chemical combination of H atoms, and the charge density of the molecular orbital which does exist in H_2. In the molecular orbital the sharing of electrons *increases the negative charge between the nuclei;* the resulting attractions decrease the total energy of the system and hold the nuclei together.

The Pauli exclusion principle applies to molecular orbitals as well as to atomic orbitals. Thus, *each molecular orbital may have a maximum of 2 electrons.*

8.2 □ SIGMA (σ) BONDS AND PI (π) BONDS

A bonding molecular orbital may be considered to be the result of the overlap of atomic orbitals. The more the atomic orbitals overlap, the greater is the charge density between the nuclei, and the more stable is the bond.

Schematic representations for bonding molecular orbitals are shown in Fig. 8.3 from (a) overlap of two *s* atomic orbitals, (b) head-to-head overlap of an *s* and a *p* atomic orbital, and (c) head-to-head overlap of two *p* atomic orbitals. The three molecular orbitals represented in Fig.

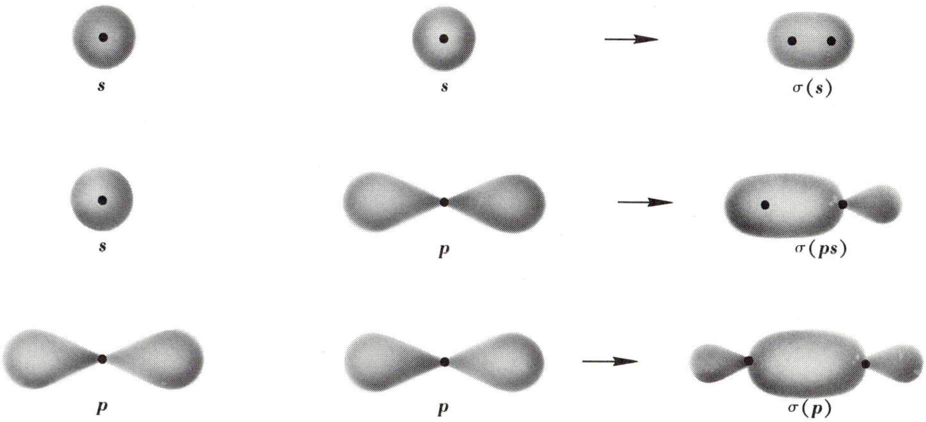

Fig. 8.3 Schematic representation of types of sigma bond.

8.3 have a symmetrical charge distribution about a line joining the nuclei, the so-called **bond axis.** Such bonds are referred to as **σ bonds.** They are more specifically designated $\sigma(s)$, $\sigma(sp)$, and $\sigma(p)$, depending on the types of atomic orbitals from which they are formed.

Two p orbitals may also overlap *laterally* (side-to-side) to form a **π bond** (Fig. 8.4). The bonding π molecular orbital consists of two electron charge clouds concentrated above and below the axis joining the nuclei. Therefore, a π bond does not exhibit the cylindrical charge symmetry of the σ bond. A π bond is never formed between two atoms unless accompanied by a σ bond. It is the σ bond that largely supplies the electron density along the axis joining the atoms. *A strong π bond can result only when the individual p orbitals are parallel.* Thus, a π bond can result from overlap of two p_x, two p_y, or two p_z orbitals, but not, for example, from a p_y and a p_z.

8.3 □ BINARY COVALENT MOLECULES AND IONS

H_2^+, [H · H]$^+$ The hydrogen molecule ion is the simplest entity of combined atoms. One electron is involved in the σ bonding of two protons. The bond is surprisingly strong (64 kcal/mole) in comparison with the bond energy of the H_2 molecule (104 kcal/mole).

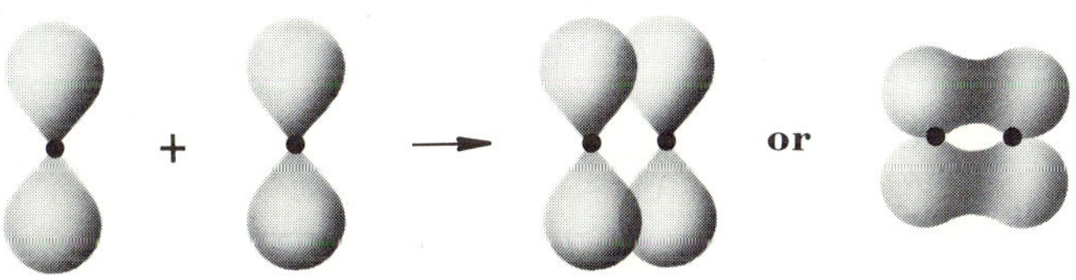

Fig. 8.4 The pi (π) bond.

Li_2, Li : Li The $2s^1$ configuration of Li, like the $1s^1$ of H, can engender a σ bond. $Li_2(g)$ does exist in the vapor state of lithium. At temperatures below 1317°C (the boiling point of lithium), however, the metallic state is more stable (see Ch. 23).

N_2, :N≡N: Nitrogen, $_7N$, $K2s^22p^3$, is represented as :N· (with one dot above and one below the N) by its Lewis symbol. Covalent bonding of the three $2p$ electrons will complete the octet for N. The three p electrons can engender one σ and two π bonds, that is, one head-to-head and two lateral overlaps. The orbital representations are analogous to those shown for the triply bonded molecule of acetylene (Fig. 8.12, p. 143).

O_2, :Ö::Ö: or :Ö:Ö: (in the second formula each O has one unpaired dot below it) Oxygen, $_8O$, $K2s^22p^4$, :Ö· (with one dot below the O), has four p electrons. Covalent bonding of two $2p$ electrons will complete the octet for O; this predicts a double bond structure for O_2. The fact that O_2 is a paramagnetic substance whose magnetic moment corresponds to two unpaired electrons is best represented by the single bond, free radical formula. We cannot write a single Lewis formula that represents both double bond and free radical characters; the actual molecule of O_2 is depicted, in part, by both formulas.

8.4 □ HYBRIDIZATION OF ATOMIC ORBITALS

We now consider bonding in molecules with more than two atoms. Methane, CH_4, will serve as the example. Any theory of bonding must be consistent with the observed chemical and physical properties of the molecule. Therefore, let us summarize some of the properties of methane:

(a) All four C—H bonds are identical; they have the same bond length (1.1093 Å) and bond energy (99.3 kcal/mole). Substitution of a Cl atom for any H atom gives only one kind of methyl chloride molecule, CH_3Cl.

(b) All H—C—H bond angles are the same; they are 109 degrees. The bond angle is the angle between the lines (axes) joining the nuclei of the atoms.

Carbon, $_6C$, $K2s^22p^2$, must provide four equivalent atomic orbitals for bonding in methane. This cannot occur when the electrons are in s and p orbitals, because the energies of these subshells are different. Any bond orbitals arising from the $2s$ atomic orbital would have a lower energy than bond orbitals arising from the $2p$ atomic orbitals. The actual equivalence of all four bonds is explained by assuming that four equivalent atomic orbitals are fabricated by blending of the $2s$ and the three $2p$ orbitals. This blending is called **hybridization.** Since these four hybrid orbitals are considered to be $\frac{3}{4}$ derived from p orbitals and $\frac{1}{4}$ from an s orbital, they are called sp^3 hybrid orbitals.

The four valence electrons of carbon are distributed among the

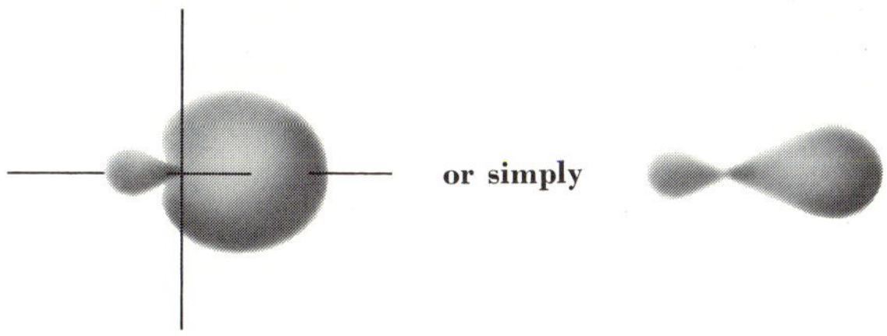

Fig. 8.5 Shape of an sp^3 hybrid atomic orbital. This is also the shape of all hybrid orbitals made up from s and p orbitals, the so-called s-p type hybrid orbitals.

hybridized orbitals in accord with Hund's rule, so that the electronic configuration of carbon in the sp^3 hybridized state is

Energy ↑	*ground state of carbon*	*sp^3 hybridized state of carbon*
	$2p$ ↓ ↓ __	$2(sp^3)$ ↓ ↓ ↓ ↓
	$2s$ ⇅	
	$1s$ ⇅	$1s$ ⇅

The shape of an sp^3 hybrid atomic orbital (Fig. 8.5) is like a p orbital, except that the two lobes are of unequal size. The atomic nucleus is near a node point. In order to simplify the use of orbital pictures for molecules, the small lobes are frequently omitted. Each sp^3 hybrid orbital overlaps with the $1s$ orbital of an H atom to form a σ bond. In fact, *bonds formed from hybrid atomic orbitals fabricated from s and p atomic orbitals are almost always σ bonds.*

Since a pair of electrons in one σ bond repels the pair of electrons in each of the other σ bonds, the theory predicts that the four σ bonds formed from the sp^3 hybrid atomic orbitals should be separated from each other in space as much as possible. This requirement is geometrically satisfied when the axis of each sp^3 orbital is directed toward the corner of a tetrahedron, with a resulting bond angle of 109 degrees (Fig. 8.6). The observed bond angle is 109 degrees and the success of

Fig. 8.6 Shape and spatial orientation of tetrahedral, sp^3 hybrid orbitals.

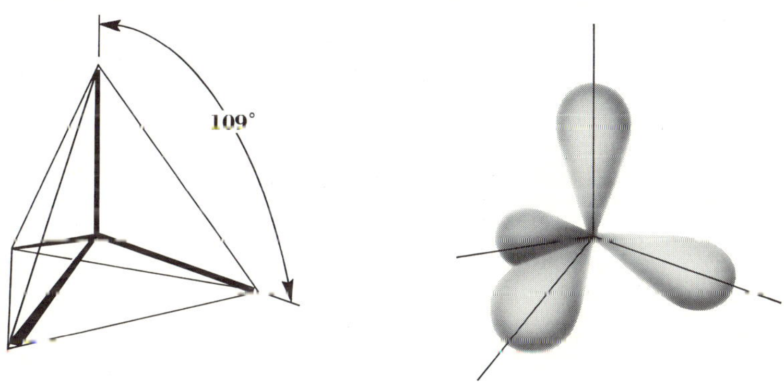

this prediction strengthens the chemist's belief in the concept of hybridization. Because of this geometry, sp^3 orbitals are called **tetrahedral hybrid orbitals.**

The concept of hybridization is also used to account for the three equivalent B—F bonds in boron trifluoride (p. 120). Each B—F bond length is 1.29 Å, and the bond energy is 154 kcal/mole. All four atoms lie in the same plane. The structure has a boron atom in the center of an equilateral triangle and a fluorine atom at each of the three corners. The F—B—F bond angle is 120 degrees.

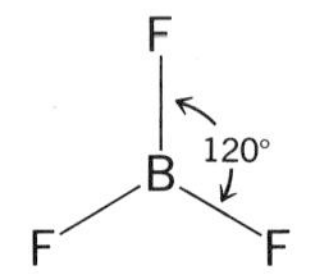

boron trifluoride

The three necessary equivalent atomic orbitals are assumed to result from a blending of the 2s and *two* of the 2p atomic orbitals of boron. The new hybrid orbitals are called sp^2:

ground state of boron *sp^2 hybridized state of boron*

The observed geometry of the molecule is in accord with the prediction that the spatial orientation of the hybrid orbitals permits the maximum bond angle. The sp^2 hybrid orbitals always have this triangular, coplanar orientation (Fig. 8.7) and hence are called **trigonal hybrid orbitals.**

A third type of hybridization of s and p atomic orbitals is suggested to explain the bonding of beryllium. For example, beryllium chloride, $BeCl_2$ is a white solid, m.p. 405°C, which readily sublimes. In the gaseous state it is a linear molecule, Cl—Be—Cl; all three atoms lie on a straight line. Each Be—Cl bond has the same bond length and energy, and so they are equivalent. It is necessary, therefore, for the Be atom to provide two degenerate* atomic orbitals to form the two equivalent σ bonds.

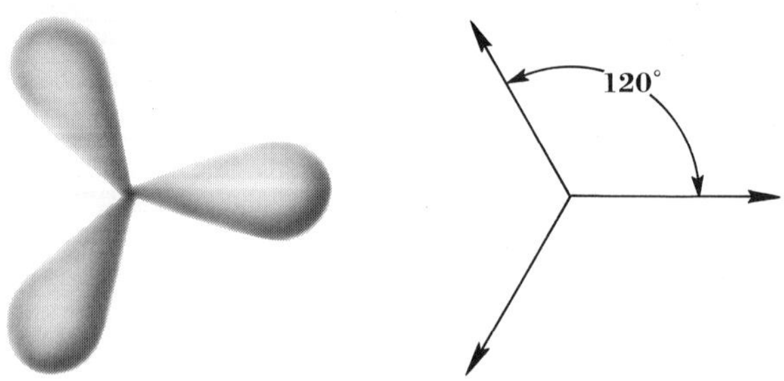

Fig. 8.7 Shape and spatial orientation of trigonal, sp^2 hybrid orbitals.

* Orbitals of equal energy are said to be "degenerate."

Fig. 8.8 Shape and spatial orientation of digonal, *sp* hybrid orbitals.

It is assumed that the two atomic orbitals which hybridize are the 2*s* and one of the 2*p* orbitals. These two hybrid orbitals are designated *sp*:

Energy ↑				
2*p*	_ _ _		2*p*	_ _
			2(*sp*)	↓ ↓
2*s*	⇅			
1*s*	⇅		1*s*	⇅

ground state of beryllium *sp hybridized state of beryllium*

Since there are only two covalent bonds to the Be atom, and no lone pairs, the maximum angle separating the axes of the two hybrid atomic orbitals should be 180 degrees (Fig. 8.8), which *is* the observed value. Because of the spatial orientation, the *sp* hybrid orbitals are also called **digonal hybrid orbitals.**

Table 8.1 summarizes the properties of the hybrid orbitals formed from *s* and *p* atomic orbitals, the so-called *s-p* type hybrids.

The bond angle can be used as a criterion for the type of atomic orbital used in bonding. For example, in ammonia, NH_3, the three N—H bonds are equivalent, and the H—N—H bond angle is 107 degrees. This is very close to the tetrahedral angle of 109 degrees, and so the N atom in ammonia is presumed to use sp^3 hybrid orbitals:

Energy ↑				
2*p*	↓ ↓ ↓		2(sp^3)	⇅ ↓ ↓ ↓
2*s*	⇅			
1*s*	⇅		1*s*	⇅

ground state of nitrogen *sp^3 hybridized nitrogen, as found in NH_3*

TABLE 8.1 ▪ PROPERTIES OF *s-p* TYPE HYBRID ATOMIC ORBITALS

Type	Name	*s* Character	*p* Character	Bond angle	Geometric arrangement	Type of bond formed
sp	Digonal	$\frac{1}{2}$	$\frac{1}{2}$	180°	Linear	Sigma
sp^2	Trigonal	$\frac{1}{3}$	$\frac{2}{3}$	120°	Coplanar triangular	Sigma
sp^3	Tetrahedral	$\frac{1}{4}$	$\frac{3}{4}$	109°	Tetrahedral	Sigma

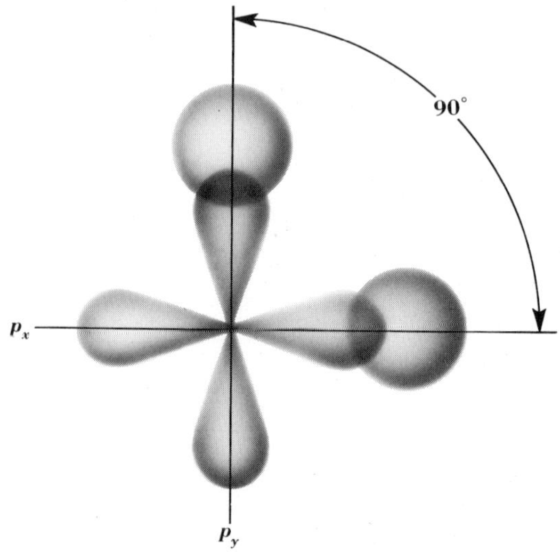

Fig. 8.9 Orbital representation of two σ bonds, each formed from a p orbital to show 90 degree bond angle.

We see that a hybrid orbital can also accommodate an unshared (lone) pair of electrons. It is noteworthy that the N atom has available three $2p$ orbitals with which to form three equivalent N—H bonds. Nevertheless, it is presumed that the N atom uses hybrid orbitals. Were the nitrogen atom to use p orbitals to form the three σ bonds with the H atoms, the bond angle would be approximately 90°, as shown in Fig. 8.9. This would be so because each p orbital is oriented along one of the three Cartesian axes, the x, y, and z axes, and the angle between any two of these axes is 90 degrees.

The two equivalent H—O bonds in water form an angle of 105 degrees. This value is much closer to the tetrahedral value of 109 degrees than it is to the 90 degree angle expected if p orbitals were used to form the two σ bonds:

Energy ↑

$2p$ ⇅ ↓ ↓
$2s$ ⇅
$1s$ ⇅

ground state of oxygen

$2(sp^3)$ ⇅ ⇅ ↓ ↓
$1s$ ⇅

sp^3 hybridized oxygen, as found in H_2O

In this respect, oxygen behaves like nitrogen in that it utilizes hybrid orbitals rather than p orbitals to form σ bonds. In general, Be, B, C, N, and O use hybrid atomic orbitals to form σ bonds and also to house lone pairs of electrons, when present.

8.5 □ MULTIPLY BONDED MOLECULES

We have seen that the triple bond in N_2 consists of a σ and two π bonds. We shall always find that *no more than one bond between two atoms can be a σ bond*, and that *all bonds in excess of one must be π bonds.* Thus, single bonds are σ bonds, double bonds are composed of one σ

and one π bond, and triple bonds are composed of one σ bond and two π bonds.

Of special interest are the multiply bonded carbon molecules, as typified by ethylene and acetylene. The geometry of these two molecules is the clue to the hybridization of the carbon atoms. Ethylene is a planar molecule in which the H—C—H and H—C—C bond angles are approximately 120 degrees:

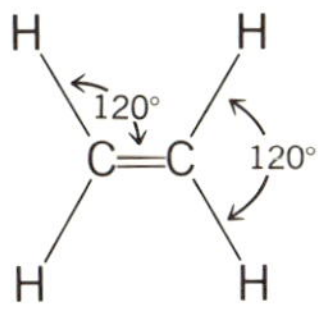

shape of ethylene, a planar molecule

Such geometry implies sp^2 hybridization for each carbon atom. The energy levels of the atomic orbitals of the sp^2 hybridized carbon atom are shown in Fig. 8.10c. The spatial arrangement of the sp^2 and p orbitals is shown in Fig. 8.11a.

The bonding of two such units with 4 H atoms is shown in Fig. 8.11b. Two of the three sp^2 hybrid orbitals of each carbon atom overlap with the 1s orbital of each of four H atoms. The third sp^2 hybrid orbital overlaps with the sp^2 hybrid orbital of the other carbon atom. This array of σ bonds, called the **skeleton** of the molecule, is coplanar (Fig. 8.11c). The p_z orbitals remaining on the adjacent carbon atoms overlap to form a π bond. The π bond is perpendicular to the skeleton of the molecule (Fig. 8.11d). The two carbon atoms and the four hydrogen atoms provide a total of 12 valence electrons. Ten of these electrons are used for the five σ bonds. The remaining two electrons comprise the π bond.

Acetylene is the simplest member of the alkyne series. The four atoms in acetylene, C_2H_2, lie in a straight line,

H—C≡C—H

acetylene,
a linear molecule

A linear geometry indicates that each carbon is using sp hybrid orbitals. The energy levels of an sp hybridized carbon atom are shown in Fig.

Fig. 8.10 Electronic structures exhibited by carbon: (a) ground state; (b) sp^3 hybridized; (c) sp^2 hybridized; (d) sp hybridized.

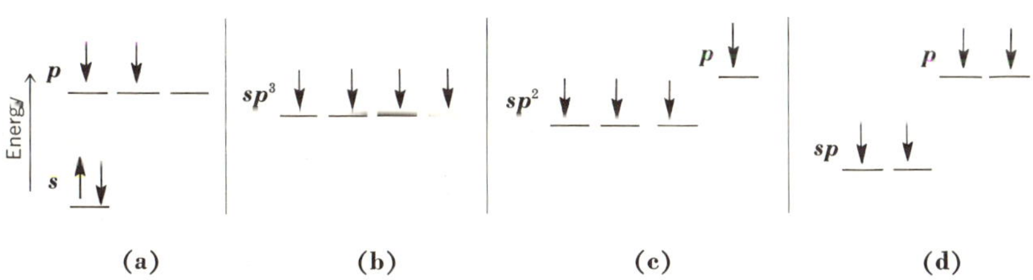

8.10d. Figure 8.12a shows the assembly of two H atoms, each with a $1s$ atomic orbital, and two C atoms, each with two sp hybrid orbitals and a p_y and p_z orbital. Each carbon atom forms two σ bonds; one by overlapping an sp hybrid orbital with the other carbon atom, and the second by overlapping an sp orbital with the s orbital of a hydrogen atom. Each carbon atom has a p_y and p_z orbital, each with one electron.

Fig. 8.11 Orbital representation of ethylene: (a) sp^2 hybrid orbitals and a p_z orbital; (b) assembly of two such units plus four H $1s$ orbitals; (c) skeleton; (d) representation of π bond. Black dots represent electrons.

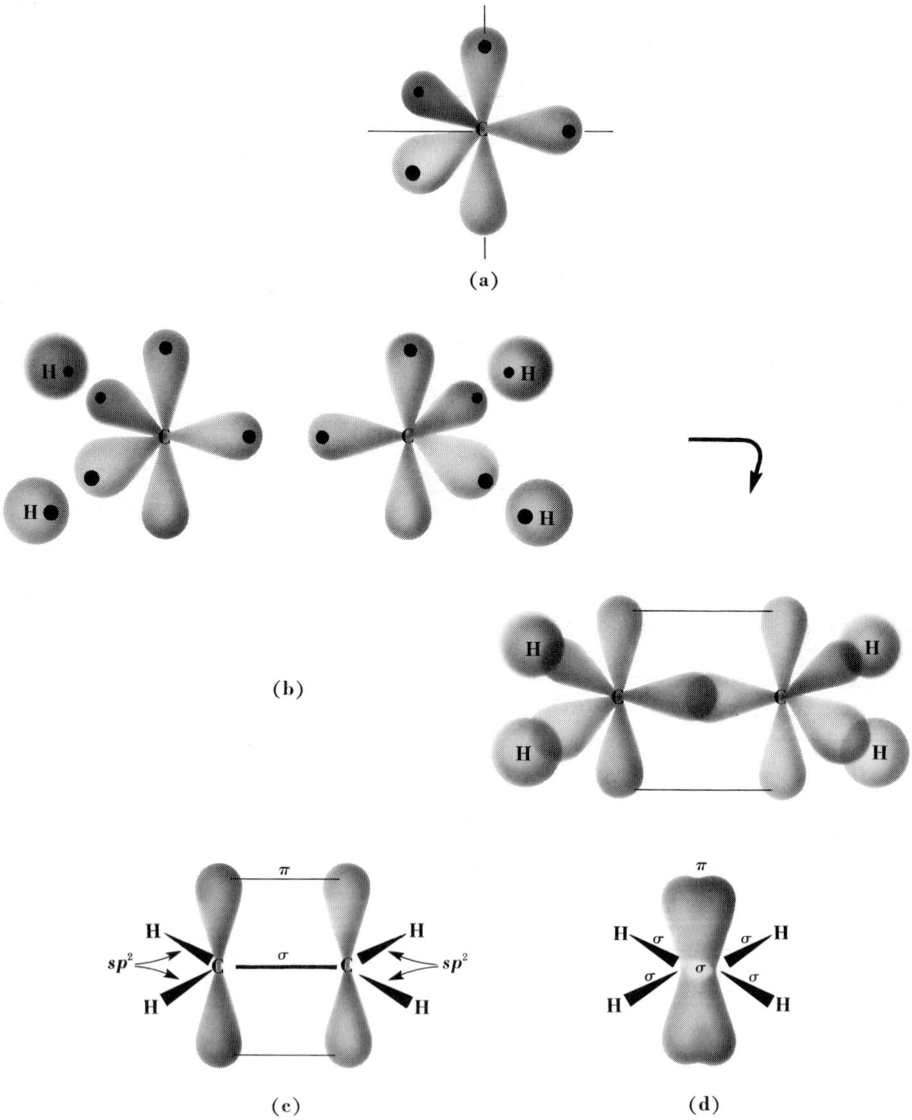

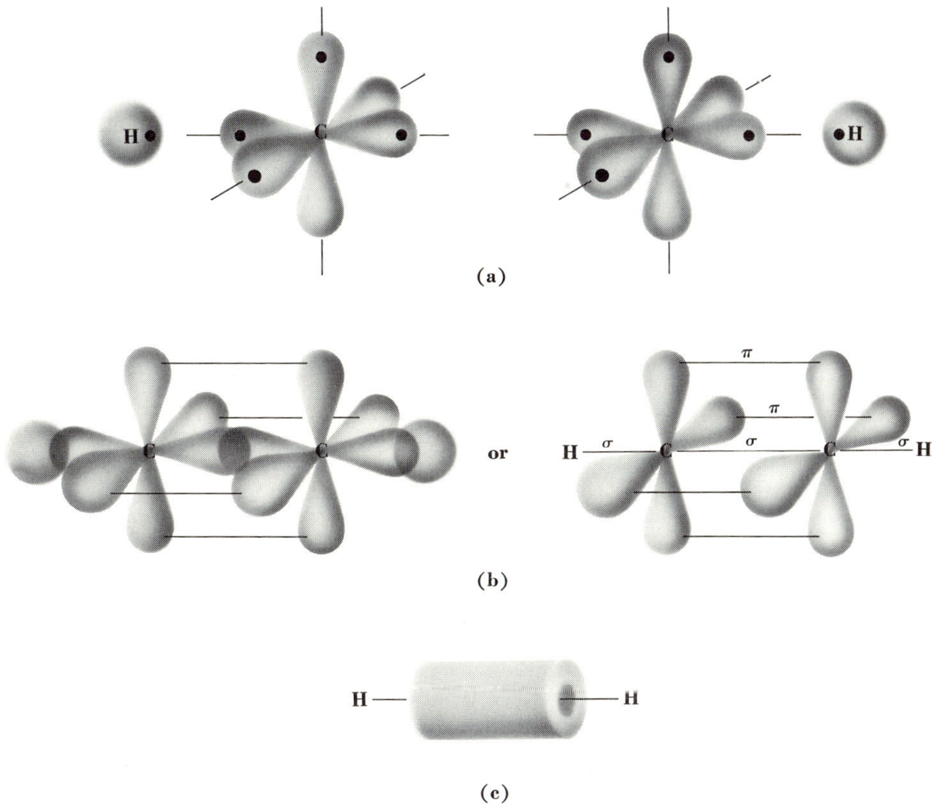

Fig. 8.12 Orbital representation of acetylene: (a) assembly of 2H and 2C atoms; (b) schematic representation showing σ-bond skeleton and π bonds; (c) representation of the acetylene-type π bond.

The p_y orbitals overlap laterally to form a π bond in the y plane, and the p_z orbitals interact similarly to form a π bond in the z plane. Therefore, the triple bond comprises one σ and two π bonds. The planes of the two π bonds are at right angles to each other and to the C—C σ-bond axis. It is believed, however, that these two π bonds coalesce into a cylindrical shape (Fig. 8.12b), called an acetylenic π bond. The acetylenic π bond is nevertheless counted as two bonds.

Carbon dioxide is an example of a molecule with two double bonds to the same carbon. It is a linear molecule in which both O═C bonds are equivalent:

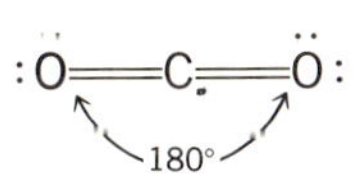

shape of a CO_2 molecule

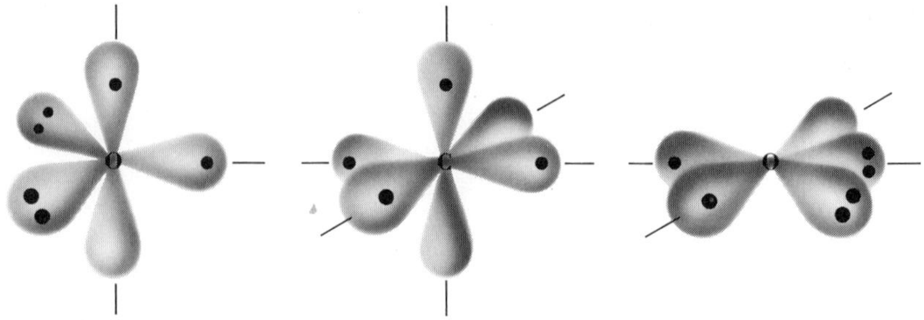

(a)

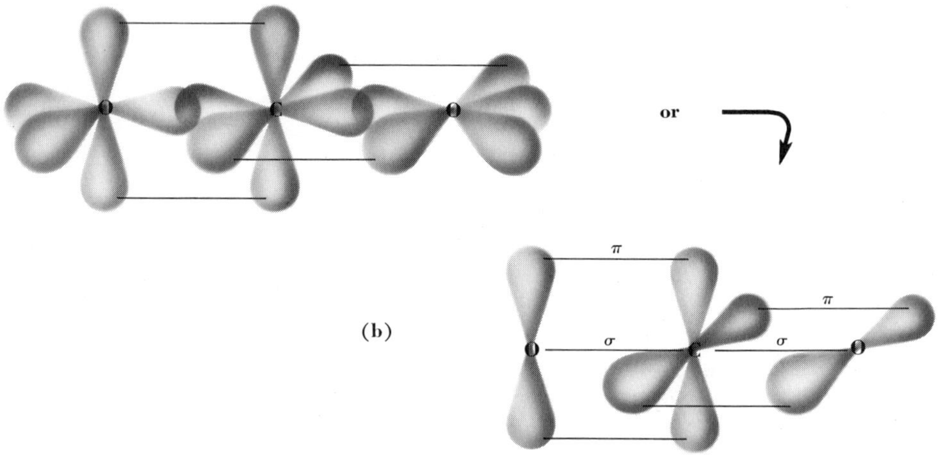

(b)

Fig. 8.13 (a) Orbital representation of the construction of a molecule of CO_2. (b) Schematic orbital representation of a CO_2 molecule.

Linearity means the O—C—O bond angle is 180 degrees, a fact indicating that the carbon uses *sp* hybridized orbitals to form a σ bond to each O atom (Fig. 8.13a). The carbon atom also has two *p* orbitals, each of which overlaps laterally with a *p* orbital of an adjacent oxygen atom to give two π bonds. Each π bond holds a pair of electrons, one from each atom, and therefore the octet requirement for carbon is satisfied. The π bonds are in planes perpendicular to each other but, since they do not involve the same two atoms, do not coalesce. If we assume the lone pairs of electrons on each oxygen atom to be contained in individual *hybrid* orbitals, then the oxygen atoms are both sp^2 hybridized.

8.6 □ HYBRID ORBITAL NUMBER

For the three molecules just discussed, the geometry was used to indicate the kind of hybridization that should be assumed for the carbon atom. However, it would be preferable if the type of hybridization could be deduced from the electronic structure, thus making it unnecessary to remember or to seek information on the geometry of each molecule. This prediction can be accomplished by utilizing the **hybrid orbital number** rule. The hybrid orbital number is the number of hybrid orbitals used by an atom, and is defined as

$$\text{hybrid orbital number} = \text{coordination number (number of } \sigma \text{ bonds)} + \text{number of pairs of unshared electrons}$$

Values of 4, 3, and 2 indicate, respectively, sp^3, sp^2, and sp hybridization. The coordination number is the number of atoms bonded to the atom being considered, and so is equal to the number of σ bonds to that atom. The basis for this rule is the assumption that *it is necessary to provide a hybrid orbital for each unshared and σ bonded pair of electrons*. Exceptions to this rule may occur for certain compounds of elements in the higher periods of the periodic table. It is observed that the tendency to use unhybridized *p* orbitals rather than hybridized orbitals increases within a group of the periodic table as the size of the atom increases.

EXAMPLE 1 Use the hybrid orbital number rule to predict the bonding of

(a) a carbanion such as $H_3C\colon^-$;
(b) a carbonium ion such as H_3C^+.

ANSWER (a) The hybrid orbital number for the C atom in $H_3C\colon^-$ is $3 + 1 = 4$; therefore the C atom uses sp^3 hybridized atomic orbitals.

(b) The hybrid orbital number for the C atom in H_3C^+ is $3 + 0 = 3$; therefore the C atom is sp^2 hybridized.

8.7 □ RELATIVE ENERGY LEVELS OF THE *s-p* TYPE OF HYBRID ORBITAL

For any quantum shell, the *s* orbital is at a lower energy level than the *p* orbital. In hybridized orbitals, the greater the *s* character, the lower is the energy:

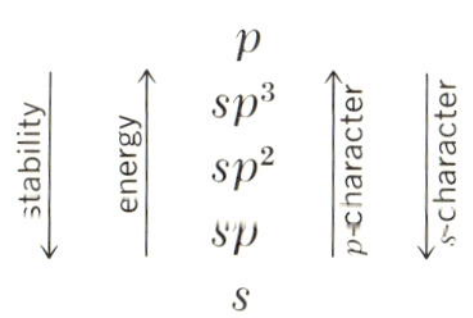

8.8 □ RESONANCE AND DELOCALIZED *p* ELECTRONS

(a) Resonance concept Let us write an electronic structure for dinitrogen oxide (nitrous oxide), N_2O. (This compound, often called "laughing gas" and used widely as a propellant in whipped cream bombs, is still used occasionally as an anesthetic in dentistry.) There are two reasonable Lewis formulas for N_2O, both of which fit the octet rule and have the required skeleton. Which structure, *if any*, is correct? This question might be resolved by comparing the calculated with the observed bond lengths.

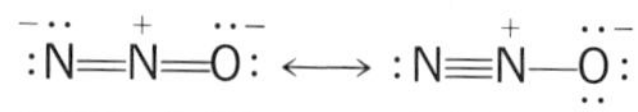

Bond lengths, Å	(calc.)	1.20	1.15	1.10	1.36
	(obs.)	1.12	1.19	1.12	1.19

(Formal charges are shown.)

The bond lengths are calculated from the single, double, and triple bond covalent radii of the individual atoms. Neither set of predicted bond lengths matches the observed values. The consequent uncertainty of assignment of electronic formula is resolved by the concept of **resonance.**

According to the resonance concept, the structure of a molecule such as N_2O cannot be accurately depicted by a single Lewis formula. Instead, the molecule is depicted by *two or more* formulas, such as those shown above for N_2O, which, taken together, serve as a better description than any single one. Thus, if you ask the question, "What kind of bond exists between the N atoms in N_2O?" each formula, taken by itself, provides an incorrect answer and predicts an incorrect bond length. The two formulas taken together, however, imply the answer, "Something between a double and a triple bond," which, while not startlingly precise, is somewhat better. The separate formulas are called **contributing** or **resonance forms,** of which the actual molecule is said to be a **resonance hybrid.** The word "hybrid" is used in an imaginary, not a biological sense, and may be exemplified by the description of a rhinoceros offered to a well-read child who had never been to the zoo as "a hybrid of a dragon and a unicorn."

A **double headed arrow,** ⟷, is written between the contributing forms to indicate resonance. There are important restrictions on the formulas one may write:

(a) The relative positions of all atoms must be the same in each formula—*only the positions of the electrons may differ.*

(b) There must be the same number of pairs of electrons in all contributing formulas. For example, :Ö::Ö: and :Ö:Ö: are not contributing forms of O_2, because the number of electron pairs is different.

EXAMPLE 2 Write contributing electronic formulas for nitrate ion, NO_3^-, showing formal charges.

ANSWER

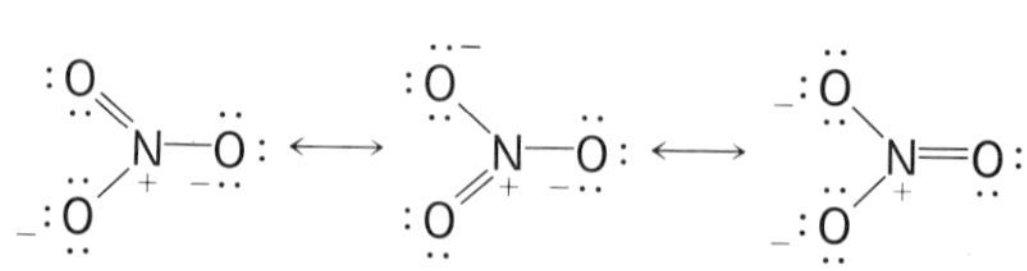

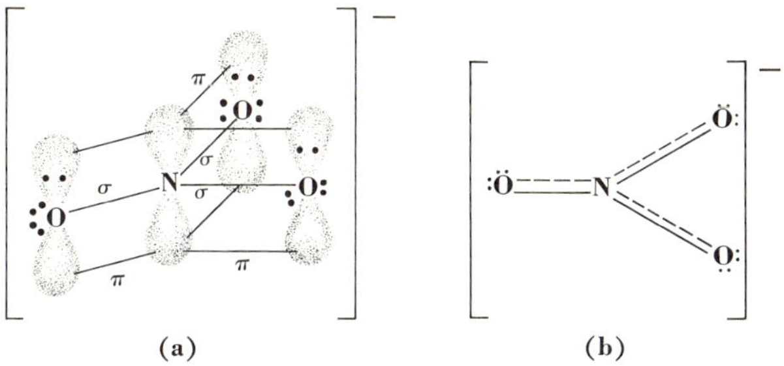

Fig. 8.14 Extended π bonding in nitrate ion.

Observe that the three contributing forms for the nitrate ion look equivalent. In each formula the N atom has a single bond to each of two O atoms, and a double bond to a third O atom. The three formulas taken together have a different connotation than any single one. Any one implies, incorrectly, two N—O bond distances of the same length and one shorter N═O bond distance. Actually, all three nitrogen-oxygen bond distances are the same. This experimental fact implies that all three oxygen atoms are indistinguishable from each other. No one oxygen atom can therefore be said to be the one that is doubly bonded to the nitrogen atom. Instead, *each O atom participates in a double bond*—a fact represented by the overall resonance hybrid.

(b) Extended π bonding If individual electronic formulas are inadequate to describe species like N_2O and NO_3^-, do we have a parallel problem with molecular orbital designations? Remember that although electrons in σ bonds and π bonds are usually shown to be localized between the bonded atoms, all the electrons of a molecule repel all other electrons and attract all the nuclei, so that the idea of localized bonds is only an approximation of reality. In an entity like NO_3^- the approximation is a particularly poor one, which we have seen engenders the dilemma, in writing formulas, that we try to circumvent by the resonance concept. The molecular orbital picture does not present this problem if we assume that the *π bonding is extended (or "delocalized"), and is not confined to individual pairs of bonded atoms.* In nitrate ion, NO_3^-, each O—N—O bond angle is 120 degrees, a fact indicating that the N atom utilizes sp^2 hybrid atomic orbitals to form the σ-bonded skeleton. The nitrogen atom has a p orbital that can act as a pivot for overlapping with a p orbital of each oxygen atom (Fig. 8.14) to give an extended π bond (Fig. 8.14a). The σ-bonded skeleton and the lone pairs of electrons on each oxygen atom account for 18 of the 24 valence electrons of the NO_3^- ion. Therefore, the remaining six electrons must be in the delocalized, extended π bond. In Fig. 8.14b, dotted lines denote extended π bonds; this convention is widely used.

PROBLEMS

1. Definitions Define and illustrate the following terms: (a) molecular orbital, (b) σ bond, (c) π bond, (d) multiple bond, (e) hybridization, (f) resonance.

2. Orbital overlap Explain why a lateral (sidewise) overlap of an s and a p orbital is nonbonding, whereas a head-to-head overlap can be bonding. Draw a schematic representation of each type of overlap.

3. Covalent molecules Write Lewis formulas for (a) F_2; (b) CO; (c) HCOCl.

4. Hybridization Show the distribution of the valence electrons in the orbitals of (a) an sp^2 hybridized N atom, (b) an sp^3 hybridized B atom, (c) an sp^2 hybridized O atom, (d) an sp hybridized N atom, (e) an sp hybridized Cl atom.

5. Hybrid orbital number rule Apply the hybrid orbital number rule (to N, C, and O atoms) to predict the molecular orbital structure of

formaldehyde, $H_2C{=}\ddot{O}{:}$
hydrogen cyanide, $H{-}C{\equiv}N{:}$
allene, $H_2C{=}C{=}CH_2$

6. Bond lengths (a) From the data in Table 3.7 on p. 56, calculate the lengths of the following bonds: O—F, O—Cl, O—Br, O—I, C—C, C—N, C—O, C—F. (b) What change is observed in the calculated bond length of A—B as (*i*) B increases in atomic weight within a group, (*ii*) B increases in atomic number within a period? Explain.

7. Relative energy levels of electrons List the following in order of *decreasing* energy: (a) The $2p$ electrons in a N atom, (b) the $2s$ electrons in a N atom, (c) the lone (unbonded) electron pairs in N_2, (d) the lone pair in ammonia, $H_3N{:}$, (e) the lone pair in formaldehyde methylimine, $H_2C{=}\ddot{N}{-}CH_3$.

8. Resonance and extended π bonding (a) Write the contributing Lewis formulas of (*i*) ozone, O_3, which does *not* have a cyclic structure; (*ii*) carbonate ion,

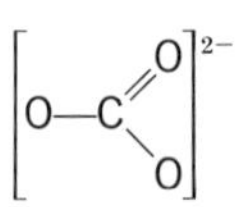

; (*iii*) formate ion,

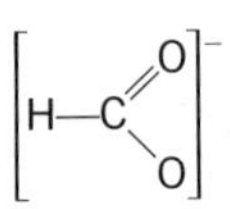

. (b) Write structures representing extended π bonding in compounds (*i*), (*ii*), and (*iii*), assuming in each case that the O atoms are sp^2 hybridized.

ADDITIONAL PROBLEMS

9. Stability of the covalent bond (a) Consider two unit charges of like sign separated by distance $2r$ in a vacuum.

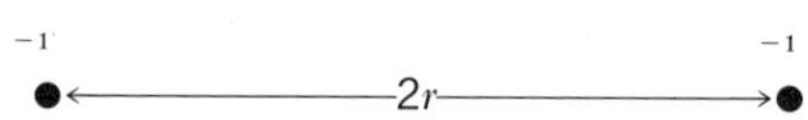

Calculate the Coulombic repulsive force in terms of r (see App. 1). Now consider a unit charge of opposite sign placed midway between the others.

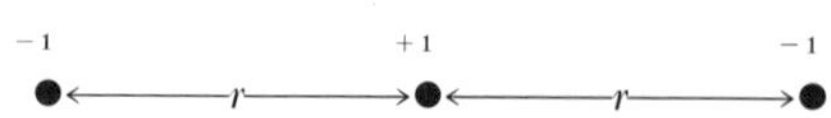

Calculate the sum of the attractive forces in terms of r. Which is greater, the total attractive force or the total repulsive force? By what ratio? Is such an alternating sequence of charged bodies more stable than the first arrangement shown? Without performing any calculation, would you predict that an extended series of alternating changes, +−+− and so on, is a stable arrangement? (b) Identify an axis along which you would expect to find alternating concentrations of positive and negative charge in (*i*) a hydrogen atom, (*ii*) a sodium chloride, Na^+Cl^-, crystal, (*iii*) a hydrogen molecule. (c) Justify the statement, "The electrostatic stabilization of the covalent bond is analogous to that of the ionic crystal and even to that of the atom itself."

10. Hybrid orbital number rule Apply the hybrid orbital number rule (to N, C, and O atoms) to predict the molecular orbital structure of

ketene, $H_2C{=}C{=}\ddot{O}{:}$
cyanic acid, $H{-}\ddot{\underset{\cdot\cdot}{O}}{-}C{\equiv}N{:}$
cyanamide, $H_2\ddot{N}{-}C{\equiv}N{:}$
diazomethane, $H_2C{=}N{=}\ddot{N}{:}$

11. Resonance (a) Write two contributing Lewis formulas consistent with the octet rule for hydrazoic acid, HN_3, for which the sequence of bonded atoms is HNNN. (b) Locate all formal charges in each formula. (c) Which if either formula correctly represents the structure of the molecule? Explain.

12. Delocalization The compound represented as

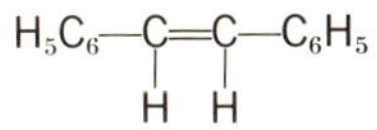

absorbs light of wavelength 3190 Å, while the compound represented as

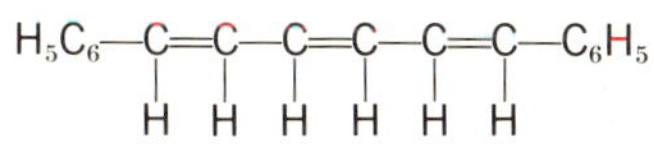

absorbs at 3770 Å. An article in the *Journal of Chemical Education* **39,** 599 (1962) explains this difference in light absorption with the language, ". . . by assuming that delocalization . . . spreads the electron system through the molecule so that it behaves as a single oscillating unit . . ." (a) In which compound is the electron system more delocalized? (b) In which compound will the delocalized electron system make a greater contribution to the stability of the compound? (c) Explain the difference in light absorption.

ANSWER **9.** (a) The ratio is 8:1.

9 THE SHAPES AND SYMMETRY OF MOLECULES

9.1 INTRODUCTION

Molecules are three-dimensional aggregates of chemically bonded atoms; they have size and shape. Experimentally, we "see" molecules by probing the substances they compose with electrons or with electromagnetic radiation like infrared and x rays, and by examining their electrical polarity. The results we obtain from such studies fix the relative positions of the atoms of a molecule; such information can be expressed entirely in terms of interatomic distances and bond angles.

We shall be concerned principally with shapes of molecules of the AB_n type, where the central atom, A, is a representative element. This means that we must examine the B—A—B bond angles.

9.2 HYBRIDIZATION OF ORBITALS OF ATOMS WITH MORE THAN AN OCTET OF VALENCE ELECTRONS

The shapes of molecules whose central atom has no more than eight valence electrons were summarized on page 139. Elements of the third and higher periods have *d* orbitals available for bonding and consequently may have more than eight valence electrons. Examples are:

Compound	Total number of valence electrons associated with central atom		
	Bonding	Unshared	Total
PCl_5	10	0	10
$:SF_4$	8	2	10
SF_6	12	0	12
$:BrF_5$	10	2	12

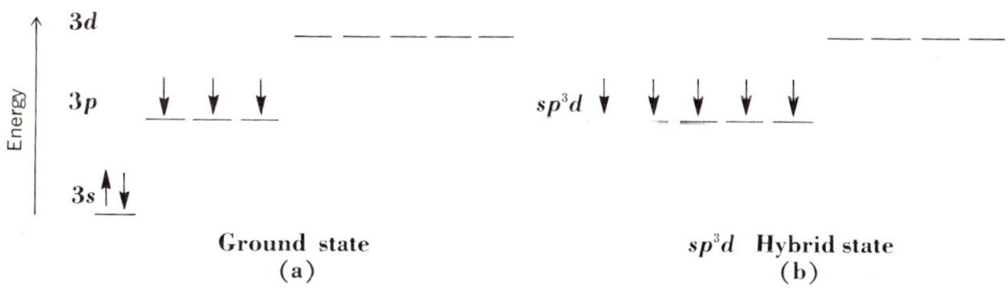

Fig. 9.1 Electronic arrangement of highest principal energy level of phosphorus: (a) ground state, (b) hybridized sp^3d state.

Because the shapes of orbitals influence bond angles, we shall consider the hybridization of orbitals of atoms with 5 or 6 pairs of valence electrons.

(a) Five pairs of electrons on the central atom Five orbitals are needed to accommodate the five pairs of electrons. It is theorized that one d orbital hybridizes with the one s and three p orbitals of the same principal quantum number to give five sp^3d hybrid orbitals. Thus, the phosphorus atom in gaseous PCl_5 uses the $3s$, three $3p$, and one $3d$ orbital to form the five sp^3d hybrid orbitals (Fig. 9.1). The PCl_5 molecule is a **trigonal bipyramid** (Fig. 9.5). Three Cl atoms lie at the corners of an equilateral triangle while a fourth Cl atom lies above and a fifth below the center of the triangle. This type of geometry indicates that the five bonds are not equivalent, because the bond angles involving the axial Cl atoms are 90 degrees while the other bond angles are 120 degrees.

(b) Six pairs of electrons on the central atom To accommodate six pairs of electrons, an atom must have available six atomic orbitals. Its outermost s and p orbitals total only four. It is theorized that this deficiency is met by hybridization of these four orbitals with two d orbitals of the same principal quantum number. In this way, six equivalent sp^3d^2 hybrid orbitals become available to accommodate the six pairs of electrons. Thus, the sulfur atom in SF_6 acquires six orbitals

Fig. 9.2 Electronic arrangement of highest principal energy level of sulfur: (a) ground state, (b) hybridized sp^3d^2 state.

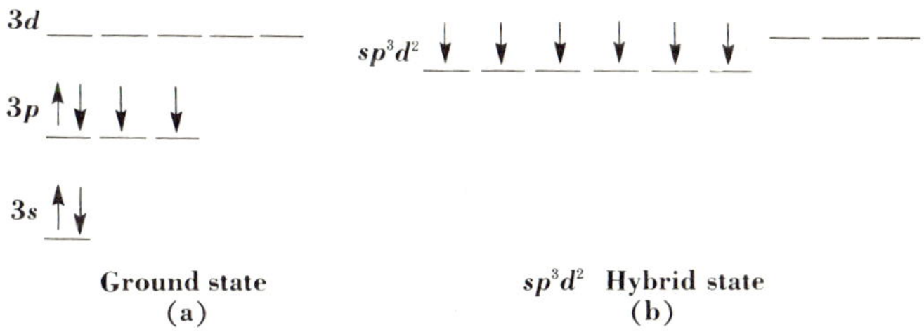

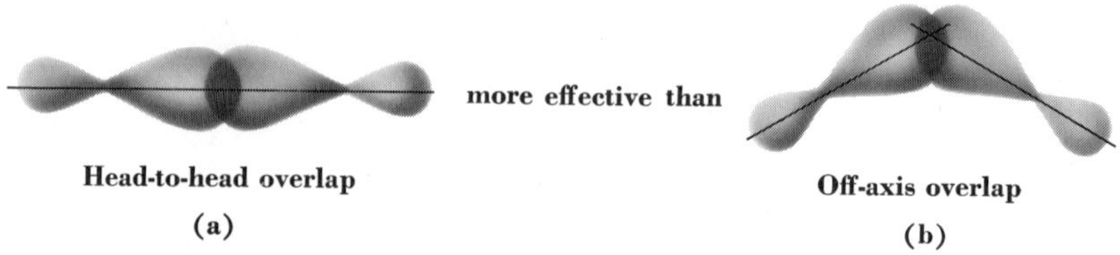

Fig. 9.3 Representation of overlap of *s-p* type atomic orbitals: (a) head-to-head (along axis); (b) off-axis.

by hybridizing its one 3*s*, three 3*p*, and two 3*d* orbitals (Fig. 9.2). Each orbital points to the corner of an octahedron (Fig. 9.5), and hence the sp^3d^2 hybrid orbitals are called **octahedral.** The angle between any two is 90 degrees.

9.3 □ PRINCIPAL FACTORS DETERMINING MOLECULAR SHAPE

(a) The sigma bonds; tendency for maximum orbital overlap The shape of a molecule of the AB_n type is determined by *the σ bonds which project from the central atom.* Since σ bonds result from overlap of atomic orbitals, it follows that the geometry of the molecule is determined essentially by the type of atomic orbitals used by the central atom, A. As long as atom B has only one σ bond, as is the case in the AB_n type molecules, the orbital it uses does *not* influence the bond angle and the molecular shape. The σ bond is strongest when the overlap of atomic orbitals is greatest; this creates the highest possible electron density in the region between the atoms. A σ bond has maximum overlap when the constituent atomic orbitals lie along the internuclear axis (Fig. 9.3a), rather than off the axis (Fig. 9.3b). Since σ bonds are formed from head-to-head overlap, the σ bond angle is defined by the atomic orbitals of the multi-σ-bonded central atom, A, rather than by the uni-σ-bonded peripheral atom, B (Fig. 9.4).

The tendency for maximum orbital overlap is a factor that predisposes an atom, as for example the nitrogen atom in ammonia and the oxygen atom in water (p. 140), to use an *s-p* type hybrid atomic orbital

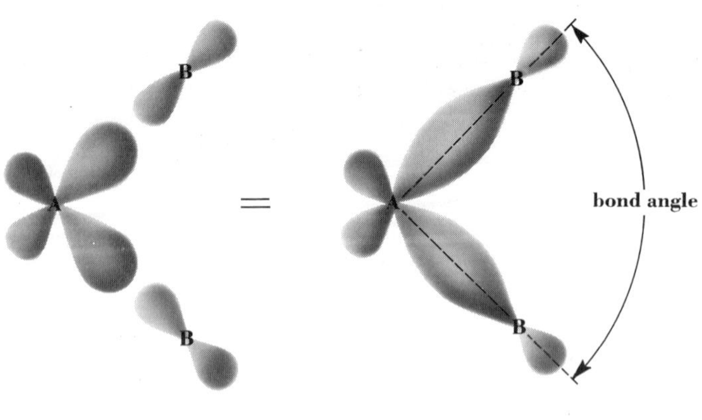

Fig. 9.4 Representation of a typical B-A-B bond angle formed by overlap of two *s-p* type atomic orbitals of A, each with a *p* orbital of atom B. (Dotted line indicates the bond axis.)

rather than an unhybridized p orbital. The following relative strengths of σ bonds formed from different atomic orbitals have been calculated from molecular orbital theory:

Orbital type	s	p	sp	sp^2	sp^3
Relative σ-bond strength	0.3	1	1.5	1.4	1.3

Evidently σ bonds formed from s-p type hybrid orbitals are stronger than those formed from a p orbital.

(b) Kinds of orbitals used; electron pair repulsion The central atom uses hybridized atomic orbitals for bonding in most cases. Some exceptions in which the central atom uses unhybridized p orbitals are discussed in Section 9.5. A hybrid atomic orbital is needed for each σ bond and for each lone pair of electrons. This requirement is the basis for the hybrid orbital number rule (p. 145). We have previously described five types of hybrid atomic orbital: sp, sp^2, sp^3, sp^3d, and sp^3d^2 (pp. 141, 151). Each type is associated with a definite spatial

Fig. 9.5 Arrangement of electron pairs to minimize repulsion: (a) two pairs, linear; (b) three pairs, plane triangle; (c) four pairs, tetrahedral: (d) five pairs, trigonal bipyramidal; (e) six pairs, octahedral.

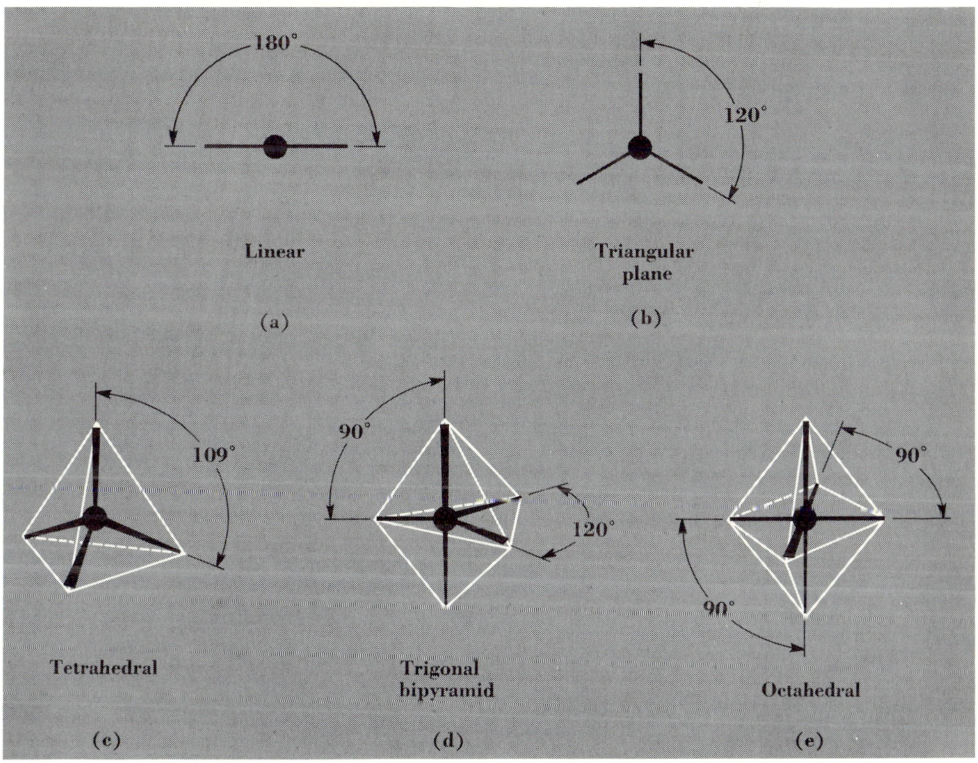

TABLE 9.1 SHAPES OF MOLECULES AND IONS OF NONTRANSITIONAL ELEMENTS WITH NO LONE PAIRS OF ELECTRONS ON CENTRAL ATOM

General type	Example	Observed shape	Hybrid orbital number	Hybrid orbital type
AB_2	$BeCl_2$, $HgCl_2$	Linear	2	*sp*
AB_3	BF_3, $GaCl_3$	Triangular	3	sp^2
AB_4	CH_4, NH_4^+, BF_4^-, $BeCl_4^{2-}$	Tetrahedral	4	sp^3
AB_5	$SbCl_5$, PCl_5	Trigonal bipyramidal	5	sp^3d
AB_6	SF_6, AlF_6^{3-}, SiF_6^{2-}	Octahedral	6	sp^3d^2

orientation which minimizes the repulsion between electron pairs in the outer shell of the central atom. These shapes are shown in Fig. 9.5 and are summarized in Table 9.1.

9.4 □ PRINCIPAL EFFECT OF A LONE PAIR OF ELECTRONS

This discussion concerns those molecules and ions in which the central atom has a lone pair, uses hybridized orbitals, and has no multiple bonds. For example, consider a molecule of the type of AB_4; the central atom, A, has an octet of bonding electrons, and the molecule is tetrahedral. In $:AB_3$ and $:\ddot{A}B_2$ the central atom, A, also has 8 electrons, but the replacement of one or more A:B bonds by a lone pair alters the shape of the molecule. Remember that molecular shape is defined by the positions of atoms, not of electrons. Thus *any* diatomic molecule, like HCl, regardless of the number of unshared pairs, has only one bond axis and is, by definition, linear.

The following paragraphs discuss specific cases of shapes of molecules or ions whose central atom has one or more lone pairs of electrons.

(a) Four electron pairs on the central atom In this group, there are two types of species, $:AB_3$, as typified by ammonia, $:NH_3$, and $:\ddot{A}B_2$, as typified by water, $:\ddot{O}H_2$. In both molecules, the central atom uses sp^3 hybrid atomic orbitals. All four orbitals are equivalent, and it is therefore immaterial which orbital houses a lone pair.

It has been experimentally determined that $:NH_3$ is a **trigonal** (triangular) **pyramid;** the three H atoms form the triangular base of the pyramid with the N atom at the apex (Fig. 9.6b). The H—N—H bond angles are 107 degrees, indicating that the orientation of the four pairs of electrons around the nitrogen atom is approximately tetrahedral. The nitrogen atom exhibits sp^3 hybridization.

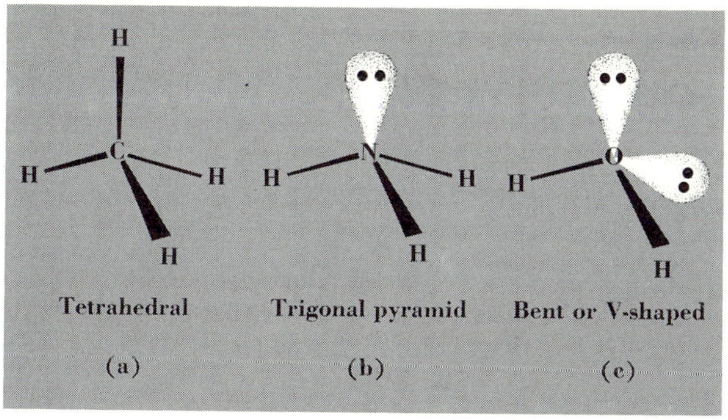

Fig. 9.6 Shapes of (a) methane, CH_4, AB_4 type; (b) ammonia, $:NH_3$, $:AB_3$ type; (c) water, $H_2O:$, $:AB_2$ type.

Water, $H_2\ddot{O}:$, is a **bent** or **V-shaped** molecule (Fig. 9.6c), with a bond angle of 105 degrees. This value is fairly close to the tetrahedral angle, and hence oxygen is believed to use sp^3 hybrid atomic orbitals.

(b) Six electron pairs on the central atom In this group, there are only two types, $:AB_5$, as typified by bromine pentafluoride, $:BrF_5$, and $:\ddot{A}B_4$, as typified by xenon tetrafluoride, $:\ddot{X}eF_4$. In the octahedral array for an AB_6 type of molecule, all six sp^3d^2 hybridized atomic orbitals are equivalent. We should expect that any of the six could accommodate the lone pair of electrons equally well. We find that the shape of $:BrF_5$ is a **square pyramid** (Fig. 9.7b); four of the fluorine atoms form the square base of the pyramid, and the fifth fluorine atom is at the apex.

When we consider $:\ddot{X}eF_4$, a new factor appears for the first time in our discussion. The two pairs of electrons can be arranged in *two* different ways about the central atom. The axes of the orbitals with the

Fig. 9.7 Structures of typical molecules whose central atom has six pairs of electrons: (a) AB_6 type, SF_6; (b) $:AB_5$ type, BrF_5; (c) $:AB_4$ type, $\ddot{X}eF_4$; (d) incorrect structure for XeF_4.

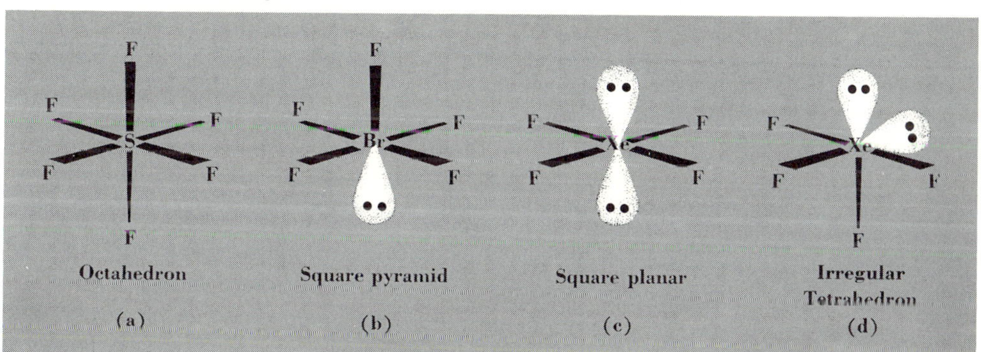

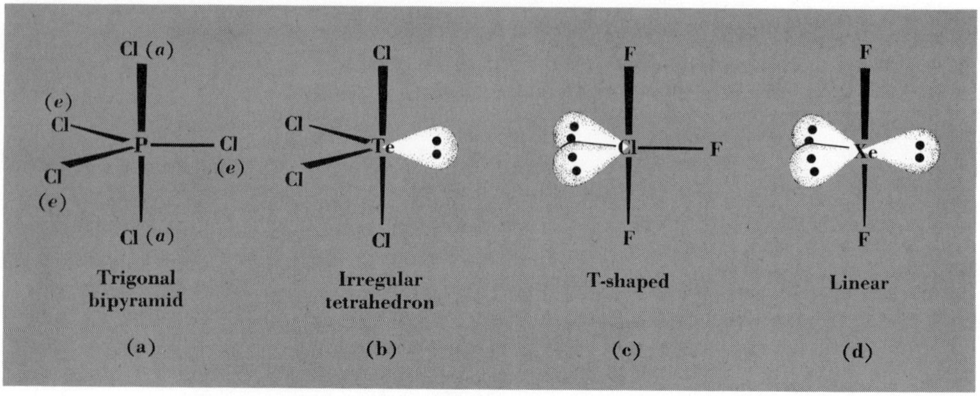

Fig. 9.8 Shapes of typical molecules with five pairs of electrons on the central atom: (a) PCl_5, AB_5 type; (b) $\dot{\text{T}}eCl_4$. : AB_4 type; (c) $:\ddot{\text{C}}lF_3$, $:\ddot{A}B_3$ type; (d) $:\ddot{\text{X}}eF_2$, $:\ddot{A}B_2$ type.

lone pairs of electrons can form either an angle of 90 degrees (Fig. 9.7d) or an angle of 180 degrees (Fig. 9.7c). The shape of xenon tetrafluoride is square-planar (Fig. 9.7c). The Xe atom is in the center and the fluorine atoms are at the corners of a square; all the atoms are thus in the same plane. Evidently the two orbitals with the lone pairs form an angle of 180 degrees with each other, rather than an angle of 90 degrees.

(c) Five electron pairs on the central atom This group has three categories of structures: (*i*) $:AB_4$, as typified by tellurium tetrachloride, $:TeCl_4$; (*ii*) $:\ddot{A}B_3$, as typified by chlorine trifluoride, $:\ddot{C}lF_3$; (*iii*) $:\ddot{A}B_2$, as typified by xenon difluoride, $:\ddot{X}eF_2$. Of the five types of hybrid

TABLE 9.2 SHAPES OF MOLECULES AND IONS HAVING LONE PAIR(S) ON CENTRAL ATOM

General type	Example			Shape	Orbital type	Electron pair arrangement
$:AB_3$	$:NH_3$	$H_3O:^+$	$:PF_3$	Trigonal pyramid	sp^3	Tetrahedral
$:\ddot{A}B_2$	$H_2\ddot{O}:$	$:\ddot{S}Cl_2$	$:\ddot{N}H_2^-$	Bent	sp^3	Tetrahedral
$:AB_4$	$:SF_4$	$:TeCl_4$		Irregular tetrahedron	sp^3d	Trigonal bipyramidal
$:\ddot{A}B_3$	$:\ddot{C}lF_3$			T-shaped	sp^3d	Trigonal bipyramidal
$:\ddot{A}B_2$	$:\ddot{X}eF_2$	$:\ddot{I}Cl_2^-$		Linear	sp^3d	Trigonal bipyramidal
$:AB_5$	$:IF_5$	$:BrF_5$		Square pyramid	sp^3d^2	Octahedral
$:\ddot{A}B_4$	$:\ddot{X}eF_4$	$:\ddot{I}Cl_4^-$		Square plane	sp^3d^2	Octahedral

atomic orbital shown in Table 9.1, the sp^3d is unique in that not all five orbitals are equivalent. There are two different sets of equivalent orbitals. The three orbitals oriented in a plane toward the corners of a triangle comprise the **equatorial (e)** group, and the two orbitals oriented perpendicular to the plane comprise the **axial (a)** group (Fig. 9.8a). The nonequivalence of the bonds is revealed by the bond distances in PCl_5: P—Cl (axial) = 2.19 Å; P—Cl (equatorial) = 2.04 Å. When bonding pairs are replaced by lone pairs, the observed shapes of the three types of molecule indicate that the lone pairs assume equatorial positions (Fig. 9.8b–d). Table 9.2 summarizes the typical shapes of molecules and ions whose central atom has at least one lone pair.

9.5 □ MOLECULES WHOSE CENTRAL ATOMS USE UNHYBRIDIZED *p* ORBITALS

Although ammonia has a bond angle (107 degrees) close to the tetrahedral value of 109.5 degrees, the hydrides of the other members of group V have bond angles close to 90 degrees. The observed values are 94 degrees for $:PH_3$, 92 degrees for $:AsH_3$, and 91 degrees for $:SbH_3$. These values are interpreted to mean that phosphorus, arsenic, and antimony use unhybridized p orbitals to form bonds to hydrogen. The lone pair in these hydrides is in an s orbital. A similar situation prevails for the hydrides of group VI. Thus, whereas the bond angle for water is 105 degrees, the bond angles are 92 degrees for H_2S, 91 degrees for H_2Se, and 90 degrees for H_2Te. As the size of the atoms in groups V and VI of the periodic table increases, the tendency to use unhybridized p orbitals for formation of a σ bond to hydrogen appears to increase. This tendency to use p orbitals seems to prevail mainly for σ bonding to hydrogen, since the bond angle in sulfur dichloride, SCl_2, is 103 degrees, and the F—P—F angle in phosphorus trifluoride, PF_3, is 104 degrees.

9.6 □ SHAPES OF MULTIPLE-BONDED MOLECULES

A multiple bond consists of σ and π bonds. We have seen that the geometry of a molecule is essentially determined by the σ bonds. The π bonds are parallel to the axis of the σ bond, and hence have no primary effect on the shape. The shape of a multiple-bonded molecule or ion therefore depends primarily on the same factors previously discussed. Table 9.3 summarizes the typical shapes of some common substances with p-p π bonds, on the basis of the number of B atoms bonded to atom A, and the number of lone pairs on atom A. Several of the substances are represented by structures showing extended π bonding (p. 147). Notice that p-p π bonds occur only if the central atom is sp or sp^2 hybridized, for only then does the central atom have a p orbital available to overlap with a p orbital of atom B.

TABLE 9.3 SHAPES OF MOLECULES AND IONS WITH p-p π BONDS

Examples	Shape	Number of σ bonds	Number of lone pairs	Hybrid orbital number	Hybrid type	Electron pair arrangement
O═C═O, H—C≡N	Linear	2	0	2	sp	Linear
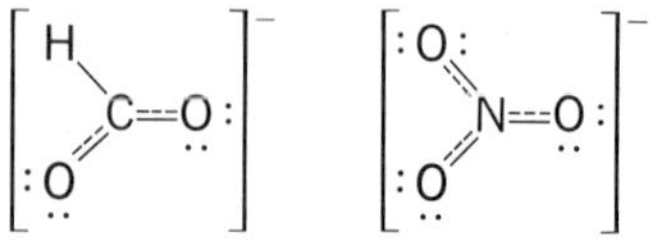	Triangular	3	0	3	sp^2	Trigonal
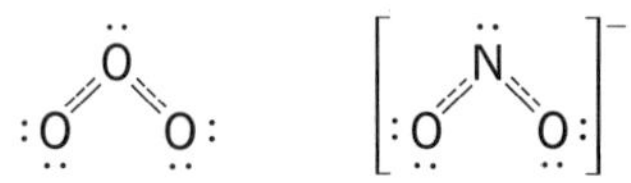	Bent	2	1	3	sp^2	Trigonal

9.7 □ MOLECULAR SYMMETRY

Symmetry is an important aspect of molecular shape. Intuitively, we recognize that a sphere and a cube are highly symmetrical bodies. In this section we describe some of the structural features, called **symmetry elements,** which make a molecule symmetrical. Visualization of these symmetry elements requires a good understanding of the typical molecular structures previously described. The use of three-dimensional models is strongly recommended for developing the ability to recognize and visualize molecular structures.

Symmetry elements include planes, a point, and lines (axes).

(a) Symmetry planes A symmetry plane cuts through the molecule so as to divide it into equivalent halves. This means that a line which starts at any atom, and is drawn perpendicular to the plane and ex-

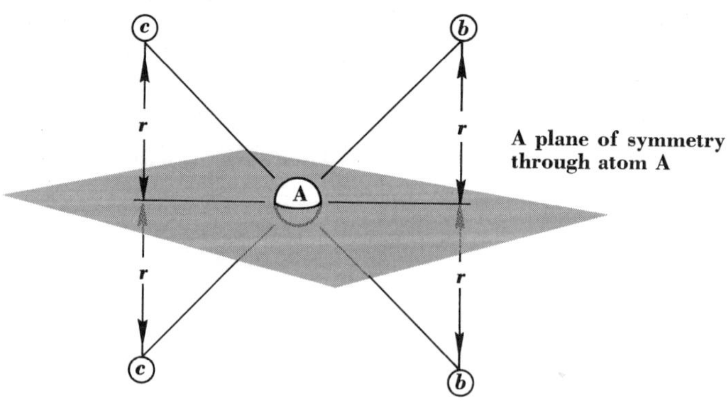

Fig. 9.9 Example of a plane of symmetry.

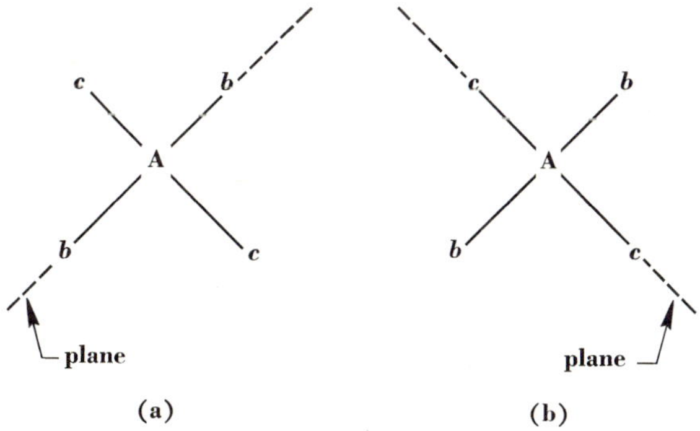

Fig. 9.10 Examples of planes of symmetry which incorporate bonds.

tended an equal distance on the other side of the plane, terminates at the same kind of atom. Figure 9.9 shows this operation for a square-planar structure of the type Ab_2c_2.

We can consider the plane as being a mirror, so that half the molecule is a mirror image of the other half. In the example cited, the plane passes through an atom, in which case it can be considered as cutting the atom into two halves that are mirror images of each other. Consequently, any planar molecule has at least one plane of symmetry, namely, the molecular plane, the plane that cuts through all its atoms. The structure in Fig. 9.9 has two planes of symmetry, the one shown and the molecular plane. Evidently a structure can have more than one plane of symmetry.

A plane can pass through more than one atom along the bonds which join them, as shown in Fig. 9.10a,b.

A plane can also pass midway between atoms without incorporating any one atom, as shown in Fig. 9.11 for ethane.

In summary, a symmetry plane divides a molecule into two identical halves. The plane may pass through one or more atoms, or midway between atoms.

Let us examine some simple types of molecule for symmetry planes. Carbon dioxide, a typical linear molecule, has an infinite number of symmetry planes through the molecular axis, the line extending from one O atom through the C atom to the other O atom. It also has a

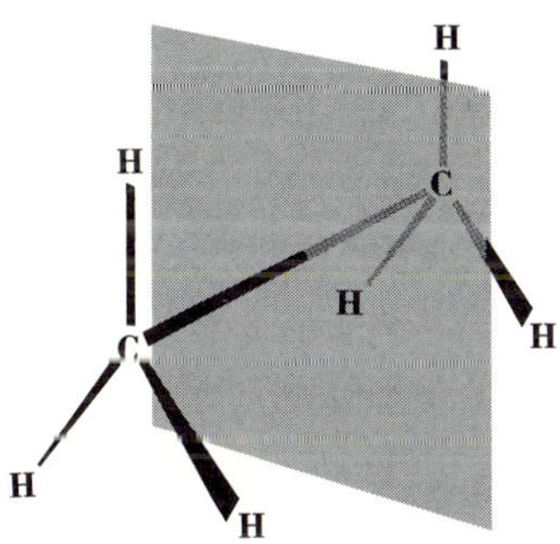

Fig. 9.11 A plane of symmetry that bisects a bond as in ethane

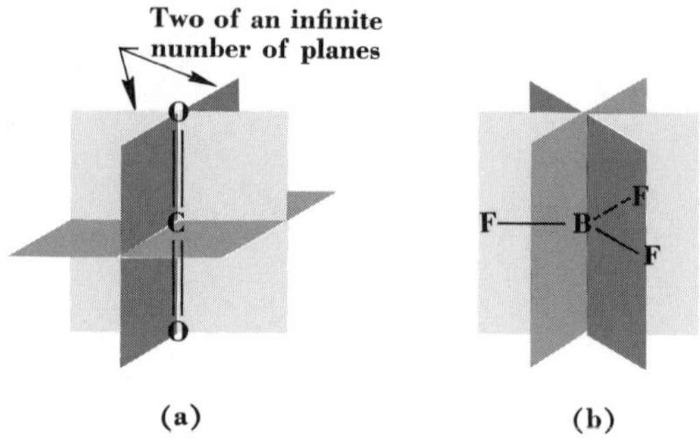

Fig. 9.12 Planes of symmetry in (a) CO_2, (b) BF_3.

plane through the carbon atom perpendicular to the molecular axis (Fig. 9.12a).

Boron trifluoride, a typical planar triangular molecule, has, in addition to the molecular plane, three planes each passing through the B atom and one of the F atoms (Fig. 9.12b).

(b) Center of symmetry The center of symmetry is a point in the center of the molecule to which lines can be drawn from all atoms so that, when each line is extended an equal distance past the center, it comes to the same kind of atom. A simple illustration of a figure with a center of symmetry is a square with points A, A and B, B diagonally opposite each other. If the positions of an A and a B are exchanged, the center of symmetry vanishes:

B—A | | A—B	B—B | | A—A	B—A | | C—B
has center of symmetry	*no center of symmetry*	*no center of symmetry*

A molecule can have only one center of symmetry, which may or may not coincide with an atom. *For a molecule to have a center of symmetry, all atoms, with the exception of the one which may coincide with the center, must exist in paired sets.* In carbon dioxide, the C atom is the center of symmetry. Of the typical molecular shapes mentioned in Tables 9.1 and 9.2, only the square planar and the octahedral forms, as exemplified by XeF_4 and SF_6, respectively, have centers of symmetry.

(c) Axes of symmetry These are elements of symmetry that can be defined in terms of rotations and reflections of the molecule with respect to the axis. Some of these elements are difficult to visualize without the construction of models, and we shall not elaborate on them in this text. The symmetry of molecules with which we shall be concerned can be examined with respect to planes and centers.

EXAMPLE 1 Determine the number and location of the symmetry elements in (a) $H_2\ddot{O}:$, (b) $:NH_3$, (c) CH_4.

ANSWER Write the structural formula showing the proper shape, and then look for the elements of symmetry:

(a)

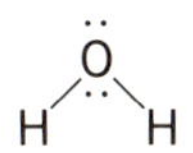

(1) Planes of symmetry: there are two present, the molecular plane, and the plane through the oxygen atom bisecting the angle.

(2) Center: not present.

(b) Consult Fig. 9.6b (p. 155).

(1) Planes of symmetry: since there is only one nitrogen atom and an odd number of H atoms, any plane of symmetry, if present, must pass through the N atom and either one or all three of the H atoms. Only in this way can the two resulting halves be identical. Since NH_3 is not a planar molecule, no plane can include all four atoms. Consequently, we seek planes including the N atom and one H atom, and bisecting the angle involving the other two H atoms. There are *three* such planes, since any one of the three H atoms can be included in a plane.

(2) Center: not present.

(c) To help our discussion, a ball and peg model of CH_4 is shown in Fig. 9.13.

(1) Planes of symmetry: any and all planes must pass through the C atom because only one C atom is present. The plane must pass through either all four H atoms, a pair of H atoms, or no H atoms. Since CH_4 is not a planar molecule, the plane cannot include all four H atoms. No plane of symmetry can be drawn through *only* the C atom, because of the tetrahedral arrangement of the four H atoms. Note that a plane drawn through C, α-H, and β-H also bisects the angle:

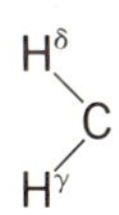

The two halves of the molecule are identical, and we have a plane of symmetry. However, there are six combinations for selecting a pair of hydrogens to lie in a symmetry plane with the C atom: α-H, β-H; α-H,

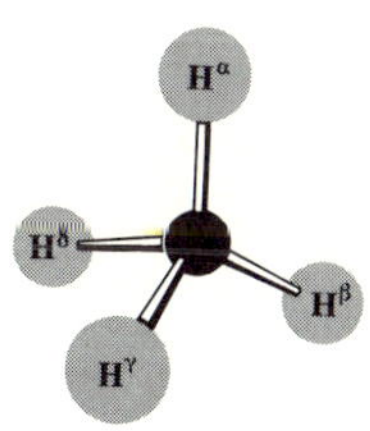

Fig. 9.13 Ball and peg model for methane. The H atoms are designated with Greek letters.

γ-H; α-H, δ-H; β-H, γ-H; β-H, δ-H; γ-H, δ-H. Therefore, methane has *six* planes of symmetry.

(2) Center: none present.

9.8 □ MOLECULAR DISSYMMETRY

Look in a mirror. There you see your mirror image. Imagine that your mirror image became a real person, and stepped into your world. Would you two be different or identical; would any differences between you be recognizable? The questions may appear to be trivial, but they conceal unexpected subtleties, and they bear important implications for molecular structure.*

Two objects are identical when all their points match. A mold, which is a confined space that just accommodates a given object, will exactly accommodate any other object that is identical with the original one. Your mirror image could *not* fit into a mold that was made for you.

If you are right-handed, your mirror image is left-handed; the positions of your wrist watch, the part in your hair, and the like, are reversed and would therefore not find their proper places in your mold. Thus, you and your mirror image are different. So are many of your corresponding molecules. For instance, one of the acids in your body is L-lactic acid, a substance in whose molecules four different atoms or groups, H, CH_3, OH, and CO_2H, are linked to a central carbon atom. L-Lactic acid molecules are not identical with their mirror images because they do not match them, atom for atom, in space. They bear to each other the same relationship that a right hand bears to a left hand. Such molecules are said to be **dissymmetric**—they *lack a plane or a point of symmetry*. The relationships are

Symmetry designation	Relationship to mirror image	Symmetry elements
Symmetric	Identical	Plane and/or point of symmetry present
Dissymmetric	Not identical	Plane and point of symmetry absent

Nonidentical mirror images are called **enantiomorphs;** the enantiomorph of L-lactic acid is D-lactic acid† (see Fig. 9.14). An example of enantiomorphism in the octahedral system is shown in Fig. 9.15.

Now we return to the question of recognition. Your mirror image, usurping your place at home, would be recognized as an impostor because his right-left configuration would be reversed. But this means that the individuals who do the recognizing must know the difference between right and left by reference to their own bodies; in other words,

* This discussion is restricted to events on a molecular level, not an atomic or subatomic one. For all we know, your mirror image may be made of antimatter, in which case his transformation to reality would be quite catastrophic.

† The symbols L and D denote **configuration,** like left-handed and right-handed.

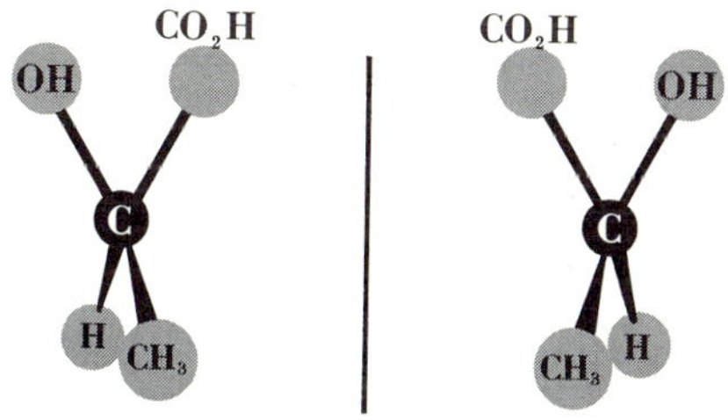

Fig. 9.14 Nonidentical mirror images. Enantiomorphism in the tetrahedral system.

they must be dissymmetric, as all of us in fact are. So it is with molecules. Hydrogen iodide, HI, reacts with L-lactic acid to remove the OH group and replace it with hydrogen.

$$CH_3CH(OH)CO_2H \xrightarrow{HI} CH_3CH_2CO_2H$$

HI is symmetric; it does not know right from left, or D-lactic acid from L-lactic acid. The reaction of HI with D-lactic acid is therefore the same in every respect as it is with L-lactic acid. But our digestive systems contain enzymes that are dissymmetric; they function normally in the digestion of foods that contain L-lactic acid, but would not do so for foods that contain D-lactic acid. As Alice said (Through the Looking-Glass, by Lewis Carroll), "Perhaps looking-glass milk isn't good to drink."

Considerations of molecular dissymmetry are thus of great importance in many chemical reactions, especially those involving living organisms (see Chapter 26).

PROBLEMS

1. Definitions Define and illustrate the following terms: (a) trigonal bipyramidal shape, (b) maximum orbital overlap, (c) plane of symmetry, (d) center of symmetry, (e) dissymmetry.

2. Valence electrons Identify the central atom and state how many bonding

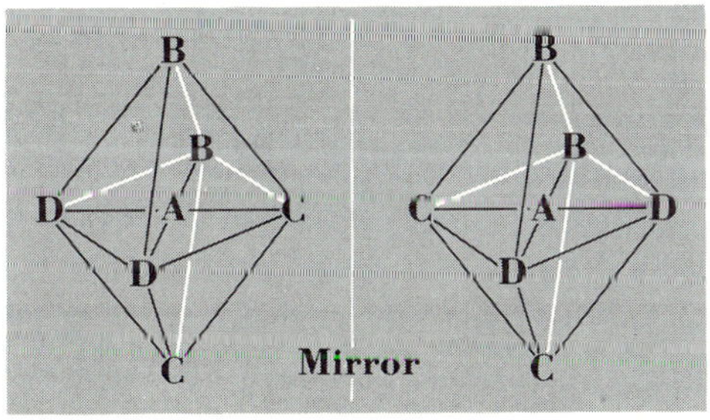

Fig. 9.15 Octahedral nonidentical mirror images.

and unshared valence electrons are associated with it, in each of the following compounds: (a) PBr_2F_3; (b) CCl_2F_2; (c) IF_7; (d) $:SeBrCl_3$; (e) $:AsH_3$.

3. Hybridization Arsenic pentafluoride, AsF_5, is a colorless gas (b.p. −53°C). Predict the probable hybrid state of valence electrons of As and the shape of the AsF_5 molecule.

4. Orbital structures Describe the orbitals (s, p, sp^2, etc.) of the central atom used for bond orbitals and lone pairs in (a) H_2Se (H—Se—H angle, 91 degrees); (b) CH_3OCH_3 (C—O—C angle, 110 degrees); (c) S_8 (S—S—S angle, 105 degrees); (d) $BeCl_4^{2-}$; (e) $(CH_3)_2Sn^{2+}$ (C—Sn—C angle, 180 degrees); (f) planar $(CH_3)_3Sn^+$.

5. Shapes of molecules and ions Draw the electronic structures and predict the orbital types and the shapes of the following ions and covalent molecules: (a) CdI_2; (b) CCl_4; (c) $BeBr_4^{2-}$; (d) $SeCl_2$; (e) SeF_4; (f) BrF_3; (g) AsF_6^-; (h) BrF_4^-; (i) ICl_2^-.

6. Shapes of multiple-bonded molecules Draw electronic structures and predict the shapes of the following molecules and ions: (a) nitrosyl chloride, ClN=O; (b) nitryl chloride, 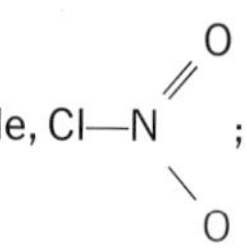; (c) acetylide ion, HC≡C:⁻; (d) ketene, $H_2C{=}C{=}O$.

7. Symmetry Indicate the symmetry planes and center (if any) in each of the following compounds or ions: (a) CBr_4; (b) $CHCl_3$; (c) CCl_2F_2; (d) $ClNH_2$; (e) ICl;

(f) 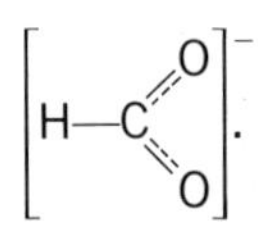.

8. Dissymmetry Consider the two substances L-lactic acid and D-lactic acid. Indicate whether they are alike or different according to each of the following criteria.

Criterion	Same or different?
(a) Oxidation with O_2	
(b) Oxidation with O_3	
(c) Reaction with D-amyl alcohol	
(d) Digestion by a dog	
(e) Solubility in water	
(f) Solubility in CCl_4	
(g) Solubility in L-2,4-dimethylheptane	
(h) Density	
(i) Recognition of a model of either the D- or L-molecule by a chemist	

ADDITIONAL PROBLEMS

9. Orbital structures and electron-pair repulsion Account for the fact that the H X—H bond angle contracts in the order $H_2S > H_2Se > H_2Te$.

10. Hybridization Sulfur tetrafluoride, SF_4, is toxic and highly reactive. The hexafluoride, SF_6, on the other hand, is nontoxic and quite inert. Predict the hybrid state of valence electrons of S in each compound. Can you suggest any reason for the marked difference in reactivity?

11. Orbital structures Show the distribution of electrons in the outer energy level of the central atom in (a) AlF_6^{3-}; (b) $SbCl_5$; (c) $O{=}PCl_3$; (d) $:ClO_3^-$.

12. Shape and hybrid orbitals (a) Draw the Lewis (electronic) structure for the covalent molecule $SnCl_2$. (b) Describe the types of orbital used by the Sn atom if the observed bond angle is 95 degrees.

13. Molecular shape The molecular structure of a spirane comprising two three-membered rings is shown. What is the angle formed between the planes of the two rings? Explain.

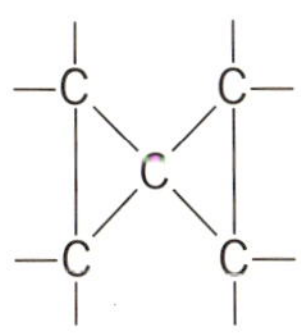

14. Symmetry There are two compounds with molecular formula $Pt(NH_3)_2Cl_2$. One of them has a center of symmetry in the molecule; the other does not. Draw reasonable structural formulas for the two compounds.

15. Symmetry There is evidence that the free radical $H_3C\cdot$ has four planes of symmetry, but the anion $H_3C:^-$ has only three. (a) Illustrate the molecular shape of the radical and of the ion. (b) What is the hybrid state of carbon in the radical? In the ion?

16. Symmetry and dissymmetry If a person who had perfect bilateral symmetry were provided with a compass that indicated only North, he could choose a direction between North and South, but not between East and West. Argue for or against this statement.

10 ■ INTERMOLECULAR FORCES

10.1 □ INTRODUCTION

The preceding chapters have been largely concerned with the structures of atoms and molecules. Molecular structure and shape are determined by the properties of the constituent atoms and by the forces (chemical bonds) that hold the atoms in a condition of chemical combination. But most chemical reactions involve events among molecules, and it is therefore important to consider the forces that molecules exert on one another, and the patterns of molecular aggregation into gases, liquids, solids, and other states of matter. In this chapter we shall discuss intermolecular forces, and in subsequent chapters the states of molecular aggregation.

10.2 □ DIPOLE-DIPOLE INTERACTION

We have seen that certain covalent bonds are polar. Because of the polarity of individual bonds, the entire molecule *may* have separated centers of positive and negative charge. Such a molecule constitutes a dipole. A crude illustration of a polar molecule is

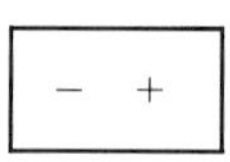

The dipole is symbolized by $\leftrightarrow$, where the arrow points toward the negative pole. Polar molecules possess a dipole moment, μ, which is the

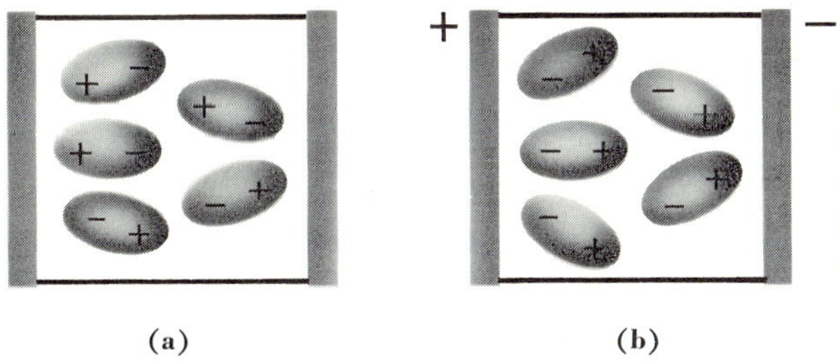

Fig. 10.1 Orientation of polar molecules in an electric field: (a) field off, (b) field on.

product of the magnitude of effective charge, q, and the distance, d, between the centers of opposite charge,

$$\mu = q \times d$$

The charge and distance are, respectively, of the order of 10^{-10} esu and 10^{-8} cm, so that the dipole moments are about 10^{-18} esu-cm/molecule. The unit 1×10^{-18} esu-cm/molecule is named the debye (D), in honor of Peter Debye, who pioneered in this field of study.

As a consequence of the dipole moment, polar molecules tend to be oriented in an electric field with the positive ends directed toward the negative electric plate and the negative ends toward the positive plate (Fig. 10.1). The orientations are far from perfect because of the randomness imposed on them by the kinetic energy of the molecules. Measurements based on these orientations are used to calculate dipole moments. In fact, it was the discovery that molecules were polar that led to the supposition that individual bonds are polar. When polar molecules attract each other they so orient themselves that the positive pole of one molecule is close to the negative pole of another molecule, as shown in Fig. 10.2. The force of this dipole-dipole attraction is inversely proportional to the distance between the molecules raised to the seventh power, $F \propto 1/r^7$. Consequently, because molecules in the gaseous state under normal pressures are far apart, their dipole-dipole attractions are very small. As the pressure on the gas increases, the molecules approach each other more closely, and, as the temperature decreases, so does the kinetic energy. The combination of these effects permits the dipole-dipole interaction to promote liquefaction and eventually solidification.

The extent of dipole-dipole interaction is one of the factors that determine the melting and boiling points of polar substances. Other factors such as molecular weight and molecular shape being equal, a

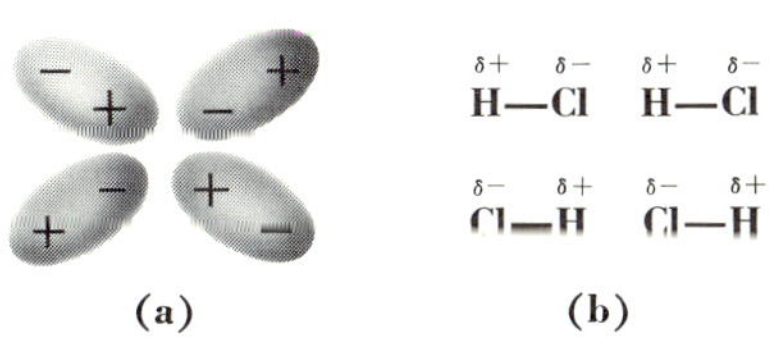

Fig. 10.2 (a) Dipole-dipole attraction; (b) exemplified by HCl.

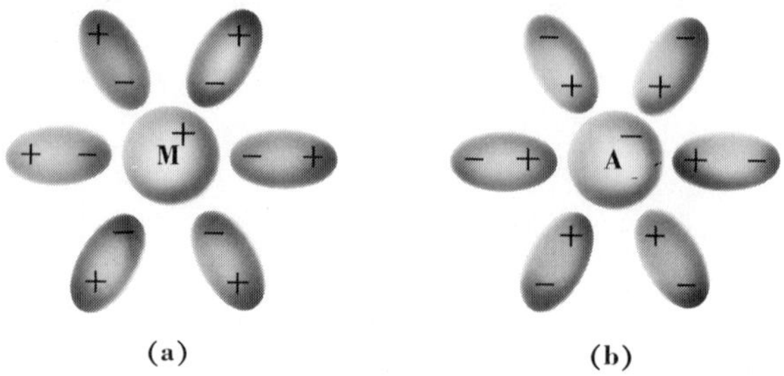

Fig. 10.3 Ion-dipole interaction: (a) cation; (b) anion.

substance with no dipole moment will have a lower boiling point and melting point than a polar molecule. Thus, the nonpolar molecules N_2 and O_2 have boiling points of $-196°C$ and $-183°C$, respectively, whereas the somewhat polar molecule NO ($\mu = 0.070$ D) boils at $-151°C$.

10.3 □ ION-DIPOLE ATTRACTIONS

Polar molecules are attracted to ions. The negative pole is attracted to cations; the positive pole is attracted to anions. This type of attraction, shown in Fig. 10.3, is called **ion-dipole** interaction. Ion-dipole interactions are involved in solution processes (Chapter 13).

10.4 □ HYDROGEN BONDING

The presence of an attractive force between certain molecules, which is greater than that expected from dipole-dipole interaction, is evidenced by examining the boiling points of binary covalent hydrides in Table 10.1. The group IV hydrides, CH_4, SiH_4, GeH_4, and SnH_4, are nonpolar and tetrahedral. As the atomic weight of the central atom increases, there is a regular increase in boiling point. In the absence of dipole-dipole interaction and major differences in shape, this direct relationship between boiling point and molecular weight is typical.

TABLE 10.1 BOILING POINTS OF BINARY COVALENT HYDRIDES: PERIODIC ARRANGEMENT

Period	IV A	V A	VI A	VII A
2	CH_4 112°K	NH_3 240°K	H_2O 373°K	HF 293°K
3	SiH_4 162°K	PH_3 185°K	H_2S 213°K	HCl 188°K
4	GeH_4 184°K	AsH_3 211°K	H_2Se 232°K	HBr 207°K
5	SnH_4 221°K	SbH_3 256°K	H_2Te 271°K	HI 238°K

(Groups; scale for each cell: 0°K, 100°K, 200°K, 300°K, 400°K)

Even in the case of the polar hydrides, those of group V, VI, and VII, this direct relationship prevails except for the hydrides of the first members of these groups. NH_3, H_2O, and HF are more polar than the other hydrides in their respective groups, because N, O, and F are the most electronegative elements. Nevertheless, their startlingly high boiling points indicate that a force stronger than a dipole-dipole interaction prevails for H_2O, HF, and NH_3.

These three polar molecules have one structural feature in common: each has at least one hydrogen atom covalently bonded to a highly electronegative atom with at least one unshared pair of electrons. Even when covalently bonded to H, the O, F, and N atoms attract the somewhat positive H atom of another molecule,

$$:\overset{\delta^-}{X}—\overset{\delta^+}{H}---:\overset{\delta^-}{X}—\overset{\delta^+}{H}$$

hydrogen bonding

The hydrogen atom increases the intermolecular attraction by bridging the two molecules. Since the very small hydrogen atom is the bridge between the two electronegative atoms, the interacting molecules can approach sufficiently close to each other to produce an attraction strong enough to be considered a *bond*, rather than just another dipole-dipole attraction. This unique type of bond is called the **hydrogen bond.** The energy* of the hydrogen bond, which varies from 3 to 10 kcal/mole, depends as a first approximation on the electronegativity of the X atom, and hence usually decreases in the order $F > O > N \approx Cl$.

One of the strongest hydrogen bonds is observed in hydrogen fluoride, whose formula should be written $(HF)_x$, since it exists as linear or cyclic aggregates, as shown in Fig. 10.4a,b.

The hydrogen difluoride ion, HF_2^-, possesses the strongest known hydrogen bond. It can be shown experimentally that the H atom is midway between the two F atoms; hence, the ion consists of two fluoride ions shielded from each other by a proton,

$$[F---H---F]^- \quad \text{or} \quad [F^- \; H^+ \; F^-]^-$$

structure of HF_2^-

If it is true that the hydrogen bond in $(HF)_x$ is stronger than the hydrogen bond in water, how can we account for the observation that water (b.p. 100°C, molecular weight 18) has a much higher boiling point than hydrogen fluoride (b.p. 19.4°C, molecular weight 20)? The answer comes from considering the geometry of the two hydrogen-bonded systems. In the case of hydrogen fluoride, any one F atom can be surrounded by only two hydrogen atoms, and so can participate in only one hydrogen bond. This limitation leads to linear (Fig. 10.4a) or

* Recall (p. 104) that the "energy of a bond" means the energy required to break the bond.

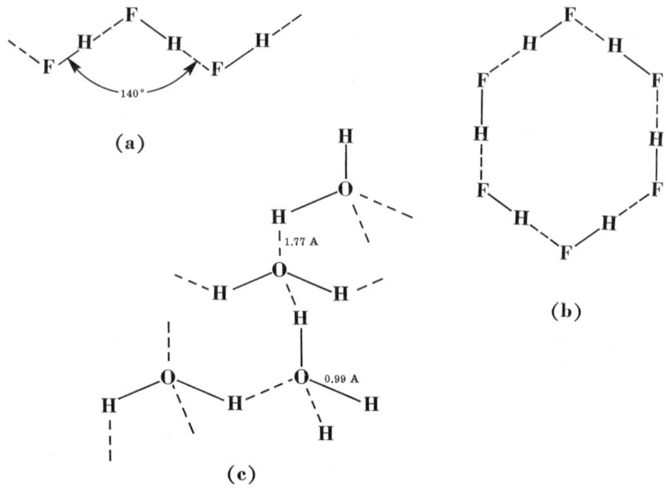

Fig. 10.4 Hydrogen-bonded structure for (a) H-F (linear); (b) $(HF)_6$ (cyclic); and (c) ice (cross-linked); broken lines indicate H bond.

cyclic (Fig. 10.4b) arrays. In ice, the oxygen atom is surrounded tetrahedrally by four hydrogen atoms, and so participates in two hydrogen bonds (Fig. 10.4c). The presence of two hydrogen bonds per H_2O molecule increases the attraction between the individual H_2O molecules. The unusually high melting point of 0°C for a molecule with a molecular weight of 18 g/mole is accounted for by the cross-linked nature of the hydrogen bonds. Since cross-linked hydrogen bonds persist to some extent in the liquid state, water also has an inordinately high boiling point. As ice melts, the hydrogen bonding becomes more random and the molecules at first are able to move closer together. The density of water increases, therefore, from 0°C to 4°C, at which temperature it is at the maximum. Above 4°C, the increase in kinetic energy of the molecules is sufficient to cause the molecules to begin to disperse, and the density steadily decreases with increasing temperature.

Hydrogen bonding also occurs between unlike molecules. For example, the very large solubility of ammonia, NH_3, and ammonium salts in water is principally due to this phenomenon. Recent studies have shown that hydrogen bonding is not restricted to strongly electronegative elements like, F, O, N, and Cl. Whenever a molecule with an H atom attached to an element more electronegative than hydrogen comes in contact with a substance possessing an atom with an unshared pair of electrons, hydrogen bonding can occur,

$$\overset{\delta^-}{\mathrm{A}}\text{—}\overset{\delta^+}{\mathrm{H}} + :\mathrm{B} \longrightarrow \overset{\delta^-}{\mathrm{A}}\text{—}\overset{\delta^+}{\mathrm{H}}\text{- - -}:\mathrm{B}$$

10.5 □ LONDON FORCES

All gases, *including the noble elements and nonpolar molecules* such as O_2, N_2, and F_2, can be liquefied. Apparently there is some attractive force even among these nonpolar molecules and atoms. Since these substances have very low boiling points, the attractive forces are relatively weak, certainly weaker in most cases than dipole-dipole, ion-dipole, or hydrogen bond forces. These weak forces are called **London forces** after Fritz London, who developed a theoretical explanation of them in 1928.

According to quantum mechanical theory, the electrons in a molecule may be regarded as in constant motion. A given molecular orbital represents an *average* charge density. Hence, at any instant, there may be an imbalance of charge distribution in the molecule. Thus, a nonpolar molecule may be momentarily *self-polarized* because of the unbalanced charge distribution. This polarized molecule induces a dipole moment in a neighboring molecule. *These induced dipoles then cause the nonpolar molecules to be mutually attracted.*

The data in Table 10.2 for noble gases and nonpolar molecules containing fewer than 20 electrons reveal some correlation between the boiling point and the number of electrons. The boiling point increases as the number of electrons increases. The London forces apparently are roughly proportional in strength to the number of electrons per molecule. It is also true that for more complicated series of nonpolar molecules, the boiling point increases with increasing molecular weight, assuming that the molecular structures are comparable. In the alkane series (Table 21.1), for example, as more carbon and hydrogen atoms, and consequently more electrons, are included in the molecule, the London forces become stronger and the boiling points increase.

In general, for similarly constituted molecules, as size increases, polarizability increases and boiling point increases.

10.6 □ LONDON FORCES AND MOLECULAR SHAPE

The correlation of boiling point with molecular weight is considerably influenced by the shape of the molecule. A comparison of the boiling points of *n*-pentane and neopentane, which have the same molecular

TABLE 10.2 BOILING POINTS OF NONPOLAR GASES RELATIVE TO NUMBER OF ELECTRONS PER MOLECULE

Substance	He	H_2	Ne	N_2	O_2	Ar	F_2	Kr	Xe
No. of electrons:	2	2	10	14	16	18	18	36	54
B.p. (°C):	269	−253	−246	−196	−183	−186	−186	−152	−107

n-pentane (a) neopentane (b)

Fig. 10.5 Representation of molecular shape of (a) *n*-pentane; (b) neopentane.

weight, reveals this dependence. Both molecules have the same kind and number of atoms and therefore the same number of electrons, yet the boiling points differ by 27°C. A molecule of *n*-pentane may be regarded as a zigzag chain (Fig. 10.5a), a molecule of neopentane as a sphere (Fig. 10.5b). For *n*-pentane, the approach between the two molecules can occur over the entire length of the chain, whereas for neopentane the approach can occur only at a tangential point. The lateral, side-to-side approach of two *n*-pentane molecules involves more contact, which increases the London forces, and therefore the boiling point of *n*-pentane is higher:

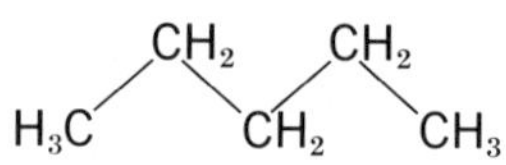

(b.p. 36.2°C)
n-pentane

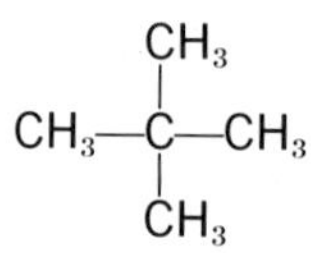

(b.p. 9.5°C)
neopentane

PROBLEMS

1. Definitions Define and illustrate the following terms: (a) dipole-dipole attraction; (b) ion-dipole attraction; (c) hydrogen bonding; (d) London forces.

2. Dipole moment Account for the following order of dipole moments observed for the hydrogen halides: HF, 1.9 D; HCl, 1.03 D; HBr, 0.74 D; HI, 0.38 D.

3. Hydrogen bonding Suggest a structure for H_3F_3, a component of liquid hydrogen fluoride.

4. Hydrogen bonding Account for the following facts. (a) Although ethyl alcohol, C_2H_5OH (b.p. 80°C), has a larger molecular weight than water (b.p. 100°C), it has a lower boiling point. (b) Salts of the HCl_2^- anion are known. (c) Mixing 50 ml each of water and ethyl alcohol gives a solution whose volume is less than 100 ml.

5. Intermolecular attraction Select the compound in each of the following pairs which you think has the higher boiling point: (a) O_2, oxygen, or H_2S, hydrogen sulfide; (b) Ar, argon, or Xe, xenon; (c) H_2S, or HCl, hydrogen chloride; (d) CH_3CH_2OH, ethanol, or CH_3OCH_3, dimethyl ether; (e) $CH_3CH_2CH_2CH_2OH$, *n*-butyl alcohol, or its isomer, 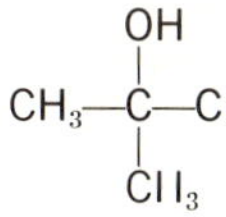, *t*-butyl alcohol; (f) 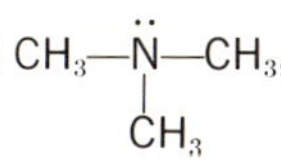, trimethylamine, or its isomer, $CH_3CH_2CH_2NH_2$, *n*-propyl amine. Explain your choice. Check your results in a handbook.

6. Hydrogen bonding Acetic acid, 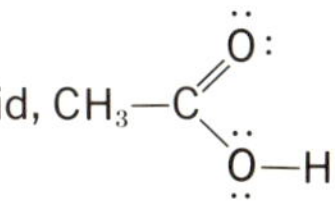, has a strong tendency to form dimers (structures made up of two individual molecules). Draw a likely structure for the dimer of acetic acid.

11 GASES

11.1 THE STATES OF AGGREGATION OF MOLECULES IN MATTER

In a sample of matter, any portion that is homogeneous, physically distinct, and separated from other parts of the sample by a definite boundary is called a **phase.** Phases of matter may be classified according to the condition of aggregation of their molecules. For the purpose of this classification, the molecules may be said (1) to be arranged (a) randomly, or (b) in an orderly manner, and (2) to attract each other (a) so little that they tend to disperse rapidly in space, or (b) enough so that they maintain a compact mass but do not maintain rigidity under the stress of gravity, or (c) so well that they maintain a rigid shape in spite of the stress imposed by gravity.

This classification yields six combinations, four of which exist in ordinary substances. These are listed and named in Table 11.1.

None of the distinctions given in Table 11.1 is absolute, but some are sharper than others. The distinction between liquid and glass, for example, depends only on assignment of degree of rigidity, and is often quite poorly defined. The randomness of molecular arrangements in liquids is far less than it is in gases.

11.2 THE PROPERTIES OF GASES

Gaseous materials are well known in our experience, and, under ordinary conditions, we have no trouble in differentiating them from solids, liquids, or glasses. A qualitative summary of the characteristic properties of gases under ordinary terrestrial environments is given in Table 11.2.

TABLE 11.1 STATES OF AGGREGATION OF MATTER

Predominant arrangement of molecules	Degree of cohesiveness	Common name of state
Random	Low (no well-defined boundary)	Gas
Random	Intermediate (well-defined boundary, but no rigidity)	Liquid
Random	High (rigid)	Glass (also called noncrystalline solid)
Orderly	Low	Does not ordinarily exist[a]
Orderly	Intermediate	Does not ordinarily exist[b]
Orderly	High	Crystalline solid

[a] Exception: solid helium near 0°K.
[b] Exception: "liquid crystals"—liquids that have some optical properties typical of crystals.

Quantitative experimental investigations of gases were begun in the seventeenth century. We shall discuss the findings of some of these experiments and then consider a general theory of gas behavior.

11.3 □ BOYLE'S LAW

Robert Boyle (1627-1691) investigated pressure—volume relationships of air by pouring successive quantities of mercury into the open arm of a J-shaped tube as shown in Fig. 11.1. After each addition, he measured (a) the volume of air in the lower section of the tube, and (b) the difference between the heights of the mercury columns in the two arms. The pressure on the trapped air is the difference between the mercury levels plus the atmospheric pressure (the height of mercury in a barometer).* Boyle's PV values were constant to within about 1% relative error.

TABLE 11.2 CHARACTERISTIC PROPERTIES OF GASES

Color	All gases are transparent; almost all are colorless; some exceptions: fluorine, chlorine (both green-yellow), bromine (red-brown), iodine (violet), nitrogen dioxide, dinitrogen trioxide (both red-brown)
Mobility	Compared with other forms of matter, gases show little resistance to flow; gases diffuse rapidly in space and through porous barriers
Response to changes of pressure, temperature	Compared with other forms of matter, gases expand greatly on heating or on reduction of applied pressure, or both

* Addition of barometer reading takes into account the atmospheric pressure on the right arm of the tube.

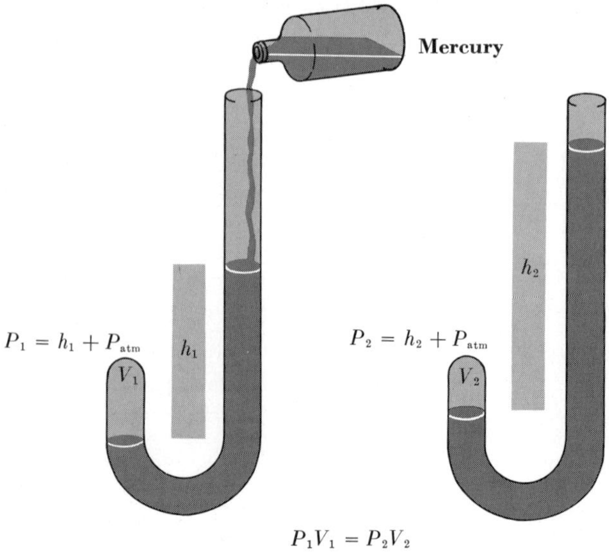

Fig. 11.1 Boyle's apparatus. P_{atm} refers to barometric pressure.

Boyle's law states that $PV = constant$, or $V \propto 1/P$.

Problems dealing with V-P relationships among gases can be solved by using the relationship derived from $PV = constant$ at constant temperature for a given quantity of gas,

$$P_2V_2 = P_1V_1 \tag{1}$$

Alternatively, we can convert the original volume to the new volume by applying as a correction factor the appropriate ratio of pressures. The form of the correction ratio is established by our "common sense" knowledge of gas behavior.

The term **"standard pressure,"** as used in calculations involving gases, means standard atmospheric pressure, or 760 torr.*

EXAMPLE 1 The pressure on 10.0 liters of gas at 760 torr is reduced to 700 torr at constant temperature. What will the new volume be?

ANSWER $V_1 = 10.0$ liters, $P_1 = 760$ torr, $P_2 = 700$ torr.

From Equation (1)

$$V_2 = V_1 \times \frac{P_1}{P_2}$$

$$= 10.0 \text{ liters} \times \frac{760 \text{ torr}}{700 \text{ torr}}$$

"Common sense" method

New volume = original volume × pressure correction

(the gas will expand because of pressure reduction, so correction is greater than unity)

$$= 10.0 \text{ liters} \times \frac{760 \text{ torr}}{700 \text{ torr}}$$

Therefore $V_2 = 10.9$ liters.

* See Appendix 1 for definitions of units.

11.4 □ THE LAW OF CHARLES AND GAY-LUSSAC; ABSOLUTE TEMPERATURE

Jacques Charles, a French physicist, measured the relationship between temperature* and volume of gases around 1787, but his data are not available. His work was confirmed in 1802 by his countryman, Joseph Gay-Lussac.

The results of such a study give a straight-line plot of gas volume versus temperature, such as appears in Fig. 11.2. This plot corresponds to Charles' law, $V \propto T$ or $V = constant \times T$.

The temperature scale designates three points: (1) the steam point, 100°C; (2) the ice point, 0°C; and (3) the temperature at the point where the extension of the straight line would reach zero gas volume. It is found experimentally that the distance x (between zero gas volume and the ice point) is close to 2.73 times as great as the distance y (between the ice point and the steam point), or

$$x = 2.73y \qquad (2)$$

But $y = 100°\text{C}$ minus 0°C, and $x = 0°\text{C}$ minus the temperature at zero gas volume. Substituting these values for x and y in Equation (2), and solving for the temperature at zero gas volume, we get

$$0°\text{C} - \text{temperature at zero gas volume} = 2.73(100°\text{C} - 0°\text{C})$$
$$\text{temperature at zero gas volume} = -273°\text{C}$$

If the molecules of the gas were points (occupying no volume), the lowest possible temperature would correspond to a total loss of volume.

Fig. 11.2 Volume-temperature relationship for a gas obtained from experiments at low constant pressure; $x = 2.73y$.

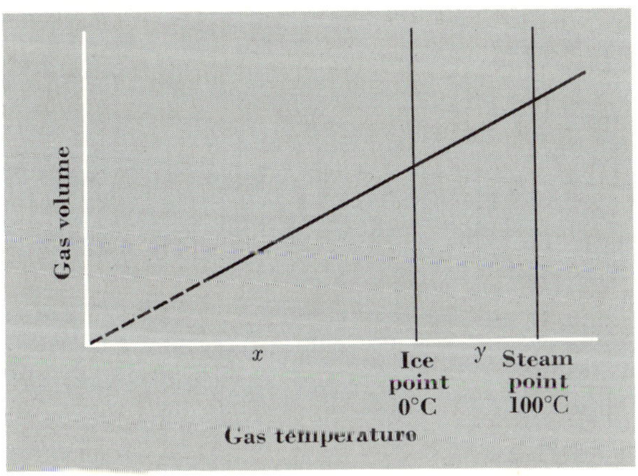

* Thermometers had been developed in Europe starting around 1600. Earliest Western concepts of gas expansion by heating probably originated in the mechanical toys of the Hellenic period, such as Hero's engine.

This temperature is called **"absolute zero"** because it is irrational to assume that any molecules might possess a negative volume. The best current value for absolute zero, obtained from volume-temperature studies at low pressures, is -273.15°C. This temperature is taken as the zero point of the **Kelvin*** **scale** of **absolute temperature,** T, and

$$°K = °C + 273.15 \tag{3}$$

Real gases do not in fact approach zero volume near 0°K. Even though the free space between molecules decreases as the temperature decreases, the molecules themselves occupy volume that is relatively independent of the temperature.

Some residual energy would be retained by molecules even at 0°K; this quantity is called the **zero-point energy.**

Charles' law can be used in a simple way to solve problems concerning V-T relationships of gases at constant pressure, If V_1, T_1 refer to one set of conditions and V_2, T_2 refer to another set of conditions of the same quantity of gas, then $V_1 = constant \times T_1$, and $V_2 = constant \times T_2$.

Dividing one equation by the other,

$$\frac{V_2}{V_1} = \frac{T_2}{T_1} \tag{4}$$

The "common sense" application of correction factors can also be used (see below).

The term **"standard temperature,"** as used in calculations involving gases, means 0°C. The term **"standard conditions"** (sc), means standard temperature and pressure, 0°C and 760 torr.

EXAMPLE 2 10.0 ml of gas at 20.0°C is cooled at constant pressure to the ice point. What will the new volume be?

ANSWER $V_1 = 10.0$ ml, $T_1 = (20 + 273)$°K, $T_2 = (0 + 273)$°K.

From Equation (4)	"Common sense" method
Solving for V_2,	New volume = original volume × temperature correction
$V_2 = V_1 \times \frac{T_2}{T_1}$	(because cooling reduces gas volume, correction factor must be less than unity)
$= 10.0 \text{ ml} \times \frac{273°K}{293°K}$	$= 10.0 \text{ ml} \times \frac{273°K}{293°K}$

Therefore $V_2 = 9.32$ ml.

* Named after Lord Kelvin (1824–1907), who suggested the scale.

11.5 □ AVOGADRO'S LAW

The hypothesis "equal volumes of all gases at the same temperature and pressure contain the same number of molecules" (p. 65) was vaguely conceived by scientists as early as about 1760; it was first clearly delineated by Amadeo Avogadro in 1811, and is therefore named in his honor. The concept is highly useful, since it teaches us that we may select equal numbers, N, of molecules simply by selecting equal gas volumes at constant temperature and pressure. The expression of Avogadro's law is $N \propto V$, or $N/V = constant$.

EXAMPLE 3 One gram of radium emits alpha particles at the rate of 11.6×10^{17} per year. Each α particle becomes a molecule of helium. The total volume of the 11.6×10^{17} molecules is 0.0430 ml at standard conditions. How many molecules are there in one liter of helium under standard conditions? In one liter of oxygen? In one liter of any gas?

$$\frac{N}{V} = \frac{11.6 \times 10^{17} \text{ molecules}}{0.043 \cancel{\text{ml}}} \times 1000 \frac{\cancel{\text{ml}}}{\text{liter}}$$
$$= 2.70 \times 10^{22} \frac{\text{molecules}}{\text{liter}}$$

Since N/V is a constant regardless of the nature of the gas, the ratio applies to helium, oxygen, or any other gas.

11.6 □ THE GAS LAW; THE MOLE; GAS DENSITIES

Boyle's, Charles', and Avogadro's laws, taken together, are expressed by the relationship $PV \propto NT$, or $PV = constant \times NT$. It is found, within the precision of Boyle's and Charles' law, that the constant is the same for all gases. To evaluate the constant, we must adopt standard values for any three of the variables; the value of the fourth variable will be fixed by this selection. The standard values are

P = one standard atmosphere.

T = 273.15°K (the ice point).

V = the volume occupied by 31.9988 g of oxygen at the standard temperature and pressure = 22.4136 liters.* (The rationale for the selection of this particular volume was explained in the discussion of molecular weights in Chapter 4.)

N = the number of molecules fixed by the selection of the above values for P, T, and V. Since the quantity is fixed it need not be measured, only named. It is called the **Avogadro number** of molecules, designated $\mathfrak{N}$, and the quantity of substance that contains this number

* This value is corrected for nonideality. It is the volume that the oxygen would have if it were an ideal gas. The measurements were actually made at a series of low pressures and extrapolated to zero pressure, where deviations from ideal behavior would disappear (see p. 191).

of molecules is called **one mole** (see p. 68). The designation for number of moles is n. The relationship between moles and molecules is

$$n(\text{moles}) \times \mathfrak{N}(\text{Avogadro number, molecules/mole}) = N(\text{molecules})$$

The ideal gas equation may now be expressed in terms of moles (n) and a constant, usually designated R:

$$PV = nRT \qquad (5)$$

To evaluate R,

$$\begin{aligned} R &= PV/nT \\ &= \frac{1.0000 \text{ atm} \times 22.4136 \text{ liters}}{1.0000 \text{ mole} \times 273.15°\text{K}} \\ &= 0.082056 \text{ liter atm/°K mole} \end{aligned}$$

PV has the dimensions of energy (see p. 97); therefore we can substitute other energy units for "liter atmospheres" and compute other expressions of the gas constant. Some of these appear in Table 11.3. Try Problem 22, p. 193, to practice the necessary calculations.

It will now be possible to execute calculations on a broader basis than is permitted by the two-variable laws described in the preceding sections. There are three approaches:

(a) Use $PV = nRT$. Conditions *must* be expressed in the units consistent with those used for R.

(b) Refer to two sets of conditions,

$$\begin{aligned} P_2V_2 &= n_2RT_2 \\ P_1V_1 &= n_1RT_1 \end{aligned}$$

Therefore

$$\frac{P_2V_2}{P_1V_1} = \frac{n_2T_2}{n_1T_1} \qquad (6)$$

(c) Use the "common sense" method. Correction factors for pressure and temperature are applied as for Boyle's and Charles' law calculations.

Methods (b) and (c) make it unnecessary to memorize the value of R.

The total mass of a given sample of a gas, w, is determined by the weight of each molecule of the gas, m, and the number of molecules,

TABLE 11.3 ▪ VALUES OF THE GAS CONSTANT

0.0821	liter atm/°K mole
62.4	liter torr/°K mole
8.31	joule/°K mole
1.986	cal/°K mole

N, or moles, n, present. Therefore, n and w are directly proportional.

$$w(\text{grams}) = n(\cancel{\text{moles}}) \times \mathfrak{N}\left(\frac{\cancel{\text{molecules}}}{\cancel{\text{mole}}}\right) m\left(\frac{\text{grams}}{\cancel{\text{molecule}}}\right)$$

Since $\mathfrak{N}$ and, for a given gas, m are constant,

$$w \propto n$$

Then

$$PV = constant \times wT$$

or

$$P = constant \times \frac{w}{V}T$$

But w/V is the density d of the gas. Therefore

$$P = constant \times dT$$

Referring to two sets of conditions for the same gas and dropping the constant, we have

$$\frac{P_2}{P_1} = \frac{d_2}{d_1}\frac{T_2}{T_1} \tag{7}$$

EXAMPLE 4 A gas occupies 10.0 liters at standard conditions. What volume will it occupy at 20.0°C and 700 torr?

ANSWER $V_1 = 10.0$ liters, $P_1 = 760$ torr, $T_1 = 273°\text{K}$, $P_2 = 700$ torr, $T_2 = 293°\text{K}$.

From Equation 6

$$V_2 = V_1 \times \frac{n_2}{n_1} \times \frac{T_2}{T_1} \times \frac{P_1}{P_2}$$

(But $n_1 = n_2$)

$$V_2 = 10.0 \text{ liters} \times \frac{293°\cancel{\text{K}}}{273°\cancel{\text{K}}} \times \frac{760 \cancel{\text{torr}}}{700 \cancel{\text{torr}}}$$

"Common sense" method

New volume = original volume × temperature correction × pressure correction
(temperature correction > 1, pressure correction > 1)

$$= 10.0 \text{ liters} \times \frac{293°\cancel{\text{K}}}{273°\cancel{\text{K}}} \times \frac{760 \cancel{\text{torr}}}{700 \cancel{\text{torr}}}$$

Therefore $V_2 = 11.7$ liters.

EXAMPLE 5 To what temperature must 1.00×10^{-4} mole of a gas be heated in a 25.0-liter container to obtain a pressure 1.00×10^{-1} torr?

ANSWER This problem is concerned with a single set of conditions, not with ratios between different sets. It will therefore be best to use Equation

5 directly. Solving for T,

$$T = \frac{PV}{nR}$$

$$= \frac{1.00 \times 10^{-1}\ \cancel{\text{torr}} \times 25.0\ \cancel{\text{liters}}}{760\ \frac{\cancel{\text{torr}}}{\cancel{\text{atm}}} \times 1.00 \times 10^{-4}\ \cancel{\text{mole}} \times 0.0821\ \frac{\cancel{\text{liter atm}}}{\cancel{\text{mole}}\ \text{deg}}}$$

$$= 401°\text{K, or } 128°\text{C.}$$

EXAMPLE 6 Using the data of Example 3 (p. 179), calculate the Avogadro number.

ANSWER The Avogadro number is the number of molecules in 22.414 liters, or one mole, of a gas. From Example 3, this is

$$2.70 \times 10^{22}\ \frac{\text{molecules}}{\cancel{\text{liter}}} \times 22.4\ \frac{\cancel{\text{liters}}}{\text{mole}} = 6.05 \times 10^{23}\ \text{molecules/mole}$$

The Avogadro number has been determined by several methods in addition to the one described by Example 3, and the best currently accepted value is $(6.0225 \pm 0.0003) \times 10^{23}$ molecules/mole.

EXAMPLE 7 10.0 g of a gas occupies 50.0 liters at 25.0°C and 600 torr. What is the density of the gas at standard conditions?

ANSWER $P_1 = 600$ torr, $P_2 = 760$ torr, $T_1 = 298°$K, $T_2 = 273°$K, $d_1 = \frac{10.0\ \text{g}}{50.0\ \text{liters}}$.

From Equation (7)

$$d_2 = d_1 \times \frac{P_2}{P_1} \times \frac{T_1}{T_2}$$

$$= \frac{10.0\ \text{g}}{50.0\ \text{liters}} \times \frac{760}{600} \times \frac{298}{273}$$

$$= 0.277\ \text{g/liter}$$

"Common sense" method

Solving for new volume at sc,

$$\text{volume}_{sc} = 50.0\ \text{liters} \times \frac{273\cancel{°\text{K}}}{298\cancel{°\text{K}}} \times \frac{600\ \cancel{\text{torr}}}{760\ \cancel{\text{torr}}}$$

$$= 36.1\ \text{liters}$$

$$\text{density}_{sc} = \frac{\text{wt.}}{\text{vol.}_{sc}} = \frac{10.0\ \text{g}}{36.1\ \text{liters}}$$

$$= 0.277\ \text{g/liter}$$

11.7 □ DALTON'S LAW OF PARTIAL PRESSURES

It follows from $PV = nRT$ that in a given volume of gas at constant temperature (V, T constant), the pressure of the gas depends only on the number of moles,

$$P = constant \times n$$

The gas constant applies to all gases, as well as to mixtures of noninteracting gases, like air. Let us consider the total pressure, P_T, of a mixture of gases n_T, that consists of n_1 moles of gas$_1$, n_2 moles of gas$_2$, and so on. Then

$$P_T = constant \times n_T$$
$$= constant \times (n_1 + n_2 + \cdot\cdot\cdot) \qquad (8)$$

Let us now separate the gases into individual containers,* each of which has the original volume. The pressure exerted in each of the new containers is called the **"partial pressure,"** $p_1, p_2, \ldots$, which is defined simply as the pressure that one component of a gas mixture would exert if it occupied the same volume alone. Then, $p_1 = constant \times n_1$, $p_2 = constant \times n_2$, and so on, and, by addition,

$$p_1 + p_2 + \cdots = constant \times (n_1 + n_2 + \cdots) \tag{9}$$

By combining Equations (8) and (9) we have

$$P_T = p_1 + p_2 + \cdots \tag{10}$$

Equation (10) is Dalton's law of partial pressures (1801), which states that, in a mixture of gases, the total pressure equals the sum of the partial pressures.

This relationship is important in measurements of gases confined by volatile liquids. The evaporation of the liquid contributes a new component to the gas and makes up a part of its total pressure in accordance with Dalton's law. Fortunately for measurement purposes, the pressure contribution from such evaporation at constant temperature reaches a constant value called the **vapor pressure of the liquid.** In laboratory practice, the confining liquid is often water, for which vapor pressures are found in Appendix 5.

EXAMPLE 8 The composition of dry air by volume is 78.1% nitrogen, 20.9% oxygen, and 1.0% of other gases. Calculate the partial pressures, in atmospheres, in a tank of dry air compressed to 10.0 atm.

ANSWER According to the data, if the nitrogen were removed to a separate container and compressed to 10.0 atm it would occupy 78.1% of the original volume. If the volume were then expanded to 100% of the original volume, the pressure would drop to 78.1% of the original pressure, or (78.1/100) × 10.0 atm, or 7.81 atm. This is the partial pressure of nitrogen in the original container. By Dalton's law, the sum of the partial pressures is the total pressure. In short,

partial pressure = proportion by volume × total pressure

$$p_{\text{nitrogen}} = \frac{78.1}{100} \times 10.0 \text{ atm} = 7.81 \text{ atm}$$

$$p_{\text{oxygen}} = \frac{20.9}{100} \times 10.0 \text{ atm} = 2.09 \text{ atm}$$

$$p_{\text{other gases}} = \frac{1.00}{100} \times 10.0 \text{ atm} = 0.10 \text{ atm}$$

EXAMPLE 9 10.0 ml of oxygen is collected over water at 20°C and 770 torr. What is the volume of the dry gas at standard conditions? The vapor pressure of water at 20°C is 17.5 torr.

* This experiment need not be performed. It is sufficient just to think of it. Werner Heisenberg called such efforts "*Gedanken*" (imaginary) experiments.

ANSWER $V_1 = 10.0$ ml, $T_1 = 293°K$. To find P_1 use the partial pressure law, $P_T = p_{oxygen} + p_{water}$. Therefore

$$P_{oxygen} = P_T - p_{water} = 770 \text{ torr} - 17.5 \text{ torr} = 753 \text{ torr}$$

P_2 and T_2 are the standard conditions (sc), 760 torr and 273°K. Then ("common sense")

$$V_2 = 10.0 \text{ ml} \times \frac{273°K}{293°K} \times \frac{753 \text{ torr}}{760 \text{ torr}}$$
$$= 9.23 \text{ ml}$$

11.8 □ THE PRESSURE OF A GAS CONFINED BY A LIQUID

A quantity of gas confined by a liquid* can be made to undergo a change of pressure simply by raising or lowering the liquid level. Such changes commonly occur in laboratory manipulations of gases in **eudiometers** (inverted graduated gas-collecting tubes, such as are pictured in Fig. 11.3).

To find the total pressure of a confined gas, note the level of the liquid with which it is in contact. If the level inside the tube equals the level outside (P_2 in Fig. 11.3), then the pressures are also equal, and the confined gas is therefore under barometric pressure. If the level inside is lower than that of the unconfined liquid, the pressure of the confined gas (P_1 in Fig. 11.3) must be greater than barometric pressure,

$$P_1 = \text{barometric pressure} + \text{pressure needed to depress confining liquid to depth } h_1$$

If the confined liquid level is elevated, then the pressure (P_3 in Fig. 11.3) of the gas must be less than barometric pressure,

$$P_3 = \text{barometric pressure} - \text{pressure needed to support confining liquid to height } h_3$$

This condition can be thought of as a rarefied gas that sucks the liquid up in the tube. The general relationship is

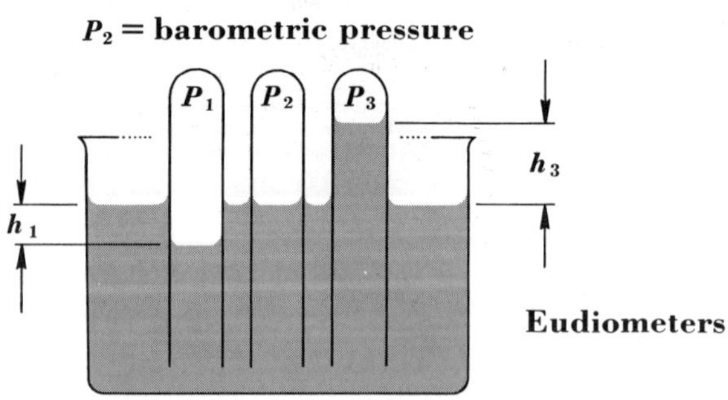

Fig. 11.3 Gas confined by a liquid.

* The vapor pressures of solids are usually too small to influence gas pressures significantly. The subject is treated in more detail in the next chapter.

$$\text{pressure of confined gas } (P) = \text{barometric pressure} \begin{cases} + \text{ pressure needed to depress liquid column} \\ \quad or \\ - \text{ pressure needed to support elevated liquid column} \end{cases} \tag{11}$$

The pressure that corresponds to the height of a liquid column is

$$\text{pressure} \left(\frac{\text{g}}{\text{cm sec}^2}\right) \text{ or } \left(\frac{\text{dynes}}{\text{cm}^2}\right) = \text{height (cm)} \times \text{density} \left(\frac{\text{g}}{\text{cm}^3}\right) \times \begin{matrix}\text{acceleration}\\ \text{of gravity}\end{matrix} \left(\frac{\text{cm}}{\text{sec}^2}\right)$$

Let us consider two liquids, densities d_1 and d_2, the heights of whose columns are h_1 and h_2, exerting pressures P_1 and P_2 respectively, under the influence of gravity, g. Then

$$P_1 = h_1 d_1 g$$

and

$$P_2 = h_2 d_2 g$$

Dividing,

$$\frac{P_1}{P_2} = \frac{h_1 d_1}{h_2 d_2} \tag{12}$$

Now, consider any column of liquid of height h_1 mm and density d_1 g/ml, that exerts a pressure P_1. We balance this pressure exactly with a column of mercury at 0°C, of height h_2 mm. The density of mercury at 0°C is 13.5951 g/ml. Since the two columns are balanced, their pressures are equal, and $P_1 = P_2$, the pressure of the mercury column. Solving Equation 12 for h_2 yields

$$h_2 = h_1 \frac{d_1}{d_2}$$

$$h_2 \text{ (mm)} = h_1 \text{ (mm)} \times \frac{d_1 \text{ (g/ml)}}{13.5951 \text{ g/ml}}$$

But h_2 (mm), which is the height of the column of mercury at 0°C in millimeters, is numerically equal to the pressure of the liquid column (P_1 or P_2) expressed in torr units (see App. 1). Under these conditions, we can use the fraction $\dfrac{1 \text{ torr}}{1 \text{ mm}}$ as a conversion factor. Thus,

$$\left(\begin{matrix}\text{pressure of a liquid}\\ \text{column (torr)}\end{matrix}\right) = \left(\begin{matrix}\text{height of liquid}\\ \text{column (mm)}\end{matrix}\right) \times \frac{\text{density of liquid (g/ml)}}{13.5951 \text{ g/ml}} \times \left(\frac{\text{torr}}{\text{mm}}\right)$$

Substituting the above expression for pressure of a liquid column in

Equation 11, we obtain

$$\begin{pmatrix}\text{pressure of}\\ \text{confined gas}\\ \text{(torr)}\end{pmatrix} = \begin{pmatrix}\text{barometric}\\ \text{pressure}\\ \text{(torr)}\end{pmatrix} \pm \left(\begin{matrix}\text{depth or height of}\\ \text{confining liquid}\\ \text{column (mm)}\end{matrix} \times \frac{\text{density of confining liquid (g/ml)}}{13.5951\text{ g/ml}}\right)\frac{\text{torr}}{\text{mm}} \tag{13}$$

EXAMPLE 10 Referring to Fig. 11.3, the confining liquid is water at 20°C, h_1 is 6.31 cm below the liquid level in the vessel, and h_3 is 15.00 cm above the liquid level. The barometer reading is 767.5 torr. Density of water at 20°C is 0.9982 g/ml. What are the values of P_1, P_2, and P_3?

ANSWER From Equation 13,

$$\begin{aligned} P_1 &= 767.5\text{ torr} + \left(63.1\text{ mm} \times \frac{0.9982\text{ g/ml}}{13.60\text{ g/ml}}\right)\left(\frac{\text{torr}}{\text{mm}}\right)\\ &= 772.1\text{ torr}\\ P_2 &= 767.5\text{ torr}\\ P_3 &= 767.5\text{ torr} - \left(150.0\text{ mm} \times \frac{0.9982\text{ g/ml}}{13.60\text{ g/ml}}\right)\left(\frac{\text{torr}}{\text{mm}}\right)\\ &= 756.5\text{ torr} \end{aligned}$$

EXAMPLE 11 Referring to Fig. 11.3, if oxygen is collected over water at 20°C in the three tubes, what is the partial pressure of the oxygen in each tube?

ANSWER From Dalton's law,

$$P = p_{\text{oxygen}} + p_{\text{water vapor}}$$

and, from Appendix 5, the vapor pressure of water at 20°C is 17.5 torr. Therefore

$$\begin{aligned} p_{\text{oxygen}} &= P - 17.5\text{ torr}\\ &= 772.1\text{ torr} - 17.5\text{ torr} = 754.6\text{ torr (first tube)}\\ &= 767.5\text{ torr} - 17.5\text{ torr} = 750.0\text{ torr (second tube)}\\ &= 756.5\text{ torr} - 17.5\text{ torr} = 739.0\text{ torr (third tube)} \end{aligned}$$

11.9 □ GAS VOLUME CORRECTIONS IN STOICHIOMETRY

One mole of a gas occupies (ideally) 22.4 liters at standard conditions. The volume of gases that are involved in calculations from equations (Chapter 5) are derived from the numbers of moles consumed or produced in the reaction, and hence such volumes must refer to standard conditions. Gas volumes at other conditions must be corrected to standard conditions. The rules are listed below. Remember that sc means standard conditions and cp means the conditions of the problem.

Rule 1 When the gas volume is *given*, convert it from cp to sc before starting the stoichiometric calculation.

$$\text{gas volume (sc)} = \text{gas volume (cp)} \times \frac{273°\text{K} \times \text{partial pressure (cp), torr}}{\text{temperature (cp), °K} \times 760 \text{ torr}}$$
$$= 0.359 \frac{°\text{K}}{\text{torr}} \times \frac{\text{gas volume (cp)} \times \text{partial pressure (cp), torr}}{\text{temperature (cp), °K}}$$

Rule 2 A gas volume that is calculated from a chemical equation refers to sc. It must then be converted to cp.

$$\text{gas volume (cp)} = \text{gas volume (sc)} \times \frac{\text{temperature (cp), °K} \times 760 \text{ torr}}{273°\text{K} \times \text{partial pressure (cp), torr}}$$
$$= 2.78 \frac{\text{torr}}{°\text{K}} \times \frac{\text{gas volume (sc)} \times \text{temperature (cp), °K}}{\text{partial pressure (cp), torr}}$$

EXAMPLE 12 What volume of Cl_2 in liters at 27°C and 810 torr is required to produce 100 g of C_2Cl_4? (see Example 1b, p. 85).

ANSWER The gas volume obtained from the stoichiometric calculation (p. 85) is 40.5 liters (sc) of Cl_2. Therefore (Rule 2)

$$\text{volume (cp)} = 2.78 \frac{\text{torr}}{°\text{K}} \times \frac{40.5 \text{ liters} \times (273 + 27)°\text{K}}{810 \text{ torr}}$$
$$= 41.8 \text{ liters}$$

EXAMPLE 13 What weight in grams of chlorine will react with 75.0 liters of C_2H_2 measured at 24.0°C and 790 torr? (See Example 1c, p. 86.)

ANSWER We must first convert the volume to sc (Rule 1):

$$\text{volume (sc)} = 0.359 \frac{°\text{K}}{\text{torr}} \times \frac{75.0 \text{ liters} \times 790 \text{ torr}}{(273 + 24.0)°\text{K}}$$
$$= 71.7 \text{ liters (sc)}$$

When we use this value in the calculation of Example 1c (p. 86), we obtain the answer 682 g Cl_2.

11.10 □ IDEAL GASES. THE KINETIC MOLECULAR HYPOTHESIS

The observed behavior of gases is fairly well approximated by a model that we call the **ideal gas.**

A model is a representation of reality, and a good model is a successful compromise between (a) simplicity and (b) faithful representation of the phenomenon to be studied.

The ideal gas consists of molecules that (a) are points in space, (b) have mass and velocity, (c) exhibit no attractive or repulsive forces among themselves or with other matter, and (d) undergo collisions with no net loss of kinetic energy.* These assumptions, taken together, are called the **kinetic molecular hypothesis.**

The ideal gas, when unconfined, will rapidly disperse into space. In a rigid container, however, the molecules will bounce off the inside walls and at all times be distributed in some arrangement within the volume of confinement. Any one molecule will move at constant speed in a straight line until it suffers a collision; the molecule will then have a new speed and direction until the next collision, and so on. Considering all the molecules in the container at any one time, some will therefore have abnormally high speeds, others will be relatively slow, and most will be moving at a rate close to some average value. The distribution of speeds is not symmetrically arranged around the average value, because there is a lower limit (at rest) but no upper one.

Figure 11.4 shows the expected shape of a curve of distribution of molecular speeds in an ideal gas. If the container is insulated from the outside, no energy enters or leaves the gas and the average speed of the molecules remains constant.

The **pressure** of the gas is the force that it exerts per unit area of its container (Fig. 11.5). This force is applied by the collisions of the molecules with the walls. Each molecule, as it travels back and forth across the container, hits the wall repeatedly. Because of the very small mass of the molecule, however, we cannot detect the force exerted during each collision. All that matters is the average force over a period of time during which there are many collisions. The average force exerted by

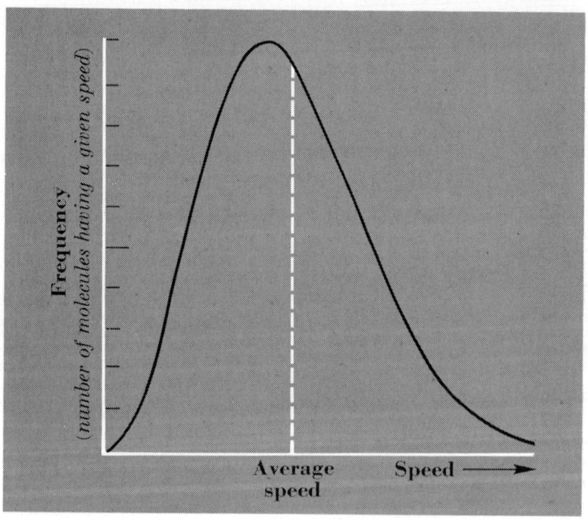

Fig. 11.4 Distribution of molecular speeds.

* This model has some inconsistency in that it describes collisions between points, which would be events of infinitesimal probability.

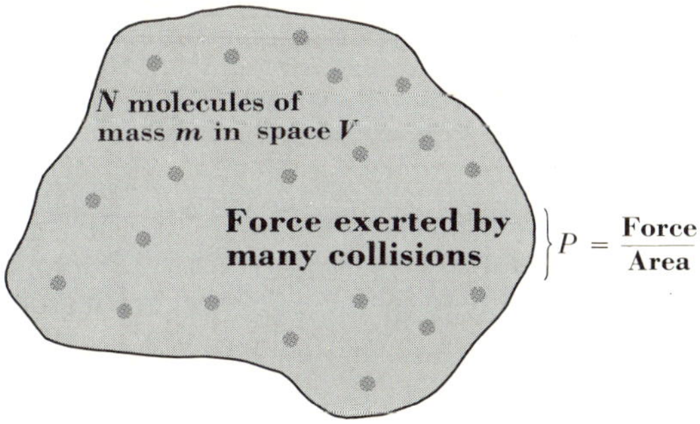

Fig. 11.5 Gas molecules confined in a container.

all the collisions of molecules with the walls of the container, divided by the inside wall area, is the pressure of the gas.

Explanation of Boyle's law For a given number of molecules with a constant average velocity, the gas pressure, which depends on the frequency of molecular collisions per unit area of the container, is inversely proportional to the average distance that a molecule must travel to reach the wall, and to the wall area that must be hit. But the product of distance times area is directly proportional to the volume of the container. Hence the pressure is inversely proportional to the volume (Boyle's law).

Explanation of Charles' law The kinetic energy of any moving particle, such as a molecule, depends on its mass and its velocity (see Appendix 1). When the average energy of gas molecules increases, they therefore travel faster and collide with the walls of the container more frequently. The effect of more frequent collisions is thus to increase the pressure in a container of fixed volume. If a constant pressure is to be maintained, the volume of the container must be increased (Boyle's law). What property of a gas depends on the kinetic energy of its molecules? Both theory and experiment lead us to believe that this property is the absolute temperature, and, indeed, that the kinetic energy is proportional to the absolute temperature. Thus, at constant pressure, the volume occupied by a fixed number of molecules in the gaseous state is directly proportional to the absolute temperature (Charles' law).

The student is invited to rationalize Avogadro's law (see Problem 11, p. 192).

11.11 ▫ DEVIATIONS FROM IDEAL BEHAVIOR

Real gases do not behave in an ideal way. The important differences between ideal and real gases, and the consequences of these differences, are given in the following tabulation.

Ideal gases	Real gases	Difference
Molecules exert no forces on each other	Molecules of gases do exert forces of attraction among themselves and to other matter. A result of such mutual attractions is the formation, at any time, of loosely bonded aggregates of two or more gas molecules. The total number of molecules, and hence of moles, is, in effect, reduced by such aggregation. The pressure exerted by the gas is therefore reduced ($P \propto N$, see discussion of Dalton's law).	The pressure exerted by a given quantity of a real gas in a rigid container at a definite temperature is less than that predicted by the ideal gas law. Therefore, since $(PV)_{\text{ideal}} = nRT$, or $\frac{(PV)_{\text{ideal}}}{nRT} = 1$ and P_{real} is less than P_{ideal}, $\frac{(PV)_{\text{real}}}{nRT} < 1$
Molecules are points that have no volume	Molecules occupy volume that is relatively independent of P and T	The volume occupied by a given quantity of a real gas at a definite T and P is greater than that predicted by the ideal gas law. Therefore, since $V_{\text{real}} > V_{\text{ideal}}$, $\frac{(PV)_{\text{real}}}{nRT} > 1$

Fig. 11.6 **Deviations from ideal gas behavior.**

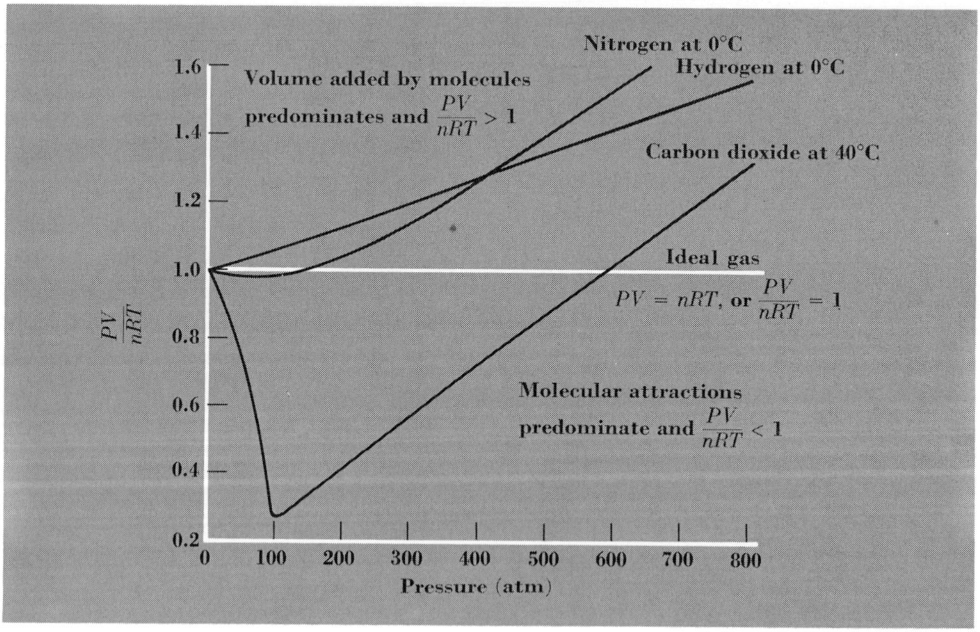

Deviations from ideal behavior are shown in Fig. 11.6. For hydrogen, the molecular attractive forces are so low that they do not depress PV below the ideal values. Instead, the volume added by the molecules yields PV products exceeding the ideal values at all pressures shown on the graph. For nitrogen, the attractive forces are great enough to yield negative PV deviations up to about 150 atm. Intermolecular attractions in carbon dioxide, even at 40°C, are obviously much more important than those in nitrogen at 0°C. For either nitrogen or carbon dioxide a pressure exists (150 atm or 600 atm, respectively) at which the two effects cancel each other and the PV product is ideal.

At low pressures, the volume of a given quantity of gas becomes large, and the volume occupied by the molecules becomes relatively small. Since the molecules become more widely separated at low pressures, their mutual attractions and tendencies to aggregate lessen. At high temperatures, the molecules have greater kinetic energy and their tendencies to aggregate lessen. Thus, high temperatures and low pressures both favor ideal gas behavior.

At one atmosphere, deviations from ideality are of the order of 1% for most gases.

PROBLEMS

1. Definitions Discuss the inadequacies and the limitations of the following definitions. Supply better ones. (a) A *liquid* is matter that takes the shape of its container. (Is this true of a liquid in a vehicle in interplanetary space?) (b) A *gas* is matter that is low in density, transparent, and expands when heated.

2. Properties of gases State whether each of the following samples of matter is or is not a gas, or whether the information is insufficient for deciding: (a) A material is in a steel tank under 100 atm pressure. When the tank is opened to the atmosphere, the material suddenly expands, increasing its volume by 10%. (b) A material, on being emitted from an industrial smokestack, rises about 25 ft into the air; viewed against a clear sky, it has a dense, white appearance. (c) 1 cm^3 of a material weighs 8.2 g. (d) When a material is released from a point 30 ft below the level of a lake, it rises rapidly to the surface, at the same time doubling its volume. (e) A material is transparent and pale green in color. (f) One cubic foot of a material contains as many molecules as one cubic foot of air at the same temperature and pressure.

3. Boyle's law A sample of neon occupies 250 ml at 1.00 standard atmosphere. Assuming that the temperature remains constant, what volume will the neon occupy (a) at 4.0 atm; (b) at 200 torr; (c) at 3.00×10^{-3} torr?

4. Kelvin scale Convert the following to the Kelvin scale: (a) −49°C; (b) −269.9°C; (c) 0.000°C; (d) 3.8×10^{-3}°C.

5. Charles' and Gay-Lussac's law The volume of gas in a cylinder at 100°C is 1.34 liters. What will its volume be if the gas temperature rises to 300°C with no change in pressure?

6. Avogadro's law A 25 ml container holds 1.2×10^{18} molecules of oxygen at a certain temperature and pressure. (a) How many molecules would remain if the container volume were compressed to 5.0 ml? (b) What volume in liters is needed to hold 6.0×10^{23} molecules at the original temperature and pressure? (c) What volume in milliliters would be occupied by 1.2×10^{8} molecules at the original temperature and pressure?

7. Ideal gas law A sample of gas occupies 0.045 ml at standard conditions. What will its volume be at 373°K and 740 torr?

8. Ideal gas law 100 ml of oxygen is collected over water at 20°C and 760 torr. Calculate the volume of the dry gas at standard conditions.

9. Dalton's law A rigid container is filled with helium at 1.20 atm. Air (approximate composition 80% N_2 and 20% O_2 by number of molecules) is pumped into the tank until the total pressure is 10.40 atm. Calculate the partial pressure of each gas in the tank.

10. Eudiometer (a) A gas is confined in a eudiometer over mercury at 0.00°C. The mercury level inside is 4.00 cm below the level outside. The barometer reads 759.1 torr. What is the pressure of the confined gas? (b) If the confining liquid is water at 4.00°C (density 1.00 g/ml) under the conditions described above, what is the pressure of the confined gas?

11. Kinetic-molecular hypothesis Offer an explanation for Avogadro's law in terms of the kinetic-molecular hypothesis.

ADDITIONAL PROBLEMS

12. Properties of gases Offer an explanation for the fact that all gases are transparent.

13. Properties of gases Suppose you were asked to supply a stated mass of a specific gas in a given rigid container at a specified pressure and temperature. Would it be likely that you could fulfill the request? Explain.

14. Boyle's law A 2.00 liter flask contains air at 25°C and 765 torr. 125 ml of mercury is injected into the flask without any transfer of air or change in temperature. What is the new pressure in the flask?

15. Boyle's law Assume that the value of a constant k in the equation $PV = k$ is 12. Plot the graph of P (abscissa) versus V (ordinate) for the values $V = 1$, 2, 3, 4, 6, and 12. What is the shape of the curve? Plot the graph of $1/P$ (abscissa) versus V (ordinate) for the same values of V. What plot is obtained?

16. Charles' and Gay-Lussac's law The device shown in Fig. 11.7 is a **gas thermometer.** (a) At the ice point, the gas volume is 1.00 liters. What would be the new volume if the gas temperature were raised from the ice point to 2.8°C? (b) Assume the cross-sectional area of the graduated arm is 1 cm². What would be the difference in height if the gas temperature went up from 0°C to 2.8°C? (c) What modifications may be made to increase the sensitivity of the thermometer?

17. Kelvin scale Convert the following to the Celsius scale: (a) 373.15°K; (b) 3.8×10^{3} °K; (c) 0.2°K; (d) 32°K.

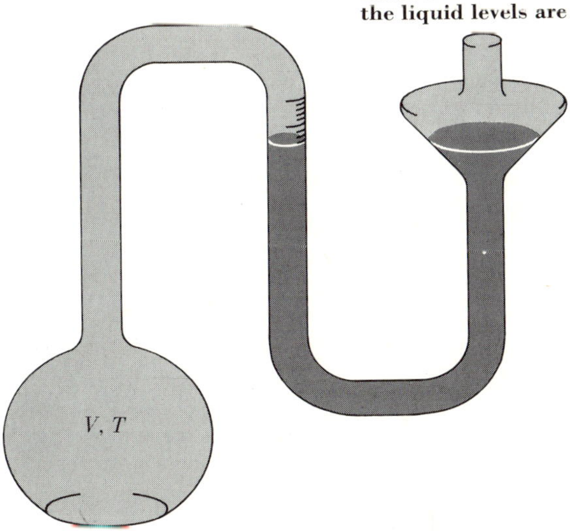

Fig. 11.7 Gas thermometer.

18. Absolute zero One gram of dry air occupies 1060 ml at 100°C and 776 ml at 0.00°C. Calculate the value of absolute zero in degrees Celsius.

19. Ideal gas law At standard conditions 22.4 liters of nitrogen weigh 28.0 g. What is the density of nitrogen in grams per liter at 120°C and 2.00 atm?

20. Ideal gas law A sample of gas occupies 6.1 liters at 20°C and 600 torr. What will its volume be at 15°C and 300 torr?

21. Ideal gas law Evaluate and assign units to the constant k for the ideal gas law expressed in the form $PV = kNT$, where $N = 6.02 \times 10^{23}$ molecules, and the other factors have the same standard values as given in Table 11.3, p. 180.

22. Values of the gas constant Evaluate the gas constant, k, in the equation $PV = kNT$, when the other factors have the following values:

	P	V	N	T
(a)	760 torr	22.4 liters	1 mole	273°K
(b)	1 atm	22.4 liters	6.02×10^{23} molecules	273°K
(c)	$PV = 542$ cal		1 mole	273°K

23. Amontons' law This law states that the pressure of a fixed quantity of gas at a fixed volume is directly proportional to the absolute temperature. (a) Show that this law follows from the ideal gas law. (b) A steel tank containing nitrogen at 10.0 atm pressure at 20°C is provided with a safety valve that will yield at a pressure of 20.0 atm. To what minimum temperature must the nitrogen be heated to blow the safety valve?

24. Dalton's law 50 ml of nitrogen gas is collected over water at 293°K and 600 torr. The nitrogen is separated from the water and dried at constant temperature. (a) What is the new volume of the dry nitrogen at standard conditions? (b) If the water vapor removed from the nitrogen were stored at 50°C and 16 torr, what volume would it occupy?

25. Dalton's law A mixture of nitrogen and oxygen at 1.00 atm pressure is stored in a 4.0 ft^3 iron container at constant temperature. The iron eventually reacts with all the oxygen, converting it to a solid oxide of negligible volume. The final pressure is 400 torr. (a) What is the final volume of the nitrogen? (b) What are the initial and final partial pressures of N_2 and O_2?

26. Eudiometer A gas is confined in a eudiometer over mercury at 0.00°C. The mercury level inside the tube is 6.00 cm above the level outside. The barometer reads 760.3 torr. (a) What is the pressure of the confined gas? (b) If the confining liquid is water at room temperature (assume 25°C, density 0.997 g/ml) and the conditions are as described above, what is the pressure of the confined gas?

27. The kinetic-molecular hypothesis Imagine that the container of Fig. 11.5 (p. 189) is about 1 ft^3 in volume and made of a hard, polished steel, that about 50 hard steel spheres of $\frac{1}{4}$ in. diameter are placed inside the container, and that all other matter is removed. The container is now shaken violently once or twice to set the steel balls in motion, after which no further outside work is done on the container and its contents.

How well or poorly is this real system represented by the ideal gas model? Be specific in your answer, discussing the following points: (a) Can the concept of temperature as a manifestation of kinetic energy be applied to this system; if the average speed of the balls is increased, can the contents of the container be said to be "hotter?" (b) Is the concept of pressure applicable to this system? If you added a certain amount of mechanical energy to the system initially at rest, and you knew the volume of the container and the number of balls, could you calculate the "pressure" from the assumptions of the kinetic-molecular hypothesis? Could a pressure be measured? (c) What are the most striking differences between the properties of this system and those of a tank of helium gas? How do these differences affect the applicability of the kinetic molecular hypothesis to this system?

28. Deviation from ideal behavior (a) Referring to Fig. 11.6 (p. 190), compare, qualitatively, the intermolecular attractive forces among nitrogen, hydrogen, and carbon dioxide at 100 atm pressure. (b) At 600 atm and 0°C, the PV product for nitrogen exceeds that for hydrogen. Does this mean that the intermolecular forces in hydrogen are the greater at this pressure? Why or why not?

29. Graham's law This law states that the rates of **effusion** (the escape of molecules through orifices) of gases are inversely proportional to the square roots of their densities, at constant temperature. Derive this law from the following assumptions: (*i*) The temperature is directly proportional to the average kinetic energy of the molecules, $T \propto \frac{1}{2}mu^3$, where m is mass and u is speed.* (*ii*) The rate of effusion of a gas is directly proportional to the average speed of its molecules. (*iii*) The density of a gas (at constant temperature and pressure) is directly proportional to the mass of its molecules.

30. Gas volume corrections in stoichiometry Refer to the appropriate problems in Chapter 5 and make the corrections requested.

* More precisely, u is the *root mean square* speed, defined as the square root of the average of the squares of the speeds, or $(\overline{u^2})^{1/2}$.

Chapter 5	Correction
(a) Problem 5, p. 90	What is the volume of nitrogen at 20°C and 772 torr?
(b) Problem 8(a), p. 91	What volume of oxygen measured at 100°C and 20.0 atm would be needed?
(c) Problem 15(a), p. 92	How many liters of methane would be needed if the gas were measured over water at 22°C and a pressure of 750 torr?
(d) Problem 18(a), p. 92	How many grams of NH_4HSO_4 are needed to produce 1.00 liter of hydrogen measured over water at 22°C and 750 torr?

ANSWERS

3. (a) 63 ml; (b) 950 ml; (c) 6.33×10^4 liters.

6. (a) 1.2×10^8 molecules; (b) 1.3×10^4 liters; (c) 2.5×10^{-9} liter.

8. 91.0 ml.

10. (a) 799.1 torr; (b) 762.0 torr.

16. (a) 1.01 liter; (b) 10 cm.

19. 1.74 g/liter.

23. 313°C.

25. (a) 4.0 ft^3; (b) initial p_{N_2} = final $p_{N_2} = 400$ torr, initial $p_{O_2} = 360$ torr, final $p_{O_2} = 0$.

30. (a) 2.36 liters; (d) 9.13 g NH_4HSO_4.

12 ■ AGGREGATED STATES OF MATTER

12.1 □ CRYSTALLINE SOLIDS; METHODS OF INVESTIGATION

The geometrical patterns of crystals have been of interest to man since earliest civilization, and more especially since the invention of the microscope. The study of such patterns is the science of **crystallography.** The internal structures of crystals can be studied more effectively, however, by measuring the degree to which they reflect or scatter radiation. Since the distances between atoms in matter are of the same order of magnitude as x-ray wavelengths, x-ray patterns may be expected to give useful information about the arrangement of atoms in crystals. This idea was experimentally verified by Max von Laue in 1912. The diffraction of x rays through crystals, the array of atoms being a diffraction grating, was found to produce patterns of reinforcement and attenuation of the radiation. X-ray diffraction patterns, as illustrated in Fig. 12.1, give a good picture of the symmetry of a crystal and of the relative densities of different crystal planes. In 1913, William Henry Bragg and William Lawrence Bragg (father and son) devised a satisfactory method for measuring the spacing of crystal planes by measuring reflectance angles (Fig. 12.2). Developments and refinements by many other investigators have made x-ray analysis a very powerful, if complicated, method of examining orderly arrangements in crystals, crystalline powders, and large molecules like those of proteins, rubber, and textiles.

12.2 □ THE SPACE LATTICE; THE UNIT CELL

A complete elucidation of the internal structure of a crystal gives all the information about the spatial arrangement of the constituent particles

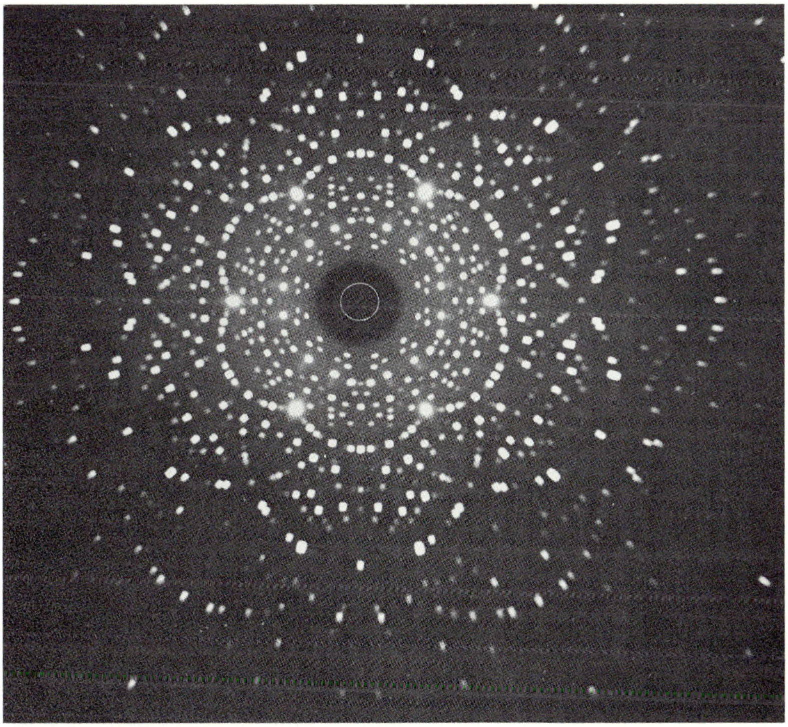

Fig. 12.1 X-ray diffraction pattern of beryl, $Be_3Al_2Si_6O_{18}$. (Courtesy Eastman Kodak Laboratories.)

Fig. 12.2 Examination of crystals by x-ray reflectance.

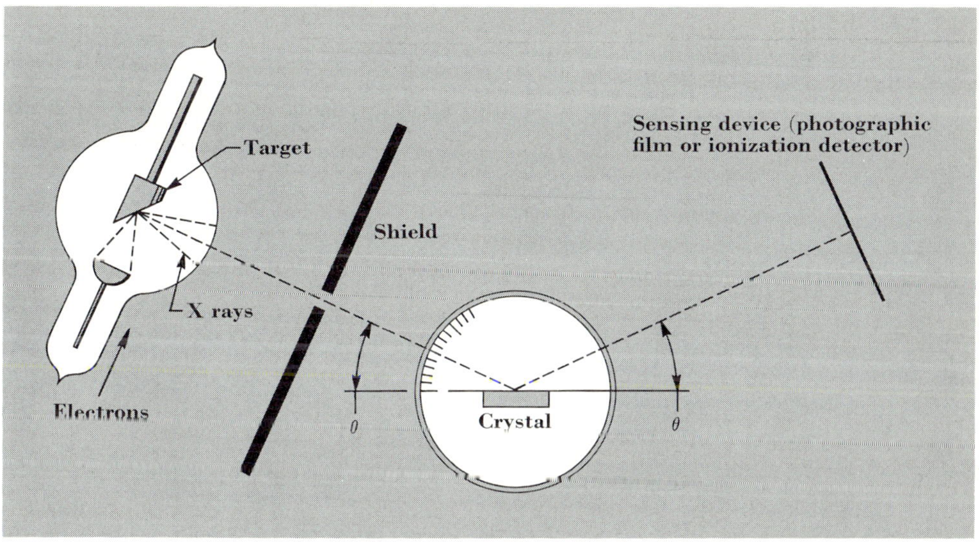

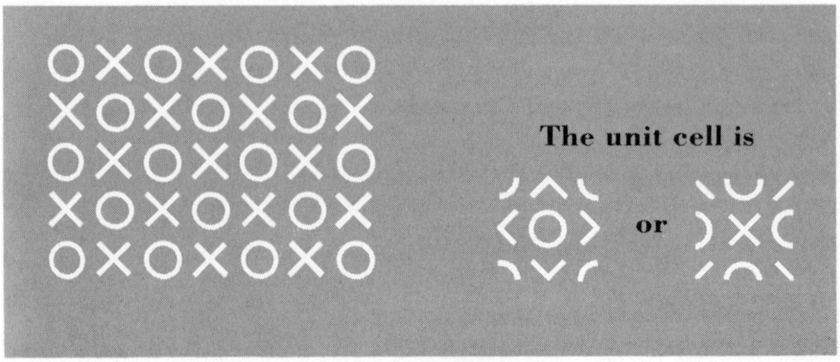

Fig. 12.3 A two-dimensional space lattice and unit cells.

(atoms, ions, or molecules). To describe such an arrangement, we may represent the particles by points or other designations and illustrate them in their relative positions in space. Since the atoms vibrate in the crystal because of their thermal energy, each designated position is actually a "center of vibration." Such a representation is called a **space lattice.**

The smallest sample that represents the entire crystal is called the **unit cell** (Fig. 12.3). The crystal may be considered to consist of an indefinite number of unit cells, each one in direct contact with its nearest neighbors, and all similarly oriented in space (not rotated or reflected). If a crystal consists of more than one kind of lattice particle (for example, sodium chloride crystals consist of lattices of sodium ions and chloride ions), we must describe either a separate space lattice for each kind of particle, or else a space lattice for the compound

Fig. 12.4 Cubic crystal systems. (a) Face-centered cube and packing model; (b) Simple cubes; (c) Body-centered cube.

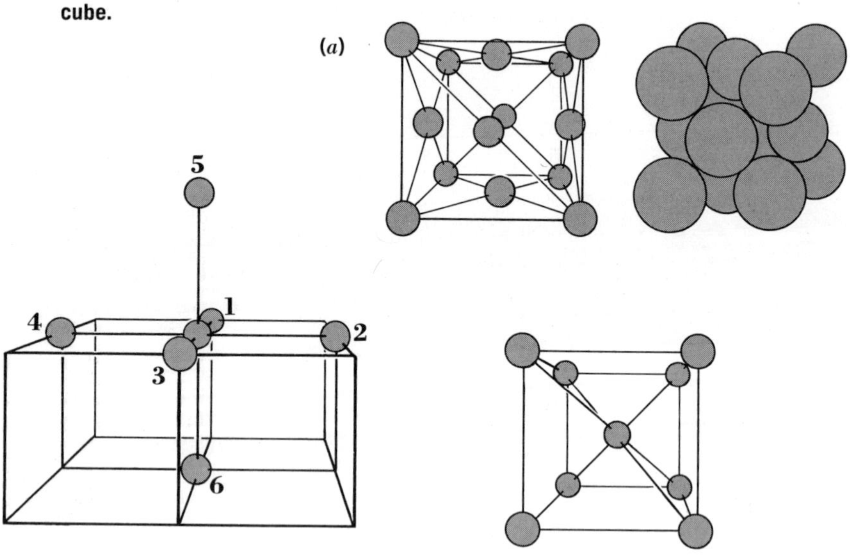

substance in which each point is identified according to the kind of particle that it represents.

The unit cells of crystal lattices are polyhedra whose faces are quadrilaterals or hexagons. Other shapes do not fill a space completely and therefore do not lend themselves to the compact orderly arrays of crystalline structures. The simplest and perhaps the most important type of lattice is the **cubic system,** which occurs in three different distributions of atomic centers: simple, body-centered, and face-centered. These appear in Fig. 12.4. Note that the *center* of each particle corresponds to a lattice point. Two questions may be asked: (a) How many particles exist within the space defined by the cubic unit cell, and (b) how many particles "touch" a given particle?

Assignment of spherical particles to the unit cell Assignment of particles to the cubic unit cell is made on the following basis:

A **corner particle** (Fig. 12.4a) is shared equally by all the unit cells that touch the same point. In two dimensions, four squares can touch a given point. In three dimensions, eight cubes can touch a given point. Therefore only $\frac{1}{8}$ of a corner particle is assigned to one unit cell.

A **face particle** (Fig. 12.4a) is shared equally between the two cells that face each other. Therefore $\frac{1}{2}$ of a face particle is assigned to one unit cell.

An **edge particle** (not shown in Fig. 12.4) is shared equally among the 4 unit cells that have a common edge. Therefore $\frac{1}{4}$ of an edge particle is assigned to one unit cell.

A **body-centered particle** (Fig. 12.4c) belongs entirely to its unit cell.

In summary:

Particle	Portion in cubic unit cell
Corner	1/8
Edge	1/4
Face	1/2
Center	1

EXAMPLE 1 How many atoms are assigned to the unit cell of the body-centered cube?

ANSWER 8 corners — 1; 1 center — 1; total = 2.

Coordination number or ligancy The number of particles touching a given particle is an important value called the **coordination number** or **ligancy.*** The ligancy for a given space lattice may be determined by inspection of the model.

* "Touching" is a macroscopic concept that is less meaningful on an atomic scale. The ligancy of a particle may therefore be defined simply as its number of nearest neighboring particles.

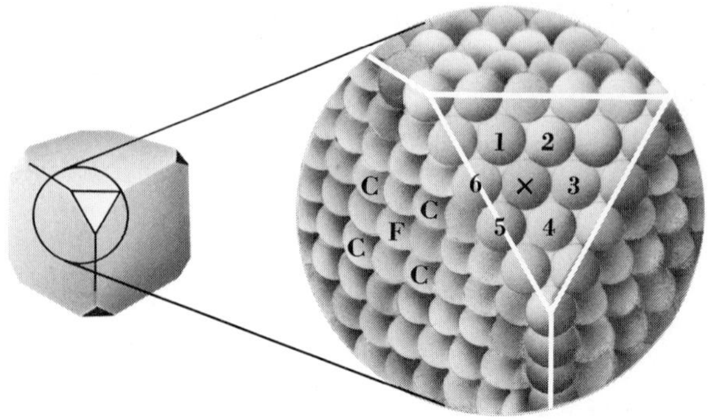

Fig. 12.5 Cubic closest packing.

In the **simple cube,** each particle touches six adjacent particles (Fig. 12.4a).

In the **body-centered cube,** each particle may be considered to be in the center of a cube and touching the eight corner particles (Fig. 12.4c).

In the **face-centered cube** the ligancy cannot readily be visualized from the model of Fig. 12.4. An extended face-centered cubic packing is shown in Fig. 12.5, with some spheres removed from the corners. Note that sphere X touches 6 other spheres (numbered 1 through 6) in a single plane that is diagonal to the side of the cube. Above this plane there is room for 3 more spheres to touch X. One would fit in the depressions bounded by 1, 2, X, another by 3, 4, X, and a third by 5, 6, X. Below this plane there are 3 additional spheres (not visible) touching X. The total number of X's nearest neighbors is therefore 12. This ligancy of 12 is called *cubic closest packing;* it is impossible to pack more uniform spheres into a given space by any other arrangement.*

In summary:

Lattice of uniform spheres	Ligancy
Simple cubic	6
Body-centered cubic	8
Face-centered cubic	12

EXAMPLE 2 The element chromium exists as a body-centered cubic lattice whose unit cell edge is 2.88 Å. The density of chromium is 7.20 g/cm³. Calculate the Avogadro number.

ANSWER The Avogadro number is the number of atoms in a gram-atom of an element (p. 68). The atomic weight of chromium is 52.0 g/g-atom.

$$\text{volume of unit cell} = (2.88\ \text{A})^3 = 23.9\ \text{Å}^3$$

$$\text{volume of 1 g-atom of chromium} = \frac{52.0\ \text{g/g-atom}}{7.20\ \text{g/cm}^3}$$

* There is another equally closely packed arrangement based on the hexagonal crystal type called **hexagonal closest packing.**

$$= 7.22 \text{ cm}^3/\text{g-atom}$$

$$\text{number of unit cells in this volume} = \frac{7.22 \text{ cm}^3/\text{g-atom}}{23.9 \text{ Å}^3/\text{unit cell}} \times (10^8 \text{ Å/cm})^3$$

$$= 3.02 \times 10^{23} \text{ unit cells/g-atom}$$

Since each body-centered cube contains 2 atoms, the total number is

$$2 \text{ atoms/unit cell} \times 3.02 \times 10^{23} \frac{\text{unit cells}}{\text{g-atom}}$$

$$= 6.04 \times 10^{23} \text{ atoms/g-atom}$$

Compound lattices Compound lattices, as previously mentioned, are more complicated than the elemental ones just described; the different kinds of atom may differ both in number and in size. The relative number and size of the different kinds of atom will, in turn, determine which types of crystal packing may be favored. Figure 12.6 illustrates the lattices and the packing models of sodium chloride and of cesium chloride. Note that the ligancies in compound lattices do not necessarily correspond to those of elemental lattices. In sodium chloride, for example, the ligancy of each ion, taken to be the number of its nearest neighbors *of opposite charge*, is 6.

12.3 □ THE TETRAHEDRON; THE OCTAHEDRON

White phosphorus is a simple cubic crystal lattice in which only alternate corners of the cube are occupied by phosphorus atoms. The shape defined by four such points is a regular **tetrahedron** (Fig. 12.7a,b). This is a body whose four faces are equilateral triangles. The body-centered tetrahedron (Fig. 12.7c,d), with a ligancy of 4, is the geometric basis of a very great number of compounds, notably those of the element carbon.

In the simple cube of uniform spheres (Fig. 12.4), the ligancy is 6. The 6 spheres that touch any given sphere define a regular **octahedron** (Fig. 12.8), a bipyramidal body with eight triangular faces. This shape is the geometric basis of many inorganic compounds.

12.4 □ LIQUIDS; GLASSES

The liquid state cannot yet be described by any concept that has the intellectual elegance of the theories of the gaseous or crystalline states. Nonetheless, important advances have been made by Henry Eyring, J. D. Bernal, and others.

Is there any order at all in liquid structure? X-ray examination shows diffuse diffraction patterns indicating some limited regularity—but greatly contrasting with the sharp patterns obtained from crystalline

Packing model

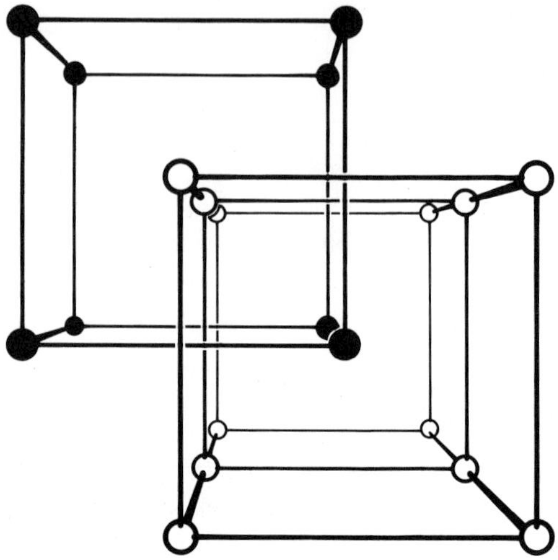

Lattice

Cesium chloride

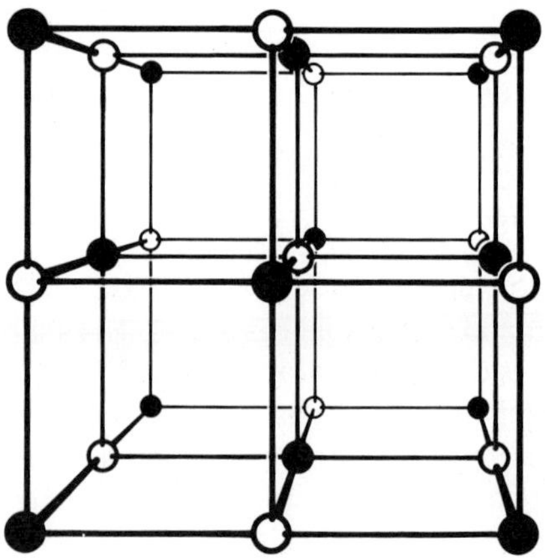

Lattice

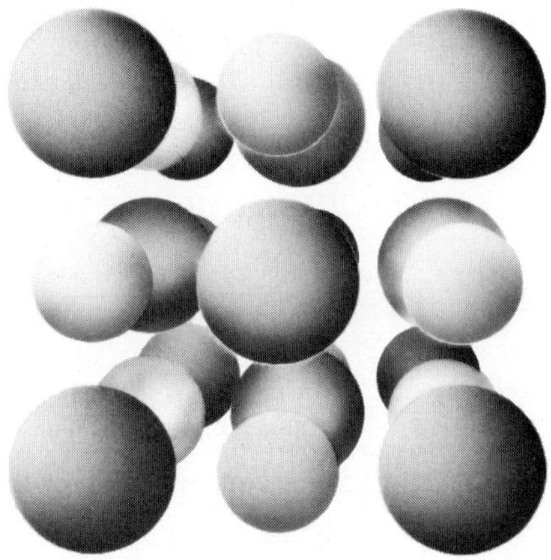

Packing model

Sodium chloride

Fig. 12.6 Compound lattices and packing models. Lighter spheres: sodium or cesium ions. Darker spheres: chloride ions.

solids. There are, of course, conditions (at or somewhat below the freezing point) under which a given substance may exist as either a liquid or a solid, with not much difference in density between the two states. Under such conditions the coordination number, or ligancy, in the liquid state cannot differ greatly from that of the solid. Thus, in a liquid, each individual atom or molecule may be thought of as occupying a lattice point and having an immediate environment that is crystalline-like.

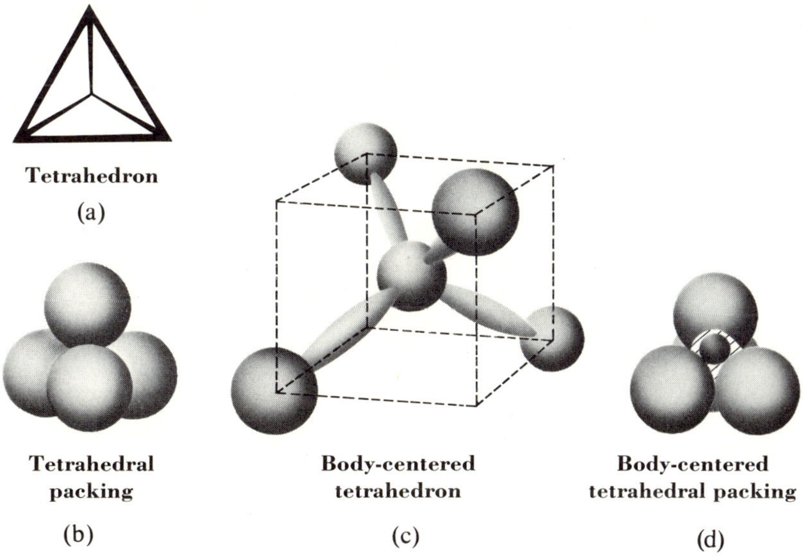

Fig. 12.7 The tetrahedron.

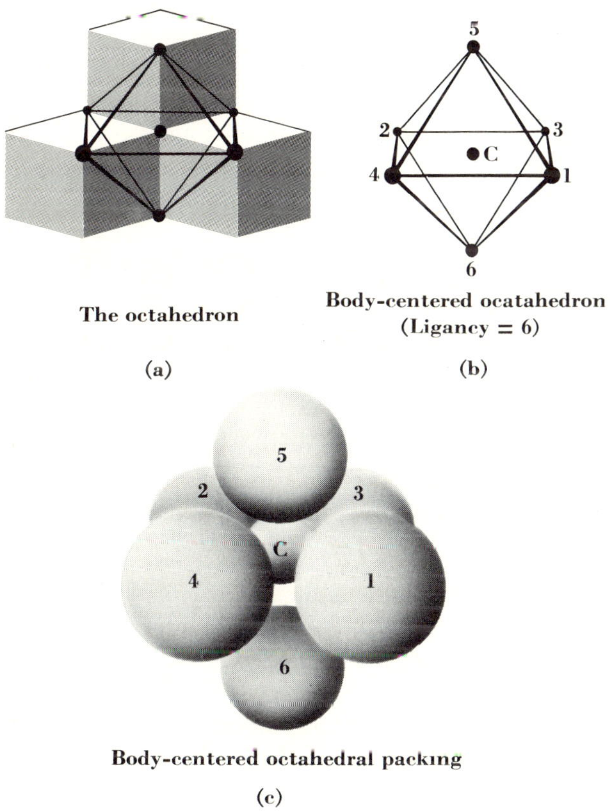

Fig. 12.8 The octahedron.

The lattice itself, however, is so defective that no long-range order exists. A highly defective lattice of this type offers little resistance to stress—hence a liquid has fluidity. Another way of looking at this is to consider that there are many alternative types of disordered structure differing little in energy among themselves; hence movement of the liquid, which involves a series of shifts from one structure to another, can be accomplished with little work. Still another way to regard the liquid state is as a crystalline-like mass interspersed with molecular-sized spaces whose arrangements are entirely random, or gaslike. Molecules may enter or leave these "holes"; this interchange produces overall randomizing and fluidity.

If a liquid is cooled until the decrease in molecular energy and the increase in density rob it of its fluidity, and if crystallization does not occur, then the substance becomes a **glass.** A glass thus has the overall randomness of a liquid. When a glass is heated, fluid properties appear gradually, since no sudden breakup of an orderly structure is required.

Examples of liquids which, on cooling, are likely to produce glasses rather than crystalline solids include the following:

(a) A molten mixture of metal silicates cools to give the material whose common name is "glass."

(b) Molten tar, on cooling, forms a glassy product. If a piece of cold "solid" tar is cracked, the vitreous (glassy) appearance of the fresh surface is apparent.

(c) Melted sugar compositions frequently fail to crystallize on cooling. Ordinary "hard candy" is a glassy solid obtained from such melts. "Rock candy," by contrast, is crystalline sugar.

12.5 □ CHANGES OF STATE

The various states of matter are frequently interconvertible. As the internal energy of matter is increased, the arrangement of its constituent particles becomes more disordered; as the energy is decreased, the opportunities for orderly arrangements increase. There are limitations, however, in both directions. Increase of energy tends to destroy particles as well as to disorder them; sometimes the destruction occurs first. For example, crystalline insulin (a protein) cannot be liquefied; less energy is required to destroy the molecules than to make them exhibit mobile flow. Therefore, for insulin, decomposition precedes melting. In the reverse direction, loss of energy may afford a better chance for an orderly pattern, but it does not insure it. Instead (as discussed earlier), a glass may form.

It will be helpful to learn the vocabulary of changes of state, as illustrated in Fig. 12.9. A change of state involves gain or loss of heat but does not involve a temperature change. Thus the **heat of fusion** of a substance is the amount of heat necessary to melt 1 g of the solid to a liquid *at the same temperature.* The energy of fusion, as illustrated in

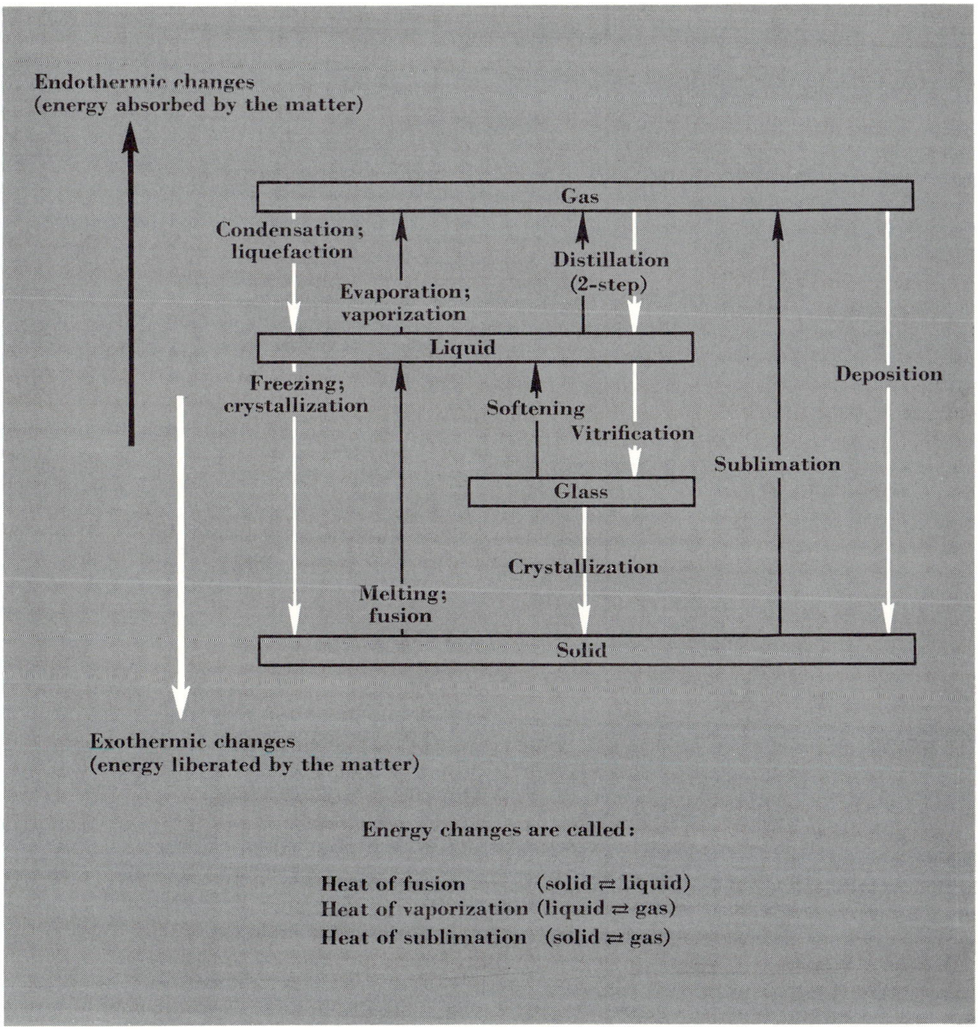

Fig. 12.9 Vocabulary of changes of state.

Fig. 12.9, is absorbed by the substance from the environment (**endothermic**); the opposite process (freezing) involves the same quantity of heat per gram of substance, but the heat is released by the substance to the environment (**exothermic**). Analogous expressions are used for the other changes of state. The heats are often referred to 1 mole, rather than 1 g, of a substance.

Any of the endothermic processes in Fig. 12.9 that proceed in the direction solid → liquid → gas produce a twofold change in the affected matter:

(1) The internal energy of the matter is increased because work is being done to overcome attractive forces.
(2) The orderliness of the molecular arrangement is decreased.

TABLE 12.1 HEATS OF FUSION AND VAPORIZATION

Substance	Melting point °C	Boiling point °C	Heat (cal/g) of Fusion	Heat (cal/g) of Vaporization
Aluminum	658	2057	94	2.5×10^3
Beeswax	62	—	42.3	—
Carbon tetrachloride	−24	76.8	4.2	46.4
Ethyl alcohol	−114	78.3	24.9	204
Mercury	−39	357	2.8	70.6
TNT	79	—	22.3	—
Water	0	100	79.7	539.6

Conversely, in any of the exothermic processes that proceed in the direction gas → liquid → solid, the internal energy is decreased and the order is increased.

Table 12.1 presents data on heats of fusion and vaporization.

12.6 □ SPONTANEOUS CHANGE; ENTROPY

Which set of processes shown in Fig. 12.9, the exo- or the endothermic ones, would be expected to be "natural" or "spontaneous"? The direction of energy flow will be influenced by the relative temperatures of the system and the environment; if the environment is cold, then heat will tend to flow out of the matter, and the gas → solid processes will be favored. The reverse will be true in a hot environment.

There is also, however, *a natural tendency for disorder*. This deceptively simple idea has far-reaching consequences. For changes of state, it means there is *always* a drive that favors the solid → gas processes, regardless of the outside temperature. Thus, it is not necessary to place solids or liquids in a warm environment for them to evaporate; evaporation will occur even when they are cooling. Examples: snow sublimes even on a cold winter day; moth-repellent crystals sublime in a closet, even though the temperature may be falling.

The attribute expressing the disorder or randomness of a system is called the **entropy.** The relationship between the entropy and the properties of a system can be arrived at either from statistical or from thermodynamic considerations.

Statistically, we may argue that disordered conditions are more probable than ordered ones because there are many more of the former. We may visualize this with the simple model of a tic-tac-toe field used by a single player (Fig. 12.10). If he makes his three marks without plan, what are his chances of obtaining an orderly arrangement? Order is characterized by symmetry, or repetition of unit parts. Let us therefore say that an orderly arrangement is one which can be folded so that the facing sides superimpose exactly on each other. The "fold line" then

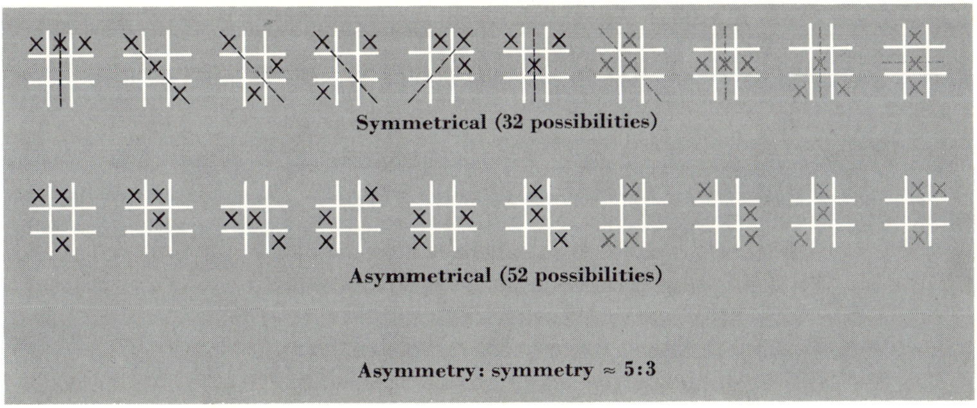

Fig. 12.10 Entropy on a tic-tac-toe field.

becomes a line of symmetry. The total possible number of plays is 84. Of these, 32 have a line of symmetry; the chances for disorder are thus 52 to 32, or about 5 to 3. With matter in the quantities usually handled in the laboratory ($\sim 10^{-3}$ to 10^{+3} g), we deal not with 3 units but with the order of 10^{23} units. Based on chance alone, the preponderances of disorder over order are overwhelming, and are expressed by the largest numbers of any physical significance to man (far larger than the numbers used, say, by astronomers). Statistically, then, there is a natural tendency to disorder. Because chance favors disorder, it follows that there is a *tendency for the entropy of a system to increase*. For change-of-state processes, then, the likely entropy change favors the solid $\rightarrow$ gas conversions.

Thermodynamically, we may argue that ordered states such as solids are usually lower in energy than disordered states such as gases. In a cold environment, heat tends to flow from a system to the outside, and the internal energy of the system therefore tends to become lower. This preference for low energy decreases, however, as the temperature increases, because at higher temperatures the tendency to disorder becomes more predominant. A mechanical illustration is afforded by a vertically vibrating tray with a shallow depression in the middle (Fig. 12.11). If some balls are placed in the tray, and the vibration is gentle,

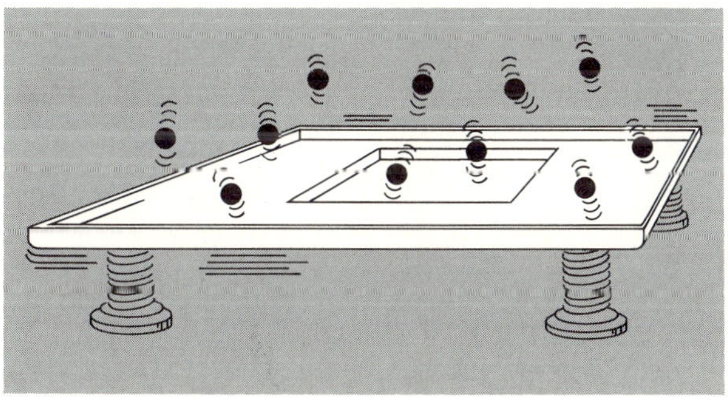

Fig. 12.11 Energy and disorder in a vibrating tray.

the balls will tend to accumulate in the depression. This accumulation is a relatively ordered arrangement. As the vibration of the tray becomes more vigorous (analogous to a rise of temperature), the balls show less and less preference for the depression. Such a dispersed arrangement is more disordered.

12.7 □ LIQUID-GAS INTERCONVERSION; VAPOR PRESSURE

Molecules of a liquid, like those of a gas (see Fig. 11.4, p. 188), exhibit a distribution of energies. The molecules that are energetic, close to the surface, and directed toward the surface, have the best chance to escape (vaporize). The vaporized molecules may be condensed if, in their kinetic path, they reach and are held by the attractive forces of the liquid. Any condensable gas is, in fact, called a **vapor.**

A high *net* rate of vaporization (evaporation) is therefore favored by (1) high temperature of the liquid, (2) small attractive forces in the liquid, (3) a large surface area, (4) low atmospheric pressure above the liquid—to decrease the number of collisions with other molecules and thereby minimize return of molecules to the liquid—and (5) a mass motion of the atmosphere (breeze) above the liquid to carry away vaporized molecules and minimize reentry. The housewife takes advantage of these factors when she spreads out her wash on a clothesline on a warm, dry, windy day.

If the enclosed space above a liquid is small enough, and if environmental factors are held constant, a condition will be reached in which the rates of vaporization and of condensation will be equal. The system is now said to be in **dynamic equilibrium,** or simply in **equilibrium.**

A system at equilibrium is defined as one in which two opposing

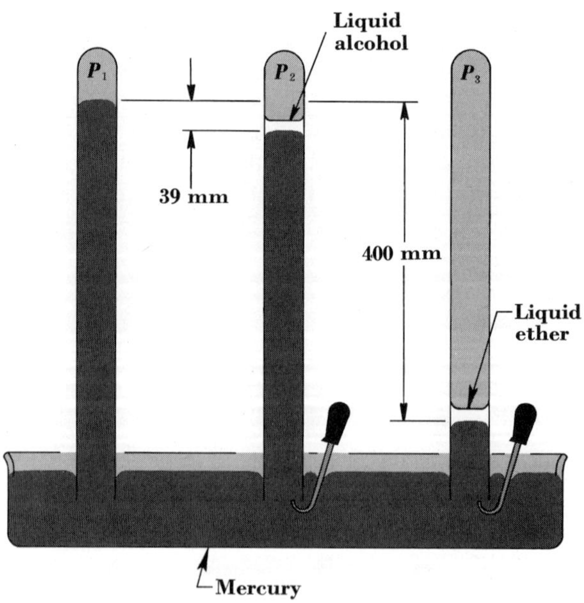

Fig. 12.12 Vapor pressure. $T = 18°C$
P_1 = Torricellian vacuum
= vapor pressure of mercury
= 10^{-3} torr ≈ 0
P_2 = vapor pressure of alcohol
= 39 torr
P_3 = vapor pressure of ether
= 400 torr

(Not to scale)

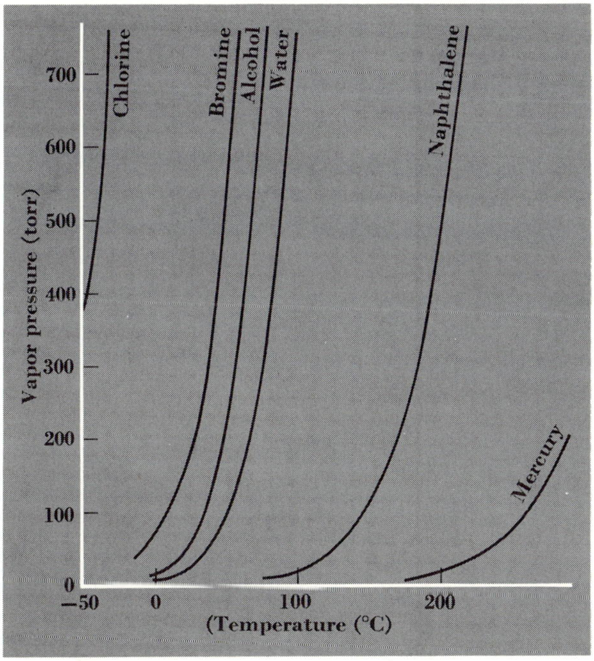

Fig. 12.13 Vapor pressures of liquids.

processes are going on simultaneously at the same rate. The effect is one of no net change.

The **vapor pressure** of a liquid is the pressure exerted by the vapor that is in equilibrium with the liquid at a definite temperature. The vapor pressure depends on the temperature and on the nature of the liquid; it is, within the limits of the ideal gas law, independent of the presence of other gases.

Vapor pressure may be measured in a number of ways. A direct method makes use of the mercury barometer, illustrated in Fig. 12.12. If a small quantity of liquid is allowed to float up to the top of the mercury column and evaporate into the Torricellian vacuum there, then the vapor establishes equilibrium with the remaining liquid and the vapor pressure in torr units is numerically equal to the depression of the mercury column in millimeters.* Figure 12.13 illustrates vapor pressure-temperature relationships for several substances.

12.8 □ CRITICAL CONSTANTS

Consider a strong, sealed vessel containing a certain quantity of liquid substance in equilibrium with its vapor. The attractive forces among the molecules constrain some of them to the liquid state. At the same time, the kinetic energy disrupts the aggregations of the more energetic molecules and thrusts them into the gaseous state. Now we heat

* Assuming temperature of 0°C, and standard gravity.

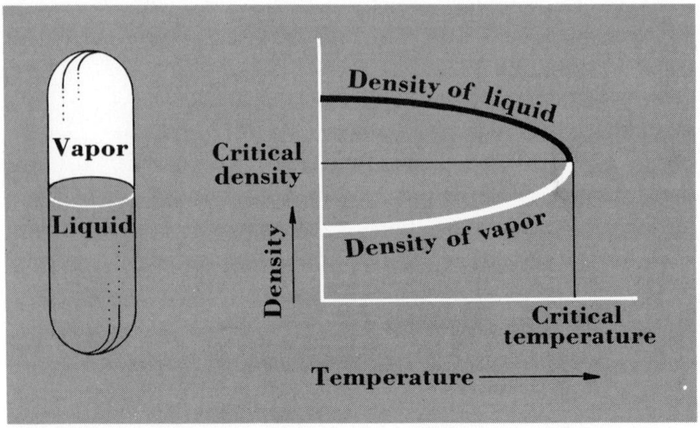

Fig. 12.14 Critical temperature.

the vessel and its contents. The molecular motions become more violent. The liquid expands, becoming less dense. More evaporation occurs, and so there are more molecules in the gaseous state, and the gas phase becomes more dense. A temperature is reached at which the attractive forces cannot overcome the random motions, the boundary between liquid and gaseous phases disappears, and the two density curves meet* (see Fig. 12.14). When this happens, the properties of the liquid and the gas become identical. The temperature and pressure under these circumstances are the critical constants, defined as follows. The **critical temperature** is *the temperature at which the properties of the liquid and vapor become identical;* it is also *the highest temperature at which a vapor can be liquefied.* The **critical pressure** is *the minimal pressure needed to liquefy a gas at the critical temperature.*

SAFETY NOTE: THE STUDENT MUST NOT ATTEMPT SUCH AN EXPERIMENT UNDER ANY CONDITIONS, BECAUSE CRITICAL PRESSURES ARE USUALLY HIGH ENOUGH TO CAUSE EVEN STRONG VESSELS TO EXPLODE.

The critical temperature of a substance is related to the magnitude of the attractive forces among its molecules. If the attractive forces are great, liquefaction can be achieved in spite of kinetic opposition at elevated temperatures. The critical temperature is, therefore, high. If attractive forces are small, liquefaction is easily opposed by thermal agitation, and the critical temperature is low. Refer to Table 12.2 for data.

12.9 □ BOILING, MELTING, and FREEZING POINTS

There are two cases of phase equilibrium that are worthy of note. One is **boiling,** the condition of vaporization of a liquid that is usually accom-

* For this to occur in an experiment, the original mass of substance divided by the volume of the vessel must correspond to the critical density.

TABLE 12.2 CRITICAL CONSTANTS

Substance	Temperature, °C	Pressure, atm
Air	−140.7	37.2
Carbon dioxide	31.1	73.0
Ethyl alcohol	243.1	63.1
Helium	−267.9	2.26
Hydrogen	−239.2	12.8
Mercury	$>$1550	$>$200
Oxygen	−118.8	49.7
Water	374.0	217.72

panied by the formation of bubbles of vapor in the body of the liquid. This can occur when the vapor pressure is equal to the total pressure of the atmosphere in contact with the liquid. Thus, the **boiling point** of a liquid is *the temperature at which the liquid is in equilibrium with its vapor at the pressure of the atmosphere in contact with the liquid.*

The boiling point of a liquid may therefore be varied by changing the atmospheric pressure. When the pressure is one standard atmosphere, the equilibrium temperature is called the **normal boiling point**.

The **melting point** and **freezing point** of a substance are equivalent; they are the temperatures at which liquid and solid are in equilibrium at standard atmospheric pressure. The difference between the two is one of approach, not of equilibrium. The melting point is reached by warming the solid, the freezing point by cooling the liquid. The effect of change of atmospheric pressure is usually so small that it need not be considered.

At the melting or freezing point, the tendencies for increase of entropy (→ liquid) and for decrease of energy (→ solid) just balance and

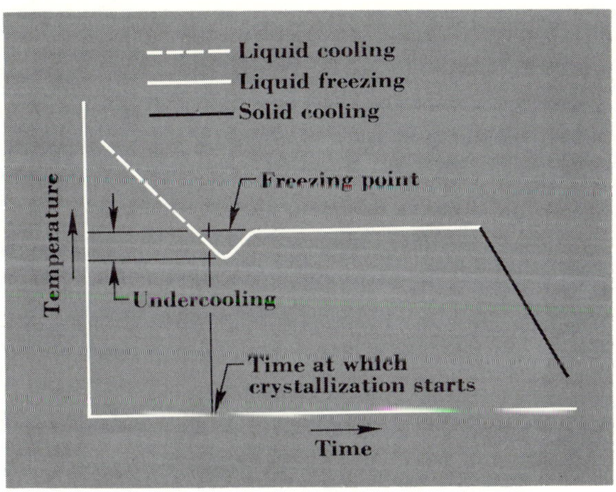

Fig. 12.15 Idealized cooling curve of a pure liquid, with undercooling.

equilibrium is established. If heat is added under conditions such that equilibrium is maintained, some solid melts without change of temperature. If heat is removed, some liquid freezes, also at constant temperature. If heat is continuously extracted from a pure substance originally in the liquid state (assume that the temperature of the entire sample is kept the same) the liquid will cool until it starts to freeze, then the liquid will freeze at constant temperature, and finally the solid will start to cool. The graphical representation of this sequence, called a **cooling curve,** is shown in Fig. 12.15. Note that the liquid may cool **below** its freezing point before crystallization occurs and solid-liquid equilibrium is established. This phenomenon is called **supercooling,** or more aptly, **undercooling.** The time direction of this curve may be reversed by warming the solid. Such a trace is called a **warming curve.** Superheating does not occur during warming of a crystal.

12.10 □ COLLOIDS; ADSORBENTS

Intermolecular forces in the solid state take on special significance at the surface of the solid. The molecules at the surface are subjected to unbalanced forces toward the center of the solid, where there are more molecules to exert attractions. Work would have to be done against these forces to move a molecule from the interior to the surface. Thus the surface molecules are in a condition of higher energy, and are

Fig. 12.16 Small particles.

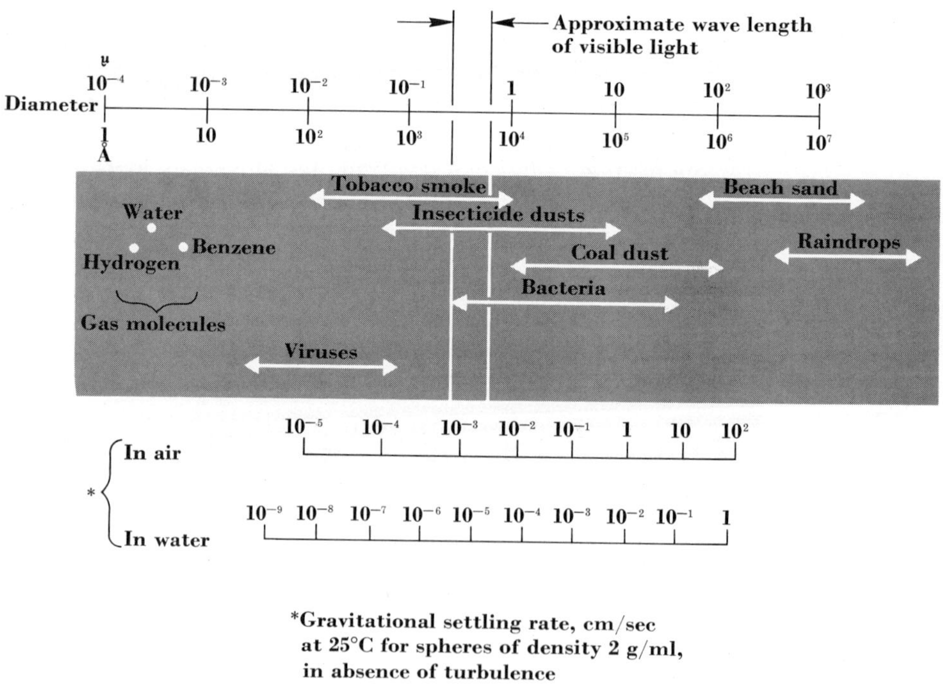

capable of attracting other molecules which might come close to them.

When the surface area of any body is very large compared with its bulk volume, the special properties of the surface molecules influence the overall properties of the body to a significant degree. Large surface-to-volume ratios exist when either

(a) the particles of the substance are very small, or
(b) the body is interlaced with small pores throughout its bulk volume.

Particles so small that surface effects assume great relative importance in determining their properties or behavior are called **colloids.** Colloidal particles dispersed in air are called **aerosols,** or **aerocolloids.** It will be helpful to refer to Fig. 12.16 to note the particle sizes of some common aerocolloids in comparison with coarse matter (such as beach sand) and with gas molecules. Particles smaller than the wavelength range of visible light cannot be seen with the ordinary microscope. The lower limit of microscopic visibility, about 0.2 micron, is frequently taken to be the upper limit of colloidal particle sizes. The lower size limit of the colloid range is taken to be the size of such known particles as protein or starch molecules and some small viruses, roughly 50 A.

Colloidal particles have the ability to scatter visible light. A thin beam of light passing through a colloid in a gas or a liquid can be seen at right angles because of the scattering. This phenomenon, called the **Tyndall effect,** was discovered in 1857, and constituted a simple (and still reliable) criterion of the colloidal state of matter. Colloids are characterized by very slow settling rates. Note (Fig. 12.16) that at 0.2 μ, settling in air is about 5×10^{-4} cm/sec and in water is several hundredfold slower.

The high energy characteristic of surface molecules may be lessened if foreign molecules move to the surface under the influence of the residual attractive forces available there. Any gas, vapor, or liquid will, therefore, adhere to some degree to any surface of a body. This phenomenon is called **adsorption,** or **sorption.** Solid bodies having large surface areas compared to their bulk volumes are therefore called **adsorbents.** Colloids, as described above, fulfill this requirement because of their small size. All colloidal matter is therefore highly adsorbent. Even large bodies, however, may have large surface-to-volume ratios if a network of pores within them provides extensive surface area. Matter of this type can be produced from coconut shells, for example, by charring and then heating them in a steam atmosphere to about 1500°C to drive out all material which can be gasified at that temperature. This process is called **activation.** The matter left behind, consisting mainly of carbon and retaining the many pores created by the activation process, is called **activated carbon.** Its surface area may be as much as about 1000 m²/g. The absorbing capacity of such material is startlingly high. For example, activated carbon at 20°C can adsorb up to $\frac{3}{4}$ its own weight of carbon tetrachloride (dry cleaning solvent) vapor from air that is saturated with the vapor at 0°C (vapor pressure = 30 torr). Other effective adsorbents may be made from metallic oxides, silicates, and metals. A good example of the latter is a highly porous form of platinum called **platinum black.**

PROBLEMS

1. Definitions Define and illustrate the following terms: (a) space lattice; (b) unit cell; (c) coordination number or ligancy; (d) body-centered cubic system; (f) compound lattice; (g) vapor pressure; (h) critical temperature; (i) critical pressure; (j) colloid, (k) adsorbent.

2. States of matter Suppose we classified the orderliness of arrangements of molecules in matter as (1) none, (2) intermediate, and (3) high, and the degree of cohesiveness as (a) low, unable to maintain a rigid shape under the stress of gravity, and (b) high, able to maintain a rigid shape under the stress of gravity. Outline the states of aggregation of matter based on this classification and, where possible, assign the appropriate common name of a state of matter to each category.

3. Crystal structure State the kinds of information about crystalline substances that may be obtained by (a) visual observation; (b) x-ray examination by the Bragg method.

4. Unit cell What fraction of a circle in each of the following locations is located within the two-dimensional square unit cell: (a) corner; (b) side; (c) center?

5. Space lattice Refer to the diagram of the sodium chloride lattice (Fig. 12.6). What is the coordination number (a) of the sodium; (b) of the chlorine? (c) Tabulate the number of corner, edge, face, or center particles of sodium atoms (smaller spheres) and of chlorine atoms (larger spheres) in the unit cell of sodium chloride. How many atoms of each are assigned to the unit cell? (d) If all the sodium ions were removed from the lattice, what would the lattice structure of the chloride ions be? If all the chloride ions were removed from the lattice, what would the lattice structure of the sodium ions be?

6. Space lattice The metal palladium exists as a face-centered cubic lattice whose unit cell edge is 3.88 Å at 20°C. (a) How many atoms are there in the unit cell? (b) Predict the density of palladium. Compare your answer with the value 12.0 g/cm^3 at 20°C (from the Handbook of Chemistry and Physics).

7. Liquid and glassy states A zinc metal rod and a glass rod are each heated in the absence of air to a temperature somewhat above 419°C. The zinc rod remains rigid, but liquid zinc begins to drip from it. The glass rod sags, but dripping does not occur. Account for these phenomena.

8. Heats of fusion and vaporization (a) How many calories of heat are needed to melt 10.0 g of solid ethyl alcohol at −114°C, warm the liquid to 78.3°C, and vaporize it at 78.3°C? The specific heat (App. 1) of ethyl alcohol is 0.58 cal/g deg. See Table 12.1 (p. 206) for other data. (b) Do you think the answer would be the same if the operations were carried out at different temperatures but started and ended at the same conditions specified in (a)?

9. Entropy (a) An atmosphere contains soot from chimneys and pollen from vegetation. Discuss the chances that, if two equivalent air samplers are placed side by side, one collects only soot, the other only pollen. Explain in terms of the entropy concept. (b) Assume that your room has a specific arrangement of clothing, books, papers, and other personal effects. For one month, whenever you have finished using any item, you replace it without plan in any location in the room. Does the room become more orderly or more disorderly? Explain in terms of the entropy concept.

10. Vapor pressure Refer to Fig. 12.13 (p. 209). (a) What are the approximate normal boiling points of chlorine, bromine, alcohol, water, and naphthalene? Estimate the absolute error in your answers. (b) At what approximate pressure is it possible to boil (*i*) bromine at 40°C, (*ii*) alcohol at 30°C, and (*iii*) water at 10°C? Estimate the absolute error in your answers.

11. Critical temperature The intermolecular attractive forces in liquid A are greater than those in liquid B. Predict which liquid will have the higher critical temperature. Explain.

ADDITIONAL PROBLEMS

12. Definitions Discuss the limitations and inadequacies of the following definitions. Improve them. (a) A *solid* is matter that retains its shape. (b) A *glass* is a transparent solid.

13. Vocabulary of changes of state Substitute appropriate words for the blank spaces to complete the following sentences: The substance sometimes known as "iso-octane" is the standard for 100 octane gasoline. Its boiling point is 99°C, and its freezing point is −107°C; it is therefore a ___ in its stable condition at 100°C and 1 atm pressure, and a ___ under the ordinary conditions of man's environment. If the ___ is distilled, it is converted first to a ___ and then to a ___. When the liquid is cooled sufficiently, the iso-octane will ___ and become a ___ if the molecules arrange themselves in an orderly lattice, but if molecular disorder is maintained, the iso-octane will ___ and become a ___. If helium gas is swept over iso-octane crystals at −110°C, the crystals will gradually disappear by the process of ___, being converted to a ___.

14. Unit cell Designate a unit cell in (a) a sheet of postage stamps; (b) black and white squares arranged in a striped pattern; (c) black and white squares arranged in a checkerboard pattern.

15. Unit cell Draw a diagram showing how the diameter of a sphere may be part of the edge of a cube. If each edge contains one such sphere, what is the total number of spheres assigned to the space within the cube?

16. Space lattice Refer to the diagram of the cesium chloride lattice (Fig. 12.6, p. 202). What is the coordination number (a) of the cesium; (b) of the chlorine? (c) How many atoms of each are assigned to the unit cell? (d) If all the cesium ions were removed, what would the lattice structure of the chloride ions be?

17. Space lattice You are given your choice of pea-sized, baseball sized, or molecular-sized spheres of the same material. You are to pack the spheres as closely as possible. Which choice will yield the most efficient filling of space?

18. Space lattice Lithium crystallizes in a cubic system whose unit cell edge is 3.50 Å in length at 20°C. The density of lithium at 20°C is 0.534 g/cm^3. How many atoms of lithium are assigned to one of its unit cells? To what particular cubic crystal system does this number correspond?

19. The liquid state A traveler in a space vehicle (zero gravity) must have a usable supply of drinking fluid. What would be the result of attempting to use earth-bound kitchen methods of storage, handling, and drinking? Suggest methods appropriate for space travel.

20. Liquid and glassy states A man wearing spiked shoes finds that he sinks slowly when he stands on a tar road or on an ice rink. Ice is crystalline; tar is not. Account for both phenomena.

21. The mesomorphic state Some soapy solids (for example, potassium oleate), on heating, melt sharply to form homogeneous, turbid liquids that have some of the optical properties of crystalline solids, and that give much sharper x-ray diffraction patterns than ordinary liquids give. On further heating, an abrupt change to a clear liquid occurs. Such turbid liquids are sometimes called "liquid crystals" and their state of aggregation is called the "mesomorphic state." Suggest a theory of molecular arrangement to account for the mesomorphic state.

22. Entropy The Haber process for fixing nitrogen from the air and the Dow process for extracting magnesium from the sea are said to be antientropic processes. These are in contrast to procedures of the type that involve the mining of a concentrated ore and the wide distribution of the final product; such procedures are said to be entropic. An example is the mining of mercury and its widespread use in industrial and commercial products. Explain how the concept of entropy is related to these considerations. Is the stockpiling of gold in Fort Knox entropic or antientropic?

23. Critical constants Large quantities of liquid oxygen are transferred to and kept at missile sites in containers that are vented to the atmosphere. Liquid carbon dioxide, on the other hand, is handled and shipped in closed steel cylinders. Explain.

24. Colloids (a) What is the surface area of a 1-cm cube? (b) If the cube is cut in half by a slice that is parallel to one set of sides, how much surface area is added? (c) How many such slicings would be needed to produce a total surface area of 10^3 m^2? (d) How much volume would each piece occupy? (e) If each sliced particle were a cube, how long would its edge be, in microns?

25. Adsorption It is observed that when air containing gasoline vapor passes through a bed of activated carbon, the gasoline vapor is removed from the air and the temperature of the carbon bed rises. Account for these phenomena.

ANSWERS **6.** (a) 4 atoms.

8. (a) 3.4×10^3 cal.

24. (a) 6 cm^2; (b) 2 cm^2; (c) 5×10^6 slices; (d) 2×10^{-7} cm^3; (e) $6 \times 10\ \mu$.

13 SOLUTIONS

13.1 INTRODUCTION

A **solution** is a *mixture of two or more substances dispersed as molecules* rather than as larger aggregates. The intermediate zone between solutions and coarse suspensions embraces the colloidal dispersions (p. 213). Although we shall not be explicitly concerned with them in this chapter, most of what we say will apply to colloids as well as to "true" solutions. Two important classes of solutions will be omitted: gaseous mixtures, such as air, and ionic solutions, in which some of the constituent particles are electrically charged. Because of the special properties and great importance of ionic solutions, Chapter 15 has been reserved for them.

13.2 LIQUID SOLUTIONS

The word "solution" suggests a liquid to most of us, and liquid solutions are indeed the most important and interesting ones. The molecules in a liquid are in intimate contact, and thus the properties of each component in a liquid solution are considerably influenced by the presence of the others. For example, the energy of a molecule is affected by the nature of the molecules surrounding it; as a result, heat is usually absorbed or emitted when a solution is formed. In some cases, there is actually a chemical reaction between the components, so that the molecules present in solution are quite different from the molecules of the pure components; the clearest examples are found among the ionic solutions. There is a twilight zone in which it is not clear whether the formation of the solution should be described as a physical change—a mere mixing—or as a chemical change. Sometimes a liquid solution can be prepared from gases alone ($HCl + H_2O$ vapors) or from solids alone ($Na + K$), but usually at least one component is a liquid, when it is pure under the conditions of the experiment. When one component is a gas or a solid and the other is a liquid, the former is called the **solute** and the latter the **solvent;** when both are liquids, and one is much more abundant than the other, the major component is considered the solvent and the minor component the solute. The reason for this nomenclature is that we think of a *solvent* as something that has the capacity to dissolve a *solute* and hold it in solution. Of course, there may be several solutes—a solution need not have only two components.

A **dilute solution** is one which contains only a small quantity of solute (or solutes) relative to the quantity of solvent. A **concentrated solution** contains a large proportion of solute. These terms are no more precise than the words "large" and "small"—as many a student has learned when he used "dilute hydrochloric acid" and found that it was 20 or 30 times more concentrated than the acid he was supposed to use.

13.3 □ SATURATION: GASES IN LIQUIDS

Consider a gas being dissolved in a liquid, with the entire gas-liquid system enclosed (Fig. 13.1a). The gas molecules are moving about in all directions. Molecules frequently strike the surface of the liquid; they may merely bounce off, but there is a chance that a gas molecule will be captured and will diffuse into the body of the liquid. Conversely, a molecule already in the liquid may happen to reach the surface with enough kinetic energy to escape and become part of the gas again. The rate at which gas molecules enter the liquid depends on the number of collisions with the liquid surface. At a fixed temperature this rate therefore depends on the pressure of the gas. The rate at which gas molecules leave the liquid, on the other hand, depends on the num-

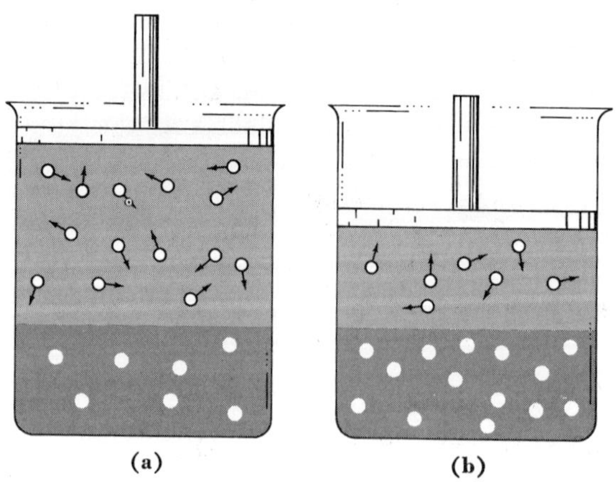

Fig. 13.1 Dissolving of a gas in a liquid at (a) low pressure, (b) higher pressure.

ber already dissolved. When the rate of escape of gas from the solution equals the rate of dissolving, the composition of the solution remains constant; no more of the gas dissolves. We say that **equilibrium** has been attained. The processes of capture and escape are still taking place, but their rates are equal, and there is no further net change in the pressure of the vapor or the composition of the solution. The solution is said to be **saturated.**

If the pressure of the gas is increased (Fig. 13.1b), the number of collisions with the surface is increased, and the rate of capture of gas molecules by the liquid therefore increases. The quantity of dissolved gas thus increases until the rate of escape has built up to equality with the new rate of capture. Thus, the **solubility** *of a gas in a liquid*—the quantity of gas that dissolves in a given quantity of liquid to form a saturated solution—*increases with rising pressure.* A carbonated beverage must be kept under pressure; when the cap or cork is removed, the pressure drops, the solubility of the gas decreases, and the beverage fizzes.

Raising the temperature nearly always decreases the solubility of a gas in a liquid. Although higher temperature results in more frequent collisions of gas molecules with the liquid surface, the principal effect of raising the temperature is that a larger fraction of the dissolved molecules have the kinetic energy needed in order to escape from the liquid. This is what we expect; the vapor pressure (p. 209) of a pure liquid increases with increasing temperature, and the pressure required to keep a gas in equilibrium with a solution of it may be regarded as the vapor pressure of the solute over that solution. When *the temperature rises*, the vapor pressure of a given solution rises, or *the solubility under a given pressure decreases.*

Table 13.1 gives the solubilities of some common gases in water, as the weight in grams of the gas which can be dissolved in 100 g of water when the total pressure of gas and water vapor is 1 atm. This is the usual manner of expressing solubility, although any of the measures of composition to be described later (p. 226) can be used. The very large solubility of SO_2 is noteworthy; this is believed to be one of the

TABLE 13.1 SOLUBILITIES OF GASES IN WATER (IN GRAMS OF GAS PER 100 G H_2O, TOTAL PRESSURE 1 ATM)

	0°C	25°C	50°C
N_2	0.00294	0.00175	0.00122
O_2	0.00694	0.00393	0.00266
H_2	0.000192	0.000154	0.000129
CO_2	0.335	0.145	0.076
SO_2	22.83	9.41	

cases in which dissolving is accompanied by a chemical reaction,

$$SO_2 + H_2O \rightleftharpoons (HO)_2SO \text{ (\textit{sulfurous acid})}$$

13.4 □ SATURATION: SOLIDS IN LIQUIDS OR LIQUIDS IN LIQUIDS

A solid dissolves in a liquid by essentially the same process by which a gas dissolves, except that we should picture the solute molecules not as colliding with the solvent, but as diffusing away from the surface of the solid and becoming surrounded by—and, to some extent, attached to—the solvent molecules. When some of the solute has gone into solution, some solute molecules will be recaptured by the solid from the solution, and the rate of this redeposition will increase as the solution becomes more concentrated. If enough solid is present, the solution will attain a composition at which dissolving and deposition are equal in rate. At this point, the solution is saturated.

The rates of dissolving and of deposition are both proportional to the surface area of the solid; if the rates are equal and the area is doubled, both rates are doubled, so that they are still equal, and the solubility is independent of the surface area. (However, very fine subdivision of a solid, finer than about 10^{-5} cm in diameter, increases the solubility.) At this point, we should note the distinction between *solubility* and *rate of dissolving*. A finely divided solid dissolves much more rapidly than large lumps—the solution becomes saturated sooner—but that does not mean that the total amount which can be dissolved is greater. Similarly, stirring accelerates the dissolving process, but, once the solution is saturated, you can stir your arm off without dissolving another milligram.

Many pairs of liquids are completely **miscible**—that is, they dissolve in each other in all proportions. Examples are water and acetic acid, water and glycerol, and benzene and toluene. In such cases, there is no such thing as a saturated solution. Other pairs of liquids, however, are only partially miscible: each dissolves in the other to some extent, giving two saturated solutions. Even so-called "immiscible" liquids—water and carbon tetrachloride, for example—usually have a measurable solubility in each other. A familiar example of partial miscibility is water and diethyl ether, $(C_2H_5)_2O$: when these two liquids are shaken together at 25°C, two layers are obtained: 94.1% water and 5.9% ether (by weight) in the bottom layer, 1.3% water and 98.7% ether in the top layer.

13.5 □ DEPENDENCE OF SOLUBILITY ON TEMPERATURE

We observed in Section 13.3 that, at constant pressure, the solubility of a gas in a liquid decreases with increasing temperature. The solubility of a solid in a liquid may change with temperature in either direction, as illustrated by the following solubilities in water, in grams of solute per 100 g H_2O:

Temperature (°C)	20	40	60	80
Sucrose ($C_{12}H_{22}O_{11}$)	204	238	287	362
Lithium carbonate (Li_2CO_3)	1.33	1.17	1.01	0.85

The typical behavior is that of sucrose; the solubility of a solid in a liquid usually increases with increasing temperature. We may think of dissolving as analogous to melting; we know that raising the temperature causes melting and, similarly, it causes more solid to dissolve.

This is not the whole story, however. Two things happen when a solid dissolves: solute molecules are pulled apart, as in melting (or even as in evaporation), which requires an input of heat; solute molecules become associated with solvent molecules, with a resultant output of heat. For most solutions of solids in liquids, the "melting" effect predominates. When a gas dissolves in a liquid, only the association effect is present (as in condensation), and heat is given out. The change of solubility with temperature can be predicted if we know whether heat is given out or absorbed when the solution is formed.* When 1 mole of Li_2CO_3 is dissolved in 220 moles of water, 3060 cal of heat is given out; when 1 mole of SO_2 is dissolved in 2000 moles of water, 8554 cal of heat is given out. In both these cases, the solubility decreases with increasing temperature. However, when 1 mole of sucrose is dissolved in a large quantity of water, 1320 cal of heat is absorbed. In this case, the solubility increases with increasing temperature. The rule is general: *if heat is given out on dissolving one component in the other, the solubility decreases with increasing temperature; if heat is absorbed, the solubility increases with increasing temperature.*

13.6 □ SUPERSATURATION

Suppose that we prepare a solution of a solid in a liquid at a high temperature, and then cool the solution. In most cases, the solubility decreases on cooling, and the solution may, at some temperature, become saturated. If we expect a crystal to appear at or just below the temperature of saturation, we may be disappointed. It is often possible to cool the solution far below this temperature without the appearance of solid. The solution existing under these conditions contains a larger percentage of solute than a saturated solution at the same temperature; it is said to be **supersaturated**. One who has learned that a saturated solution contains the largest possible proportion of solute, at a given temperature, may feel that a solution in which this proportion is exceeded is somehow freakish. But there is nothing wrong with a supersaturated solution, any more than with a book standing on end; in each case, there exists another situation which is more stable,

* Strictly speaking, the heat effect in question is that associated with the addition of solute to a nearly saturated solution. There are rare instances in which heat is emitted on forming the dilute solution and absorbed on adding solute to a concentrated solution (for example, NaOH in H_2O), or conversely.

namely, saturated solution plus excess solid, or the book lying flat. Just as the book needs a push in order to reach its more stable position, the supersaturated solution may not disgorge its excess solute without a little encouragement. Supersaturation can be observed in any kind of solution, provided a finite solubility exists, but the phenomenon is most commonly observed when one component is a crystalline solid.

It is not easy to initiate the growth of a crystal. The embryonic crystal must be formed by several molecules or ions that come together in the correct configuration and stay that way long enough for other particles to deposit on them. Such an event is not impossible, but remains improbable until the solution is considerably supersaturated. Supersaturation is much more likely to be relieved by deposition of solute molecules on a dust particle, the container wall, or any other solid present. The best crystallization nucleus, of course, is a fragment of the solute itself, or of something else with a similar crystal structure. When such a "seed" is introduced, crystals usually form very rapidly, leaving a saturated solution.

13.7 ▫ SOLUBILITY AND MOLECULAR STRUCTURE

When a solid or a liquid dissolves in a liquid, the molecules of each kind must lose some of their like neighbors, obtaining in exchange neighbors of the other kind. When the solute is a gas, its molecules have no near neighbors, but they lose something else by dissolving: their freedom to move about. The stronger the attractive forces between unlike molecules, the greater the solubility. For a component that is, when pure, a solid or liquid, strong forces between its molecules will inhibit its solubility, unless the molecules of the other component can exert a comparable attraction.

It is a time-honored, although ambiguous, maxim among chemists that "like dissolves like," and this is as we should expect; forces between similar molecules are comparable to those between identical molecules, a condition under which the new molecules can replace the previously neighboring molecules with impunity. The most important kind of similarity is the presence or absence, in the molecule, of local concentrations of electric charge. The water molecule, for example, is highly polar. Water molecules, in the solid or the liquid, hang firmly together (see p. 170). If another molecule is to break into this happy arrangement, it, too, must have positive and negative charges to attract the opposite charges in the water molecule; best of all, it may have sites where hydrogen bonding to H_2O can occur. Examples are

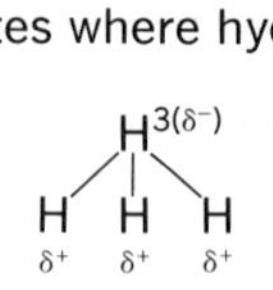

ammonia

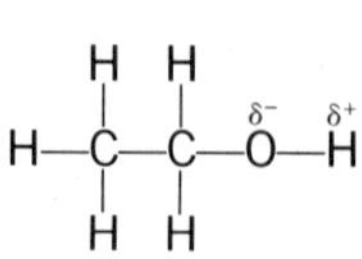

ethanol

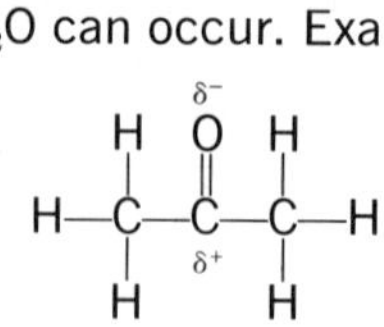

acetone

Ammonia is a gas so soluble that, under ordinary conditions, nearly 700 volumes will dissolve in 1 volume of water; the other two compounds are liquids completely miscible with water. On the other hand, 2,2,4-trimethylpentane (iso-octane), a typical constituent of gasoline, has no strong inhomogeneities in its electric charge distribution; its structure is

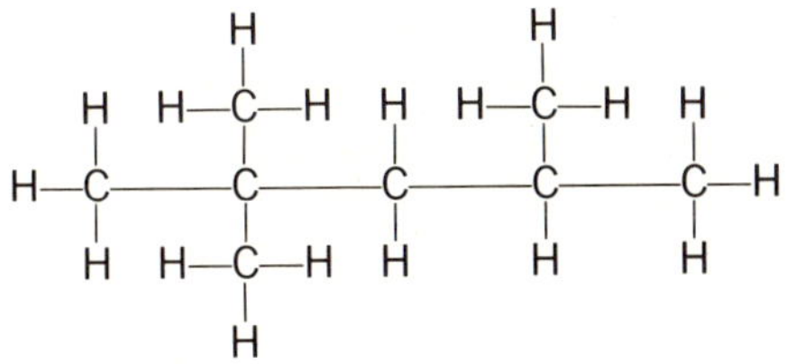

The C—H bond is only slightly polar. It should not surprise us that water and gasoline are not appreciably miscible. But gasoline is a very good solvent for oils, greases, and waxes, which also have nearly nonpolar molecules; neither substance has very strong intermolecular forces in these cases.

The most striking examples of the relation between charge distribution and solubility will be considered in Chapter 15, where we shall see that such highly polar liquids as H_2O and NH_3 (which boils at $-33°C$) can hold ions (charged particles) in solution, a feat that gasoline could never hope to imitate.

13.8 □ DETERGENCY

There is no reason why a single molecule cannot have a part that is polar, or even charged—and therefore has a strong affinity for water—and another part consisting of a long hydrocarbon chain. Such a molecule or ion—the most important examples are ions—should be somewhat soluble both in oil and in water. More important, it can facilitate the dispersion of oil into water, or of water into oil. A salt containing such ions may be useful as a **detergent,** or cleansing agent, because of its ability to bring the oily, greasy components of dirt into colloidal dispersion in water.

Molecules at the surface of a phase have higher energy than molecules deeper within the phase (see p. 212). When two immiscible liquids, or a liquid and a solid that does not dissolve in it, are in contact, the area of contact counts as liquid surface. Work must therefore be done to create additional surface area. The work required, per unit increase in area, is called the **interfacial tension.**

Detergent ions find an especially congenial environment at the interface between oil and water, for they can have their hydrocarbon ends in the oil and their ionic ends in the water—the best of both worlds (Fig. 13.2). The oil and water molecules are likewise attracted to the corresponding ends of the detergent ions, the interfacial tension being thereby greatly reduced. It thus becomes possible to detach oil or

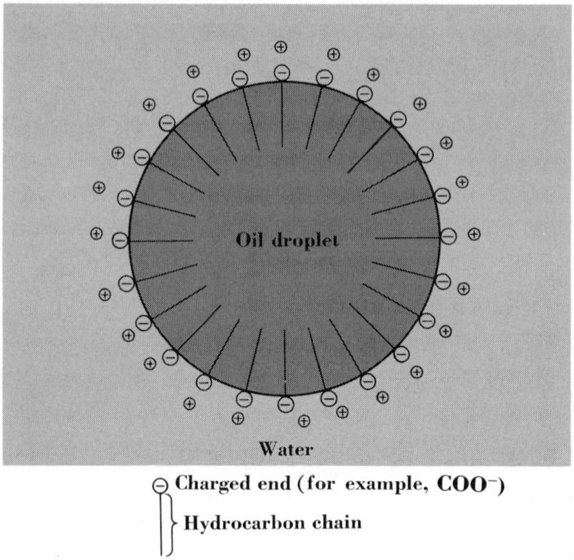

Fig. 13.2 Detergent ions at an oil-water interface.

grease from a soiled surface and bring it into a finely divided state, dispersed throughout the water, with the surface of each globule protected by a layer of detergent ions.

The most common detergents have either —COO^- or —OSO_3^- as their ionic parts. Three examples are shown below.

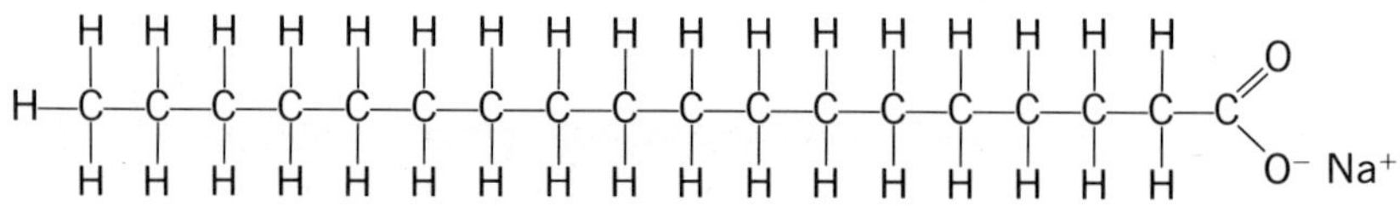

sodium stearate

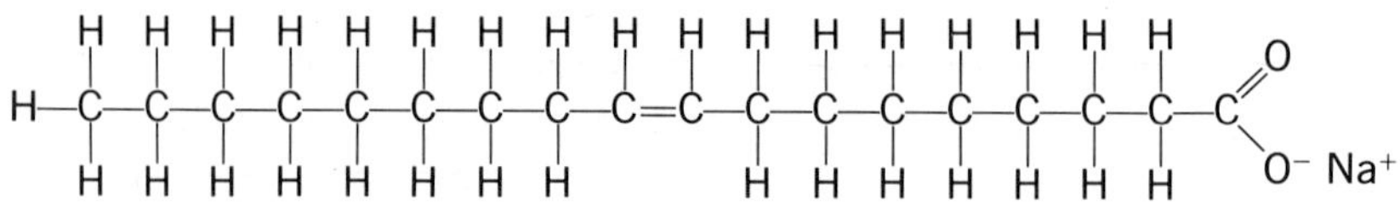

sodium oleate

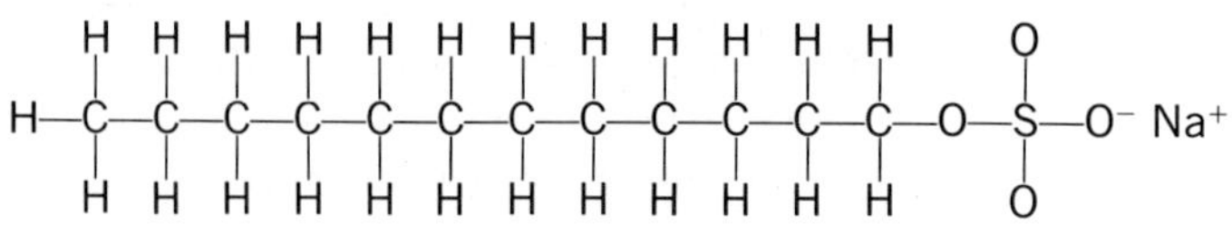

sodium lauryl sulfate

The word **soap** is used technically to refer to any salt in which the negative ion contains the carboxylate group, —COO^-, and a long hydrocarbon chain, about $C_{15\text{-}17}$. Thus, sodium stearate and sodium oleate are soaps, but sodium lauryl sulfate is not; it is a nonsoap detergent or synthetic detergent ("syndet"). Carboxylate salts of metals other than the alkali metals are usually only slightly soluble in water and therefore not useful as detergents, but are nevertheless called "soaps." Some of them find industrial application as lubricants. However, insoluble soaps are more often nuisances. Natural waters usually contain Ca^{2+}, Mg^{2+}, Fe^{2+}, and perhaps other ions. Such water is described as "hard," and these positive ions react with, for example, the stearate ion, forming nearly insoluble salts. The result is that soap is wasted—the first portion is consumed in reacting with the troublesome metal ions—and the insoluble soap clings to the fabric, the sides of the bathtub, or whatever other surface is available. The greatest advantage of the synthetic detergents is the solubility of the salts they form with Ca^{2+}, Mg^{2+}, and so on. They are much better than soaps for use with hard water.

13.9 □ SOLID SOLUTIONS

A solid solution is usually prepared by mixing the liquids and freezing the liquid solution. Although solid solutions are not at all uncommon, the number of pairs of substances that show appreciable miscibility as solids is much less than the number of pairs that mix in the liquid state. The regular arrangement of molecules in a crystal will tolerate only those replacements that resemble the solvent molecules in their type of intermolecular forces, and have nearly the correct size and shape to fit in without causing serious distortion of the lattice. Solid solutions are most frequently found among metals; all metals are made of spherical ions and free electrons, with essentially the same type of bond (Ch. 23) holding them together, and ions of similar size should be able to replace each other. Thus, copper and nickel form crystals with the same structure, and the radii of the atoms (half the internuclear distances) are not too different: 1.278 and 1.246 Å, respectively. These two metals are completely miscible in the solid state. The same is true of silver and gold, with radii of 1.444 and 1.442 Å, respectively. But copper and silver are only partially miscible. Another large class of solid solutions consists of ionic crystals, where ions of similar size, shape, and charge can replace each other indiscriminately. Examples are K_2SO_4 and $(NH_4)_2SO_4$; $FeCO_3$ and $MnCO_3$. The radii of the positive ions are K^+, 1.33 Å; NH_4^+, 1.48 Å; Fe^{2+}, 0.76 Å; Mn^{2+}, 0.80 Å. In $CaCO_3$, on the other hand, only a small amount of Mg^{2+} (0.65 Å) can replace Ca^{2+} (0.99 Å).

In the solid solutions described in the preceding paragraph, each solute particle replaces a solvent particle at its proper lattice position. In another kind of solid solution, small atoms fit into the crevices between large ones without changing the arrangement of the large atoms, except perhaps for some distension of the lattice. Many metals, notably palladium, can absorb hydrogen, which is believed to be present as interstitial H atoms.

13.10 □ MEASURES OF COMPOSITION FOR SOLUTIONS

In discussing solutions, we must be able to specify their compositions—that is, the relative amounts of the several components. Composition is expressed in a number of ways. Let us assume that we have a solution of two components, A and B. It is usual to represent the solvent by A and the solute by B. We adopt the following notation:

w_A, w_B = mass ("weight") in grams of A or B in the solution
n_A, n_B = number of moles of A or B
V_A, V_B = volume of pure A or B, in liters
V = total volume of the solution, in liters

The most important measures of composition for liquid solutions are the following.

(a) **Weight fraction** (more properly "mass fraction") of B is

$$w_B/(w_A + w_B)$$

Weight percentage is, of course, 100 times the weight fraction. These are the simplest and perhaps most useful measures; nothing need be known about the components except their masses, which are easily determined experimentally, and which always add up to the mass of the solution.

(*b*) **Weight ratio** is w_B/w_A; more commonly, 100 w_B/w_A is given. Solubilities are often tabulated in terms of this measure, which must not be confused with weight percentage.

(*c*) **Volume fraction** of B is V_B/V, the volume of pure B divided by the volume of the solution. It should be noted that, in general, $V \neq V_A + V_B$; there is often an appreciable change in volume when two substances form a solution. Volume fraction (or percentage) is confined to liquid-liquid solutions, and is more commonly used in commerce than in scientific work. The "proof number" of an ethanol-water solution is twice the volume percentage of the ethanol; 100 proof whiskey contains a quantity of ethanol that would occupy, if pure, half the volume of the whiskey.

(*d*) **Mole fraction** of B, denoted by X_B, is the ratio of the number of moles of B to the total number of moles,

$$X_B = \frac{n_B}{n_A + n_B}$$

Let $\mathcal{N}$ be Avogadro's number; then $\mathcal{N}n_A$ and $\mathcal{N}n_B$ are the numbers of molecules of A and of B, respectively, in the solution, and

$$X_B = \frac{\mathcal{N}n_B}{\mathcal{N}n_A + \mathcal{N}n_B}$$

A mole fraction is a number fraction—the mole fraction of B is the fraction of all the molecules in the solution that are B molecules. It is obvious from the definition that $X_A + X_B = 1$ or, for more than two components, $X_A + X_B + X_C + \cdots = 1$.

(*e*) **Molality** of B is

$$m_B = \frac{n_B}{(w_A/1000)} = \frac{\text{moles B}}{\cancel{\text{g A}} \times \dfrac{1\text{ kg A}}{1000\ \cancel{\text{g A}}}} = \frac{\text{moles B}}{\text{kg A}}$$

that is, the number of moles of B dissolved in one kilogram (1000 g) of A; w_A is in grams, and $w_A/1000$ is the number of kilograms of A in the solution.

(*f*) **Molarity,** or **concentration** of B is n_B/V, with V in liters; it is the *number of moles of B per liter of solution.* The molarity of B is represented by [B], M_B, or c_B. When we know the molarity of a solution, we can measure out a certain volume of it and calculate the number of moles of B in that volume. Against this virtue must be set the disadvantage, not shared by molality or mole fraction, that the molarity changes on a mere change of temperature, because of the thermal expansion or contraction of the solution.

The following statements are equivalent: the molarity of B is 0.1; the molarity of the solution with respect to B is 0.1; the solution is 0.1 *M* (0.1 molar) with respect to B. The same forms of language are used in specifying molality.

It is customary to refer to the molality or molarity of a solution with respect to the solute (B), not the solvent. These measures of composition are most useful for dilute solutions.

EXAMPLE 1 A solution of 20.0% ethanol, C_2H_5OH, and 80.0% water, by weight, has density 0.966 g/ml at 25°C. Find (a) the mole fraction, (b) the molality, (c) the molarity of ethanol in this solution.

ANSWER The molecular weights are 18.016 g/mole for water and 46.07 g/mole for ethanol. We fix our attention on a definite but arbitrary quantity of solution, say 100 g. In this 100 g of solution, there are 20.0 g ethanol and 80.0 g water; the numbers of moles of the components are

$$n_{\text{water}} = \frac{80.0\text{ g}}{18.02\ \dfrac{\text{g}}{\text{mole}}} = 4.44\text{ moles}$$

$$n_{\text{ethanol}} = \frac{20.0\text{ g}}{46.1\ \dfrac{\text{g}}{\text{mole}}} = 0.434\text{ mole}$$

The volume of the 100 g of solution (needed only in calculating the molarity) is

$$\frac{100.0\text{ g}}{0.966\text{ (g/ml)}} = 103.5\text{ ml} = 0.1035\text{ liter}$$

(a) $$X_{\text{ethanol}} = \frac{0.434}{4.44 + 0.434} = \frac{0.434}{4.87} = 0.0891 \text{ (mole fraction)}$$

(b) $$m_{\text{ethanol}} = \frac{0.434 \text{ mole } C_2H_5OH}{0.080 \text{ kg } H_2O} = 5.43 \frac{\text{mole}}{\text{kg}} \text{ (molality)}$$

(c) $$[C_2H_5OH] = \frac{0.434 \text{ mole}}{0.1035 \text{ liter}} = 4.19 \frac{\text{mole}}{\text{liter}} \text{ (molarity)}$$

A solution of known molarity can be prepared without knowing its density or the weight of solvent used, with the aid of a volumetric flask (Fig. 13.3). The solute must be available in a state of high purity, in order that a known mass will correspond to a definite number of moles. An accurately measured mass of solute (usually a solid) is dissolved in the solvent, and the solution is tranferred completely (with rinsing) to the flask; or, if the solute flows freely and dissolves easily, it may be introduced directly into the flask through a dry funnel with a wide aperture. Solvent is then added to slightly below the mark on the neck. When the solute is entirely dissolved and the solution well mixed, solvent is added carefully to the mark, and the solution mixed again. We now have a known volume of solution containing a known number of moles of solute, and can easily calculate the molarity. If we need to

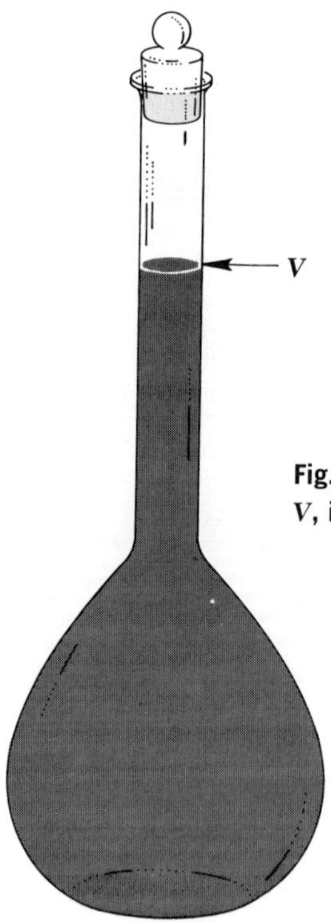

Fig. 13.3 Volumetric flask. The capacity, up to the mark *V*, is accurately known.

know the molality, weight percentage, or some other measure of composition, we shall need the density of the solution.

EXAMPLE 2 (a) A solution was prepared by dissolving 86.53 g of sodium carbonate, Na_2CO_3, in water in a 1000-ml volumetric flask at 20°C, adding water to the mark, and mixing. Find the molarity of the solution. (b) The density of the solution, at 20°C, is 1.0816 g/ml. Find its molality.

ANSWER (a) The molecular weight of Na_2CO_3 is 105.99 g/mole. The number of moles used is

$$\frac{86.53 \not{g}}{105.99 (\not{g}/\text{mole})} = 0.8164 \text{ mole}$$

The volume of the solution is 1.000 liter; the molarity is therefore 0.8164 mole/liter.

(b) The mass of 1 liter of solution is

$$1.0816 \frac{\text{g}}{\not{\text{ml}}} \times 1000 \frac{\not{\text{ml}}}{\text{liter}} = 1081.6 \frac{\text{g}}{\text{liter}}$$

We know that this quantity of solution contains 86.53 g Na_2CO_3, and it must therefore contain 995.1 g H_2O:

$$1081.6 - 86.53 = 995.1 \text{ g } H_2O$$

There is 0.8164 mole Na_2CO_3 in 995.1 g H_2O; in 1000 g H_2O, there is

$$0.8164 \text{ mole} \times \frac{1000 \text{ g/kg}}{995.1 \text{ g}} = 0.8203 \text{ mole/kg}$$

The molality is 0.8203.

(g) **Normality** of B, represented by N_B, is the number of equivalents of B per liter of solution. The abbreviation for equivalent is eq.

$$N_B = \frac{\text{equivalents}_B}{V} \qquad (V \text{ is in liters})$$

The number of equivalents of a substance is its quantity in grams divided by the **equivalent weight** expressed in grams per equivalent (g/eq),

$$\text{equivalents (eq)} = \frac{\text{mass of substance } (\not{g})}{\text{equivalent weight } (\not{g}/\text{eq})}$$

Note that this expression is analogous to that which defines moles in terms of molecular weight (p. 68).

The equivalent weight of a substance is the weight that, in a particular chemical reaction, is involved in the exchange of one mole of unit charges. The equivalent weight is therefore the molecular weight divided by a whole number—the number of unit charges exchanged in a particular reaction per molecule of the substance. The units of this whole number are equivalents per mole. Equivalent weights are therefore not constants; they depend on the nature of the reaction.

$$\text{equivalent weight}\left(\frac{\text{g}}{\text{eq}}\right)=\frac{\text{molecular weight (g/mole)}}{\text{whole number of unit charges exchanged per molecule (eq/mole)}}$$

EXAMPLE 3 Calculate the equivalent weight of the first species represented in each of the following equations.

(a) $H_3PO_4 + Na^+OH^- \longrightarrow Na^+H_2PO_4^- + HOH$
(phosphoric acid)
(b) $H_3PO_4 + 2Na^+OH^- \longrightarrow Na^+{}_2HPO_4^{2-} + 2HOH$
(c) $H_3PO_4 + 3Na^+OH^- \longrightarrow Na^+{}_3PO_4^{3-} + 3HOH$
(d) $Cr_2O_7^{2-} + 3SO_2 + 5H^+ \longrightarrow 2Cr^{3+} + 3HSO_4^- + H_2O$
(e) $2La^{3+} + 3C_2O_4^{2-} \longrightarrow La_2(C_2O_4)_3$

ANSWER (a) The H_3PO_4 (98.00 g/mole) transfers 1 mole of charge (H^+).

$$\text{equivalent weight}=\frac{\text{molecular weight (g/mole)}}{1\text{ (eq/mole)}}=\frac{98.00\text{ (g/mole)}}{1\text{ (eq/mole)}}$$
$$=98.00\text{ g/eq}$$

(b) The H_3PO_4 transfers 2 moles of H^+.

$$\text{equivalent weight}=\frac{98.00\text{ (g/mole)}}{2\text{ (eq/mole)}}=49.00\,\frac{\text{g}}{\text{eq}}$$

(c) The H_3PO_4 transfers 3 moles of H^+.

$$\text{equivalent weight}=\frac{98\text{ (g/mole)}}{3\text{ (eq/mole)}}=32.67\,\frac{\text{g}}{\text{eq}}$$

(d) One $Cr_2O_7^{2-}$ ion acquires 6 electrons (see Sect. 15.7) to become $2Cr^{3+}$. The same result can be calculated from the total change in oxidation number. The oxidation number of Cr in $Cr_2O_7^{2-}$ is +6 (see p. 126), and in Cr^{3+} it is +3. This is a change of 3 unit charges per Cr atom or 6 unit charges for the two Cr atoms in $Cr_2O_7^{2-}$.

$$\text{equivalent weight}=\frac{216\text{ (g/mole)}}{6\text{ (eq/mole)}}=36.0\,\frac{\text{g}}{\text{eq}}$$

(e) The La^{3+} transfers its charge of 3+ to become associated with the $C_2O_4^{2-}$ ion in an ionic solid.

$$\text{equivalent weight}=\frac{138.9\text{ (g/mole)}}{3\text{ (eq/mole)}}=46.3\,\frac{\text{g}}{\text{eq}}$$

Calculations of the normality of solutions are strictly analogous to those for molarity, except that equivalents are used instead of moles, and equivalent weights instead of molecular weights.

EXAMPLE 4 Give directions for preparing 500 ml of a 0.500 N solution of phosphoric acid that is to be used in a reaction with NaOH to produce Na_3PO_4 in a nonaqueous solution.

ANSWER In this reaction H_3PO_4 has 3 eq/mole and its equivalent weight is therefore 32.67 g/eq. (See Example 3c.) From the definition of normality, the number of equivalents needed is

$$\begin{aligned}\text{equivalents} &= NV\\ &= 0.500\ \frac{\text{eq}}{\text{liter}} \times 0.500\ \text{liter}\\ &= 0.250\ \text{eq}\end{aligned}$$

Thus, the weight of H_3PO_4 needed is

$$0.250\ \text{eq} \times 32.67\ \frac{\text{g}}{\text{eq}} = 8.17\ \text{g}$$

We therefore place 8.17 g of phosphoric acid in a 500 ml volumetric flask and fill the flask to the mark with solvent.

The utility of expressing concentrations in normality derives from the fact that electrical charge is conserved in chemical reactions. Consequently, numbers of equivalents in a chemical reaction, which correspond directly to numbers of electrical charges, are also conserved, and *equal numbers of equivalents of substances are involved in a chemical reaction*. Therefore, for any two substances, A and B, in any reaction, we may write

$$\text{eq}_\text{A} = \text{eq}_\text{B}$$

and since eq $= NV$, we may also write

$$N_\text{A}V_\text{A} = N_\text{B}V_\text{B}$$

Thus, when substances react *in solution*, and their concentrations are expressed in normality, the above equation can be solved to determine what volume of one solution will react completely with a given volume of another.

EXAMPLE 5 (a) What volume of 0.0100 N NaOH will just react completely with (neutralize) 15.1 ml of 0.0250 N H_2SO_4? (b) What weight of NaOH will just neutralize 200 ml of 10.0 N H_2SO_4?

(a) $0.0100\ N \times V_\text{NaOH} = 0.0250\ N \times 15.1\ \text{ml}$

$$V_\text{NaOH} = \frac{0.0250\ N}{0.0100\ N} \times 15.1\ \text{ml} = 37.8\ \text{ml}$$

Note that if we use milliliters as the units of volume on both sides of the equation, the expression is dimensionally consistent and it is unnecessary to convert to liters.

(b) The number of equivalents of H_2SO_4, and therefore also of NaOH, is

$$\frac{200\ \text{ml}}{10^3\ \text{ml/liter}} \times 10.0\ \text{eq/liter} = 2.00\ \text{eq}$$

The equivalent weight of NaOH is $\frac{40.0\ \text{g/mole}}{1\ \text{eq/mole}}$, or 40.0 g/eq. The mass (weight) needed is therefore

$$2.00\ \text{eq} \times 40.0\ \text{g/eq} = 80.0\ \text{g}$$

13.11 □ VAPOR PRESSURES OF SOLUTIONS; RAOULT'S LAW

We saw in Chapter 12 that a solid or liquid, placed in an enclosed space, evaporates to some extent. The partial pressure of the vapor at equilibrium is the **vapor pressure** of the substance.

Many substances have no appreciable vapor pressure at ordinary temperatures—most metals and most salts, for example. However, their vapor pressures can be measured at high temperatures, and there is every reason to believe that, on cooling, the vapor pressure becomes smaller and smaller without every becoming exactly zero.

Each component of a solid or liquid solution has a certain vapor pressure, which is *less than the vapor pressure of the pure substance* at the same temperature. The other component apparently occupies part of the surface, and thus decreases the rate at which molecules pass into the vapor. In order to obtain quantitative information about the vapor pressures of solutions, one may prepare a series of solutions of two components, measure the total vapor pressure of each solution at a fixed temperature, and then analyze samples of the vapor to obtain the partial pressure (p. 183) of each component. The vapor pressures are then plotted against one of the measures of composition given on p. 226. The most interesting graphs are obtained when mole fraction (X) is chosen. Examples of such graphs are given in Fig. 13.4. When the mole fraction of component A is zero, its vapor pressure is, of course, zero; when the mole fraction is 1, its vapor pressure p_A is that of the pure substance ($p_A{}^0$). In between, the vapor pressure of each component rises with increasing mole fraction. In Fig. 13.4, the observed (experimental) values of p_{CCl_4} are close to the dashed straight line, which is a plot of $X_{CCl_4}p^0_{CCl_4}$—the calculated value which p_{CCl_4} would have if it were directly proportional to the mole fraction X_{CCl_4} of CCl_4. Likewise, $p_{C_6H_6}$ is close to the dashed line which represents $X_{C_6H_6}p^0_{C_6H_6}$.

There are many liquid solutions in which *the partial vapor pressure of each component is nearly equal to the mole fraction of that component times the vapor pressure of the pure component,*

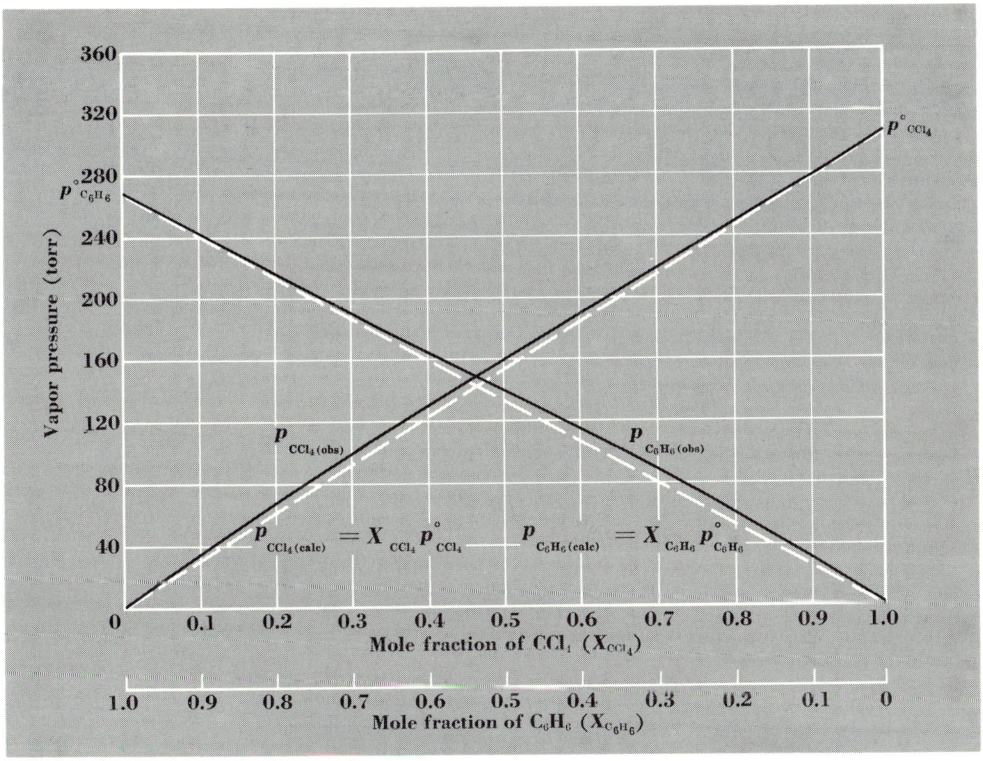

Fig. 13.4 Vapor pressures of benzene-carbon tetrachloride solutions at 49.99°C. $p°C_6H_6$ and $p°CCl_4$ are the vapor pressures of pure benzene and pure carbon tetrachloride, respectively. The solid curves represent the experimental values; the dashed lines represent the vapor pressures predicted by Raoult's law.

$$p_A = X_A p_A^0 \tag{1}$$

This relation was first pointed out by François Marie Raoult in 1886, and is called **Raoult's law.** It is a good approximation for solutions in which the different molecules are very similar in size and polarity. It seems that, in such cases, the fraction of the surface occupied by A molecules is X_A—the same as the number fraction of A molecules in the bulk of the solution—and the rate of escape of A molecules into the vapor is only X_A times as large as in pure A, while the rate of return is unaffected. A solution in which both (or all) components have vapor pressures as given by Raoult's law is called an **ideal solution.** (An "ideal" solution has nothing to do with an "ideal" gas, except that each is described by an especially simple law.) Probably there are no exactly ideal solutions, but many solutions are nearly ideal. C_6H_6 and CCl_4 are an example of a pair of liquids that form an almost ideal solution.

13.12 □ VAPOR PRESSURE DEPRESSION

In any *dilute* solution, the vapor pressure of the *solvent* A is approximately equal to the mole fraction of A times the vapor pressure of pure A: $p_A = X_A p_A{}^0$. This equation holds for the solvent even when the solution is not ideal. Let us assume that there is only one solute, B. Then

$$X_A = 1 - X_B$$
$$p_A = (1 - X_B)p_A{}^0$$

or

$$p_A{}^0 - p_A = \Delta p_A = X_B p_A{}^0 = \frac{n_B}{n_A + n_B} p_A{}^0$$

where n_A and n_B are the numbers of moles of the two components. If the solution is dilute, n_B is much less than n_A, and we can simplify this equation to

$$\Delta p_A = p_A{}^0 - p_A \approx \frac{n_B}{n_A} p_A{}^0 \qquad (2)$$

The quantity Δp_A is the vapor pressure depression caused by the addition of the solute to the solvent. Any property of a solution that is approximately proportional to the number of moles (or molecules) of solute per unit quantity of solvent, independently of the identity of the solute, is called a **colligative property;** vapor pressure depression is an example of such a property.

EXAMPLE 6 By how much is the vapor pressure of 100 g of water depressed when 2.00 g of sucrose ($C_{12}H_{22}O_{11}$) is dissolved in it at 25°C?

ANSWER We first find the numbers of moles of the two substances (the molecular weight of sucrose is 342.3 g/mole):

$$n_{\text{water}} = \frac{100 \text{ g}}{18.0 \text{ (g/mole)}} = 5.55 \text{ moles}$$

$$n_{\text{sucrose}} = \frac{2.00 \text{ g}}{342 \text{ (g/mole)}} = 0.00584 \text{ mole}$$

The vapor pressure of pure water at 25°C is 23.8 torr $= p^0_{\text{water}}$. Then

$$\Delta p = p^0_{\text{water}} - p_{\text{water}} = \frac{0.00584 \text{ mole}}{5.55 \text{ moles}} \times 23.8 \text{ torr} = 0.0250 \text{ torr}$$

13.13 □ BOILING-POINT ELEVATION AND FREEZING-POINT DEPRESSION

Two other colligative properties* can be understood with the aid of Fig. 13.5. The black lines represent temperatures and pressures at which two phases of a typical substance (call it A) are in equilibrium

* Osmotic pressure is also a colligative property of a solution. It is the pressure exerted on a solution by its solvent when they are separated by a semipermeable membrane.

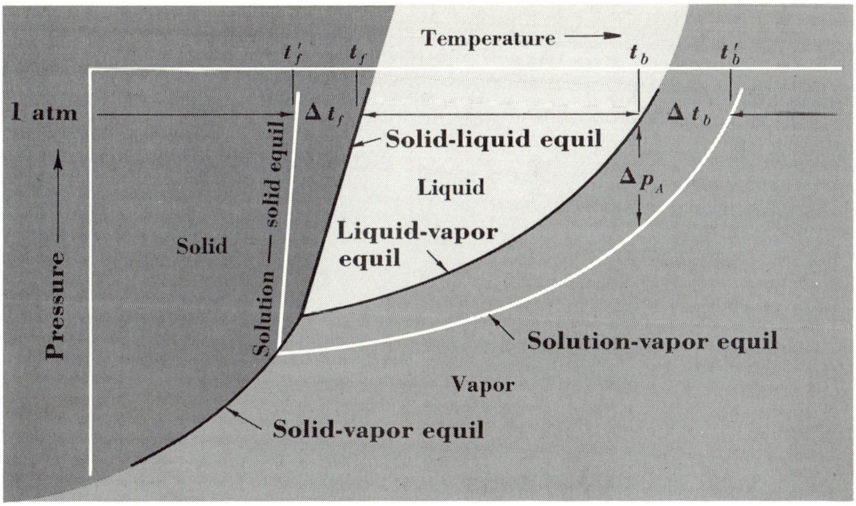

Fig. 13.5 Phase diagram for pure solvent (black curves) and solution (white curves).

with each other. The line labeled "liquid-vapor equilibrium" gives, by definition, the vapor pressures of the liquid at different temperatures; the curves of Fig. 12.13 (p. 209) represent similar equilibria. The three black lines, taken together, are called the **phase diagram** of the substance. Now, let us add a solute (call it C) that has a negligible vapor pressure to the solvent A. The vapor pressure of the liquid is lowered by the presence of the solute. The white line labeled "solution-vapor equilibrium" gives the vapor pressures of the solution at different temperatures; the difference between the two values, Δp_{A}, at any temperature, is the vapor pressure depression.

The boiling point of the pure liquid, t_b, is the temperature at which the vapor pressure equals 1 atm. The solution must be heated to a higher temperature before its depressed vapor pressure can be brought to 1 atm; that is to say, its boiling point, t'_b, is greater than that of the pure liquid. The difference, $t'_b - t_b$, or Δt_b, is called the **boiling point elevation.** Since we are considering only the case of a nonvolatile solute, the vapor is composed entirely of A, regardless of whether the liquid is pure A or a solution.

The freezing point of the pure liquid, t_f, is the temperature at which solid and liquid are in equilibrium at 1 atm. The solution must be cooled to a lower temperature before it reaches equilibrium with the solid; that is to say, its freezing point, t'_f, is lower than that of the pure liquid. The difference, $t_f - t'_f$, or Δt_f, is called the **freezing point depression.** We consider only a solute which does not form a solid solution with the solvent A. Then the solid is composed entirely of A, regardless of whether the liquid is pure A or a solution.

The value of Δt_b depends on the degree of separation between the liquid-vapor line and the solution-vapor line in the region near 1 atm pressure. We can see from Fig. 13.5 that this separation is related to Δp_A. Over a small range of pressure (near 1 atm) this relationship approximates a direct proportionality, and

$$\Delta t_b \propto \Delta p_A$$

and, from Equation 2, assuming the solution is dilute,

$$\Delta t_b \propto \frac{n_C}{n_A} p_A{}^0$$

By convention, we refer to molality (which is proportional to the mole ratio) and

$$\Delta t_b \propto m_C$$

or

$$\Delta t_b = K_b m_C$$

The constant, K_b, depends both on Δp_A and (see Fig. 13.5) on the slope of the liquid-vapor equilibrium line at 1 atm. The latter, and therefore K_b, are characteristic of the solvent.

A similar expression, derived for t_f, is

$$\Delta t_f = K_f m_C$$

Constants for some common solvents are listed in Table 13.2.

EXAMPLE 7 Find the freezing and boiling points of the water-sucrose solution described in Example 5 (2.00 g sucrose in 100 g water).

ANSWER This solution has 0.00584 mole sucrose dissolved in 100 g water. It therefore contains 0.0584 mole sucrose per 1000 g water, and this is its molality; $m = 0.0584$ mole/kg. Then the freezing-point depression is

TABLE 13.2 ▪ BOILING-POINT ELEVATION AND FREEZING-POINT DEPRESSION CONSTANTS

Solvent	Boiling point (°C) at 1 atm	K_b $\left(\frac{\text{deg kg}}{\text{mole}}\right)$	Freezing point (°C)	K_f $\left(\frac{\text{deg kg}}{\text{mole}}\right)$
Acetic acid (CH_3COOH)	118.1	3.07	16.6	3.90
Benzene (C_6H_6)	80.1	2.53	5.51	4.90
Carbon tetrachloride (CCl_4)	76.8	5.03	− 22.8	31.8
Ethanol (C_2H_5OH)	78.5	1.22	−117.3	1.99
Water	100	0.512	0	1.86

$$t_f - t_f' = K_f m = 1.86 \frac{\text{deg } \cancel{\text{kg}}}{\cancel{\text{mole}}} \times 0.0584 \frac{\cancel{\text{mole}}}{\cancel{\text{kg}}} = 0.109°$$

and the freezing point is

$$0° - 0.109° = -0.109°\text{C}$$

The boiling-point elevation is

$$t_b' - t_b = K_b m = 0.512 \frac{\text{deg } \cancel{\text{kg}}}{\cancel{\text{mole}}} \times 0.0584 \frac{\cancel{\text{mole}}}{\cancel{\text{kg}}} = 0.0299°\text{C}$$

and the boiling point is

$$100° + 0.0299° = 100.0299°\text{C}$$

13.14 □ DETERMINATION OF MOLECULAR WEIGHTS

The strongest reason for interest in colligative properties is that they provide a means of measuring the number of moles of solute present, n. If we also know the mass of the solute, w, we can calculate its molecular weight, w/n. Freezing-point depression is the easiest of the properties to measure experimentally, except with solutes of very high molecular weight.

EXAMPLE 8 The freezing point of pure benzene was measured as 5.49°C. With the same benzene and equipment, the freezing point of a solution containing 0.100 g naphthalene in 10.0 g benzene was found to be 5.10°C. Find (a) the molecular weight of naphthalene; (b) the boiling point of a solution containing 3.21% (by weight) naphthalene in ethanol.

ANSWER (a) The freezing-point depression is

$$t_f - t_f' = 5.49 - 5.10 = 0.39°$$

$$t_f - t_f' = K_f m$$

$$K_f = 4.90 \frac{\text{deg kg}}{\text{mole}} \quad \text{(Table 13.2, p. 236)}$$

The molality of naphthalene is

$$m = \frac{t_f - t_f'}{K_f} = \frac{0.39 \cancel{\text{deg}}}{4.90 (\cancel{\text{deg}} \text{ kg/mole})} = 0.080 \frac{\text{mole}}{\text{kg}}$$

The weight of naphthalene per kilogram of benzene is

$$w = \frac{0.100 \text{ g naphthalene}}{10.0 \cancel{\text{g}} \text{ benzene}} \times 1000 \frac{\cancel{\text{g}}}{\text{kg}} = 10.0 \frac{\text{g}}{\text{kg}}$$

The molecular weight is therefore

$$\frac{w}{m} = \frac{10.0\ \dfrac{\text{g}}{\cancel{\text{kg}}}}{0.080\ \dfrac{\text{mole}}{\cancel{\text{kg}}}} = 125\ \frac{\text{g}}{\text{mole}}$$

$$\left(\text{actual value, } 128\ \frac{\text{g}}{\text{mole}}\right).$$

(b) From the statement of the problem, 100 g of solution contains 3.21 g naphthalene and 96.79 g ethanol. The weight of naphthalene per kilogram of ethanol is

$$\frac{3.21\text{ g naphthalene}}{96.79\ \cancel{\text{g}}\text{ ethanol}} \times 1000\ \frac{\cancel{\text{g}}}{\text{kg}} = 33.2\ \frac{\text{g}}{\text{kg}}$$

$$m = \frac{33.2\ (\cancel{\text{g}}/\text{kg})}{125\ (\cancel{\text{g}}/\text{mole})} = 0.266\ \frac{\text{mole}}{\text{kg}}$$

From Table 13.2 (p. 236),

$$K_{b\ \text{ethanol}} = 1.22\ \frac{\text{deg kg}}{\text{mole}}; \qquad t_b = 78.5°\text{C}$$

Then

$$t_b' - t_b = K_b m = 1.22\ \frac{\text{deg kg}}{\text{mole}} \times 0.263\ \frac{\text{mole}}{\text{kg}} = 0.321\text{ deg}$$

and

$$t_b' = 0.321°\text{C} + 78.5°\text{C} = 78.8°\text{C}$$

PROBLEMS

1. **Definitions** Define and illustrate the following terms: (a) saturated solution; (b) solubility; (c) miscibility; (d) supersaturation; (e) solid solution.

2. **Saturation** State whether each of the following solutions is unsaturated, saturated, or supersaturated. Refer to Table 13.1 (p. 219) for data, and assume that the total pressure in each case is 1 atm. (a) 1.75×10^{-5} g N_2 per g H_2O at 25°C; (b) 0.076 g CO_2 per 100 g H_2O at 60°C; (c) 0.145 g CO_2 per 100 g H_2O at 24°C; (d) 22.83 g SO_2 per 100 g H_2O at −1°C.

3. **Solubility and temperature** Saturated solutions of lithium nitrate, $LiNO_3$, in water contain 34.8% $LiNO_3$ at 0°C and 66.0% $LiNO_3$ at 70°C. A solution made from 50.0 g $LiNO_3$ and 50.0 g water was cooled from 70° to 0°C. What mass of $LiNO_3$ precipitated, assuming no water evaporated and there was no supersaturation?

4. **Solubility and molecular structure** From each of the following pairs select the compound that you would expect to be more soluble in water. Explain (you may wish to refer to Table 7.1 of electronegativities, p. 119).

(a) C_2H_5F and C_2H_5I;
(b) C_3H_7OH and C_3H_7SH;
(c) 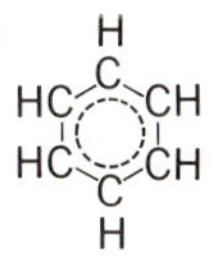 and 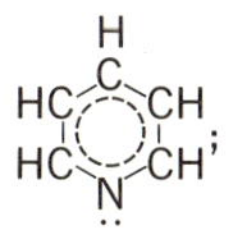;
(d) NH_3 and AsH_3.

5. Molarity Solutions were prepared by dissolving 25.0 g of each of the following solutes in water, and adding water to the mark in a 250-ml volumetric flask. Find the molarity of each solution: (a) KCl; (b) $NaNO_3$; (c) sucrose, $C_{12}H_{22}O_{11}$.

6. Composition The label on a bottle of "concentrated sulfuric acid" gives the following information: density, 1.84 g/ml; 98% H_2SO_4 by weight. Calculate (a) the mole fraction; (b) the molarity; and (c) the normality (assuming the transfer of $2H^+$) of H_2SO_4 in this solution.

7. Raoult's law The vapor pressures of toluene, C_7H_8, and hexachlorobenzene, C_6Cl_6, at 0°C are 111 torr and 309 torr, respectively. Find the total vapor pressure at 0°C of a solution prepared by mixing 5 g C_7H_8 and 20 g C_6Cl_6. Assume that Raoult's law is applicable.

8. Raoult's law Glucose, $C_6H_{12}O_6$, has practically no vapor pressure at ordinary temperatures. The vapor pressure of pure water at 40°C is 55.3 torr. A solution contains 50.0 g glucose and 1000 g water. Find the vapor pressure of this solution at 40°C.

9. Determination of molecular weight An aqueous solution contains 0.34 g of an unknown solute and 50.0 g H_2O. It freezes at −0.092°C. Find (a) the molecular weight of the solute, (b) the boiling point of the solution.

10. Antifreezes Of the commonly used antifreezes, the lower-boiling, or so-called "temporary," one is methanol, CH_3OH, and the higher-boiling, or "permanent," is ethylene glycol, $(CH_2OH)_2$. At one time ethyl alcohol, C_2H_5OH, was also used. (a) Estimate the weight of each of these solutes that must be added to 1 kg of water to prevent the water from freezing at −10°C. List the three antifreezes in order of the degree to which they will lower the freezing point of a given quantity of water per unit weight of antifreeze. (b) Assume that a mixture of water and methanol whose freezing point is −10°C deposits pure ice on freezing. If an automobile whose radiator is filled with this mixture is kept at a temperature of −11°C for an indefinitely long time, will the liquid in the radiator freeze solid? Explain.

ADDITIONAL PROBLEMS

11. Solubility and temperature The solubility of ammonium sulfamate ("ammate" weed killer, $NH_4SO_3NH_2$) in water is 134 g per 100 g H_2O at 0°C and 357 g per 100 g H_2O at 50°C. A solution was prepared by dissolving $NH_4SO_3NH_2$ in 300 ml H_2O at 50°C until no more would dissolve. The solution was then cooled to 0°C. How many grams of ammonium sulfamate were originally dissolved? How many grams precipitated, assuming no evaporation of water, and no supersaturation?

12. Equivalent weights Calculate the number of equivalents per mole and the equivalent weight of the first substance represented in each of the following equations:

(a) $Ca(OH)_2 + 2H^+ \longrightarrow Ca^{2+} + 2H_2O$
(b) $Na_2SO_4 + H^+ \longrightarrow Na^+HSO_4^- + Na^+$
(c) $2KMnO_4 + 5SnCl_2 + 16HCl \longrightarrow 2MnCl_2 + 2KCl + 5SnCl_4 + 8H_2O$
(d) $Hg_2^{2+} + 2Cl^- \longrightarrow Hg_2Cl_2$

13. Molarity You need 100 ml of a 0.15 M solution of NaCl at 20°C. Write *specific* laboratory instructions for preparing this solution. Is it necessary to know what mass of water this solution contains? (Yes or no.)

14. Composition measures A bottle of concentrated hydrochloric acid has the following information on the label: "Assay 38% HCl. Specific gravity 1.18." Find (a) the molarity of HCl in this solution: (b) the normality of this solution for use in the reaction $HCl + NH_3 \longrightarrow NH_4^+Cl^-$; (c) the volume of the concentrated acid that must be used in order to prepare 100 ml of 3.0 M solution by dilution with water.

15. Composition measures An aqueous solution of acetic acid, CH_3COOH, is 0.168 molal with respect to CH_3COOH and has density 1.00 g/ml at 20°C. Find (a) the mole fraction, (b) the weight fraction, (c) the molarity of acetic acid in this solution, and (d) the normality of the solution in a reaction in which acetic acid is used to prepare barium acetate, $Ba^{2+}(CH_3COO^-)_2$.

16. Composition measures You are given a supply of distilled water and of solid, anhydrous sodium oxalate, $Na_2C_2O_4$. Its solubility is 3.7 g $Na_2C_2O_4$ in 100 g water at 20°C, and is greater at higher temperatures. Give specific laboratory directions for making: (a) 1 liter of saturated $Na_2C_2O_4$ solution at 25°C; (b) 500 ml of a supersaturated solution of $Na_2C_2O_4$ at 20°C; (c) 500 ml of 0.2 M $Na_2C_2O_4$ at 20°C; (d) 1 liter of 0.10 N $Na_2C_2O_4$ to be used for a reaction to produce $Ca^{2+}C_2O_4^{2-}$.

17. Composition measures The molarity of HCl in an aqueous solution is 6.05 moles/liter. The density of the solution is 1.10 g/ml. Find (a) the molality, (b) the weight fraction, (c) the weight percent, and (d) the mole fraction of HCl in this solution.

18. Composition measures A solution is 14.0% HCl and 86.0% H_2O by mass. Its density is 1.07 g/ml. Find (a) the mole fraction, (b) the molality, and (c) the molarity of HCl in this solution.

19. Molarity What mass of each of the following compounds should be weighed out in order to prepare a 0.200 molar solution with the aid of a 1000-ml volumetric flask? (a) $AgNO_3$; (b) $MgSO_4(H_2O)_7$; (c) acetamide, H_3CCONH_2.

20. Raoult's law At 25°C, the vapor pressure of liquid n-butane, C_4H_{10}, is 1823 torr; the vapor pressure of liquid n-pentane, C_5H_{12}, is 521 torr. Assume that solutions of these two compounds are ideal. (a) Find the total vapor pressure at 25°C of a solution containing 20.0% n-butane and 80.0% n-pentane, by weight. (b) Find the mole fraction of n-pentane in a solution of n-butane and n-pentane having a total vapor pressure of 1520 torr at 25°C.

21. Raoult's law A recent perfume advertisement states, "Traditional perfumes blend florals, herbs and oils with volatile alcohol. When mixed with air, the alcohol takes off, taking the precious perfume with it." It continues, "Taji is a hauntingly beautiful blend of all precious essences and oils with no volatile alcohol. Touch Taji to your skin and it warms up with you. As your pulse beats, it beats out its fragrance. Instead of coming on strong, then leaving you flat, Taji spends the evening with you. . . ." Restate this advertisement in technical language. Are the implications of these statements consistent with the predictions of Raoult's law with respect to the behavior of volatile substances in solution? Explain your answer.

22. Boiling-point elevation Separate solutions of two nonvolatile solutes, B and C, in the same solvent A were prepared. The molecular weight of B is 50. The boiling points of the solvent and the solutions were as follows:

Solute	Weight of solute (per 100 g A)	Boiling point
None		70.00°C
B	1.00 g	70.40°C
C	0.50 g	70.30°C

Find (a) the boiling-point elevation constant of A; (b) the molecular weight of C.

23. Determination of molecular weight A solution of 8.00 g acetic acid in 100 g benzene freezes at 2.10°C. A solution of 8.00 g acetic acid in 100 g water freezes at −2.50°C. Find the molecular weight of acetic acid from each of these data. What can you conclude about the state of the acetic acid molecules dissolved in each of these solvents? How are these results related to the properties of the water and benzene molecules?

24. Determination of molecular weight A certain hydrocarbon (a compound containing C and H only) contains 6.24% H by weight. When 3.00 g of this hydrocarbon is dissolved in 100 g benzene, the resulting solution freezes 1.146°C below the freezing point of pure benzene. Find the molecular weight and the molecular formula of the hydrocarbon.

25. Determination of molecular weight You are asked to determine the molecular weight of an unknown volatile solid.

Solvent	Solubility of unknown (g per 100 g solvent) at freezing point	Solubility of unknown (g per 100 g solvent) at boiling point
Water	0.2	3.1
Benzene	7.0	7.8

Indicate and justify the procedure and the solvent you would select for the most precise measurement of molecular weight of the unknown.

ANSWERS

3. 15.2 g

5. (a) 1.34 *M*; (b) 1.18 *M*; (c) 0.292 *M*.

8. 55.0 torr.

10. (b) Hint: What is the effect of removing ice on the concentration of antifreeze in the liquid?

12. (a) 37.05 g/eq; (c) 31.61 g/eq.

14. (a) 12 *M*; (b) 12 *N*; (c) 25 ml.

17. (a) 6.96 moles/kg; (b) 0.201; (c) 20.1%; (d) 0.111.

20. (a) 830 torr; (b) 0.233.

22. (a) 2.0 deg kg/mole; (b) 33 g/mole.

23. 115 g/mole in benzene, 60 g/mole in water.

14 ■ CHEMICAL EQUILIBRIUM

14.1 □ INTRODUCTION

Consider a mixture of hydrogen and iodine, in a closed vessel, at a temperature such that both are gases. They will react to form hydrogen iodide, also a gas,

$$H_2 + I_2 \longrightarrow 2HI \tag{1}$$

If we had started with pure HI at the same temperature, it would have decomposed to give hydrogen and iodine,

$$2HI \longrightarrow H_2 + I_2 \tag{2}$$

Now, the HI that is formed in reaction 1 acts like any other HI: as soon as some of it has been formed, the decomposition (reaction 2) sets in. The H_2 and I_2 formed in reaction 2 will react to give HI, as in reaction 1. Thus, in any mixture of the three gases H_2, I_2, and HI, reactions 1 and 2 will both take place. This fact is expressed by saying that reaction 1 or 2 is **reversible,** and writing it with a double arrow,

$$H_2 + I_2 \rightleftharpoons 2HI$$

or

$$2HI \rightleftharpoons H_2 + I_2$$

If reaction 1 is faster than reaction 2, there is a net formation of HI, so that its concentration increases with time, while the concentrations of H_2 and I_2 decrease. As we shall see in Chapter 19, the rate of a reaction usually increases as the concentrations of reactants increase, and decreases as they decrease. In the present case, the falling concentrations of H_2 and I_2 result in a slowing of reaction 1, while the rising concentration of HI speeds up reaction 2. If reaction 2 is faster, the loss of HI will slow it down, and the gain of H_2 and I_2 will speed up reaction 1. Thus, the faster reaction becomes slower, and the slower becomes faster. A state of **equilibrium** will eventually be reached when the two reactions, forward and reverse, are occurring at the same rate; from then on, there will be *no further net change* in the composition of the reaction mixture, unless some external disturbance is imposed, such as a change in temperature or volume of the container, or the addition or removal of one of the substances involved in the reaction. The concept of equilibrium is the same here as for a saturated solution; in that case, the two processes of dissolving and coming out of solution were equal in rate at equilibrium. Now we are concerned with two chemical reactions, one the reverse of the other, which proceed at the same rate when the various concentrations have adjusted themselves to the proper values.

The reaction $H_2 + I_2 \rightleftharpoons 2HI$ was one of the first gas reactions studied in detail with respect to rate and equilibrium. Some of the

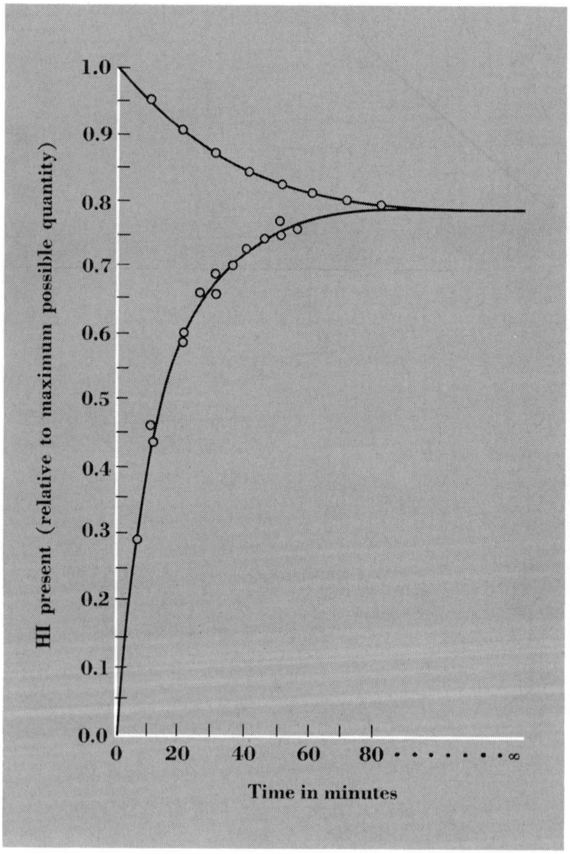

Fig. 14.1 Experimental data on the reaction $H_2(g) + I_2(g) \rightleftharpoons 2HI(g)$ at the boiling point of sulfur, 445°C. The horizontal axis represents the time elapsed from the beginning of the reaction. The vertical axis represents the composition of the reaction mixture, from pure HI at the top to $H_2 + I_2$ at the bottom. [After M. Bodenstein, *Z. phys. Chem.* 13, 111 (1894)]

original experimental data, published by Max Bodenstein in 1894, are shown in Fig. 14.1. In some experiments, he started with various quantities of pure HI, and the quantity initially present is taken as one unit. In other experiments, he started with equal numbers of moles of H_2 and I_2; the quantity of HI that would be formed by complete combination of these starting materials is again called one unit. The vertical axis gives the fraction of this maximum quantity of HI that was present when the sealed tube was opened and its contents analyzed; the length of time (in minutes) that the reaction had been in progress is shown on the horizontal axis. The two curves show clearly that the proportions of the three substances approach the same final values, whether one starts with HI (*upper curve*) or with H_2 and I_2 (*lower curve*).

In this particular case, equilibrium is attained in an hour or two; in other reactions, it may take a tiny fraction of a second, and there are many reactions that would be nowhere close to equilibrium after a million years. Some reactions settle down to equilibrium after formation of only minute amounts of the products; in other reactions, only minute amounts of the reactants remain unreacted at equilibrium. Sometimes, as in the HI reaction, appreciable quantities of both reactants and products are present in the equilibrium mixture.

14.2 □ EQUILIBRIUM IN GASES; THE EQUILIBRIUM CONSTANT

Let the ideal gases A, B, C, and D undergo, at constant temperature, the reaction

$$a\text{A} + b\text{B} \rightleftharpoons c\text{C} + d\text{D}$$

where a, b, c, and d are the stoichiometric coefficients (small integers or simple fractions). We define the **reaction quotient in terms of molar concentrations** as

$$Q_c = \frac{[\text{C}]^c[\text{D}]^d}{[\text{A}]^a[\text{B}]^b}$$

where [A], . . . , [D] are the concentrations in moles per liter of the gases in the reaction mixture. The **law of chemical equilibrium** states that, at each temperature, there is associated with this reaction a number K_c, called the **equilibrium constant in terms of concentrations,** such that at equilibrium, $Q_c = K_c$. Thus at equilibrium we can write

$$K_c = \frac{[\text{C}]^c[\text{D}]^d}{[\text{A}]^a[\text{B}]^b}$$

The condition for equilibrium* can be written on inspection of the chemical equation for the reaction. The molar concentrations of the products, raised to powers equal to the stoichiometric coefficients are

* This condition is derivable from the assertion that *a system in equilibrium can do no work*, that, indeed, work must be done on the system to displace it from equilibrium. The derivation is beyond the scope of this book; we shall content ourselves with the results.

multiplied to form the numerator. The molar concentration of the reactants are treated similarly to form the denominator. The value for the equilibrium constant is a characteristic of the given reaction and varies only with changes in temperature. For this reason, in tables, the temperature is specified when the value of K_c for a reaction is given. An infinite number of variations are possible for the separate values of [A], [B], [C], and [D] at equilibrium, but the value of K_c remains constant.

When $Q_c < K_c$, [C] and [D] are smaller than they would be at equilibrium, for given values of [A] and [B]. Alternatively, we may say that [A] and [B] are greater than they would be at equilibrium, for given values of [C] and [D]. The only possible direction of the reaction, under these conditions, is the direction that makes [C] and [D] larger, and—since A and B are consumed when C and D are formed —makes [A] and [B] smaller. This process is described by saying that the reaction "goes from left to right." Similarly, when $Q_c > K_c$, [C] and [D] are too great, or [A] and [B] are too small; in this case, the only possible direction of the reaction is from right to left—the formation of more A and B at the expense of C and D. In general, the reaction can proceed in the direction making Q_c *more nearly* equal to K_c. We have not said that the reaction will take place in that direction; there may be no reaction at all, or there may be an imperceptibly slow reaction. All we know is that the reaction will not go in the other direction.

An example of a reversible gas reaction is

$$2SO_2(g) + O_2(g) \rightleftharpoons 2SO_3(g)$$

This reaction is the crucial step in the most common process for making sulfuric acid, $(HO)_2SO_2$. The reaction is extremely slow under ordinary conditions, but proceeds quite rapidly above 500°C in the presence of a catalyst, such as platinum or divanadium pentoxide (V_2O_5). The equilibrium condition in this case is

$$\frac{[SO_3]^2}{[SO_2]^2[O_2]} = K_c$$

It is easier to measure the pressure than the molar concentration of a gas. From the ideal gas law, the pressure of a gas is related to its molar concentration (moles per liter) as follows:

$$P_A = \frac{N_A}{V_A}RT = [A]RT$$

Accordingly, the equilibrium constants for gaseous reactions are frequently written in terms of the partial pressures of the gases. For the sulfur dioxide-sulfur trioxide equilibrium, we write

$$\frac{P_{SO_3}^2}{P_{SO_2}^2 P_{O_2}} = K_p$$

where P_{SO_3}, P_{SO_2}, and P_{O_2} are the partial pressures of SO_3, SO_2, and O_2 respectively. K_p is called the **equilibrium constant in terms of pressures.**

It is necessary to agree on the units in which the partial pressures are to be expressed. Atmospheres are most commonly used.

14.3 □ CHANGE OF *K* WITH FORM OF EQUATION

Instead of writing

$$2SO_2 + O_2 \rightleftharpoons 2SO_3$$

with the equilibrium condition

$$\frac{[SO_3]^2}{[SO_2]^2[O_2]} = K_c$$

we could write, no less correctly,

$$4SO_2 + 2O_2 \rightleftharpoons 4SO_3$$

Let us represent by K' the equilibrium constant for the reaction written in this form; the equilibrium condition then becomes

$$\frac{[SO_3]^4}{[SO_2]^4[O_2]^2} = K_c'$$

Obviously, $K_c' = K_c^2$. When the chemical equation is multiplied through by a constant n, the equilibrium constant is raised to the nth power. Thus, for

$$SO_2 + \tfrac{1}{2}O_2 \rightleftharpoons SO_3$$

the equilibrium condition is

$$\frac{[SO_3]}{[SO_2][O_2]^{1/2}} = K_c''$$

and

$$K_c'' = K_c^{1/2} = \sqrt{K_c}$$

If the reaction is written backward

$$2SO_3 \rightleftharpoons 2SO_2 + O_2$$

the equilibrium condition becomes

$$\frac{[SO_2]^2[O_2]}{[SO_3]^2} = K_c''' = \frac{1}{K_c}$$

Reversing the chemical equation changes the equilibrium constant to its reciprocal. To speak of "the equilibrium constant for the formation of SO_3 from SO_2 and O_2" is ambiguous—one should write the chemical equation.

14.4 □ COMBINATION OF EQUILIBRIA

It is possible to find the equilibrium constant for a reaction when the constants for certain other reactions are known. One may be interested, for example, in the reaction

$$SO_2(g) + CO_2(g) \rightleftharpoons SO_3(g) + CO(g) \tag{3}$$

but may find that data on it are not readily available. However, the equilibrium constants for the reactions

$$2SO_2(g) + O_2(g) \rightleftharpoons 2SO_3(g) \tag{4}$$

and

$$2CO_2(g) \rightleftharpoons 2CO(g) + O_2(g) \tag{5}$$

are well known. We can write for them the equilibrium conditions

$$K_4 = \frac{[SO_3]^2}{[SO_2]^2[O_2]} \qquad \text{and} \qquad K_5 = \frac{[CO]^2[O_2]}{[CO_2]^2}$$

respectively. (The constants are labeled by the same numbers as the equations. To avoid cumbersome notation, we omit the subscript c in this section.) In a mixture of all five gases—SO_2, SO_3, CO, CO_2, and O_2—these two conditions must be satisfied simultaneously at equilibrium. If we multiply the left side of one equation by the left side of the other, and similarly multiply the right sides together, we obtain another equation,

$$K_4K_5 = \frac{[SO_3]^2}{[SO_2]^2[O_2]} \times \frac{[CO]^2[O_2]}{[CO_2]^2}$$

or

$$K_4K_5 = \frac{[SO_3]^2[CO]^2}{[SO_2]^2[CO_2]^2} \tag{6}$$

But the equilibrium condition for Equation 3 is

$$K_3 = \frac{[SO_3][CO]}{[SO_2][CO_2]} \tag{7}$$

On comparing Equations 6 and 7, we see that

$$K_3 = (K_4K_5)^{1/2}$$

Of course, K_3 is the same whether or not O_2 is present.

The procedure can be simplified if we observe that, when two reactions are added to obtain a third reaction, their equilibrium constants are multiplied to give the equilibrium constant of the third reaction. Thus, we may add reactions 4 and 5,

$$\begin{array}{rcll} 2SO_2(g) + O_2(g) & \rightleftharpoons & 2SO_3(g) & (4) \\ 2CO_2(g) & \rightleftharpoons & 2CO(g) + O_2(g) & (5) \\ \hline 2SO_2(g) + 2CO_2(g) & \rightleftharpoons & 2SO_3(g) + 2CO(g) & (8) \end{array}$$

Without writing out the equilibrium conditions, we can see, from the way the reaction quotient is constructed, that $K_4K_5 = K_8$. That $K_8 = K_3^2$ is obvious from the rule (p. 247) about multiplying the equation through by a constant. In general, when a number of reactions are added, their equilibrium constants are multiplied together to give the equilibrium constant of the overall reaction—care being taken, of course, to use the equilibrium constants for the reactions as they are actually written, with all reversals, doublings, halvings, and so on taken into account. The process is similar to the calculation of a heat of reaction with the aid of Hess's law (p. 101), except that equilibrium constants are multiplied together, while heats of reaction are added. Changing the sign of ΔH, when the equation is reversed, corresponds to taking the reciprocal of K. Multiplying ΔH by 2, when each term in the equation is doubled, corresponds to squaring K.

14.5 □ PRINCIPLE OF LE CHATELIER

The composition of a mixture at equilibrium depends, in general, on the quantities of the reacting substances that are initially brought together, on the volume of the container in which they are placed, and on the temperature. The only one of these factors affecting the actual value of the equilibrium constant is the temperature. However, a change in volume, or the addition or removal of matter, may also require a change in the proportions of the reacting substances in order that the equilibrium condition shall continue to be satisfied under the new conditions. Before we study quantitative calculations on equilibrium mixtures, we shall develop some useful rules enabling us to predict qualitatively the direction in which the composition of an equilibrium mixture will change.

(a) Change of quantity If a mixture of SO_2, O_2, and SO_3 is in equilibrium and more O_2 is added, the initial effect is to make $[SO_3]^2/[SO_2]^2[O_2]$ smaller than K_c because the denominator has become larger. If equilibrium is restored, the numerator must increase, or the denominator must decrease, or both. Actually, both will happen: SO_3 can be produced only by consuming SO_2 and O_2. When SO_3 is thus produced at the expense of SO_2 and O_2, the reaction $2SO_2 + O_2 \rightleftharpoons 2SO_3$ is said to *shift to the right*. Similarly, adding SO_2 causes a shift to the right; adding SO_3 causes a shift to the left (formation of more SO_2 and O_2). The general rule is that *increasing the concentration (or partial pressure) of one substance in an equilibrium mixture causes the reaction to take place in that direction which consumes some of the material added.* Similarly, decreasing the concentration of a substance causes the production of more of that substance.

(b) Change of volume Another way in which we can disturb an equilibrium mixture of gases is to change the volume of the container. We recall that, for each component, $[\mathrm{A}] = n_\mathrm{A}/V$, where n is the number of moles of that component present. Thus, for the reaction $2SO_2 + O_2 \rightleftharpoons 2SO_3$,

$$Q_c = \frac{(n_{SO_3}/V)^2}{(n_{SO_2}/V)^2(n_{O_2}/V)} = \frac{n_{SO_3}^2}{n_{SO_2}^2 n_{O_2}} \times V$$

Increasing the volume, in this case, makes $Q_c > K_c$, and the numbers of moles must change to restore equilibrium. To decrease Q_c, more SO_2 and O_2 must form, at the expense of SO_3. Decreasing the volume, conversely, results in the formation of more SO_3. The rule is that *increasing the volume favors that reaction which produces more moles of gas.* $2SO_2 + O_2$ is 3 moles of gas, $2SO_3$ is 2 moles of gas, and the equilibrium $2SO_2 + O_2 \rightleftharpoons 2SO_3$ is therefore shifted to the left when the volume increases, to the right when it decreases.

(c) Change of total pressure Given a reaction mixture at a constant temperature, the only way to change the total pressure without adding or removing material is to change the volume. A decrease in pressure resulting from an increase in volume favors that reaction which produces more moles of gas. Thus for the reaction $2SO_2 + O_2 \rightleftharpoons 2SO_3$, a decrease in pressure causes more SO_2 and O_2 to be formed. Conversely, an increase in pressure resulting from a decrease in volume causes a shift in equilibrium to produce fewer moles of gas. Thus an increase in pressure favors the formation of SO_3. Increasing the pressure by adding an inert gas has no effect on the equilibrium.

(d) Change of temperature An equilibrium constant is constant only as long as the temperature does not change. A change in temperature results in a change, often quite large, in the equilibrium constant. *Increasing the temperature causes reaction to occur in that direction which results in absorption of heat; decreasing the temperature causes reaction to occur in that direction which results in emission of heat.* In the formation of SO_3, heat is given out,

$$2SO_2 + O_2 \longrightarrow 2SO_3 \qquad \Delta H = -47 \text{ kcal}$$

Therefore, raising the temperature decreases K_c, shifting the equilibrium to the left; lowering the temperature increases K_c. However, the rate of attainment of equilibrium is increased by raising the temperature, regardless of the effect that temperature has on the final state of equilibrium. An increase in temperature speeds up both the formation and the decomposition of SO_3, but speeds up the decomposition more than the formation, so that less SO_3 is present when the two rates become equal.

The rules given in the preceding discussion are summarized by **Le**

Chatelier's principle*: *when a disturbance is imposed on a system in equilibrium, the equilibrium shifts in such a way as to undo, in part, the effect of the disturbance.* Adding a reactant results in partial consumption of that reactant; decreasing the volume (and thus increasing the total pressure) favors the formation of those substances which would, at a given pressure and temperature, occupy a smaller volume (and thus exert a smaller pressure in a given volume); raising the temperature results in a reaction which absorbs heat and makes the rise in temperature, for a given quantity of heat added, less than it would otherwise be.

We saw some illustrations of Le Chatelier's principle when we studied solubility. If heat is absorbed in the process, solute + solvent $\longrightarrow$ solution, raising the temperature shifts the equilibrium to the right—that is, increases the solubility—whereas, if heat is given out, raising the temperature shifts the equilibrium to the left. If the solute is a gas, decreasing the volume of the container increases the pressure and forces more gas into solution; that is, the equilibrium, gas + solvent $\rightleftharpoons$ solution, has been shifted to the right: from the side on which gas appears (the left) to the side on which no gas appears.

One thing that does *not* cause a shift in equilibrium is the addition or removal of a catalyst (Sect. 19.2). A catalyst changes the rate of approach to equilibrium, but affects the rates of the forward and reverse reactions in the same way, so that the composition of the *equilibrium mixture is unchanged* by its presence.

14.6 □ HETEROGENEOUS EQUILIBRIUM

When chemical equilibrium is established in one phase—a mixture of gases, a liquid solution—we have a case of **homogeneous equilibrium.** When more than one phase is involved—gas and solid, for example, or liquid and solid—the equilibrium is said to be **heterogeneous.** A saturated solution in equilibrium with solute is a familiar instance of heterogeneous equilibrium. For the present we shall concern ourselves with gas-solid reactions.

For this discussion we use K_p rather than K_c. We have studied the formation of HI from H_2 and I_2, at a temperature at which I_2 is a gas. At room temperature, on the other hand, I_2 is a solid, although its vapor pressure is high enough that the violet color can be seen in an enclosed space above solid iodine. At 25°C, the vapor pressure of I_2 is 4.04×10^{-4} atm. The equilibrium constant of the reaction

$$H_2(g) + I_2(g) \rightleftharpoons 2HI(g)$$

* After Henry Le Chatelier (1850–1936), who formulated the principle in 1888. Essentially the same principle was recognized by F. Braun in 1886, as a result of his work on solubility.

is 871 (with pressures in atmospheres) at 25°C,

$$\frac{p_{HI}^2}{p_{H_2}p_{I_2}} = 871$$

But now, as long as solid I_2 is present, $p_{I_2} = 4.04 \times 10^{-4}$ atm; it can have no other value at 25°C, provided that equilibrium is established. The equilibrium condition can be rewritten

$$\frac{p_{HI}^2}{p_{H_2}} = 871p_{I_2} = 871 \times 4.04 \times 10^{-4} = 0.352$$

and this equation may be regarded as the equilibrium condition for the reaction where I_2 is a solid.

$$H_2(g) + I_2(c) \rightleftharpoons 2HI(g) \qquad (9)$$

We say that, for this reaction, $K_p = 0.352$ at 25°C. The constant pressure of $I_2(g)$ has been incorporated into the equilibrium constant to give a new constant, just as constant as the old one, provided that the temperature is fixed and solid is present to maintain the equilibrium vapor pressure. *Whenever a reaction involves pure solids or pure liquids,* their vapor pressures are handled in this way, and *they do not appear in the reaction quotient.*

When K_c is used, the pure solids or pure liquids, likewise, do not appear in the equilibrium expression. For reaction 9 we write

$$\frac{[HI]^2}{[H_2]} = K_c$$

The vapor pressures need not be known; if we measure p_{H_2} and p_{HI} at equilibrium, in the presence of $I_2(c)$, K_p for the solid-gas reaction can be calculated immediately. Similarly, if we measure $[H_2]$ and $[I_2]$ at equilibrium, in the presence of solid I_2, K_c can be calculated.

The same rule applies even when the vapor pressure is too small to be measured. Steam will oxidize red-hot iron, forming magnetite (Fe_3O_4) and hydrogen. The reaction also proceeds in the other direction if we start with H_2 and Fe_3O_4.

$$4H_2O(g) + 3Fe(c) \rightleftharpoons 4H_2(g) + Fe_3O_4(c) \qquad (10)$$

For this reaction we write

$$\frac{p_{H_2}^4}{p_{H_2O}^4} = K_p \qquad \text{or} \qquad \frac{[H_2]^4}{[H_2O]^4} = K_c$$

The application of Le Chatelier's principle to heterogeneous reactions deserves a few remarks. When determining the effect of changes in volume or pressure on the equilibrium, *only the moles of gaseous substances are considered.* For example, in reaction 10,

$$4H_2O(g) + 3Fe(c) \rightleftharpoons 4H_2(g) + Fe_3O_4(c)$$

there is the same number of moles (4) of gas on each side of the equation. Hence, changes in volume or pressure have no effect on this equilibrium.

However, in the reaction $C(c) + CO_2(g) \rightleftharpoons 2CO(g)$ there is a different number of moles of gas on each side of the equation. There is 1 mole of CO_2 on the left side and 2 moles of CO on the right side. An increase in volume (decrease in pressure) favors the right side and formation of CO whereas a decrease in volume (increase in pressure) favors the left side and formation of CO_2.

Adding or subtracting a solid has no effect, as long as some of each solid involved in the reaction is still present. The vapor pressure of a pure substance is independent of the quantity present. Changing the quantity of a gas, however, shifts the equilibrium in the direction that counteracts the change, just as in an all-gas reaction.

The principles in this section are applicable to any reaction involving *pure solids or liquids;* these substances *are ignored in writing the equilibrium condition.*

14.7 □ EQUILIBRIUM IN SOLUTIONS

One of the first reactions to be studied in detail was the formation of an ester and water from an alcohol and an acid,

$$\underset{\substack{\textit{ethanol} \\ \text{(an alcohol)}}}{C_2H_5OH(l)} + \underset{\textit{acetic acid}}{CH_3COOH(l)} \rightleftharpoons \underset{\substack{\textit{ethyl acetate} \\ \text{(an ester)}}}{CH_3COOC_2H_5(l)} + H_2O(l) \qquad (11)$$

The four liquids form one solution over a wide range of compositions. The equilibrium condition is

$$\frac{[\text{ester}][\text{water}]}{[\text{ethanol}][\text{acid}]} = K_c \qquad (12)$$

An experimental test of Equation 12 is shown in Table 14.1, based on work done by Marcellin Berthelot and Péan de Saint-Gilles in 1862. They did not calculate equilibrium constants—for the law of chemical equilibrium was first stated two years later, by the Norwegians Cato Guldberg and Peter Waage—but gave data from which K_c can be calculated. Their procedure was to seal the reactants in glass tubes and keep them at 100°C until equilibrium was attained, about 4 days. Acetic acid was then determined by reaction with base, and the quantities of the other substances could be calculated in a way to be studied later. The last column gives the reaction quotient in terms of molar concentrations,

$$Q_c = \frac{[\text{ester}][\text{water}]}{[\text{ethanol}][\text{acid}]}$$

At equilibrium, Q_c should have the same value, K_c, for all the solutions;

TABLE 14.1 DETERMINATION OF THE EQUILIBRIUM CONSTANT OF $C_2H_5OH(l) + CH_3COOH(l) \rightleftharpoons CH_3COOC_2H_5(l) + H_2O(l)$ AT 100°C

Initial number of moles per liter		Final numbers of moles per liter				
C_2H_5OH	CH_3COOH	C_2H_5OH	CH_3COOH	$CH_3COOC_2H_5$	H_2O	Q_c
1	1	0.335	0.335	0.665	0.665	3.94
1	2.9	0.12	2.02	0.88	0.88	3.19
2	1	1.172	0.172	0.828	0.828	3.40

if we remember that chemistry was still a young science in 1862, the results are not bad.

Frequently homogeneous equilibria are studied in dilute solutions. One component is the solvent present in an overwhelming amount; the other components are solutes present in very small quantities. For example, the reverse of reaction 11, called ester hydrolysis, is best achieved in a large excess of water. In this way the equilibrium is shifted to the left as predicted by Le Chatelier's principle.

In a liter of a dilute aqueous solution of ethyl acetate, acetic acid, and ethanol, practically the entire liter of solution is H_2O. Under these conditions,

$$[H_2O] = \frac{1000 \text{ g/liter}}{18 \text{ g/mole}} = 55.5 \text{ moles/liter}$$

Provided the solution remains dilute, this molar concentration of water, the solvent, is a constant even when the quantities of the other three components vary. We can now write Equation 12

$$\frac{[\text{ester}]55.5}{[\text{acid}][\text{alcohol}]} = K_c$$

and by dividing through by 55.5, we get a new equation,

$$\frac{[\text{ester}]}{[\text{acid}][\text{alcohol}]} = \frac{K_c}{55.5} = K_c'$$

with a new constant, K_c'. This new constant can be determined experimentally. The important feature of this equation is the absence of the concentration of the solvent—in this case H_2O.

We thus have a simple rule for writing the equilibrium condition for a reaction in dilute solution: *the molarity of each product is written in the numerator, and the molarity of each reactant in the denominator, each molarity raised to the appropriate power, except that the solvent is omitted.* It should be noted however, that the value of the equilibrium constant depends on the nature of the solvent. If we were to add some seemingly innocuous solute—sodium chloride, sugar, acetone, for example—in large amount, the nature of the solvent would be changed, and we would no longer be justified in using the K_c that was applicable when the solution was mostly water.

14.8 ▫ EQUILIBRIUM CALCULATIONS

The most important calculation involving equilibria is, of course, the calculation of the equilibrium constant. When the values at equilibrium of all the molarities or partial pressures are known, this calculation involves merely a substitution of the numbers into the reaction quotient. More commonly, however, what we know is the initial quantity of each reactant or product, and the final quantity of some one substance. Conversely, we may know the equilibrium constant at some fixed temperature, and wish to find the equilibrium quantities of the reacting substances when the initial quantities are given. Methods of performing such calculations as these are the subject of this section.

In the reaction

$$2SO_2 + O_2 \rightleftharpoons 2SO_3$$

the equation tells us that, for every mole of O_2 consumed, 2 moles of SO_2 must also be consumed, and 2 moles of SO_3 must be produced. Conversely, if the reaction goes from right to left, 2 moles of SO_2 and 1 mole of O_2 must be produced for every 2 moles of SO_3 consumed. We can start with any amounts; let us introduce 3.00 moles SO_3, with no SO_2 or O_2, into a container of one liter capacity. We do not know how much of the SO_3 will have been consumed when equilibrium is attained, nor do we know how much SO_2 and O_2 will have been formed, but we do know the ratios of these quantities. The balanced equation says that 1 mole of SO_3 gives 1 mole of SO_2 and $\frac{1}{2}$ mole of O_2. Let x be the number of moles per liter of O_2 produced; then $2x$ is the number of moles per liter of SO_2 produced, and $2x$ is the number of moles per liter of SO_3 consumed. The numbers of moles present at equilibrium are SO_2, $2x$; O_2, x; SO_3, $3.00 - 2x$. A summary follows:

	$2SO_2$ +	O_2	$\rightleftharpoons 2SO_3$
Initial molarity	0	0	3.00
Change in molarity	$+2x$	$+x$	$-2x$
Molarity at equilibrium	$0 + 2x$	$0 + x$	$3.00 - 2x$

Suppose that we find, by experiment, that the $[O_2]$ at equilibrium is 0.95 mole per liter. Then $x = 0.95$.

We find K_c by substituting these values into the equilibrium expression

$$\frac{[SO_3]^2}{[SO_2]^2[O_2]} = K_c \qquad \text{or} \qquad \frac{1.10^2}{1.90^2 \times 0.95} = K_c = 0.35$$

When the value of K_c and the initial molarities are known the composition of the equilibrium mixture can be found.

EXAMPLE 1 K_c for the reaction $H_2 + I_2 \rightleftharpoons 2HI$ at 490°C is 45.9. Initially $[H_2] = [I_2] = 0.500$ and $[HI] = 0.00$. Find the concentrations of H_2, I_2, and HI when equilibrium has been established.

ANSWER Let x equal moles per liter of H_2 consumed. According to the chemical equation x is also the number of moles per liter of I_2 consumed. Furthermore for every mole of H_2 or I_2 consumed 2 moles of HI must appear.

Therefore $2x$ is equal to the number of moles per liter of HI formed to reach equilibrium. A summary follows.

	H_2	+	I_2	$\rightleftharpoons$	2HI
Initial molarity	0.500		0.500		0.00
Change in molarity	$-x$		$-x$		$+2x$
Molarity at equilibrium	$0.500 - x$		$0.500 - x$		$0.00 + 2x$

Substitution of the molarities at equilibrium in the equilibrium expression

$$\frac{[HI]^2}{[H_2][I_2]} = K_c \text{ gives } \frac{(2x)^2}{(0.500 - x)(0.500 - x)} = 45.9$$

By taking the square root of both sides, this quadratic equation* is simplified to give

$$\frac{2x}{0.500 - x} = 6.77$$

$$x = 0.386$$

Therefore at equilibrium

$$[HI] = 2x = 0.772 \text{ mole/liter}$$
$$[H_2] = [I_2] = 0.500 - x = 0.114 \text{ mole/liter}$$

You can check your answers by substituting them back into the equilibrium equation and solving for K_c, which should equal 45.9.

Equilibrium can be approached from either direction. Let us consider what happens when we start only with HI.

EXAMPLE 2 We start with 1.000 mole per liter of HI at 490°C. What are the molarities of each component at equilibrium?

ANSWER Let x be the number of moles of HI decomposing to establish equilibrium. Half as many moles of H_2 and I_2 are produced as moles of HI

* Generally we must solve these quadratic equations using the method described in Appendix 2. Two roots must be obtained but one can be discarded because it is physically impossible—it may be negative.

decompose. Hence from x moles of HI, there are formed $0.500x$ moles each of H_2 and I_2. A summary follows.

	H_2	+	I_2	$\rightleftharpoons$	$2HI$
Initial molarity	0		0		1.000
Change in molarity	$+\frac{1}{2}x$		$+\frac{1}{2}x$		$-x$
Molarity at equilibrium	$0 + 0.500x$		$0 + 0.500x$		$1.000 - x$

Substitution into the equilibrium expression

$$\frac{[HI]^2}{[H_2][I_2]} = K_c \quad \text{gives} \quad \frac{(1.000 - x)^2}{(0.500x)(0.500x)} = 45.9$$

from which $x = 0.228$. Therefore at equilibrium $[H_2] = [I_2] = 0.500x = 0.114$ mole/liter and $[HI] = 1.000 - x = 0.772$ mole/liter.

PROBLEMS

1. Form of equilibrium condition Write the equilibrium condition for each of the following reactions:

(a) $CO(g) + \frac{1}{2}O_2(g) \rightleftharpoons CO_2(g)$

(b) $2Cl_2(g) + 2H_2O(g) \rightleftharpoons 4HCl(g) + O_2(g)$

(c) $C(c) + CO_2(g) \rightleftharpoons 2CO(g)$

(d) $NH_4HS(c) \rightleftharpoons NH_3(g) + H_2S(g)$

(e) $C_6H_5COOH(l) + C_2H_5OH(l) \rightleftharpoons C_6H_5COOC_2H_5(l) + H_2O(l)$
(Assume that the solution is not dilute)

(f) same equation as (e), but assume C_2H_5OH is the solvent.

2. Principle of Le Chatelier For the reaction

$$2Cl_2(g) + 2H_2O(g) \rightleftharpoons 4HCl(g) + O_2(g) \qquad \Delta H = +27 \text{ kcal}$$

The four gases Cl_2, H_2O, HCl, and O_2 are mixed and the reaction is allowed to come to equilibrium. State and explain the effect (increase, decrease, no change) of the operation in the left column (below) on the equilibrium value of the quantity in the right column. Each operation is to be considered separately. Temperature and volume are constant except when the contrary is stated.

(a) Increasing the volume of the container	Number of moles of H_2O
(b) Adding O_2	Number of moles of H_2O
(c) Adding O_2	Number of moles of O_2
(d) Adding O_2	Number of moles of HCl
(e) Decreasing the volume of the container	Number of moles of Cl_2
(f) Decreasing the volume of the container	Partial pressure of Cl_2
(g) Decreasing the volume of the container	K_c
(h) Raising the temperature	K_c
(i) Raising the temperature	Partial pressure of HCl
(j) Adding He, an inert gas	Number of moles of HCl
(k) Adding catalyst	Number of moles of HCl

3. Combination of equilibria Given the following equilibrium constants at 1476°K:

$$CO(g) + \tfrac{1}{2}O_2(g) \rightleftharpoons CO_2(g) \qquad K_c = 2.5 \times 10^5$$
$$C(\text{graphite}) + CO_2(g) \rightleftharpoons 2CO(g) \qquad K_c = 1.67 \times 10^3$$

Find K_c for the reaction

$$C(\text{graphite}) + O_2(g) \rightleftharpoons CO_2(g)$$

at 1476°K.

4. Calculation of equilibrium constant At 503°K, 3.00 moles NO, 2.00 moles Cl_2, and 5.00 moles ClNO were introduced into a 1.0 liter container. When the reaction

$$2NO(g) + Cl_2(g) \rightleftharpoons 2ClNO(g)$$

had come to equilibrium, there were 6.12 moles of ClNO in the container. Find (a) the number of moles of Cl_2 present at equilibrium; (b) the number of moles of NO present at equilibrium; (c) K_c for the reaction.

5. Dilute-solution equilibrium The following equilibrium is established in the presence of water and a certain enzyme:

$$\underset{\text{fumaric acid}}{H_2C_4H_2O_4(aq)} + H_2O \rightleftharpoons \underset{\text{malic acid}}{H_2C_4H_4O_5(aq)}$$

One liter of solution was prepared from water and 0.100 mole of pure malic acid. At equilibrium, the solution contained 0.022 mole of fumaric acid, in addition to some unchanged malic acid. Find the equilibrium constant K_c for the reaction, treating the solution as dilute.

6. Gas equilibrium At 1557.8°K, the equilibrium constant of the reaction

$$Br_2(g) \rightleftharpoons 2Br(g)$$

is $K_c = 1.0 \times 10^{-3}$. Find the percentage of Br_2 that is dissociated into atoms when 0.010 mole of Br_2 is confined in a 1.0-liter container and heated to 1557.8°K.

7. Gas equilibrium; K_p The equilibrium constant K_c of the reaction

$$H_2(g) + I_2(g) \rightleftharpoons 2HI(g)$$

is 66.5 at 633°K and 50.7 at 713°K. (a) Is ΔH for this reaction positive or negative? (b) What is K_p for the reaction at 633°K? (c) Find K_c for the reaction

$$\tfrac{1}{2}H_2(g) + \tfrac{1}{2}I_2(g) \rightleftharpoons HI(g)$$

at 633°K. (d) Pure HI is placed in a 1 liter container and heated to 713°K. What fraction of the HI is decomposed to H_2 and I_2 at equilibrium?

ADDITIONAL PROBLEMS

8. Combination of equilibria Given the following equilibrium constants at 1362°K:

$$H_2(g) + \tfrac{1}{2}S_2(g) \rightleftharpoons H_2S(g) \qquad K_p = 0.80$$
$$H_2(g) + Br_2(g) \rightleftharpoons 2HBr(g) \qquad K_p = 7.2 \times 10^4$$

Find K_p for the reaction

$$Br_2(g) + H_2S(g) \rightleftharpoons 2HBr(g) + \frac{1}{2}S_2(g)$$

9. Solution equilibrium When 1.00 mole per liter of benzoic acid, C_6H_5COOH, and 3.00 moles per liter of ethanol, C_2H_5OH, are mixed and kept at 200°C (under sufficient pressure to prevent vaporization) until equilibrium is attained, it is found that 87% of the acid is consumed by the reaction

$$C_6H_5COOH(l) + C_2H_5OH(l) \rightleftharpoons C_6H_5COOC_2H_5(l) + H_2O(l)$$

(a) Find the equilibrium constant K_c for this reaction. (b) Find the percentage of the acid that would be consumed if 1.00 mole benzoic acid and 4.00 moles ethanol were mixed and treated in the same way.

10. Principle of Le Chatelier Ammonia, NH_3, is prepared by the Haber process as described by the equation

$$N_2(g) + 3H_2(g) \rightleftharpoons 2NH_3(g) \qquad \Delta H = -22 \text{ kcal}$$

Suggest four ways in which the concentration of NH_3 at equilibrium can be increased.

11. Equilibrium calculation K_c for the reaction $H_2 + I_2 \rightleftharpoons 2HI$ at 490°C is 45.9. Find the number of moles per liter of H_2, I_2, and HI at equilibrium when 6.0 moles of HI are placed initially in a 2.0 liter reaction chamber.

12. Heterogeneous equilibrium; K_p The equilibrium constant for the reaction

$$H_2(g) + Br_2(l) \rightleftharpoons 2HBr(g)$$

is $K_p = 4.5 \times 10^{18}$ at 25°C. The vapor pressure of liquid Br_2 at this temperature is 0.28 atm. Assume that the gases do not dissolve in the liquid. (a) Find K_p at 25°C for the reaction

$$H_2(g) + Br_2(g) \rightleftharpoons 2HBr(g)$$

(b) How will the equilibrium in (a) be shifted by a decrease in the volume of the container if (*i*) liquid Br_2 is absent, (*ii*) liquid Br_2 is present? Explain why the effect is different in the two cases.

ANSWERS

3. 1.04×10^{14}.

5. 3.5.

7. (b) 66.5; (c) 8.15; (d) 0.219.

9. (a) 2.73; (b) 90.3%

12. (a) 1.6×10^{19}

15 IONIC SOLUTIONS

15.1 ELECTRICAL CONDUCTANCE

As long ago as the eighteenth century, certain aqueous solutions were found to have a special property not shared by the pure solutes, by water, or by most other liquids: these solutions easily conduct electricity. Solutes that show this property are of the kinds known as acids, bases, and salts. The first rough measurements of the conductivities of such solutions were made by Henry Cavendish in 1777: he compared the lengths of different solutions through which the discharge of a battery of Leyden jars gave equal shocks. More recent data, obtained by more accurate and less heroic methods, on the conductivities of solutions, in comparison to other conductors and insulators, are shown in Table 15.1.

The solutions are rather poor conductors in comparison to the metals, although far from being in the same category as the insulators. It is noteworthy that the conductivity of a solution increases sharply with increasing temperature, whereas the conductivity of a metal decreases as the temperature increases.

There is another striking difference between the solutions and the metals in behavior as conductors. When a current is passed through a copper wire, the copper is entirely unaffected by this process. The same wire can carry a current for years or decades. Of course, the wire will become hot, and may melt if too large a current passes through, but such an effect is attributable to the heat and not directly to the electricity. On the other hand, when a direct current* is passed through

* With an alternating current, the reaction that occurs in each half-cycle is undone in the next half-cycle, unless the frequency is so low that gas can escape before the current reverses.

TABLE 15.1 CONDUCTIVITIES AT ROOM TEMPERATURE (in $ohm^{-1}\ cm^{-1}$)

Metals		Aqueous solutions		Insulators	
Ag	6.1×10^5	0.1 *M* HCl	3.5×10^{-2}	H_2O (pure)	6×10^{-8}
Cu	5.8×10^5	0.1 *M* KCl	1.1×10^{-2}	Maple wood	3×10^{-11}
Fe	1.0×10^5	0.1 *M* NaCl	9.2×10^{-3}	Glass	10^{-14}
Hg	1.0×10^4	0.1 *M* CH_3COOH	4.7×10^{-4}	Hard rubber	10^{-15}
C[a]	1.2×10^3	0.1 *M* NH_3	3.1×10^{-4}		

[a] Graphite. Carbon is not a metal, but the electrical properties of graphite are similar to those of metals.

a solution of, say, NaCl, gases are evolved at the electrodes (the metallic conductors in contact with the solution), and the chemical composition of the solution near the electrodes changes. It seems that the mere passage of an electric current into, and out of, such a conducting solution must be accompanied by a chemical reaction—gas evolution is a common, but not universal, feature of these reactions. Such a pair of reactions (one at each electrode), caused by electricity, is called **electrolysis.** A solute that can be dissolved in a suitable solvent (usually water) to give an electrically conducting solution is called an **electrolyte.** Many electrolytes conduct not only when dissolved, but also when melted.

15.2 COLLIGATIVE PROPERTIES OF SOLUTIONS OF ELECTROLYTES

Another peculiarity of solutions of electrolytes was discovered in 1884 by François Marie Raoult, of vapor-pressure fame. He observed that the freezing-point depression in an aqueous solution of NaCl is almost twice what one would expect from the molality of the solution. The atomic weights of Na and Cl are 23.0 and 35.5, respectively; the molecular weight of NaCl clearly cannot be less than 58.5 g/mole. If we dissolve 58.5 g of NaCl in 1000 g of water, the solution is 1.00 molal, and we expect it to freeze approximately 1.86 degrees below the freezing point of pure water (p. 236). The actual freezing point is −3.37°C. In the case of zinc chloride, $ZnCl_2$, the freezing-point depression is nearly three times as great as we would calculate from its formula weight. The more dilute the solution, the more nearly is the depression equal to twice or three times the calculated value. The ratio of the observed to the calculated value of the freezing-point depression is called the **van't Hoff factor,** and is represented by i. Table 15.2 illustrates how i depends on molality for several typical electrolytes.

Similar magnifications are found for the other colligative properties: boiling-point elevation, vapor-pressure depression, and osmotic pressure are all multiplied by the same integral factor in dilute solutions of an electrolyte. The conclusion is inescapable: what is present in the

TABLE 15.2 FREEZING-POINT DEPRESSIONS OF AQUEOUS SOLUTIONS OF ELECTROLYTES

Molality m $\left(\frac{\text{moles}}{\text{kg } H_2O}\right)$	Calculated freezing-point depression °C (1.86 m)	van't Hoff factor $i = \frac{(\Delta t_f)_{\text{measured}}}{(\Delta t_f)_{\text{calculated}}}$		
		NaCl	$ZnCl_2$	$MgSO_4$
0.00100	0.00186	1.97		1.82
0.0100	0.0186	1.94	2.77	1.53
0.100	0.186	1.87	2.66	1.21
1.00	1.86	1.81	2.80	1.09

solution is not NaCl molecules, or $ZnCl_2$ molecules, but individual Na and Cl particles, or Zn and Cl particles. In a $MgSO_4$ solution, however, the colligative properties are multiplied by 2, not by 6, so that this substance must dissociate not into Mg, S, and four O's, but into two fragments—probably Mg and SO_4, since the SO_4 group turns up in many compounds and is undoubtedly a stable entity.

15.3 IONIC CONDUCTION

The remarkable properties of solutions of electrolytes can be understood if the solute is assumed to be present in the form of charged particles—**ions.** When an electric field is applied to a liquid containing ions, the positive ions will move toward the negative electrode and the negative ions will move toward the positive electrode (Fig. 15.1). Such a motion of charges constitutes an electric current. The temperature dependence of the conductivity of a solution is consistent with this

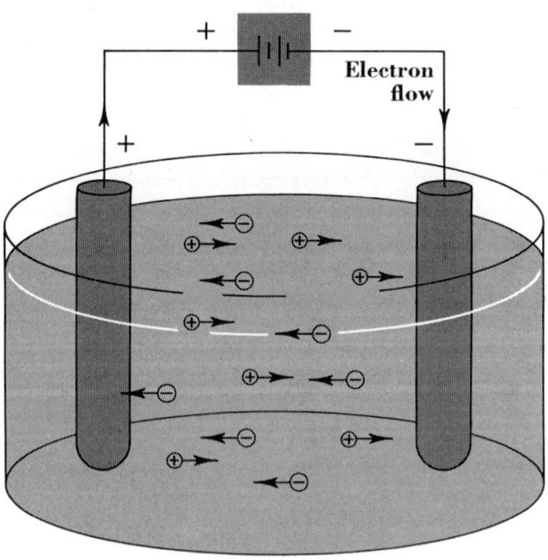

Fig. 15.1 The motion of ions in an electric field.

picture; since the viscosity* of a liquid decreases with increasing temperature, the ions can more easily make their way through the solution at a higher temperature, and the conductivity should increase with increasing temperature, as it does.

When an ion reaches an electrode, its journey ends. In the simplest cases, a negative ion gives up one or more electrons to the electrode to which it moved, and a positive ion accepts one or more electrons from the other electrode. (Other possibilities will be discussed later, p. 269.) When, for example, a solution of $CuBr_2$ is electrolyzed by a direct current between unreactive electrodes such as platinum, copper is deposited on one electrode, while bromine, as a red gas or liquid, appears at the other. These processes are represented, respectively, as

$$Cu^{2+} + 2e^- \longrightarrow Cu \qquad \text{and} \qquad 2Br^- \longrightarrow Br_2 + 2e^-$$

Such a transfer of electrons to or from a chemical species changes it into a new species, and thus constitutes a chemical reaction—specifically, an **oxidation-reduction reaction.** The chemical changes associated with electrolytic conduction are thus consistent with the theory that the solute is present as ions. It should be emphasized that every part of the solution remains electrically neutral during conduction.

15.4 ▫ SOLVATION OF IONS

When Svante Arrhenius proposed his **theory of electrolytic dissociation** in 1887, most chemists greeted it with less than enthusiasm. The principal objection was that forces between charged particles are very strong, and it seemed implausible that, in solution, these particles would be separated from each other and free to move independently. An explanation is suggested by the fact that the solvents in which electrolytic conduction is observed—such as water, liquid NH_3, and liquid SO_2—consist of molecules with large dipole moments (p. 166). An ion attracts the oppositely charged ends of such solvent molecules almost as strongly as it attracts another ion. The ions are said to be **solvated,** or, when the solvent is water, **hydrated** (see p. 168),

$$Cu^{2+} + 4H_2O \rightleftarrows Cu(H_2O)_4^{2+}$$

Solvation sometimes involves the formation of covalent bonds between the ion and the adjacent solvent molecules, in addition to simple electrostatic attraction. Solvation is an exothermic process, corresponding to a decrease in energy. The effect of solvation is that the state in which the ions are separated by solvent becomes nearly as low in energy as the state in which they are combined to form a crystalline ionic solid. For each ionic crystal, there is a certain concentration of its ions (in a given solvent at a given temperature) at which the solution and the

* The **viscosity** of a liquid is its resistance to flow.

crystal are equally stable; this concentration is the solubility (p. 219) of the substance, and corresponds to a state of equilibrium in a process such as

$$KCl(c) \rightleftharpoons K^+(aq) + Cl^-(aq)$$

or

$$Ag_2SO_4(c) \rightleftharpoons 2Ag^+(aq) + SO_4^{2-}(aq)$$

Nonpolar (and slightly polar) molecules, on the other hand, like benzene, C_6H_6, carbon tetrachloride, CCl_4, carbon bisulfide, CS_2, and *n*-octane, $CH_3(CH_2)_6CH_3$, provide relatively little solvation effect. Ionic salts seldom have appreciable solubility in these solvents.

15.5 □ COVALENT ELECTROLYTES

Ionic crystals are not the only solutes that yield ions in solution. Some covalent molecules react with the solvent to give ions, usually by the transfer of a proton from one molecule to the other. An example is hydrogen chloride, which is a gas under ordinary conditions and consists of covalently bonded molecules. When it is dissolved in water, the reaction

$$HCl + H_2O \longrightarrow H_3O^+ + Cl^-$$

produces the **hydronium** or **hydrogen** ion, H_3O^+,* and the chloride ion, Cl^-. Similarly, another gas, ammonia, reacts with water, not by giving it a proton, but by taking one away,

$$NH_3 + H_2O \rightleftharpoons NH_4^+ + OH^-$$

A proton donor, as HCl, is called an **acid;** a proton acceptor, as NH_3, is a **base.** Acids and bases will be discussed in detail in Chapter 17.

15.6 □ NET IONIC EQUATIONS

As soon as we recognize that ions exist in solution as more or less independent chemical species, it becomes natural to write chemical equations in terms of ions, as we have written them in this and previous chapters. In order to decide when to write an equation in ionic form, we must recognize which substances exist primarily as ions—that is, which substances are strong electrolytes (p. 274). The custom is *to represent each substance involved in a reaction by the formula corresponding to the form in which most of it exists.* Thus, we write the reaction between hydrochloric acid and ammonia, in aqueous solution,

$$H_3O^+ + Cl^- + NH_3 \longrightarrow NH_4^+ + Cl^- + H_2O \qquad (1)$$

* One theory is that the proton (H^+) is equidistant from the oxygen ends of three water molecules: $[H(OH_2)_3]^+$. Another theory is that three protons are covalently bonded to an oxygen atom, and each is hydrogen-bonded to the oxygen end of a water molecule: $[OH_3(OH_2)_3]^+$. We shall represent this ion by H^+ or H_3O^+, without claiming that either formula depicts its actual structure.

The strong acid HCl is represented as $H_3O^+ + Cl^-$, since it is almost completely ionized. Ammonium chloride, NH_4Cl, is a completely ionized salt, and is therefore written as the two distinct species, $NH_4^+ + Cl^-$. Ammonia and water, on the other hand, are weak electrolytes (p. 276)—that is, they are ionized only to a small extent—and are therefore represented by their molecular formulas, NH_3 and H_2O. If the acid were acetic acid, which is weak, the equation would be

$$CH_3COOH + NH_3 \longrightarrow NH_4^+ + CH_3COO^-$$

The weak acid is represented by its molecular formula, CH_3COOH; the product, ammonium acetate, is a typical strong electrolyte and is written in terms of ions.

Inspection of Equation 1 shows that a further simplification is possible. Cl^- appears on both sides, which means that it was present initially and is not changed by the reaction. Any species remaining unchanged can be omitted from a chemical equation. Equation 1 can therefore be replaced by

$$H_3O^+ + NH_3 \longrightarrow NH_4^+ + H_2O \qquad (2)$$

Equation 2 is called a **net ionic equation** because superfluous species have been omitted, leaving only those that actually participate in the reaction. We know that negative ions must be present to keep the solution electrically neutral, but Equation 2 does not tell us what they are, nor do we necessarily care. The reaction is the same whether the acid is hydrochloric, nitric, hydrobromic, or perchloric, as long as it is a strong acid. Only when the solution is evaporated, to yield a solid product, does the identity of the acid make a difference; the solid product obtained will then be NH_4Cl, NH_4NO_3, NH_4Br, or NH_4ClO_4, depending on which acid is used.

SAFETY NOTE: AMMONIUM PERCHLORATE IS AN EXPLOSIVE, AND THE STUDENT SHOULD NEVER EVAPORATE ITS SOLUTIONS TO DRYNESS.

Solids, including ionic crystals, and gases are represented by their molecular formulas. For example, when solutions of the salts KCl and $AgNO_3$ are mixed, a white precipitate of silver chloride, AgCl, is formed. We write the formulas of the soluble salts in ionic form, but AgCl, which is present almost entirely as the solid, is represented as AgCl,

$$K^+ + Cl^- + Ag^+ + NO_3^- \longrightarrow AgCl(c) + K^+ + NO_3^-$$

This equation is not yet a net equation. K^+ and NO_3^- appear on both sides and may be omitted,

$$Cl^- + Ag^+ \longrightarrow AgCl(c)$$

This form of the equation emphasizes that the only real chemical change is the combination of Ag^+ and Cl^- ions. A reaction involving the other two ions will occur only if the final solution is evaporated; when it becomes sufficiently concentrated, KNO_3 will precipitate,

$$K^+ + NO_3^- \longrightarrow KNO_3(c)$$

15.7 □ BALANCING OXIDATION-REDUCTION EQUATIONS BY THE ION-ELECTRON METHOD

Equations for oxidation-reduction (redox) reactions are often difficult to balance. The method to be described in this section is especially appropriate for reactions involving ions in solution. The principle underlying this method is that the reaction can be resolved, on paper, into two parts, or **half-reactions:** a loss of electrons (**oxidation**) and a gain of electrons (**reduction**). In reality, most oxidation-reduction reactions probably do not proceed in this way, except in an electrolytic or galvanic cell (Chapter 16), where electrons must be lost at the anode and gained at the cathode. (Even then, we do not necessarily know what species are involved in the electrode processes.) The method is essentially a bookkeeping device enabling us to deal with the oxidation and with the reduction separately, and finally to combine the results into a balanced equation.

The steps are

(a) Write the skeleton (unbalanced) partial ionic equations.
(b) Balance each equation with respect to atoms.
(c) Balance each equation with respect to electrons.
(d) Balance the overall exchange of electrons.
(e) Add the partial equations.

In aqueous solution, we assume the presence of water (even if it is not stated in the problem). If the solution is acidic, we also assume the presence of H^+ (or H_3O^+) ions, and, if it is basic, the presence of OH^- ions. These entities are then assumed to be sources of hydrogen and oxygen in balancing the equations.

	Source of hydrogen	Source of oxygen	Procedure
In acid solution	H^+ (or H_3O^+)	H_2O	To the side deficient by x O atoms, add x H_2O. Then add $2x$ H^+ to opposite side.
In basic solution	H_2O	OH^-	To the side deficient by x O atoms, add $2xOH^-$. Then add xH_2O to opposite side.

As an example, we take the reaction

$$Cu(c) + HONO_2(aq) \longrightarrow Cu(NO_3)_2(aq) + H_2O + NO(g)$$

which occurs when copper is added to dilute nitric acid.

Step (a). Partial ionic equations When we write the equation in ionic form, $HONO_2$ (or HNO_3) and $Cu(NO_3)_2$ are strong electrolytes and are therefore represented as ions,

$$Cu + H^+ + NO_3^- \longrightarrow Cu^{2+} + 2NO_3^- + H_2O + NO$$

Now, we must observe which species are involved in oxidation and reduction processes (although it is not yet necessary to decide which is which). We write separate equations for each. It is obvious that Cu becomes Cu^{2+} and that NO_3^- somehow becomes NO.

$$Cu \longrightarrow Cu^{2+} \tag{3}$$

$$NO_3^- \longrightarrow NO \tag{4}$$

Step (b). Balance atomically Equation 3 is balanced atomically, but Equation 4 is not. Where did the missing oxygens go? Unless we are told differently, we assume that they become one of the components of water (see tabulation, p. 266). Since the solution is acidic, we assume that the hydrogen atoms of the water came from H^+ ions in the solution. Thus we write

$$NO_3^- + 4H^+ \longrightarrow NO + 2H_2O \tag{5}$$

This is now balanced atomically.

Step (c). Balance electrically Equation 3 has a charge of 0 on the left and +2 on the right. This discrepancy is remedied by adding two electrons (e^-) to the right side.

$$Cu \longrightarrow Cu^{2+} + 2e^- \ (\textit{oxidation}) \tag{6}$$

This is an oxidation because electrons are lost. Equation (5) has $-1 + 4 = +3$ on the left, and zero charge on the right. We add three electrons to the left to balance it electrically.

$$NO_3^- + 4H^+ + 3e^- \longrightarrow NO + 2H_2O \ (\textit{reduction}) \tag{7}$$

This is a reduction because electrons are gained.

Step (d). Balance the overall exchange of electrons *The number of electrons given out in the oxidation must be equal to the number of electrons taken up in the reduction.* In this case, two electrons were lost in the oxidation of one Cu atom, and three electrons were gained in the reduction of one NO_3^- ion. It follows that, for every three Cu atoms oxidized, two NO_3^- ions must be reduced, so that six electrons are lost and six gained. We multiply Equation 6 by 3, and Equation 7 by 2.

$$3Cu \longrightarrow 3Cu^{2+} + 6e^- \tag{8}$$

$$2NO_3^- + 8H^+ + 6e^- \longrightarrow 2NO + 4H_2O \tag{9}$$

Step (e). Addition The two partial Equations 8 and 9 may now be added algebraically, to give the overall equation that is balanced atomically and electrically.

$$3Cu + 2NO_3^- + 8H^+ \longrightarrow 3Cu^{2+} + 2NO + 4H_2O \tag{10}$$

Note that an oxidation or reduction reaction always involves a change in oxidation number of an element. If there is no change in oxidation number, no oxidation or reduction has occurred. For example, the transformation

$$Cr_2O_7^{2-} \longrightarrow CrO_4^{3-}$$

preserves the oxidation numbers of Cr (+6) and O (−2), and is therefore neither an oxidation nor a reduction. The equation can be balanced without the introduction of electrons.

$$Cr_2O_7^{2-} + H_2O \longrightarrow 2CrO_4^{3-} + 2H^+$$

EXAMPLE 1 Balance the equation

$$Cu^+ + H_2O_2 \longrightarrow Cu^{2+} + H_2O$$

in acid solution.

ANSWER (a) Write skeleton partial equations,

$$Cu^+ \longrightarrow Cu^{2+}$$
$$H_2O_2 \longrightarrow H_2O$$

(Any other products would have been specified.)

(b) Balance with respect to atoms,

$$Cu^+ \longrightarrow Cu^{2+}$$
$$H_2O_2 + 2H^+ \longrightarrow 2H_2O \text{ (balanced for H and O)}$$

(c) Balance electrically,

$$Cu^+ \longrightarrow Cu^{2+} + e^-$$
$$H_2O_2 + 2H^+ + 2e^- \longrightarrow 2H_2O$$

(d) Balance the electron exchange,

$$2Cu^+ \longrightarrow 2Cu^{2+} + 2e^-$$
$$H_2O_2 + 2H^+ + 2e^- \longrightarrow 2H_2O$$

(e) Add partial equations,

$$H_2O_2 + 2H^+ + 2Cu^+ \longrightarrow 2Cu^{2+} + 2H_2O$$

EXAMPLE 2 Balance the equation

$$NaClO + NaBr \longrightarrow NaBrO_3 + NaCl$$

in basic solution.

ANSWER (a) $Na^+ + ClO^- + Na^+ + Br^- \longrightarrow Na^+ + BrO_3^- + Na^+ + Cl^-$

$ClO^- \longrightarrow Cl^-$

$Br^- \longrightarrow BrO_3^-$

(b) $ClO^- + H_2O \longrightarrow Cl^- + 2OH^-$

$Br^- + 6OH^- \longrightarrow BrO_3^- + 3H_2O$

(c) $ClO^- + H_2O + 2e^- \longrightarrow Cl^- + 2OH^-$

$Br^- + 6OH^- \longrightarrow BrO_3^- + 3H_2O + 6e^-$

(d) $3ClO^- + 3H_2O + 6e^- \longrightarrow 3Cl^- + 6OH^-$

$Br^- + 6OH^- \longrightarrow BrO_3^- + 3H_2O + 6e^-$

(e) $3ClO^- + Br^- + 3H_2O + 6OH^- \longrightarrow 3Cl^- + BrO_3^- + 3H_2O + 6OH^-$

$3ClO^- + Br^- \longrightarrow 3Cl^- + BrO_3^-$

15.8 ▫ ELECTRODE PROCESSES

We observed (p. 263) that the passage of a current through an electrolytic conductor is always accompanied by reactions at the electrodes. At one electrode, electrons are being supplied to the electrolyte, and some molecule or ion must accept them and be **reduced.** The electrode at which this reduction takes place is the **cathode.** At the other electrode, electrons in equal number are being removed from the electrolyte, and something is being **oxidized;** this electrode is the **anode.** Reactions of several types are possible at each electrode; the following list covers the most common types, but is not meant to be exhaustive.

Anode reactions Anode reactions include the following:

(a) The metal of which the electrode is made may be oxidized to positive ions,

$$Cu(c) \longrightarrow Cu^{2+} + 2e^-$$

(b) A constituent of the solution, neutral or charged, may be oxidized and remain in solution,

$$Fe^{2+} \longrightarrow Fe^{3+} + e^-$$

$$\underset{\textit{ethanol}}{C_2H_5OH} + 2H_2O \longrightarrow \underset{\textit{acetaldehyde}}{CH_3\overset{\overset{\displaystyle O}{\|}}{C}H} + 2H_3O^+ + 2e^-$$

(c) A negative ion in the solution may become a neutral molecule and escape as a gas,

$$2Cl^- \longrightarrow Cl_2(g) + 2e^-$$

(d) A gas in contact with the electrode may be oxidized to a positive ion,

$$H_2(g) + 2H_2O \longrightarrow 2H_3O^+ + 2e^-$$

(e) In an aqueous solution, water may be oxidized to gaseous oxygen,

$$6H_2O \longrightarrow O_2(g) + 4H_3O^+ + 4e^-$$

Since water is the predominant species in a dilute aqueous solution, reaction (e) is always a possibility.

Cathode reactions (a) A positive ion in the solution may be reduced to the element, which deposits on the electrode. (b) A constituent of the solution may be reduced and remain in solution. (c) A gas in contact with the electrode may be reduced to a negative ion. (The reverses of the anode reactions under (a), (b), and (c) above will serve as illustrations.) (d) In an aqueous solution, water may be reduced to gaseous hydrogen,

$$2H_2O + 2e^- \longrightarrow H_2(g) + 2OH^-$$

As with the production of O_2, this reaction is always a possibility.

In general, several reactions may take place simultaneously at each electrode. In a $CuCl_2$ solution, for example, the cathode process will be some combination of the reactions

$$Cu^{2+} + 2e^- \longrightarrow Cu(c)$$

and

$$H_2O + e^- \longrightarrow \tfrac{1}{2}H_2(g) + OH^-$$

The decision as to which will predominate cannot always be made without a specific experiment.

15.9 □ ELECTROLYSIS OF FUSED SALTS

When an ionic crystal melts, the liquid is still composed of ions, now having a mobility lacking in the crystal. These ions, like the ions in a liquid solution, can move under the influence of an electric field, and a fused ionic salt is therefore an electrolytic conductor. Solid ionic crystals also have appreciable conductivity, especially at high temperatures, indicating that the ions can migrate to some extent even in the solid.

Electrode reactions impossible in the presence of water become possible when the electrolyte is a fused salt. Thus, the alkali and alkaline earth metals are prepared by the electrolysis of their fused chlorides, chosen because of their abundance and relatively low melting points. The reaction at the anode is

$$Cl^- \longrightarrow \tfrac{1}{2}Cl(g) + e^-$$

and at the cathode

$$Na^+ + e^- \longrightarrow Na(l)$$

in the case of NaCl. If water were present, H_2, not Na, would be the product.* The melting point of NaCl is 801°C, so that Na, which melts at 97.5°C, is produced as a liquid.

15.10 □ FARADAY'S LAWS

When there is only one reaction at an electrode, there are two simple and significant relations between the quantity of charge passed through the circuit and the quantity of matter produced or consumed by the electrode process. These relations are called **Faraday's laws,** after Michael Faraday, who discovered them in 1832–33. Before stating these laws, we shall illustrate them with a familiar reaction, the electrolytic decomposition of water. An electrolyte, usually sulfuric acid, $(HO)_2SO_2$, is added to make the water a conductor, but the sulfate ion (SO_4^{2-} or $O_2SO_2^{2-}$) is very stable and is not oxidized at the anode. The only possible reactions are

$$3H_2O \longrightarrow \tfrac{1}{2}O_2(g) + 2H_3O^+ + 2e^-$$

at the anode, and, in acid solution,

$$H_3O^+ + e^- \longrightarrow \tfrac{1}{2}H_2(g) + H_2O$$

or, in basic solution,

$$H_2O + e^- \longrightarrow \tfrac{1}{2}H_2(g) + OH^-$$

at the cathode. Let us pass through the solution the same current for different lengths of time, or different currents for the same time. When the current or the time is doubled, the mass and volume of H_2 and of O_2 are doubled. This result is not very surprising, and had been suggested by several chemists before Faraday's careful experiments established it beyond doubt. It is an instance of Faraday's **first law:**

(1) *The mass of product produced (or reactant consumed) in a given electrode reaction is proportional to the quantity of charge (current × time) that has passed through the circuit.*†

Since the two electrodes are in series, the same current, and therefore the same total charge, must pass through them. We find that the ratio of the mass of H_2 to the mass of O_2 is 2.016:16.00—just the ratio of the molecular weight of H_2 to half the molecular weight of O_2. If another electrolysis apparatus is in series with this one, and the only reaction at the cathode of the second apparatus is the reduction of silver ion to metallic silver (which is plated on the electrode), the forma-

* Even with an aqueous solution, Na can be obtained as a product by using Hg as the cathode.

† Charge is measured in *coulombs* (C); current is measured in *amperes* (A). One ampere is one coulomb per second.

tion of 16.00 g of O_2 is associated with the deposition of 215.74 g of Ag; 16 is half the molecular weight of O_2, and 215.74 is twice the atomic weight of Ag. Equally well, we may say

1 gram-atom (107.87 g) of Ag is produced by the same quantity of charge that liberates
$\frac{1}{4}$ mole or $\frac{1}{2}$ gram-atom (8.00 g) of O_2 and
$\frac{1}{2}$ mole or 1 gram-atom (1.008 g) of H_2.

The same charge will plate out

$\frac{1}{2}$ gram-atom (31.77 g) of Cu or
$\frac{1}{3}$ gram-atom (8.99 g) of Al;

and it will liberate

35.453 g ($\frac{1}{2}$ mole or 1 gram-atom) of Cl_2.

These results illustrate Faraday's second law, which is most easily stated if we first adopt a definition: one **faraday** is *the quantity of electric charge that will deposit exactly one gram-atom of silver.* Silver is chosen as the standard element because it lends itself to accurate weighing. Then Faraday's **second law** is

(2) *The mass (in grams) of a substance produced or consumed in an electrode reaction, when one faraday passes through the circuit, is equal to the atomic or molecular weight of the substance divided by a small integer.*

One faraday is equal to 96,487 C. This quantity is taken as a fundamental unit because the production (or removal) of one gram-atom or one mole of any substance always requires 1, or 2, or 3, and so on, faradays, never a fractional part of a faraday. One mole is Avogadro's number of atoms or molecules; the fact that this number of molecules is produced by an integral number of faradays suggests strongly that one molecule is produced by the addition or removal of an integral number of atoms of electricity, or **electrons.** *A faraday is therefore Avogadro's number of electronic charges—one mole of electrons.*

In this section, we shall call that quantity of an element or compound produced or consumed by one faraday an **equivalent;** the mass in grams of this quantity is the **equivalent weight,** and is equal to the molecular or atomic weight divided by the number of electrons transferred per molecule, atom, or ion. With knowledge of chemical formulas, we can determine the number of electrons transferred, if we know the identities of the reactants and products, by writing the partial equation for the electrode reaction. Thus, the partial equation

$$2H_2O + 2e^- \longrightarrow H_2 + 2OH^-$$

show that two electrons are needed to produce one molecule of H_2, so that two faradays are needed for one mole of H_2, or one faraday for $\frac{1}{2}$ mole; the equivalent weight is $\frac{1}{2}$ the molecular weight. For O_2, we see from the partial equation

$$6H_2O \longrightarrow O_2 + 4H_3O^+ + 4e^-$$

that four electrons are removed for every molecule of O_2 that is formed, or four faradays for every mole, or one faraday for every $\frac{1}{4}$ mole, and the equivalent weight is $32.00/4 = 8.00$ g. When Cu is deposited on, or removed from, an electrode, the reaction is

$$Cu^{2+} + 2e^- \longrightarrow Cu$$

or the reverse; its equivalent weight is therefore $\frac{1}{2}$ its atomic weight. When Fe^{3+} is reduced to Fe at a cathode,

$$Fe^{3+} + 3e^- \longrightarrow Fe$$

its equivalent weight is $\frac{1}{3}$ its atomic weight. But another possible cathode reaction is

$$Fe^{3+} + e^- \longrightarrow Fe^{2+}$$

Here, the equivalent weight is equal to the atomic weight. We see that the equivalent weight of a species is not simply a property of itself, but depends on the reaction in which it participates.

EXAMPLE 3 An aqueous solution of gold(III) chloride, $AuCl_3$, was electrolyzed with a current of 0.500 A until 1.200 g of Au had been deposited on the cathode. There was no other reaction at the cathode. At another electrode in series with this one, the only reaction was the evolution of O_2. Find (a) the number of moles, (b) the volume at standard conditions, and (c) the mass of O_2 liberated; (d) the number of coulombs passed through the circuit; (e) the duration of the electrolysis.

ANSWER The atomic weight of Au is 197.0 g/g-atom. The cathode reaction is

$$Au^{3+} + 3e^- \longrightarrow Au$$

so that the equivalent weight of Au is

$$\frac{197.0 \text{ g}/\cancel{\text{g-atom}}}{3 \text{ eq}/\cancel{\text{g-atom}}} = 65.67 \frac{\text{g}}{\text{eq}}$$

The number of equivalents deposited is

$$\frac{1.200 \cancel{\text{g}}}{65.67 \frac{\cancel{\text{g}}}{\text{eq}}} = 0.01827 \text{ eq}$$

This is also the number of faradays that have passed through the circuit. (a) The number of equivalents of O_2 is 0.01827. From the partial equation

$$6H_2O \longrightarrow O_2 + 4H_3O^+ + 4e^-$$

we see that one mole of O_2 is 4 equivalents, so that

$$\frac{0.01827\ \cancel{eq}}{4\ (\cancel{eq}/\text{mole})} = 0.00457\ \text{mole}$$

of O_2 is liberated.

(b) $$0.00457\ \cancel{\text{mole}} \times 22.4\ \frac{\text{liters}}{\cancel{\text{mole}}} = 0.102\ \text{liter (sc)}$$

(c) $$0.00457\ \cancel{\text{mole}} \times 32.00\ \frac{\text{g}}{\cancel{\text{mole}}} = 0.146\ \text{g}$$

(d) $$0.01827\ \cancel{\text{faraday}} \times 9.65 \times 10^4\ \frac{\text{Coul}}{\cancel{\text{faraday}}} = 1.76 \times 10^3\ \text{C}$$

(e) The charge (q) in coulombs is the current (I) in amperes times the time (t) in seconds,

$$q = It$$

The current is given as

$$0.500\ \text{A} = 0.500\ \frac{\text{C}}{\text{sec}}$$

We have found that the charge is

$$1.76 \times 10^3\ \text{C}$$

Then

$$t = \frac{q}{I} = \frac{1.76 \times 10^3\ \cancel{\text{C}}}{0.500\ (\cancel{\text{C}}/\text{sec})} = 3.52 \times 10^3\ \text{sec}$$

15.11 ▫ STRONG ELECTROLYTES

Electrolytes that exist as ionic crystals in the solid state are believed to exist almost entirely as ions in solution as well. Such electrolytes are relatively good conductors of electricity and are called **strong electrolytes.** The ions in solution, though they can move somewhat independently, do not behave as though they were in complete isolation in empty space. Each ion has neighbors predominantly of the opposite charge; it is attracted by these ions moving in the opposite direction, and its progress is thus hindered (Fig. 15.2). These effects are, of course, greater at higher concentrations.

The colligative properties of strong electrolytes show deviations from the values to be expected on the basis of complete ionization. In a 1 m solution of either NaCl or $MgSO_4$ in water, there should be 2 moles

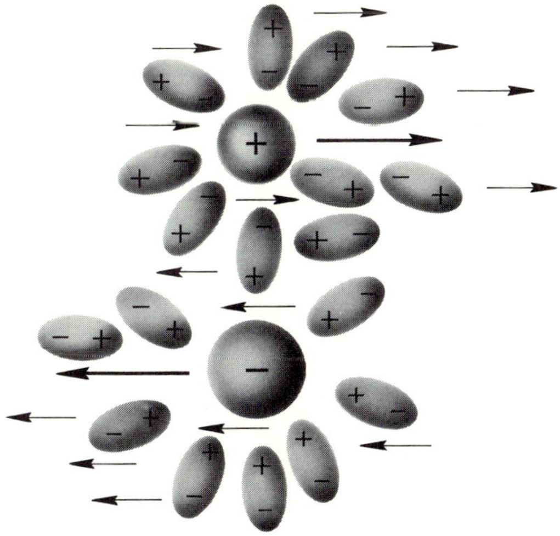

Fig. 15.2 A schematic diagram of two oppositely charged ions moving in opposite directions through a polar solvent. Each ion is attempting to carry its sheath of solvent molecules with it.

of ions per kilogram of water, giving a freezing-point depression of $2 \times 1.86 = 3.72$ deg, but, as Table 15.2 (p. 262) shows, the depression is only 1.81×1.86 deg for NaCl, and 1.09×1.86 deg for $MgSO_4$. These diminutions in the colligative properties are accounted for, at least qualitatively, by the concept of interionic attraction, which restricts the mobility of the ions and thus makes them behave as if they were not all there.

In addition to the ionic salts (including the soluble hydroxides, as NaOH), some acids behave as strong electrolytes in aqueous solution: HCl, HBr, HI, $(HO)_2SO_2$ (sulfuric acid), $HONO_2$ (nitric acid), and $HOClO_3$ (perchloric acid) are the most common of them. A few other compounds that are covalent in the pure state ionize almost completely in solution;

$$AlCl_3(c) + 6H_2O \longrightarrow Al(H_2O)_6^{3+} + 3Cl^-$$

represents a familiar case.

15.12 ▫ CONDUCTANCES OF INDIVIDUAL IONS

The electrical **conductance** of a body is a measure of its ability to conduct electricity; it is the reciprocal of the resistance.* We picture an electrolyte as consisting of two intermingled conductors: the positive ion and the negative ion. We can attribute a certain conductance to each of these ions, and the conductance of the solution is therefore

* Resistance is measured in *ohms* (Ω), conductance in $ohms^{-1}$.

the sum of the conductances of the positive ions and of the negative ions.

As explained above, the mobility of each ion in solution is somewhat restrained by oppositely charged ions and by solvent molecules. But if the solution were sufficiently dilute, each ion would in effect be isolated from other ions and its conductance in a given solvent would therefore not be affected by them. From measurements of conductances in solutions of various dilutions, it is possible to calculate the conductance that one equivalent of an ion would have if it were dissolved in an "infinitely dilute" solution, that is to say, if no other ions were close enough to interfere. Such values are called **ionic conductances at infinite dilution,** abbreviated λ_0, and are shown in Table 15.3. The conductance at infinite dilution of any strong electrolyte, abbreviated Λ_0, can be calculated simply by adding the ionic conductances. These conductances all refer to *one equivalent* (one Avogadro number of + or − charges), and hence the value is usually called **equivalent conductance.** The value for an ion of multiple charge is shown by a designation like $\frac{1}{2}Ba^{2+}$ to emphasize that it refers to the conductance of one equivalent of the ion.

EXAMPLE 4 Calculate the equivalent conductance at infinite dilution, Λ_0, of Na_2SO_4.

ANSWER.

$$\Lambda_0,\ \tfrac{1}{2}Na_2SO_4 = \lambda_0,\ Na^+ + \lambda_0,\ \tfrac{1}{2}SO_4^{2-}$$
$$= 50.11 + 79.8 = 129.9\ \text{cm}^2/\Omega\ \text{eq}$$

15.13 □ WEAK ELECTROLYTES

Acetic acid, dissolved in water, yields hydronium ions and acetate ions by the reaction

$$CH_3COOH + H_2O \longrightarrow H_3O^+ + CH_3COO^-$$

If we could calculate the equivalent conductance of an extremely

TABLE 15.3 IONIC CONDUCTANCES AT INFINITE DILUTION AT 25°C (cm²/Ω eq)*

Ag^+	61.92	Br^-	78.4
H_3O^+	349.82	CH_3COO^-	40.9
K^+	73.52	Cl^-	76.34
Na^+	50.11	NO_3^-	71.44
NH_4^+	73.4	OH^-	198
$\frac{1}{2}Ba^{2+}$	63.64	$\frac{1}{2}SO_4^{2-}$	79.8
$\frac{1}{2}Ca^{2+}$	59.50		
$\frac{1}{2}Mg^{2+}$	53.06		

* The calculation of equivalent conductance involves the dimensions of the cell.

dilute solution of acetic acid by adding the ionic conductances of H_3O^+ and CH_3COO^- (Table 15.3), we would get

$$\Lambda_0 = 349.82 + 40.9 = 390.7 \text{ cm}^2/\Omega \text{ eq}$$

The observed conductance of acetic acid is nowhere near this value, however, even in the most dilute solutions for which measurements have been made. The most likely explanation is that we have in this case a partial dissociation into ions. An electrolyte dissociated only to a small extent into ions at equilibrium is known as a **weak electrolyte.** Most acids are of this kind; the exceptions were mentioned on p. 275. The most common weak bases* are ammonia, NH_3, and many compounds related to it by the replacement of one or more hydrogen atoms by other atoms or groups of atoms; examples are methylamine, H_3CNH_2, hydroxylamine, H_2NOH, and hydrazine, H_2NNH_2, an important rocket fuel. There are a relatively few **weak salts** which exist, even in dilute solution, predominantly as undissociated molecules: cadmium chloride, $CdCl_2$; lead acetate, $(CH_3COO)_2Pb$; mercuric chloride, $HgCl_2$, for example.

15.14 □ DEGREE OF DISSOCIATION

(a) From equivalent conductance The equivalent conductance of completely ionized acetic acid at infinite dilution would be 390.7 cm²/Ω eq, as we saw in the preceding paragraph. The observed equivalent conductance at 25°C of a 0.050 M solution of acetic acid is 7.358 cm²/Ω eq. We may say that the solution is only 7.358/390.7 = 1.88% as good a conductor as it would be if the acid were completely ionized. If we adopt the approximation that each ion carries current as efficiently in the 0.05 M solution as in an infinitely dilute solution, the number of ions present in the solution is 1.88% of the number that would be present if ionization were complete—in other words, 1.88% of the acetic acid molecules have reacted with water to become H_3O^+ and CH_3COO^- ions, and 98.12% of the acetic acid is still present as neutral molecules. The fraction of a weak electrolyte that is present as ions is called the **degree of dissociation** or **degree of ionization** of the electrolyte; it depends, of course, on the concentration of the solution, the identity of the solvent, and the temperature.

(b) From colligative properties The degree of dissociation of a weak electrolyte can also be calculated from the colligative properties of its solution. Taking the freezing-point depression as an example, we recall (from p. 236) that the depression Δt_f is given by $\Delta t_f = K_f m$, where $K_f = 1.86$ deg kg/mole for H_2O, and m is the total molality of all solute particles. Suppose that a solution of a weak acid HA is prepared with molality 0.70 mole per kilogram H_2O, and that the degree of dissociation

* Here we are considering only those weak acids and bases which are neutral molecules, not ions.

of the acid 0.020, or 2.0%. Then the reaction

$$HA + H_2O \rightleftharpoons H_3O^+ + A^-$$

gives $0.020 \times 0.70 = 0.0140$ mole of H_3O^+ and 0.0140 mole of A^- per kilogram of water, and $0.70 - 0.0140$ mole of undissociated HA remains. The total molality of all solute species is

$$m = 0.70 - 0.0140 + 0.0140 + 0.0140$$
$$= 0.714 \frac{\text{mole}}{\text{kg } H_2O}$$

The freezing-point depression then is

$$\Delta t_f = K_f m = 1.86 \frac{\text{deg } \cancel{\text{kg}}}{\cancel{\text{mole}}} \times 0.714 \frac{\cancel{\text{mole}}}{\cancel{\text{kg}}} = 1.33 \text{ deg}$$

If there were no dissociation, the depression would be $1.86 \times 0.70 = 1.30$ deg. Conversely, the degree of dissociation can be calculated from measurements of freezing-point depression.

EXAMPLE 5 The freezing point of a 0.10 m aqueous solution of hydrofluoric acid, HF, is -0.198°C. Find the degree of dissociation of the acid.

ANSWER The total molality is

$$m = \frac{\Delta t_f}{K_f} = \frac{0.198 \text{ deg}}{1.86 \text{ deg kg/mole}} = 0.106 \frac{\text{mole}}{\text{kg}}$$

Let x = the number of moles of HF per kilogram of solvent that are dissociated at equilibrium. Then,

$$\begin{aligned} \text{moles of undissociated HF} &= 0.100 - x \\ \text{moles of } H_3O^+ &= x \\ \text{moles of } F^- &= \underline{\quad x} \\ \text{total number of moles} &= 0.100 + x \end{aligned}$$

Then, $m = 0.106 = 0.100 + x$

$$x = 0.006$$

The degree of dissociation is the fraction of acid dissociated, or

$$\frac{0.006}{0.100} = 0.06$$
$$= 6\%$$

Because ionization of a weak electrolyte causes only a small change in the freezing-point depression, the degree of dissociation calculated from this or other colligative properties is not very accurate, unless the temperatures are measured with extreme precision. The degree of dissociation of 0.10 m HF, calculated from conductance measurements at 0°C, is 9.05%; this figure is more reliable than the 6% obtained from freezing-point depression.

PROBLEMS

1. Definitions Define and illustrate the following terms: (a) ionic conduction; (b) solvation; (c) electrolyte; (d) equivalent conductance.

2. Ionization A 0.00200 m aqueous solution of $KCo(NH_3)_2(NO_2)_4$ freezes at −0.00736°C. How many moles of ions are present in a solution containing one mole of this compound?

3. Balancing equations Balance each of the following equations by the ion-electron method. Show the balanced partial equations for oxidation and reduction, and the ionic equation for the overall reaction.

(a) $Pb(c) + PbO_2(c) \longrightarrow Pb^{2+}$ (in acid solution)
(b) $Cr_2O_7^{2-} + H_2S(aq) \longrightarrow Cr^{3+} + S_8(c)$ (in acid solution)
(c) $Zn(c) + NO_3^-(aq) \longrightarrow Zn^{2+} + NH_4^+$ (in acid solution)
(d) $ClO_3^- + I_2 \longrightarrow IO_3^- + Cl^-$ (in acid solution)
(e) $MnO_4^- \longrightarrow MnO_4^{2-} + O_2$ (in basic solution)
(f) $HCHO$ (*formaldehyde*) $+ Ag(NH_3)_2^+ \longrightarrow HCOO^-$ (*formate ion*) $+ Ag(c) + NH_3$ (in basic solution)

4. Faraday's laws An aqueous solution of $AgNO_3$ was electrolyzed between inert electrodes for 30.0 minutes. The only cathode reaction was the deposition of 0.123 g of Ag. (a) What was the reaction at the anode? (NO_3^- ion is not changed by the electrolysis.) How many (b) moles, (c) grams, (d) liters (sc) of the gaseous product were evolved at the anode? (e) How many coulombs passed through the solution? (f) What was the average current in amperes?

5. Degree of dissociation (a) Find the equivalent conductance of ammonia, NH_3, at infinite dilution at 25°C, assuming that it is completely ionized,

$$NH_3 + H_2O \longrightarrow NH_4^+ + OH^-$$

(b) The degree of dissociation of NH_3 in an 0.100 M solution is 1.33%. Calculate the equivalent conductance of this solution.

6. Colligative properties A 0.202 m aqueous solution of chloroacetic acid, $CH_2ClCOOH$, is 24% ionized,

$$CH_2ClCOOH + H_2O \rightleftharpoons H_3O^+ + CH_2ClCOO^-$$

Calculate the freezing point of the solution.

ADDITIONAL PROBLEMS

7. Ionization The value of the van't Hoff factor i for hexammine cobalt(III) chloride, $[Co(NH_3)_6]Cl_3$, was determined to be 3.9 when the molality of its solution is about 0.001. (a) How many moles of ions are present in a solution containing one mole of this compound? (b) Calculate the freezing point of an 0.0010 m aqueous solution of this compound.

8. Conductivity and temperature Would you expect the conductivity of a fused salt to increase or to decrease as the temperature increases? Explain.

9. Balancing equations Balance each of the following equations by the ion-electron method. Show the balanced partial equations, and the overall ionic equation.

(a) $Cu(c) + HONO_2(aq,$ concentrated$) \longrightarrow Cu^{2+}(NO_3^-)_2 + NO_2(g)$ (in acid solution)
(b) $H_2S(g) + H_2SO_3(aq) \longrightarrow S_8(c)$ (in acid solution)
(c) $Zn(c) + CrO_4^{2-} \longrightarrow ZnO_2^{2-} + CrO_2^{2-}$ (in basic solution)
(d) $MnO_4^-(c) + S_2O_3^{2-} \longrightarrow MnO_2(c) + SO_4^{2-}$ (in basic solution)

10. Faraday's laws Two electrolytic cells are connected in series. In the first, Ag^+ is reduced to Ag at the cathode and oxygen is liberated from water at the anode. In the second cell hydrogen is liberated from water at the cathode, and Zn is oxidized to Zn^{2+} at the anode. If 1.00 g of Ag is deposited in the first cell, find the volumes (sc) of hydrogen and oxygen liberated, and the weight of Zn removed, at the other electrodes.

11. Faraday's laws A current of 0.500 A is used for the electrolytic decomposition of water. (a) How long (in hours) will it take to produce 1.00 liter (sc) of H_2? (b) What weight of O_2 will be produced in the same time?

12. Faraday's laws Iridium was plated from an aqueous solution containing Ir^{x+} for 2.00 hours with a current of 0.0650 A. The iridium deposit on the cathode weighed 0.311 g. (a) Find the equivalent weight exhibited by iridium in this process. (b) What is the charge of the iridium ion?

13. Faraday's laws In a chemical plant, fused NaCl is electrolyzed with a current of 500 A. (a) How many kilograms of Na will be produced every 24 hours? (b) How many liters (sc) of Cl_2 will be produced?

14. Electrode processes The antiknock agent tetraethyl lead, $(CH_3CH_2)_4Pb$, is made in one commercial process by the electrolysis of ethylmagnesium chloride, CH_3CH_2MgCl, in ether solution, using a lead anode. The CH_3CH_2MgCl dissociates into negative ethyl ions, $CH_3CH_2^-$, and positive $MgCl^+$ ions. The ethyl ions combine directly with the anode to form tetraethyl lead. At the cathode, the $MgCl^+$ ions are reduced to $MgCl_2$ and metallic magnesium. Write the electronic equation for (a) the anode reaction; (b) the cathode reaction. (c) Add the two.

15. Ionic conductance The equivalent conductance of thallium(I) perchlorate, $TlClO_4$, at infinite dilution at 25°C is 142.7 cm²/Ω eq. Experiments on the fraction of the current carried by each of the two ions indicate that the individual ionic conductances of Tl^+ and ClO_4^- are in the ratio 1.099 to 1. Find the ionic conductances of Tl^+ and ClO_4^-.

16. Degree of dissociation The equivalent conductance at infinite dilution at 25°C of sodium propionate, C_2H_5COONa, is 85.92 cm²/Ω eq. (a) Find the equivalent conductance at infinite dilution of the propionate ion. (b) The equivalent conductance of 1.0 M propionic acid, C_2H_5COOH, is 1.4 cm²/Ω eq. Find the percent dissociation of this acid.

17. Freezing-point depression The freezing point of a 0.1000 m aqueous solution of formic acid, HCOOH, is −0.194°C. Calculate the degree of ionization of the acid.

ANSWERS

4. (b) 2.85×10^{-4} mole; (c) 0.912 g; (d) 0.638 liter; (e) 110 C; (f) 0.0611 A.

5. (a) 271 cm^2/Ω eq; (b) 3.60 cm^2/Ω eq.

7. (a) 4; (b) $-0.0073°C$.

11. (a) 0.48 hour; (b) 0.71 g.

15. 74.7, 68.0 cm^2/Ω eq.

16. (a) 35.81 cm^2/Ω eq; (b) 0.36%.

16 GALVANIC CELLS AND THE DRIVING FORCE OF CHEMICAL REACTIONS

16.1 INTRODUCTION

We saw in the last chapter that an electric current can cause chemical reactions. The process can be reversed, and electricity obtained from a chemical reaction. This discovery, by Alessandro Volta in 1796, provided for the first time a source of continuous electric current, and thus made possible the great electrical discoveries of the early nineteenth century, including the electrochemical observations that we have described. Today, the electrochemical cell, or battery of cells, has been generally superseded by the generator as a source of electricity. However, cells still have a place as portable sources of relatively small amounts of power. What is more important, from the point of view of chemistry, is that the electrical work that can be obtained from a chemical reaction provides a direct measure of the driving force of the reaction, and, not quite so directly, of its equilibrium constant.

One of the most familiar chemical reactions is the reaction between zinc and a soluble copper salt,

$$\mathrm{Zn}(c) + \mathrm{CuSO_4}(aq) \longrightarrow \mathrm{Cu}(c) + \mathrm{ZnSO_4}(aq)$$

or, in ionic form,

$$\mathrm{Zn}(c) + \mathrm{Cu^{2+}} \longrightarrow \mathrm{Cu}(c) + \mathrm{Zn^{2+}}$$

The products in this reaction have lower enthalpy than the reactants; that is, ΔH is negative, or the reaction is exothermic. Merely adding a piece of zinc to a $CuSO_4$ solution results in the liberation of heat, but yields no work, in mechanical or electrical form, aside from a minute amount that the reaction mixture will do if it expands against atmospheric pressure. Of course, the liberated heat can be partially converted to work (as in a steam engine), but this process is notoriously inefficient and does not help us to measure the maximum available work—the quantity in which we are especially interested.

16.2 □ ELECTRICITY FROM A CHEMICAL REACTION

To see how a reaction can be so harnessed as to produce electrical work directly, we first recall that an oxidation-reduction reaction can be resolved into two partial reactions (p. 266),

$$Zn \longrightarrow Zn^{2+} + 2e^- \qquad (\textit{oxidation})$$
$$Cu^{2+} + 2e^- \longrightarrow Cu \qquad (\textit{reduction})$$

If we can arrange for these partial reactions to occur in two physically separated places, then the electrons will have to flow from the place where the oxidation occurs (the **anode**) to the place where the reduction occurs (the **cathode**). This flow of electrons constitutes a current, which can be used to drive a motor and lift a weight, or otherwise to do work.

The most obvious way to separate the partial reactions is to insert the Zn into a solution that does not contain Cu^{2+}—it may contain Zn^{2+}, or it may be plain water—and to insert a piece of Cu (or some unreactive metal) into a solution containing Cu^{2+} (Fig. 16.1). When this is done, a minute amount of Zn can go into solution as Zn^{2+}, leaving a negative

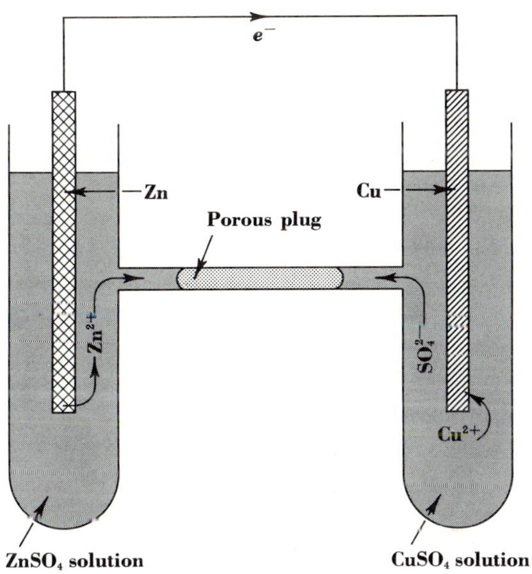

Fig. 16.1 A galvanic cell with liquid junction.

charge on the Zn (because of the electrons left behind) and a positive charge in the solution. These charges stop the process almost immediately; for every Zn^{2+} ion entering the solution, it or another Zn^{2+} ion will be driven back to the Zn electrode by the attraction of the negative electrode and the repulsion of the positive solution. The fact that Zn^{2+} can pick up two electrons to become an atom of Zn indicates the reversibility of the reactions. All such partial reactions are reversible. Similarly, a few Cu^{2+} ions may deposit on the Cu electrode as Cu atoms, but this process will make the electrode positive and the solution negative and will thus stop itself. The deadlock can be broken if we connect the two electrodes by a wire, to allow the electrons from the Zn to flow to the electron-deficient Cu electrode,* and bring the two solutions into contact, to allow the charges of the solutions to be neutralized by a flow of ions between them (Fig. 16.1). In other words, the circuit must be closed. The solutions should be connected through a narrow tube or a porous plug; we must not allow Cu^{2+} ions to reach the Zn electrode and take electrons from the Zn there, thus bypassing the wire and defeating our purpose. We can now see that the solution in contact with the Zn electrode must contain an electrolyte to make it a conductor—pure water is almost an insulator and would leave an open circuit.

An apparatus in which chemical energy is converted to electrical energy is known as a **galvanic cell,** after Luigi Galvani, whose discovery in 1786 of the effects of electricity on frog muscles prepared the way for Volta's work.

16.3 □ A GALVANIC CELL WITH ONE SOLUTION

Another galvanic cell, perhaps simpler (though less familiar) than that just described, is shown in Fig. 16.2. Two electrodes are immersed in a solution of $ZnCl_2$: one is simply a piece of zinc, and the other is a piece of silver, with a porous coating of silver chloride. Because of the very low solubility of AgCl, the solution contains only a trace of Ag^+ ion.

If the two solids, Zn and AgCl, are brought into contact, the reaction

$$Zn(c) + 2AgCl(c) \longrightarrow 2Ag(c) + Zn^{2+} + 2Cl^-$$

will take place very slowly. The slowness results merely from the difficulty of establishing sufficiently intimate contact between two solids. But the reaction has a strong tendency to take place, in the sense that equilibrium is attained only when the reaction has gone nearly to completion. In the apparatus of Fig. 16.2, if there is no wire joining the electrodes, this reaction will occur so slowly that it can be ignored. The only way it can occur is by a reaction between Zn and the minute

* It may be that both the Cu and the Zn give up positive ions to the solution and become negatively charged, or they may both become positively charged by deposition of positive ions from the solution; no one can be sure. We do know that the Cu is more positive than the Zn, or the Zn is more negative than the Cu.

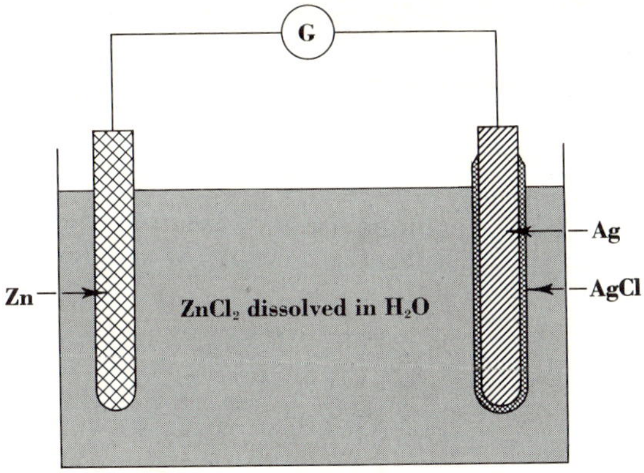

Fig. 16.2 A galvanic cell without liquid junction. *G*, galvanometer.

amount of Ag^+ in the solution,

$$Zn(c) + 2Ag^+ \longrightarrow Zn^{2+} + 2Ag(c)$$

followed by dissolution of more AgCl to replenish the Ag^+,

$$AgCl(c) \rightleftharpoons Ag^+ + Cl^-$$

However, when the two electrodes are joined by a conductor, the same reaction can proceed by an alternative route, which is much faster. Zn can enter the solution as Zn^{2+} ion at the left electrode,

$$Zn(c) \longrightarrow Zn^{2+} + 2e^-$$

The two electrons left behind on the Zn can travel through the conductor to the right electrode,* where the reaction

$$2e^- + 2AgCl(c) \longrightarrow 2Ag(c) + 2Cl^-$$

occurs. The overall process is

$$Zn(c) + 2AgCl(c) \longrightarrow 2Ag(c) + Zn^{2+} + 2Cl^-$$

the same reaction which, although spontaneous, was extremely slow until the circuit was closed. As in the Zn-Cu cell, the cell works because the reactants are kept apart and can react only by sending electrons through an external conductor.

16.4 □ ELECTRICAL WORK; ELECTROMOTIVE FORCE

When one gram-atom of Zn is converted to Zn^{2+}, as shown above, 2 faradays (p. 272) of charge, or $2 \times 9.65 \times 10^4$ C, pass through the external circuit. They can be harnessed to an electric motor, which

* More precisely, the electrons left on the Zn push other electrons toward the Ag electrode. The electrons that come out at one end of a conductor need not be the same ones that entered at the other end. It is not even meaningful to ask whether they are the same ones.

does mechanical work, or their energy can be dissipated as heat in a resistor. The quantity of work that can be obtained as a given amount of charge passes from one point to another is determined by the **difference of potential** between the two points. Potential is usually measured in *volts*.

$$\underset{\text{[in joules, J]}}{\text{work}} = \underset{\text{[in volts, V]}}{(\text{potential at } A - \text{potential at } B)} \times \underset{\text{[in coulombs, C]}}{\text{charge}}$$

A volt, then, is a joule per coulomb. The potential difference between the two electrodes in a galvanic cell is called the **electromotive force, emf, potential,** or **voltage** of the cell. The emf of the cell in Fig. 16.2 depends on the temperature and the concentration of the $ZnCl_2$ solution; if the temperature is 25°C and the concentration is suitably chosen (about 1.5 mole $ZnCl_2$ per liter), the emf of this particular cell is 1.00 V. This means that when one gram-atom of zinc is consumed, and two faradays ($2 \times 9.65 \times 10^4$ C) flow through the circuit, the work this electricity will do is

$$\begin{aligned} 2 \times 9.65 \times 10^4 \text{ C} \times 1.00 \text{ V} &= 1.93 \times 10^5 \text{ J} \\ &= 1.93 \times 10^5 \not{\text{J}} \times \frac{1 \text{ cal}}{4.184 \not{\text{J}}} \\ &= 4.61 \times 10^4 \text{ cal} \end{aligned}$$

This is the amount of work that can be obtained from the chemical reaction

$$\text{Zn}(c) + 2\text{AgCl}(c) \longrightarrow 2\text{Ag}(c) + \text{Zn}^{2+} + 2\text{Cl}^-$$

when one gram-atom of Zn reacts with 2 moles of AgCl, and is the most precise and realistic measure of the driving force of the reaction.

The reader may have noted by now that it was not necessary to have $ZnCl_2$ initially present in the solution; all we need is an electrolyte to make the solution a conductor. $ZnCl_2$ will appear as a product of the reaction. However, the emf depends on the concentrations of Zn^{2+} and Cl^- ions, so that the initial situation is more definite if we start with a known concentration of $ZnCl_2$, instead of just letting some unknown concentration appear when the reaction begins.

Emf can be thought of as electrical pressure analogous to water pressure. Water in a reservoir high in the mountains exerts a pressure on a faucet at a distance below. The greater the difference in height between the reservoir and the faucet, the greater the water pressure. In a galvanic cell, the greater the electron-forming ability of the anode and the greater the electron-consuming ability of the cathode, the greater is the emf.

16.5 □ MEASUREMENT OF ELECTROMOTIVE FORCE

Not all the work theoretically available from a chemical reaction, the so called **maximum work** or **work**$_{max}$, can actually be realized as work.

Instead of work, some of the cell energy appears as heat because of the resistance of the solution and of the wires. Another difficulty is that the electrode reactions cause local changes of concentration in the solution, and these changes are always such as to decrease the emf. Such inefficiency is inevitable in any device that transforms one kind of energy into another. But we are interested not so much in obtaining work from the reaction as in measuring the maximum work, for, as we shall see, this is what measures the driving force. To find this *maximum work*, we must therefore measure the potential difference between the electrodes *when no current is flowing through the cell*. The device used for this purpose is called a **potentiometer.**

We shall not describe the details of the potentiometer. We may regard it simply as a galvanic cell, the emf of which can be varied continuously and read from a dial. The emf of the potentiometer tries to make a current flow through the cell. There is a chemical reaction that also tries to make a current flow through the cell, and we connect the cell so that these two currents will be in opposite directions. The potentiometer is then adjusted until the current is zero. At this point, the known emf of the potentiometer is equal to the emf of the cell. The process is analogous to measuring a mass by balancing it against weights (a variable mass) until the gravitational forces on the two masses are equal.

16.6 □ FREE ENERGY AND ENTROPY

When a potentiometer is used to measure the emf of a cell, the measurement is made with no current flowing. There is, therefore, no loss because of resistance, and no creation of local concentration differences. The emf thus determined is a measure of the ideal maximum amount of work (in joules per coulomb of charge passed through the circuit) that can be obtained from the chemical reaction taking place in the cell. It is called the **reversible emf,** because the cell process can be reversed by a very small change in the opposing potential. Assume that the concentration of $ZnCl_2$, in the cell of Fig. 16.2, is such that the reversible emf is exactly 1 V. Then the maximum work available when the cell reaction

$$\text{Zn} + 2\text{AgCl} \longrightarrow 2\text{Ag} + \text{Zn}^{2+} + 2\text{Cl}^-$$

occurs is

$$1.00\ \text{V} \times 9.65 \times 10^4 \frac{\text{C}}{\cancel{\text{faraday}}} \times 2 \frac{\cancel{\text{faradays}}}{\text{g-atom}} = 1.93 \times 10^5\ \text{V C/g-atom Zn} = 1.93 \times 10^5\ \text{J/g-atom Zn}$$

This maximum work is called *minus* the **free energy change,** $-\Delta G$, in the chemical reaction written above. The symbol Δ (*delta*) signifies "change": specifically, final value minus initial value. ΔG is the total free energy of 2 g-atoms Ag, 1 mole Zn^{2+}, and 2 moles Cl^-, minus the total free energy of 1 g-atom Zn and 2 moles AgCl:

$$\Delta G = 2G_{Ag} + G_{Zn^{2+}} + 2G_{Cl^-} - G_{Zn} - 2G_{AgCl} = -193{,}000 \text{ J}$$

Here, ΔG is negative, $-193{,}000$ J, because the cell has done work, and its ability to do further work has declined. Free energy—"free" in the sense of "available"—is a measure of this ability to do work. The decrease in free energy is not equal to the electrical work actually done when current is drawn from the cell, for the work done is always less than that which might be done, and it is the possibility, not the performance, that is given by the free energy change.

If Zn and AgCl simply react with each other, not in a galvanic cell, the total heat liberated is minus the enthalpy change, $-\Delta H$ (p. 97). For this reaction, $-\Delta H = 2.33 \times 10^5$ J, which is the amount of heat emitted in the reaction. The maximum work available from the cell is 1.93×10^5 J. The maximum work is not equal to the heat emitted when the reactants merely mix and do no work (other than work against atmospheric pressure). The difference is the amount of energy that can be obtained only as heat and not as work. This difference in energy is equal to *minus* the **entropy change** multiplied by the absolute temperature, $-T\,\Delta S$.

To summarize:

$$\begin{pmatrix}\text{The heat emitted when}\\ \text{work}_{max} \text{ is done}\end{pmatrix} = \begin{pmatrix}\text{heat emitted when}\\ \text{no work is done}\end{pmatrix} - \text{work}_{max}$$

$$(-T\,\Delta S) \quad = \quad (-\Delta H) \quad - (-\Delta G)$$

Solving for ΔS, we obtain

$$\Delta S = \frac{\Delta H - \Delta G}{T}$$

The presence of T in this equation becomes plausible if we understand that the entropy of a system is a measure of its randomness (or disorder) (p. 206). If energy is added to a system, the vigor of molecular motion and hence the opportunities for disordered arrangements increase; therefore the entropy increases. At low temperatures the opportunity for increased disorder is great; therefore a given increase in energy produces a large increase in entropy. At high temperatures (when there is a higher likelihood of disorder) the effect of change in energy on change of entropy becomes small. Thus, for any given value of the difference $\Delta H - \Delta G$, the entropy change, ΔS, will be smaller at higher temperatures, greater at lower temperatures.

For the Zn + AgCl reaction at 25°C, we obtain ΔS as follows:

$$\Delta S = \frac{-2.33 \times 10^5 - (-1.93 \times 10^5) \text{ J}}{298 \text{ deg}}$$

$$= -1.3 \times 10^2 \text{ J/deg}$$

It is also possible for $work_{max}$ to be *greater* than the heat that can be given out on mixing the reactants. In such a case, heat is absorbed when the reaction occurs slowly enough in a galvanic cell, and ΔS is positive. An example is the reaction

$$Pb(c) + Hg_2Cl_2(c) \longrightarrow PbCl_2(c) + 2Hg(l)$$

for which $\Delta H = -9.43 \times 10^4$ J, and $\Delta G = -1.034 \times 10^5$ J.

Entropy is one of the most important, and also one of the most subtle, concepts in physical science. A system tends to arrive at the least orderly configuration—the configuration of greatest entropy—simply because there are many more ways to be disorderly than to be orderly. A system also tends to achieve the lowest possible energy. These tendencies are in conflict, because the least orderly configurations have the very highest energies; in general, an orderly arrangement (as in a crystal) has low energy, while a disorderly arrangement (as in a gas) has high energy. A balance is struck between low energy and high entropy: the system tends to arrive at that state in which $E - TS$ is a minimum. When the pressure is kept constant, enthalpy (p. 97) replaces energy, and *the system tends to arrive at that state in which $H - TS$ is a minimum*. The function $H - TS$ is nothing other than the free energy G. *The most stable state of the system, at a fixed temperature and pressure, is the state of minimum G.* A process in which ΔG is negative (G decreases) leads to a more stable state, and this process can occur without the application of work; indeed, with sufficient ingenuity, work can be obtained from the process, in an amount not exceeding $-\Delta G$. A process with negative ΔG is described as "spontaneous," but the word is rather unfortunate, for the process may not actually occur at a measurable rate. The reaction

$$2H_2(g) + O_2(g) \longrightarrow 2H_2O(l)$$

has $\Delta G = -113$ kcal at 25°C and is therefore "spontaneous," but the reaction does not occur at room temperature in the absence of a catalyst. The reverse reaction,

$$2H_2O(l) \longrightarrow 2H_2(g) + O_2(g)$$

has $\Delta G = +113$ kcal, which means that there is no hope of decomposing water to its elements at 25°C unless work, in the amount of at least 113 kcal for every 2 moles H_2O, is done. This information about free energy does not guarantee that water can be decomposed at all. We know that it can be, by electrolysis, and the electrical work required will be somewhat more than 113 kcal per 2 moles because of the inefficiency to be expected in any process.

The free energy change in a reaction is thus a measure of the **driving force** of the reaction. "Drift" might be a better word than "drive," for molecules behave randomly, not purposefully. The only direction in which the reaction can proceed, at constant temperature and pressure, is the direction of lower free energy, unless work is done on the reacting system. The reversible emf, $\mathscr{E}$, associated with a reaction is really the free energy change in different units, and with the sign reversed. Let n be the number of faradays which pass through the circuit when the reaction occurs as written, and let $\mathscr{F}$ be the Faraday constant, 9.65×10^4 C/faraday. Then the free energy change is

$$-\Delta G = n\mathscr{F}\mathscr{E}$$

The minus sign appears because, for a "spontaneous" reaction, ΔG is negative, while $\mathscr{E}$, by convention, is positive. $\mathscr{E}$ is in joules per coulomb (volts); ΔG is in joules or, more often, calories or kilocalories:

$$\Delta G = -n\ (\text{faradays}) \times \mathscr{F}\left(\frac{\text{C}}{\text{faraday}}\right) \times \mathscr{E}\left(\frac{\text{J}}{\text{C}}\right)$$
$$= -9.65 \times 10^4\ n\mathscr{E}\ \text{J}$$
$$\Delta G = -n\ (\text{faradays}) \times 9.65 \times 10^4 \left(\frac{\text{C}}{\text{faraday}}\right) \times \mathscr{E}\left(\frac{\text{J}}{\text{C}}\right) \times \frac{1\ \text{cal}}{4.184\ \text{J}}$$
$$= -2.305 \times 10^4\ n\mathscr{E}\ \text{cal}$$

16.7 ▫ SALT BRIDGES; CONVENTIONAL NOTATION FOR CELLS

Salt solutions, called **salt bridges,** are frequently used to permit migration of ions between the two solutions. A Zn-Cu cell, modified by inclusion of a KCl salt bridge, is shown in Fig. 16.3. In this case Cl^- flows to the anode to neutralize the positive charge created by the formation of the Zn^{2+}. The K^+ migrates to the cathode to neutralize the excess of SO_4^{2-} created when the Cu^{2+} ions are reduced to Cu. Negative ions are called anions because they migrate to the anode. Positive ions are called cations because they move to the cathode.

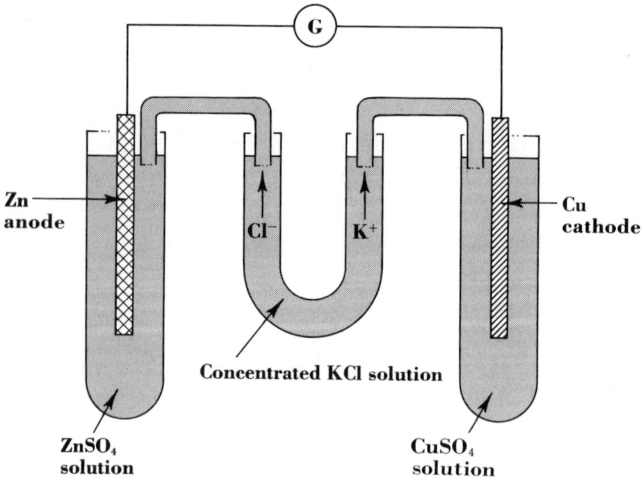

Fig. 16.3 A galvanic cell with a salt bridge. *G*, galvanometer.

One often wishes to describe a galvanic cell without taking the trouble to draw an actual picture of it. We shall use the following notation for this purpose:

The material of one electrode (normally a metal) is listed first, by chemical symbol. Any other substance, not part of the solution, which is in contact with the electrode comes next; then the solution (assumed aqueous unless otherwise specified) is described by giving the formula(s) and concentration(s) of the solute(s). Phase boundaries (between solid and liquid, liquid and gas, or solid and gas) are indicated by | . If there is a second solution, its description is separated from the first by ¦ if the two solution are in direct contact, or by ‖ if they are joined by a salt bridge. After the solution(s), the second electrode is specified, this time with the metal last.

The cells previously discussed may be represented thus:

Fig. 16.1 Zn | $ZnSO_4$(__M) ¦ $CuSO_4$(__M) | Cu
Fig. 16.2 Zn | $ZnCl_2$(__M) | AgCl | Ag
Fig. 16.3 Zn | $ZnSO_4$(__M) ‖ $CuSO_4$(__M) | Cu

The anode is written on the left, and the cathode on the right, a rule that may be remembered by observing the alphabetical order in each of the following pairs: anode-cathode, oxidation-reduction, left-right. When this rule is obeyed and the left electrode is really the anode, the emf is positive. If, for any reason, the cell description is reversed, and the left electrode turns out to be the cathode, the sign of the emf is also reversed. A negative emf signifies that the cell has been written backward, and that the reaction goes in the direction opposite to that which was assumed.

16.8 □ THE EFFECT OF CONCENTRATION ON EMF; NERNST EQUATION

It has been mentioned previously that the partial cell reactions can proceed in either direction. The extent to which such reactions proceed in one direction or the other depends on the molar concentrations of all soluble components or, for gaseous components, on their partial pressures. Thus a change in concentration affects the emf of the cell. For example, in the Zn-Cu cell (p. 282) an increase in $[Zn^{2+}]$ in the solution bathing the Zn electrode shifts the anode reaction to the left, which is the direction opposed to the formation of electrons. The electron-forming ability of the anode is decreased and the emf of the cell drops. On the other hand, an increase in $[Cu^{2+}]$ in the cathode solution drives the cathode reaction to the right, the direction favoring electron consumption. Such an effect increases the emf.

The **concentration cell** illustrates especially well the ability of partial reactions to proceed in either direction. In such a cell, the same two partial reactions occur but with at least one species produced and consumed at two different concentrations. There is thus no *net* chemical reaction—merely a transfer of matter from one concentration to another. An example is

$$\text{Cu} \mid \text{CuSO}_4(aM) \,\vdots\, \text{CuSO}_4(bM) \mid \text{Cu}$$

The reactions here are

$$\begin{array}{lrcl} \text{Anode} & \text{Cu} & \longrightarrow & \text{Cu}^{2+}(aM) + 2\text{e}^- \\ \text{Cathode} & \text{Cu}^{2+}(bM) + 2\text{e}^- & \longrightarrow & \text{Cu} \\ \hline \text{Overall} & \text{Cu}^{2+}(bM) & \longrightarrow & \text{Cu}^{2+}(aM) \end{array}$$

The cathode reaction is more likely to occur when $[Cu^{2+}]$ is larger. The anode reaction occurs in the half of the cell with the smaller $[Cu^{2+}]$. Hence, the emf is positive if $b > a$, in which case the cell process involves the transfer of Cu^{2+} from a more concentrated solution to a more dilute solution.

The quantitative relationship between concentrations* and emf of a galvanic cell is given by the **Nernst equation,** named after Walther Nernst, who developed it in 1889. The equation is

$$\mathscr{E} = \mathscr{E}^\circ - \frac{2.303RT}{n\mathscr{F}} \log Q$$

where $\mathscr{E}$ is the emf of the cell in volts; $\mathscr{E}^\circ$ is a number that depends on the temperature and on what the cell is; R is the gas constant, 1.987 cal/deg mole; T is the absolute temperature; $\mathscr{F}$ is Faraday's constant, 9.6487×10^4 C/faraday; n is the number of moles of electrons (faradays) transferred through the external circuit when the cell reaction takes place as written; Q is the reaction quotient (p. 245) for the reaction as written, and consists of the molar concentrations of each soluble product in the numerator and of each soluble reactant in the denominator, each raised to a power equal to the coefficient of that species in the equation. For a gas, the partial pressure is used. At 25°C, the Nernst equation becomes

$$\mathscr{E} = \mathscr{E}^\circ - \frac{0.05915}{n} \log Q$$

The Nernst equation for the reaction

$$\text{Zn}(c) + \text{Cu}^{2+} \longrightarrow \text{Cu}(c) + \text{Zn}^{2+}$$

at 25°C is

$$\mathscr{E} = \mathscr{E}^\circ - \frac{0.059}{2} \log \frac{[\text{Zn}^{2+}]}{[\text{Cu}^{2+}]}$$

and for

$$2\text{MnO}_4^- + 6\text{H}_3\text{O}^+ + 5\text{SO}_3^{2-} \longrightarrow 2\text{Mn}^{2+} + 9\text{H}_2\text{O} + 5\text{SO}_4^{2-}$$

* For better accuracy, activities should be used. An **activity** is a modified concentration needed because of the attraction between oppositely charged ions. In very dilute solutions, activities equal concentrations.

is

$$\mathscr{E} = \mathscr{E}^\circ - \frac{0.059}{10} \log \frac{[Mn^{2+}]^2[SO_4^{2-}]^5}{[MnO_4^-]^2[H_3O^+]^6[SO_3^{2-}]^5}$$

Water, being the solvent, does not appear in the Nernst equation. The value of n is best determined by balancing the chemical equation using the ion-electron method (p. 266) and finding the number of moles of electrons transferred in the reaction as written. Notice that when the molarities of all participating substances are 1.0,* $\mathscr{E} = \mathscr{E}^\circ$. $\mathscr{E}^\circ$ is called the **standard emf.**

16.9 □ HALF-CELL POTENTIALS; THE HYDROGEN HALF-CELL

A galvanic cell always requires two electrodes. Oxidation must occur at one electrode (the anode), and reduction at the other (the cathode). There is no way in which we can measure the potential difference between an electrode and a solution without introducing another electrode; what we always measure, as the emf of a cell, is a potential difference between two electronic conductors. Only the emf of a complete cell can be measured.

However, it is useful, as a bookkeeping device, to think of a cell as consisting of two half-cells, with each half-cell contributing its share to the emf of the entire cell. A half-cell consists of an electrode and the solution with which it is in contact. If a cell is made of two half cells containing different solutions, we consider the solutions to be joined by a salt bridge. *It is possible to assign a potential to each half-cell,* such that the emf of the entire cell is equal to the difference between the half-cell potentials. It is this fact that makes useful the resolution of an emf into half-cell potentials. Instead of tabulating the standard emf, $\mathscr{E}^\circ$, for every cell, we need tabulate it only for every half-cell, and $\mathscr{E}^\circ$ for any cell that can be made by combining two of these half-cells is obtained immediately by subtraction.

Let us return to our analogy of water pressure. How is the difference in height between two points determined? To calculate geographical height, or altitude, a reference, called sea level, is selected and assigned a height of zero. All points on the earth's surface are assigned a value either above or below sea level. The difference in height between any two points is obtained by subtracting the value of the lower point from that of the higher point.

A similar procedure is followed to get half-cell potentials. The half-cell chosen as the reference and assigned a standard half-cell potential equal to zero (the sea level) is the hydrogen half-cell, Pt | H_2(1 atm) | HCl(aq)(1 M) || shown in Fig. 16.4. Gaseous hydrogen at 1 atm pressure is allowed to bubble over a "platinized platinum" electrode—a Pt electrode coated with finely divided Pt. The Pt electrode is bathed by a 1 M HCl solution.* The Pt serves as a catalyst for the reaction

$$\tfrac{1}{2}H_2(g) + H_2O \longrightarrow H_3O^+ + e^-$$

* More accurately, all species should have unit activity.

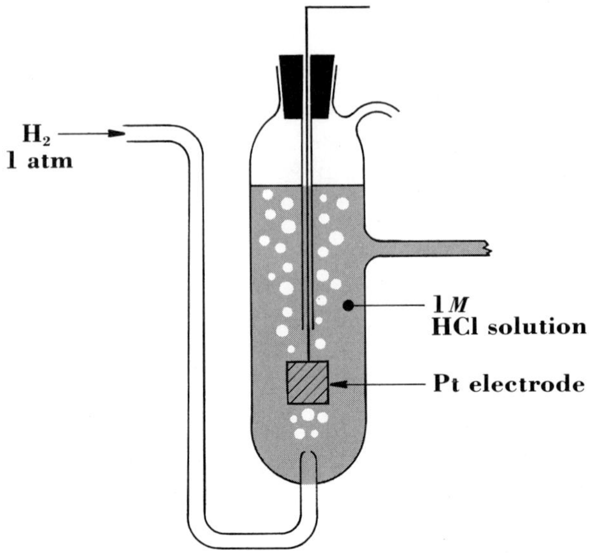

Fig. 16.4 The hydrogen half-cell.

TABLE 16.1 REDUCTION POTENTIALS IN AQUEOUS SOLUTION AT 25°C (ARRANGED ALPHABETICALLY BY CHEMICAL SYMBOLS)

Half-cell		Half-reaction	$\mathscr{E}°$ (volts)
Ag	$\mid Ag^+$	$Ag^+ + e^- \rightleftharpoons Ag$	+0.7996
Ag	$\mid AgCl(c) \mid Cl^-$	$AgCl(c) + e^- \rightleftharpoons Ag + Cl^-$	+0.2221
Ag	$\mid AgI(c) \mid I^-$	$AgI(c) + e^- \rightleftharpoons Ag + I^-$	−0.1519
Ag	$\mid NH_3(aq), Ag(NH_3)_2^+$	$Ag(NH_3)_2^+ + e^- \rightleftharpoons Ag + 2NH_3(aq)$	+0.373
Al	$\mid Al^{3+}$	$Al^{3+} + 3e^- \rightleftharpoons Al$	−1.67
Au	$\mid Au^+$	$Au^+ + e^- \rightleftharpoons Au$	+1.7
Au	$\mid Au^{3+}$	$Au^{3+} + 3e^- \rightleftharpoons Au$	+1.42
[a]Ba	$\mid Ba^{2+}$	$Ba^{2+} + 2e^- \rightleftharpoons Ba$	−2.90
Pt	$\mid Br_2(l) \mid Br^-$	$Br_2(l) + 2e^- \rightleftharpoons 2Br^-$	+1.0650
[a]Ca	$\mid Ca^{2+}$	$Ca^{2+} + 2e^- \rightleftharpoons Ca$	−2.76
Cd	$\mid Cd^{2+}$	$Cd^{2+} + 2e^- \rightleftharpoons Cd$	−0.4024
Pt	$\mid Cl_2(g) \mid Cl^-$	$Cl_2(g) + 2e^- \rightleftharpoons 2Cl^-$	+1.3583
Pt	$\mid Cr^{3+}, HCrO_4^-, H_3O^+$	$HCrO_4^- + 7H_3O^+ + 3e^- \rightleftharpoons Cr^{3+} + 11H_2O$	+1.195
Cr	$\mid Cr^{3+}$	$Cr^{3+} + 3e^- \rightleftharpoons Cr$	−0.74
Cu	$\mid Cu^{2+}$	$Cu^{2+} + 2e^- \rightleftharpoons Cu$	+0.3460
[a]Pt	$\mid F_2(g) \mid F^-$	$F_2(g) + 2e^- \rightleftharpoons 2F^-$	+2.85

(*Continued on next page*)

TABLE 16.1 (*Continued*)

Half-cell	Half-reaction	$\mathscr{E}°$ (volts)
Fe $\mid Fe^{2+}$	$Fe^{2+} + 2e^- \rightleftharpoons Fe$	−0.441
Pt $\mid Fe^{2+}, Fe^{3+}$	$Fe^{3+} + e^- \rightleftharpoons Fe^{2+}$	+0.7701
Pt $\mid H_2(g) \mid H_3O^+$	$2H_3O^+ + 2e^- \rightleftharpoons H_2(g) + 2H_2O$	0.0000
Pt $\mid H_2(g) \mid OH^-$	$2H_2O + 2e \rightleftharpoons H_2(g) + 2OH^-$	−0.8277
$Hg(l) \mid Hg_2^{2+}$	$Hg_2^{2+} + 2e^- \rightleftharpoons 2Hg$	+0.7986
$Hg(l) \mid Hg^{2+}$	$Hg^{2+} + 2e^- \rightleftharpoons Hg$	+0.852
Pt $\mid I_2(c) \mid I^-$	$I_2(c) + 2e^- \rightleftharpoons 2I^-$	+0.5356
[a]K $\mid K^+$	$K^+ + e^- \rightleftharpoons K$	−2.9241
[a]Li $\mid Li^+$	$Li^+ + e^- \rightleftharpoons Li$	−2.9595
Mg $\mid Mg^{2+}$	$Mg^{2+} + 2e^- \rightleftharpoons Mg$	−2.375
Pt $\mid Mn^{2+}, MnO_4^-, H_3O^+$	$MnO_4^- + 8H_3O^+ + 5e^- \rightleftharpoons Mn^{2+} + 12H_2O$	+1.50
[a]Na $\mid Na^+$	$Na^+ + e^- \rightleftharpoons Na$	−2.7132
Ni $\mid Ni^{2+}$	$Ni^{2+} + 2e^- \rightleftharpoons Ni$	−0.23
Pt $\mid O_2(g) \mid H_3O^+$	$O_2(g) + 4H_3O^+ + 4e^- \rightleftharpoons 6H_2O$	+1.229
Pt $\mid O_2(g) \mid OH^-$	$O_2(g) + 2H_2O + 4e^- \rightleftharpoons 4OH^-$	+0.401
Pb $\mid Pb^{2+}$	$Pb^{2+} + 2e^- \rightleftharpoons Pb$	−0.1263
Pb $\mid PbSO_4(c) \mid SO_4^{2-}$	$PbSO_4(c) + 2e^- \rightleftharpoons Pb + SO_4^{2-}$	−0.336
Pt $\mid PbSO_4(c) \mid PbO_2(c) \mid H_3O^+, HOSO_3^-$	$PbO_2(c) + 3H_3O^+ + HOSO_3^- + 2e^- \rightleftharpoons PbSO_4(c) + 5H_2O$	+1.685
Pt $\mid Pt^{2+}$	$Pt^{2+} + 2e^- \rightleftharpoons Pt$	+1.2
Sn $\mid Sn^{2+}$	$Sn^{2+} + 2e^- \rightleftharpoons Sn$	−0.1406
Zn $\mid Zn^{2+}$	$Zn^{2+} + 2e^- \rightleftharpoons Zn$	−0.7628

[a] These half-cells cannot be prepared because of a vigorous reaction between one constituent and water. Their potentials have been determined indirectly.

The standard potential, $\mathscr{E}°$, of any other half-cell (C) is derived from the standard emf of the cell formed by combining the half-cell C, on the left, with the reference H_2 half-cell on the right. If the emf of the resulting cell is positive, C is really the anode. If the emf of the cell, with C on the left is negative, the cell is represented incorrectly and C is really the cathode. The words "anode" and "cathode" apply strictly to the electrodes (the metallic conductors). However, we shall also refer to the half-cell (electrode + solution) in which oxidation occurs as the anode, and the half-cell in which reduction occurs as the cathode.

For example, the standard emf of the cell

$$Zn \mid Zn^{2+}(1\ M) \parallel HCl(aq)(1\ M) \mid H_2(1\ \text{atm}) \mid Pt$$

is +0.7628 V. The positive voltage for the cell as written means that the $Zn \mid Zn^{2+}$ half-cell is the anode relative to the hydrogen half-cell. By convention the $\mathscr{E}°$ *for a cell is* $\mathscr{E}°$ *for the cathode* (right half-cell) *minus* $\mathscr{E}°$ *for the anode* (left half-cell).

$$\mathscr{E}°_{\text{cell}} = \mathscr{E}°_{\text{cathode}} - \mathscr{E}°_{\text{anode}} \text{ or } = \mathscr{E}°_{\text{right}} - \mathscr{E}°_{\text{left}} \tag{1}$$

For the above cell, $\mathscr{E}°_{\text{cathode}} = 0$ by arbitrary assignment and

$$\mathscr{E}°_{\text{cell}} = 0 - \mathscr{E}°_{\text{anode}} = +0.7628$$

Then $\mathscr{E}°_{\text{anode}} = -0.7628$. This is the standard potential assigned to the $Zn \mid Zn^{2+}$ half-cell as listed in Table 16.1.

Let us now consider the cell

$$Pt \mid Hg \mid Hg^{2+}(1\ M) \parallel HCl(aq)(1\ M) \mid H_2(1\ \text{atm}) \mid Pt$$

whose emf is determined as -0.852. The negative sign indicates that the cell was written incorrectly and that the $Hg \mid Hg^{2+}$ half-cell is actually the cathode. If we write the cell thus,

$$Pt \mid H_2(1\ \text{atm}) \mid HCl(aq)(1\ M) \mid Hg^{2+}(1\ M) \mid Hg \mid Pt$$

we would say that the emf is $+0.852$ V—not because the cell has changed, but because our representation of it has been reversed. Substituting into equation 1 gives

$$0.852 = \mathscr{E}°_{\text{cathode}} - 0$$

and

$$\mathscr{E}°_{\text{cathode}} = 0.852$$

which is the standard potential for the $Hg \mid Hg^{2+}$ half-cell. We see that a half-cell has a *negative* potential if it functions as an *anode* when combined with the reference half-cell; a *positive* potential, if it is a *cathode* relative to the reference half-cell. All half-cell reactions in Table 16.1 are written as reductions and, therefore, the listed $\mathscr{E}°$ values are called **standard reduction potentials.**

In the cells just considered, the nature of the anion, whether, for example, it is Cl^-, NO_3^-, or SO_4^{2-}, does not affect the $\mathscr{E}°$.

The standard potential of the half-cell $Cu \mid Cu^{2+}$ may be determined by comparing it with any half-cell for which we now know the potential. If the Zn half-cell is chosen, we prepare the cell

$$Zn \mid ZnCl_2 \parallel CuSO_4 \mid Cu$$

using a salt bridge to join the two unlike solutions. (Any soluble salts of Cu^{2+} and Zn^{2+} could be used in place of $CuSO_4$ and $ZnCl_2$.) The reaction of this cell, as written, is

$$Zn(c) + Cu^{2+} \longrightarrow Zn^{2+} + Cu(c)$$

At 25°C, $\mathscr{E}°$ for this cell is found by experiment to be $+1.1088$ V. Then

$$+1.1088 = \mathscr{E}°_{\text{right}} - \mathscr{E}°_{\text{left}} = \mathscr{E}°_{\text{right}} - (-0.7628)$$

and $\mathscr{E}_r° = +0.3460$ V. By experiments of this kind, the standard potential of every possible half-cell can be determined. A list of a relatively few such potentials is given in Table 16.1. When an inert electrode is needed, we have indicated Pt, but other substances, notably graphite, could be used, except when the catalytic activity of platinum is important (as in the reactions involving gases).

EXAMPLE 1 Find $\mathscr{E}^\circ$ at 25°C of the cell

$$\mathrm{Cu} \mid \mathrm{CuCl_2}(aq) \mid \mathrm{Cl_2}(g) \mid \mathrm{Pt}$$

ANSWER First we write the anode and cathode reactions, and the overall cell process. Beside each half-reaction, we write $\mathscr{E}^\circ$ from the table on p. 294 regardless of the actual direction of the reaction:

		$\mathscr{E}^\circ$ (reduction potential, volts)
Anode:	$\mathrm{Cu} \longrightarrow \mathrm{Cu^{2+}} + 2e^-$	+0.3460
Cathode:	$\mathrm{Cl_2} + 2e^- \longrightarrow 2\mathrm{Cl^-}$	+1.3583
	$\mathrm{Cu} + \mathrm{Cl_2} \longrightarrow \mathrm{Cu^{2+}} + 2\mathrm{Cl^-}$	

Then we calculate $\mathscr{E}^\circ$ for the cell by subtracting the standard emf for the left half-cell from that for the right half-cell:

$$\mathscr{E}^\circ = 1.3583 - 0.3460 = +1.0123\ \mathrm{V}$$

The plus sign indicates that the left electrode is the anode.

Since the anode reaction is opposite in direction to the half-cell reaction listed in Table 16.1, many persons prefer to *reverse the sign* of $\mathscr{E}^\circ$ for the anode (oxidation) half-reaction, and then *add* the $\mathscr{E}^\circ$'s,

$$\begin{array}{r} -0.3460 \\ +1.3583 \\ \hline +1.0123 \end{array}$$

Obviously, there is no difference between $\mathscr{E}^\circ_r - \mathscr{E}^\circ_l$ and $\mathscr{E}^\circ_r + (-\mathscr{E}^\circ_l)$, but much difficulty results from confusing the two procedures. The best advice is: adopt one convention or the other and stick to it. Either copy the emf values *directly from the table* and *subtract*, or *reverse the sign* of the *anode* emf and *add*.

16.10 □ EMF, K, AND ΔG

The Nernst equation, in its general form (p. 292), is

$$\mathscr{E} = \mathscr{E}^\circ - \left(\frac{2.303RT}{n\mathscr{F}}\right) \log Q$$

where Q is the reaction quotient. Now, if current is drawn from the cell until the cell reaction has come to equilibrium, the emf must become zero, for a system at equilibrium can do no work. In other words, no net reaction, no emf. *The condition of equilibrium for the cell reaction is, then,* $\mathscr{E} = 0$, or

$$\mathscr{E}^\circ - \left(\frac{2.303RT}{n\mathscr{F}}\right) \log Q = 0$$

or

$$\log Q = \frac{n\mathscr{F}\mathscr{E}^\circ}{2.303RT}$$

At equilibrium, $Q = K$, therefore

$$\frac{n\mathscr{F}\mathscr{E}^\circ}{2.303RT} = \log K$$

or at 25°C

$$\frac{n\mathscr{E}^\circ}{0.059} = \log K$$

This law is a consequence of the Nernst equation for any reaction that can take place in a galvanic cell. For example, in the cell

$$Zn \mid Zn^{2+} \parallel Cu^{2+} \mid Cu$$

the reactions are

$$\begin{array}{rcl} Zn & \rightleftharpoons & Zn^{2+} + 2e \\ Cu^{2+} + 2e & \rightleftharpoons & Cu \\ \hline Zn + Cu^{2+} & \rightleftharpoons & Zn^{2+} + Cu \end{array}$$

The reaction quotient is

$$Q = \frac{[Zn^{2+}]}{[Cu^{2+}]}$$

and at equilibrium

$$K = \frac{[Zn^{2+}]}{[Cu^{2+}]}$$

where we have omitted the solids.

At 25°C, $\mathscr{E}^\circ = 1.1088$ V and $n = 2$; then

$$\log K = \frac{2 \times 1.10}{0.059} = 37.5$$

$$K = 3.2 \times 10^{37}$$

That is to say, when an aqueous solution of Zn^{2+} and Cu^{2+} is in contact with solid Zn and Cu, the equilibrium condition is

$$K = \frac{[Zn^{2+}]}{[Cu^{2+}]} = 3.2 \times 10^{37}$$

The equilibrium constant of any reaction can be determined if we can resolve it into two half-reactions for which the potentials are given in a table. The potential for the oxidation half-reaction is subtracted from the potential for the reduction half-reaction. From the resultant emf $\mathscr{E}°$, we calculate K as shown above.

We can now relate the equilibrium constant to the ΔG value for any given reaction. We have already seen (p. 290) that

$$-\Delta G = n\mathscr{F}\mathscr{E}$$

When all species involved in the reaction have standard values,* $\mathscr{E} = \mathscr{E}°$ and $\Delta G = \Delta G°$. This special value of ΔG is called the **standard free energy change** for the reaction,

$$-\Delta G° = n\mathscr{F}\mathscr{E}°$$

At equilibrium

$$2.303RT \log K = n\mathscr{F}\mathscr{E}°$$

and therefore

$$\Delta G° = -2.303RT \log K$$

When $\Delta G°$ is negative ($\mathscr{E}°$ positive), K is greater than 1; when $\Delta G°$ is positive ($\mathscr{E}$ negative), K is less than 1.

Equilibrium constants can thus be calculated equally well from $\mathscr{E}°$ values or from $\Delta G°$ values, depending on the form in which the data are available. $\Delta G°$'s for several reactions can be combined to yield $\Delta G°$ for another reaction, in exactly the same way that ΔH's are combined by Hess's law (p. 101).

16.11 □ PREDICTING THE DIRECTION OF A REACTION

Often we wish to know whether a reaction can occur when certain chemicals are present in certain amounts. A rigorous way to answer this question is to find the equilibrium constant K for the reaction in question and to calculate the reaction quotient Q for the reaction, using the initial concentrations of the species present. Then if $Q < K$, the reaction may go from left to right; if $Q > K$, from right to left.

EXAMPLE 2 A solution contains Ag^+ and CrO_4^{2-} ions, with $[Ag^+] = 10^{-3}$ mole/liter and $[CrO_4^{2-}] = 10^{-4}$ mole/liter. For the reaction $Ag_2CrO_4(c) \rightleftharpoons 2Ag^+ + CrO_4^{2-}$, $K = 1.15 \times 10^{-12}$. Will solid Ag_2CrO_4 dissolve in or precipitate from this solution?

ANSWER For the given reaction we have

$$Q = [Ag^+]^2[CrO_4^{2-}] = (10^{-3})^2(10^{-4}) = 10^{-10}$$

* By standard values, we mean 1.0 molar concentrations and for gases 1 atm pressure. More accurately, for $\Delta G°$ or $\mathscr{E}°$ to be used all species should have unit activity.

This is greater than $K = 1.15 \times 10^{-12}$, and the reaction will go to the left; in other words, Ag_2CrO_4 will precipitate.

A simpler procedure makes use of the $\mathscr{E}°$ for a cell designed for the reaction in question. *If* $\mathscr{E}° > 0$, *the reaction goes from left to right; if* $\mathscr{E}° < 0$, *it goes from right to left.* The rule becomes unreliable when $\mathscr{E}°$ is close to 0, and it applies only when all concentrations are near 1 mole/liter.

EXAMPLE 3 Will Fe reduce Fe^{3+} to Fe^{2+}?

ANSWER The half-reactions in which we are interested are

		$\mathscr{E}°$ (volts) (reduction potentials)
Oxidation:	$Fe \longrightarrow Fe^{2+} + 2e^-$	-0.441
Reduction:	$(Fe^{3+} + e^- \longrightarrow Fe^{2+}) \times 2$	$+0.7701$

For the overall reaction

$$Fe + 2Fe^{3+} \longrightarrow 3Fe^{2+}$$

the standard emf is

$$\mathscr{E}° = +0.7701 - (-0.441) = +1.211$$

Since $\mathscr{E}° > 0$, the reaction is possible when concentrations are near 1. A practical consequence of this result is that a solution of a ferrous (Fe^{2+}) salt can be protected from atmospheric oxidation by keeping solid iron in the bottle; any Fe^{3+} formed will be reduced to Fe^{2+}.

16.12 ▫ STRENGTHS OF OXIDIZING AND REDUCING AGENTS

The potentials in Table 16.1 (p. 294) are called **reduction potentials,** because they measure the tendency of a half-reaction to occur, in the direction corresponding to reduction or electron consumption.

The more positive is the $\mathscr{E}°$, the more strongly the reaction tends to proceed as a reduction and the more the higher oxidation state tends to convert to the lower oxidation state. A substance in the higher oxidation state is potentially an oxidizing agent and the corresponding substance in the lower oxidation state is potentially a reducing agent. In terms of the standard reduction potentials we can make useful generalizations about oxidizing and reducing agents.

$F_2 + 2e^- \rightleftharpoons 2F^-$; $\mathscr{E}° = 2.85$	$Li^+ + e^- \rightleftharpoons Li$; $\mathscr{E}° = -2.96$
(a) Reaction has most positive $\mathscr{E}°$.	(a) Reaction has most negative $\mathscr{E}°$.
(b) Reaction has greatest tendency to proceed as a reduction.	(b) Reaction has least tendency to proceed as a reduction.

(c) F_2 is the best oxidizing agent.	(c) Li^+ is the poorest oxidizing agent.
(d) The reverse reaction (oxidation) has least tendency to go.	(d) The reverse reaction (reduction) has the greatest tendency to go.
(e) F^- is the poorest reducing agent.	(e) Li is the best reducing agent.

The use of $\mathscr{E}°$ values enables predictions to be made about the course of redox reactions providing the concentrations of components are not too far from one molar.

16.13 □ SOME PRACTICAL CELLS

(a) Dry cell Figure 16.5 shows a typical dry cell. The anode is Zn; the cathode is C, which is chemically inert under ordinary conditions. A paste consisting of graphite, MnO_2, NH_4Cl, and H_2O surrounds the cathode. Adjacent to the Zn is a layer of paste, usually made from wheat flour, containing H_2O, NH_4Cl, and $ZnCl_2$, but not MnO_2. The cell is "dry" only in the sense that it contains paste rather than free liquid; it could not function if water were absent. The electrode reactions are complicated and subject to some controversy. One representation is

$$\text{Anode} \qquad Zn \longrightarrow Zn^{2+} + 2e^-$$
$$\text{Cathode} \qquad 2NH_4^+ + 2MnO_2 + 2e^- \longrightarrow Mn_2O_3 + 2NH_3(aq) + H_2O$$

(b) Lead-acid storage cell The lead-acid storage cell, in its charged state, consists of an electrode of spongy lead and an electrode of finely

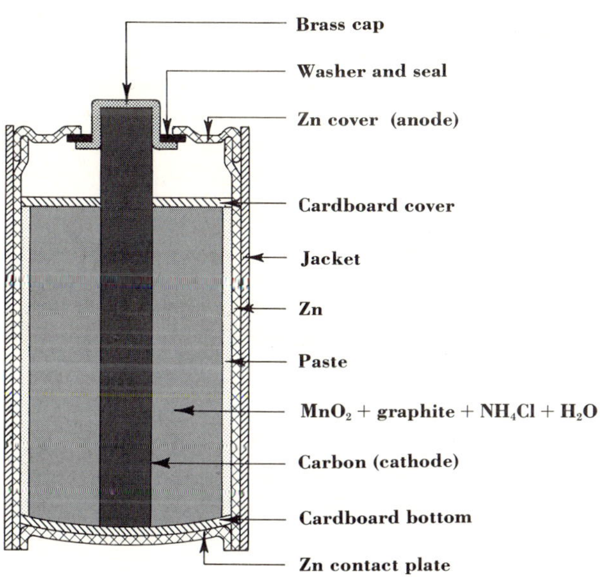

Fig. 16.5 A dry cell as used in flashlights.

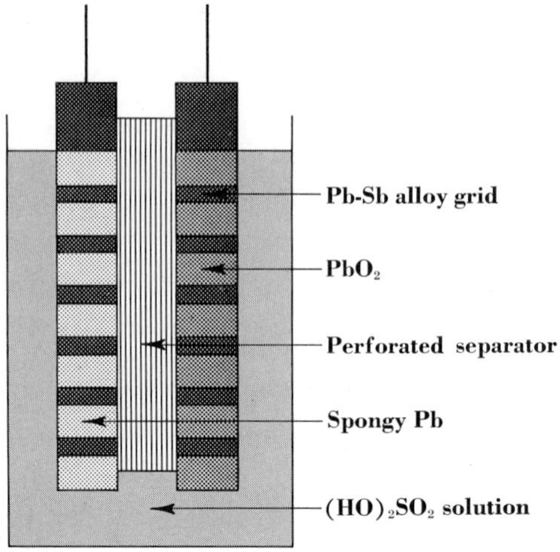

Fig. 16.6. A lead-acid storage cell.

divided solid PbO_2, each material supported by a grid of Pb-Sb alloy (stronger and more resistant to corrosion than pure Pb). The electrolyte is an aqueous solution of sulfuric acid (Fig. 16.6). The reactions, on discharging, are

$$\begin{array}{rcl} Pb(c) + HOSO_3^- + H_2O & \longrightarrow & PbSO_4(c) + H_3O^+ + 2e^- \\ PbO_2(c) + 3H_3O^+ + HOSO_3^- + 2e^- & \longrightarrow & PbSO_4(c) + 5H_2O \\ \hline Pb(c) + PbO_2(c) + 2H_3O^+ + 2HOSO_3^- & \longrightarrow & 2PbSO_4(c) + 4H_2O \end{array}$$

The cell can be recharged by passing current in the reverse direction, with electrons entering at the Pb electrode and leaving at the PbO_2 electrode. On charging, both electrode reactions are simply reversed. Sulfuric acid is consumed in the discharging process, and regenerated on charging; therefore, the percentage of acid in the solution provides an indication of the degree to which the cell is charged. This percentage is most easily estimated by measuring the density of the solution, for the density increases as the acid content increases. The solution in a charged cell usually contains 30–40% $(HO)_2SO_2$, corresponding to a density of 1.2–1.3 g/ml. The emf of the cell is roughly 2 V.

A **battery** is a number of galvanic cells connected together, in either series or parallel.

(c) Fuel cells At present, the chemical energy of fuels such as coal, petroleum, and natural gas is converted into electrical energy by burning the fuel in air (combustion) and using the liberated heat to drive a gas turbine. The operating gas turbine is coupled to a generator which produces the electrical energy. On the average, this sequence is only about 30% efficient. The efficiency of converting chemical to electrical energy would be increased substantially (theoretically to 100%) if the reaction between fuel and oxygen were made the basis of an electrochemical cell. In recent years, such electrochemical cells, called **fuel cells,** have been designed. The fuels most commonly used, H_2, CO, or CH_4, undergo oxidation at the anode. The reduction of oxygen occurs at the cathode. The different electrolytes used are aqueous KOH, molten mixtures of carbonates of alkali metals (group IA), solid oxides, or special membranes saturated with water.

In a typical case using hydrogen as the fuel, the half-cell reactions are

Cathode*	$\frac{1}{2}O_2 + H_2O + 2e^- \longrightarrow 2OH^-$
Anode	$H_2 + 2OH^- \longrightarrow 2H_2O + 2e^-$
Overall reaction	$\frac{1}{2}O_2 + H_2 \longrightarrow H_2O$

The electrolyte in this cell is aqueous KOH. During its operation the reactants can be supplied continuously.

Fuel cells now find use for special purposes where cost is of secondary importance, such as supplying electrical energy for space capsules. The use of fuel cells for large-scale power generators is not yet practical, but this situation may change as the result of intensive research now underway.

16.14 ▫ CORROSION

We may define corrosion as any undesired process in which a metal is converted to one of its compounds. We shall consider primarily the corrosion of iron, the most widely used and one of the more easily corroded metals.

A familiar fact about the corrosion, or rusting, of iron is that both oxygen and water are necessary. Perfectly dry iron does not rust at ordinary temperatures, and iron immersed in water does not rust if the water is free of O_2. Furthermore, the presence of acids and salts† in the water greatly accelerates the process of corrosion. These observations suggest that electrochemical processes may be involved in corrosion. The generally accepted theory is that Fe is oxidized to Fe^{2+} by H_2O or H_3O^+,

$$Fe \longrightarrow Fe^{2+} + 2e^- \qquad (2)$$

$$2H_3O^+ + 2e^- \longrightarrow 2H + 2H_2O \qquad (3)$$

If the H_3O^+ concentration is large, so that reactions 2 and 3 are rapid,

* The actual cathode reaction takes place in steps, the hydroperoxide anion, HO_2^-, being an intermediate

$$O_2 + H_2O + 2e^- \longrightarrow HO_2^- + OH^-$$

$$HO_2^- \xrightarrow{\text{catalyst}} \tfrac{1}{2}O_2 + OH^-$$

Net cathode reaction $\quad \frac{1}{2}O_2 + H_2O + 2e^- \longrightarrow 2OH^-$

† Salts which decrease the concentration of H_3O^+ (for example, Na_2CO_3) are exceptional in that they inhibit corrosion.

the next step is $2H \rightarrow H_2(g)$, and bubbles of H_2 appear. In corrosion by natural waters, however, dissolved O_2 usually reacts with the H atoms before the atoms pair off to form H_2,

$$2H + \tfrac{1}{2}O_2(aq) \longrightarrow H_2O \qquad (4)$$

The oxygen and water also react with the Fe^{2+} ions liberated in reaction 1, producing hydrated ferric oxide (rust),

$$4Fe^{2+} + O_2 + (12 + 2x)H_2O \longrightarrow 2Fe_2O_3(H_2O)_x + 8H_3O^+ \qquad (5)$$

The H_3O^+ consumed in reaction 3 is regenerated in reaction 5, so that H_3O^+ is behaving as a catalyst: it accelerates the reaction without being consumed. CO_2 is nearly always present in natural waters, and increases the H_3O^+ concentration by the reaction $CO_2 + 2H_2O \leftrightarrows H_3O^+ + HCO_3^-$, thus contributing to corrosion.

It is not necessary that reactions 2 and 3 should occur at the same place on the iron. Electrons liberated by reaction 2 can flow through the iron to another point at which H atoms are formed. Such a process, indeed, is conducive to rapid corrosion, for Fe^{2+} ions can more easily diffuse away from a place where the surface is not coated with H atoms. The corrosion of an iron or steel pipe is greatly accelerated if it is attached to a copper pipe; reaction 2 occurs on the Fe, reaction 3 mostly on the Cu. Any metal less active (a poorer reducing agent) than Fe will behave in the same way as Cu. Even when only Fe is present, some points on the surface ("anodic sites") will be more chemically reactive than others ("cathodic sites"), because of mechanical strain or small variations in composition. Thus, reaction 2 takes place at the anodic sites, which will be corroded, and reaction 3 occurs at the cathodic sites, which will not be corroded. The circuit through the metal must be completed by ionic conduction through the solution, which explains why an electrolyte accelerates corrosion.

If another metal—Zn or Mg, for example—more active than Fe is in contact with the Fe, the more active metal will be oxidized in preference to the Fe, and the Fe in the vicinity will be protected from corrosion. The Fe is said to be made **cathodic;** the active metal is called a "sacrificial anode," for it is allowed to corrode away in order to protect the Fe. The same result can be achieved by imposing a negative electric potential on the iron and a positive potential on other electrodes placed in the water or the ground nearby. These methods are often used to protect water tanks, boilers, ship hulls, and submerged or buried pipes.

The most obvious way to prevent corrosion is to coat the metal with an impervious layer of paint, grease, or another metal. Even this procedure has its pitfalls. When only part of the surface of a piece of Fe is exposed to O_2, the unexposed part, surprisingly, corrodes faster than the exposed part, apparently because the coating of rust formed in the presence of O_2 provides some protection, whereas reaction 2 proceeds without hindrance on the area that is wet but protected from air. If a coating of paint has holes in it, and especially if it does not adhere firmly, corrosion may actually be worse at the half-protected places than it would be if no paint had been applied. Plating with another metal is most helpful when the surface metal is more active (more easily oxidized) than the underlying metal—such as Zn on Fe (**galvanizing**). Then, if there is a small hole in the coating, the Zn serves as a sacrificial anode and protects the Fe. On the other hand, if the Fe is plated with a metal less active than itself, such as Sn or Cu, a small hole will result in very rapid corrosion, for the Fe will behave as a sacrificial anode and will protect the plating—hardly a desirable result.

Many metals start to corrode by forming a thin, continuous, adherent layer of oxide, which prevents further corrosion. Examples are Al, Co, Cr, and Ni. Such metals are resistant to corrosion under ordinary conditions, even though Al and Cr, for example, are more active than Fe. An especially impervious coating can be produced on many metals, even on Fe, by treating the metal with a strong oxidizing agent, such as concentrated HNO_3 or Pb_3O_4 ("red lead"), or by using it as the anode in an electrolysis, with high current density. The metal is said to become **passive.** Unfortunately, the passivity of Fe is easily destroyed, merely by touching the passive metal with a piece of normal (active) Fe. The resistance of Pt and related elements to electrolytic oxidation is much greater than one would expect from their oxidation potentials; this is also a case of passivity. Passivity is not always desirable, especially not when an attempt is being made to oxidize the metal electrolytically to its ion.

PROBLEMS

1. Redox reactions Which of the following reactions can be utilized to supply an electric current? Write the anode and cathode reactions for those that can be.

(a) $2Fe^{3+} + H_2S + 2H_2O \longrightarrow 2Fe^{2+} + S + 2H_3O^+$

(b) $AgCl + 2NH_3 \longrightarrow Ag(NH_3)_2^+ + Cl^-$

(c) $2MnO_4^- + 5Cu + 16H_3O^+ \longrightarrow 5Cu^{2+} + 2Mn^{2+} + 24H_2O$

(d) $Pb(OH)_2 + 2H_3O^+ + SO_4^{2-} \longrightarrow PbSO_4 + 4H_2O$

(e) $SO_2 + 5H_2O + I_2 \longrightarrow 2I^- + HOSO_3^- + 3H_3O^+$

2. Cell construction For each of the following reactions, draw a diagram of a galvanic cell in which the reaction occurs. Show the composition of the anode, the cathode, and the solution or solutions. Indicate on your diagram the direction of migration of each kind of ion in the cell, and the direction of flow of electrons in the external circuit.

(a) $2Cr(c) + 3HgCl_2(aq) \longrightarrow 2Cr^{3+} + 6Cl^- + 3Hg(l)$

(b) $Ni^{2+} + H_2(g) + 2OH^- \longrightarrow Ni(c) + 2H_2O$

(c) $Fe(c) + Hg_2Cl_2(c) \longrightarrow Fe^{2+} + 2Hg(l) + 2Cl^-$

3. Calculation of emf For each of the following cells: (a) write the equation for the cell process; (b) find $\mathscr{E}°$ at 25°C.

(*i*) Mg | $MgSO_4(aq)$ ‖ $NaBr(aq)$ | $Br_2(l)$ | Pt
(*ii*) Fe | $FeSO_4(aq)$ ‖ $FeSO_4(aq)$, $Fe_2(SO_4)_3(aq)$ | Pt

4. Emf and equilibrium constant For each of the following reactions, (a) devise a galvanic cell in which the reaction occurs; (b) find $\mathscr{E}°$ for the cell at 25°C; (c) write the Nernst equation for the cell; (d) find the equilibrium constant for the given reaction.

(*i*) $Ni + Sn^{2+} \longrightarrow Ni^{2+} + Sn$
(*ii*) $5Hg(l) + 2MnO_4^- + 16H_3O^+ \longrightarrow 5Hg^{2+} + 2Mn^{2+} + 24H_2O$
(*iii*) $2H_2(g) + O_2(g) \longrightarrow 2H_2O(l)$

5. Direction of reaction Ascertain, with as little calculation as possible, whether each of the following reactions will go as written or in the reverse direction at 25°C, when all molar concentrations are equal to 1:

(a) $2Cr + 3Cu^{2+} \longrightarrow 2Cr^{3+} + 3Cu$
(b) $Zn + Mg^{2+} \longrightarrow Zn^{2+} + Mg$
(c) $Fe + Cl_2(g) \longrightarrow Fe^{2+} + 2Cl^-$
(d) $2Fe^{2+} + Cl_2(g) \longrightarrow 2Fe^{3+} + 2Cl^-$
(e) $4Ag + O_2(g) + 4H_3O^+ \longrightarrow 4Ag^+ + 6H_2O$
(f) $4Ag + O_2(g) + 4H_3O^+ + 4Cl^- \longrightarrow 4AgCl(c) + 6H_2O$

6. Emf and work The emf of a galvanic cell is 1.30 V. (a) How much work, in joules, is done by this cell when 10 C of charge passes from one electrode to the other? (b) When a current of 0.10 A flows, how much power (in watts) is the cell producing (1 W = 1 J/sec)? (c) How much work, in joules and in calories, does the cell do when one faraday passes through the circuit? Assume that the emf of the cell is unaffected by the rate at which current is drawn.

7. Reduction potentials Given Ba, K^+, Br^-, Hg, H_3O^+, and H_2O. Select from these six (a) the strongest oxidizing agent, (b) the strongest reducing agent, (c) the weakest oxidizing agent, and (d) the weakest reducing agent.

8. Corrosion Which of the following methods of storing used steel wool is best? Which is most conducive to rusting? Explain. (a) Keep it immersed in plain tap water. (b) Keep it immersed in soapy water (a basic solution). (c) Just leave it in the sink.

ADDITIONAL PROBLEMS

9. Entropy (a) From information on p. 289, calculate the heat absorbed, when the maximum work is done, for the reaction

$$Pb(c) + Hg_2Cl_2(c) \longrightarrow PbCl_2(c) + 2Hg(l)$$

occurring as written (1 g-atom Pb, etc.) in a galvanic cell. (b) Find the entropy change (in calories per degree) in this reaction at 25°C. (c) How much heat (in calories) is absorbed or liberated when the reaction occurs on simple mixing of the reactants?

10. Form of equation Compare (a) ΔG, (b) K, and (c) $\mathscr{E}$ for reaction (*i*) with the corresponding quantity for (*ii*) and for (*iii*):

$$(i)\quad Zn + 2Ag^+ \longrightarrow Zn^{2+} + 2Ag$$
$$(ii)\quad \tfrac{1}{2}Zn + Ag^+ \longrightarrow \tfrac{1}{2}Zn^{2+} + Ag$$
$$(iii)\quad Zn^{2+} + 2Ag \longrightarrow Zn + 2Ag^+$$

Explain how each of these three quantities is affected when the equation is reversed, halved, or doubled.

11. Cell reaction and standard emf For each of the following cells (*i–v*), (a) write the equations for the anode and cathode processes, assuming that the left electrode is the anode; (b) write the equation for the overall cell process; (c) find the standard emf $\mathscr{E}°$ at 25°C; (d) state whether the cell reaction will actually occur as written in (b) or in the reverse direction when all the concentrations are 1.00 *M*.

(*i*) $Pt \mid Br_2(l) \mid HBr(aq) \mid H_2(g) \mid Pt$
(*ii*) $Al \mid AlCl_3(aq) \parallel CdSO_4(aq) \mid Cd$
(*iii*) $Pt \mid Pb(NO_3)_2(aq) \parallel FeCl_2(aq), FeCl_3(aq) \mid Pt$
(*iv*) $Mg \mid MgCl_2(aq) \mid Hg_2Cl_2(c) \mid Hg(l)$
(*v*) $Pt \mid I_2(c) \mid KI(aq) \mid AgI(c) \mid Ag$

12. Equilibrium constant Find the equilibrium constant at 25°C for each of the following reactions:

(a) $2H_2O + 2Ag + 4NH_3(aq) \rightleftharpoons H_2(g) + 2OH^- + 2Ag(NH_3)_2^+$
(b) $2H_2O + 2Ag \rightleftharpoons H_2(g) + 2OH^- + 2Ag^+$
(c) $Ag(NH_3)_2^+ \rightleftharpoons Ag^+ + 2NH_3(aq)$

Use the results of (a) and (b) in answering (c). (Review combination of equilibria, p. 248.)

13. Equilibrium constant Find the equilibrium constant at 25°C of each of the reactions in Problem 5.

14. Equilibrium constant From the results of parts (e) and (f) of the preceding problem, find the equilibrium constant (see p. 248) for the reaction $AgCl(c) \rightleftharpoons Ag^+ + Cl^-$.

15. Reduction potentials What new substances will be formed and in what approximate amounts when the following are mixed in the presence of water?

(a) Ag^+ (1 mole), Cu (1 mole), and Zn (1 mole)
(b) Ag^+ (2 moles), Cu (1 mole), and Zn (1 mole)
(c) Ag^+ (2 moles), Cu (1 mole), and Zn (3 moles)
(d) Cl_2 (1 mole), Br_2 (1 mole), and I^- (1.5 mole)

16. Reduction potentials Explain the following facts in terms of reduction potentials.

(a) FeI_3 cannot be prepared.
(b) Chlorine obtained by the electrolysis of a solution of NaCl invariably contains some O_2.
(c) Fe is added to solutions of Fe^{2+} to prevent air oxidation of Fe^{2+} to Fe^{3+}.

(d) The Au^+ state is generally obtained by reducing Au^{3+}, but not by oxidizing Au^0.

17. Dry cell Why is the dry cell so designed that Zn and MnO_2 do not come into contact? What reaction might occur if they were in contact? How would this reaction affect the usefulness of the cell?

18. Fuel cells (a) Find the $\mathscr{E}°$ of the H_2-O_2 fuel cell (p. 303) at 25°C. (b) Assume that the available emf is 75% of the value found in (a). How many joules of work can be obtained when one mole of H_2 is consumed? (c) What volume of H_2 (at standard conditions) is needed to produce 1 kilowatt-hour? (1 kWh = 3.600×10^6 J.)

19. Storage cell A galvanic cell must be so constructed that not all the reactants can come into contact and react directly. (a) How is this rule violated in the design of the lead-acid storage cell? (b) What reaction, resulting from this violation, helps to account for the fact that a charged storage cell slowly discharges itself on standing idle?

20. Corrosion Would you recommend plating Fe, to protect it from corrosion, with any of the following metals: Au; Cr; Ba; Pb? Why or why not? (Ignore the cost of the metal in making your decision.)

21. Corrosion A magnesium bar weighing 20 lb is attached to a buried iron pipe by way of a copper wire, to protect the pipe from corrosion. An average current of 0.30 A flows between the bar and the pipe. (a) What reaction occurs at the surface of the bar? Of the pipe? In which direction do electrons flow through the copper wire? (b) How many years will it take for the Mg bar to be entirely consumed? (There are 3.15×10^7 seconds in a year.) (c) What reaction(s) will occur if the bar is not replaced after the time calculated in (b)?

22. Emf and concentration Given the cell $Zn \mid Zn^{2+}(aq) \parallel Cu^{2+}(aq) \mid Cu$. State, qualitatively, the change in the emf of the cell when (a) H_2S is added to the anode compartment, (b) H_2S is added to the cathode compartment. Note: ZnS and CuS are both insoluble.

ANSWERS

3. (*i*) (b). +3.440 V.

4. (*i*) (b). +0.09 V; (d) 1.1×10^3.

9. (a) 9.1×10^3 J; 2.17×10^3 cal; (b) 7.3 cal/deg; (c) 2.25×10^4 cal liberated.

11. (c) (*i*) −1.065 V; (*ii*) +1.27 V.

12. (a) 2.5×10^{-41}; (b) 9.5×10^{-56}; (c) 6.1×10^{-8}.

13. (a) 2×10^{110}; (b) 3.1×10^{-55}.

21. (b) 7.6 years.

17 ACIDS AND BASES

17.1 EARLY DEFINITIONS

Acidic and basic solutions were first defined in terms of several easily recognizable chemical and physical properties. Thus, solutions were called acidic if they had a sour taste, changed the color of indicators (turned blue litmus red), and reacted with active metals to release hydrogen. Solutions were called basic if they felt slippery, tasted bitter, and changed red litmus to blue. Acidic and basic solutions were said to neutralize each other to produce water and a salt. With time, structural features that might account for the observed behavior received attention. Svante Arrhenius, in 1887, made the first attempt to give a structural definition of acids and bases. It was based on the knowledge that (1) pure molecular acids, such as anhydrous sulfuric acid and 100% (glacial) acetic acid, are nonelectrolytes, yet their aqueous solutions conduct a current, and (2) substances recognized as acids possess at least one hydrogen atom. Arrhenius reasoned that a pure acid, such as HCl, is a neutral molecule that dissolves in water to give H^+ and Cl^-; the acidic solution is known as hydrochloric acid. The abundance of the H^+ ion was thought to impart the familiar acidic properties to the solution.

Since all bases at the time of Arrhenius were known to possess the hydroxyl group, OH, and to conduct a current when dissolved in water, he assumed that a base is a substance that dissociates in water to give an OH^- ion and a cation: $NaOH \longrightarrow Na^+ + OH^-$. The characteristic behavior of a base was attributed to the abundance of the OH^- ion.

Equilibria were thought to exist between the molecules and the ions in solution. For a strongly electrolytic acid such as hydrochloric acid, and a strongly electrolytic base such as sodium hydroxide, the values of the constants for the equilibria are large numbers. Consequently, for dilute solutions, ionization is practically complete, as indicated by the relative lengths of the arrows in the equations,

$$HCl \rightleftharpoons H^+ + Cl^- \qquad K > 10^2$$
$$NaOH \rightleftharpoons Na^+ + OH^- \qquad K > 10^2$$

For a weakly electrolytic acid such as acetic acid, CH_3COOH,* and a weakly electrolytic base such as ammonia, the equilibrium constants are small and ionization is very incomplete,

$$CH_3COOH \rightleftharpoons H^+ + CH_3COO^- \qquad K \approx 10^{-5}$$
$$NH_3{\cdot}H_2O \rightleftharpoons NH_4^+ + OH^- \qquad K \approx 10^{-5}$$

In terms of the Arrhenius theory, we would define an acid-base reaction as one occurring specifically between H^+ and OH^- to form water, coincident with the formation of a salt,

$$\textit{in water} \qquad \underbrace{H^+ + Cl^-}_{acid} + \underbrace{Na^+ + OH^-}_{base} \longrightarrow H_2O + \underbrace{Na^+ + Cl^-}_{salt}$$

17.2 □ BRÖNSTED-LOWRY CONCEPT OF ACID-BASE REACTIONS

One limitation of Arrhenius's definition is its failure to deal with acid-base reactions that occur without the intervention of water. Hydrogen chloride and ammonia do not ionize in benzene, yet react to form ammonium chloride, the same salt obtained from an aqueous solution. Also, one would not expect an H^+ ion to maintain its identity in water because of its exceedingly large cationic charge density (5.4×10^{15}), as indicated in Table 7.2 (p. 129). It would be strongly attracted to any molecule or ion having a free pair of electrons. It could exist in an isolated state, as in an electric discharge tube, but certainly not in the regions of high electron density surrounding the oxygen atoms of water molecules.

A major improvement on Arrhenius' definition of acids and bases came in 1923 when J. N. Brönsted and M. Lowry independently proposed a more general theory. Their theory states that *an acid-base reaction involves a proton transfer; the acid is the proton donor and the base is the proton acceptor.* The loss of a proton is often called **protolysis** and the transfer is called a **protolytic reaction.** This theory emphasizes the interdependence of an acid and a base; one cannot react without the other. The stronger the acid, the more readily it donates a proton; the stronger the base, the more readily it accepts a proton.

(a) Conjugate acid-base pairs In aqueous solution an acid, HA, ionizes

* The CH_3CO group is abbreviated Ac, standing for *acetyl*. Therefore acetic acid is often written AcOH or HOAc.

by transferring its proton to water, the base,

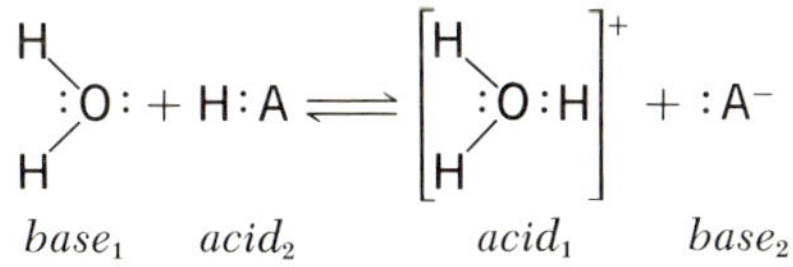
(1)

The hydronium ion is further hydrated, but for simplicity it will be written as H_3O^+. In some equations it will be further simplified by writing H^+. In the reverse reaction, H_3O^+ is the acid that donates a proton to A^-, the base. The reaction of an acid and a base *always* leads to the formation of a new acid and a new base. The product acid, H_3O^+, arises when the reactant base, H_2O, accepts a proton; H_3O^+ and H_2O are said to comprise a **conjugate acid-base pair.** Similarly, the product base, A^-, results when the reactant acid, HA, loses a proton; HA and A^- also comprise a conjugate acid-base pair. One conjugate acid-base pair in the reaction is designated by the subscript 1 and the other pair by the subscript 2, as demonstrated in Equations 1 and 2,

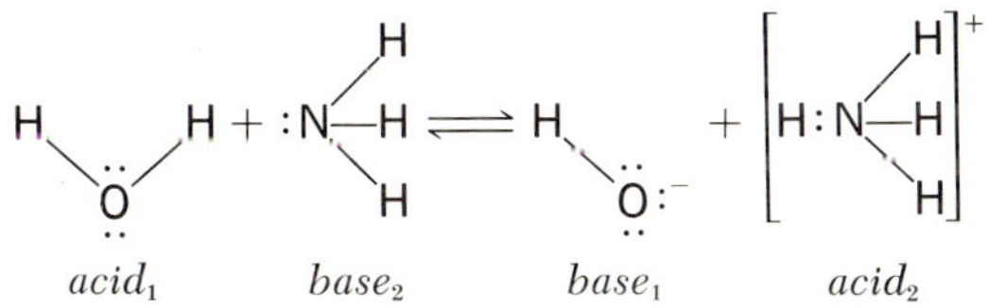
(2)

In the reaction of ammonia, $:NH_3$, with water, Equation 2, NH_3 is the base and water is the acid. The ammonium ion is the conjugate acid of NH_3, and hydroxide ion, OH^-, is the conjugate base of H_2O. In the reaction between NH_3 and HCl in benzene, hydrogen chloride, the acid, donates a proton to ammonia, the base. This produces chloride ion, the conjugate base of HCl, and ammonium ion, the conjugate acid of NH_3,

$$\underset{base_1}{H_3N:} + \underset{acid_2}{HCl} \xrightleftharpoons{\text{in benzene}} \underset{acid_1}{[H_3N{:}H]^+}\underset{base_2}{Cl^-}(c)$$

Electronic structures have been used in the preceding equations to reveal an important structural feature: *the base in each case has an unshared pair of electrons that can form a coordinate covalent bond with the transferred proton.* For this reason, the base gains in positive charge when it is converted to its conjugate acid, for example,

$$NH_3 \xrightarrow{H^+} NH_4^+$$

$$OH^- \xrightarrow{H^+} H_2O$$

$$O^{-2} \xrightarrow{H^+} OH^-$$

The reverse is true of the acid.

There is an interdependent relationship between the strengths of a given conjugate pair. In general, *the stronger an acid, the weaker its conjugate base; the stronger a base, the weaker its conjugate acid.* Thus, from observing the relative positions of the equilibria of Equations 3 and 4,

$$\underset{\substack{acid_1 \\ \left(\substack{stronger \\ than\ H_3O^+}\right)}}{HCl} + \underset{\substack{base_2 \\ \left(\substack{stronger \\ than\ Cl^-}\right)}}{H_2O} \rightleftharpoons \underset{\substack{acid_2 \\ \left(\substack{weaker \\ than\ HCl}\right)}}{H_3O^+} + \underset{\substack{base_1 \\ \left(\substack{weaker \\ than\ H_2O}\right)}}{Cl^-} \qquad K \approx 10^2 \quad (3)$$

$$\underset{\substack{acid_1 \\ \left(\substack{weaker\ than \\ H_3O^+}\right)}}{CH_3COOH} + \underset{\substack{base_2 \\ \left(\substack{weaker\ than \\ CH_3COO^-}\right)}}{H_2O} \rightleftharpoons \underset{\substack{acid_2 \\ \left(\substack{stronger \\ than\ CH_3COOH}\right)}}{H_3O^+} + \underset{\substack{base_1 \\ \left(\substack{stronger \\ than\ H_2O}\right)}}{CH_3COO^-} \qquad K \approx 10^{-5} \quad (4)$$

it is evident that hydrochloric acid is a stronger acid than acetic acid. From the relative extent of the reverse reactions, it is evident that acetate ion, CH_3COO^-, is a stronger base than Cl^-, because of its greater tendency to accept a proton from H_3O^+.

In every acid-base reaction, the two existing bases—as H_2O and Cl^- in Equation 3—are always competing for the proton. The stronger base prevails and is thereby converted to its conjugate acid. The weaker base fails to acquire the proton and thus remains unchanged. Consequently, *the equilibrium favors the formation of the weaker base.* Thus in Equation 3, Cl^- is a weaker base than H_2O, which must also mean that HCl is a stronger acid than H_3O^+. Also, in Equation 4, CH_3COO^- is a stronger base than water, which means that H_3O^+ is a stronger acid than CH_3COOH. We see that acid-base reactions "run downhill" *to favor the formation of the weaker base and weaker acid.* (It is noteworthy that a similar relationship prevails in electron transfer reactions of oxidant-reductant pairs. Thus, fluorine, the strongest oxidant, has the strongest tendency to accept electrons, while F^- ion has the least tendency to lose electrons, and is the weakest reductant known. The equilibrium of a redox reaction also tends to favor the weaker oxidant and reductant.)

(b) Relative acidity and basicity *Strengths of acids must be compared toward the same base;* the base of choice is water. Dilute aqueous solutions of strong acids, those with $K > 2$, ionize practically completely so that the only acid present is H_3O^+:

$$\underset{\text{hydrochloric acid}}{HCl} + H_2O \longrightarrow H_3O^+ + Cl^-$$

$$\underset{\text{nitric acid}}{HONO_2} + H_2O \longrightarrow H_3O^+ + NO_3^-$$

$$\underset{\text{perchloric acid}}{HOClO_3} + H_2O \longrightarrow H_3O^+ + ClO_4^-$$

strong acids of different strength ($K > 2$)*; all three stronger than* H_3O^+ — *only acid present in all three cases*

Consequently, under these conditions all such acids are equally strong. With strong acids water is a strong enough base to "level out" or obscure differences in acid strength. This behavior is called the **leveling effect.**

Since weak acids ionize incompletely, equilibrium is established and an equilibrium constant, K_a, can be calculated.

$$HA + H_2O \rightleftharpoons H_3O^+ + A^-$$

$$K_a = \frac{[H_3O^+][A^-]}{[HA]}$$

The magnitude of K_a, the ionization constant, is a measure of acidity—the larger the value for K_a, the stronger is the acid. Thus, acetic acid, $K_a = 1.8 \times 10^{-5}$, is more acidic than hydrocyanic acid, $K_a = 7.2 \times 10^{-10}$.

To avoid the use of exponential expressions, a new term, pK_a, is defined, such that $pK_a = -\log K_a$.

EXAMPLE 1 Find the pK_a for (a) acetic acid, and (b) hydrocyanic acid. (See App. 2 for use of logarithms.)

ANSWER

(a) For acetic acid:

$$\begin{aligned} pK_a &= -\log(1.8 \times 10^{-5}) \\ &= -(\log 1.8 + \log 10^{-5}) \\ &= -(0.25 - 5) \\ &= 4.8 \end{aligned}$$

(b) For hydrocyanic acid:

$$\begin{aligned} pK_a &= -\log(7.2 \times 10^{-10}) \\ &= -(\log 7.2 + \log 10^{-10}) \\ &= -(0.86 - 10.00) \\ &= 9.1 \end{aligned}$$

The weaker acid thus has the larger pK_a value.

The same principles apply for comparing basicities. The bases B: and OH^- compete for the proton,

$$B: + H\!-\!\ddot{\underset{..}{O}}\!-\!H \rightleftharpoons [B:H]^+ + [:\ddot{\underset{..}{O}}H]^-$$

When the base, B:, is a weaker base than OH^-, the equilibrium does *not* favor the formation of OH^- ions. The weak ionization of ammonia illustrates this point,

$$\underset{\text{base}_1}{:NH_3} + \underset{\text{acid}_2}{H\!-\!\ddot{\underset{..}{O}}\!-\!H} \rightleftharpoons \underset{\text{acid}_1}{[H:NH_3]^+} + \underset{\text{base}_2}{[:\ddot{\underset{..}{O}}H]^-}$$

base$_1$ (*weaker than* OH^-) — *acid*$_2$ (*weaker than* NH_4^+) — *acid*$_1$ (*stronger than* H_2O) — *base*$_2$ (*stronger than* NH_3)

The expression for the equilibrium constant is

$$K_b = \frac{[NH_4^+][OH^-]}{[NH_3]} = 1.8 \times 10^{-5}$$

When the base, B:, is a molecule (has no charge) it is invariably a weaker base than OH^-.

17.3 □ AMPHOTERISM; AUTOPROTOLYSIS (SELF-IONIZATION)

Water behaves either as an acid or as a base:

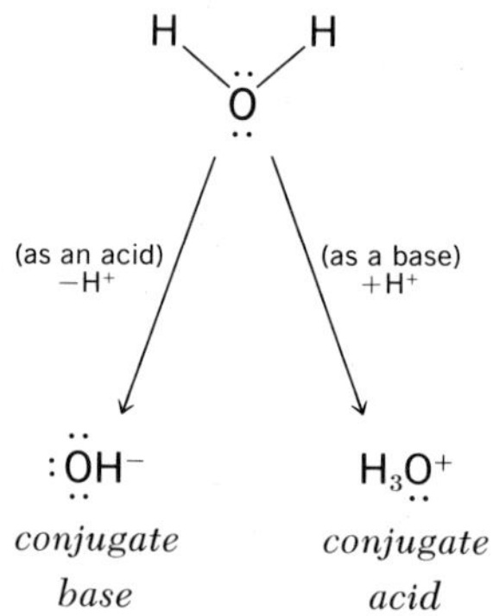

It is an example of a group of substances that are said to be **amphiprotic** or **amphoteric.** A substance with an H atom attached to a fairly strongly electronegative atom with an unshared pair of electrons may be amphiprotic; examples are $:NH_3$ and $C_2H_5\ddot{O}H$. Since water possesses dual acid-base behavior, it undergoes **self-ionization (autoprotolysis):**

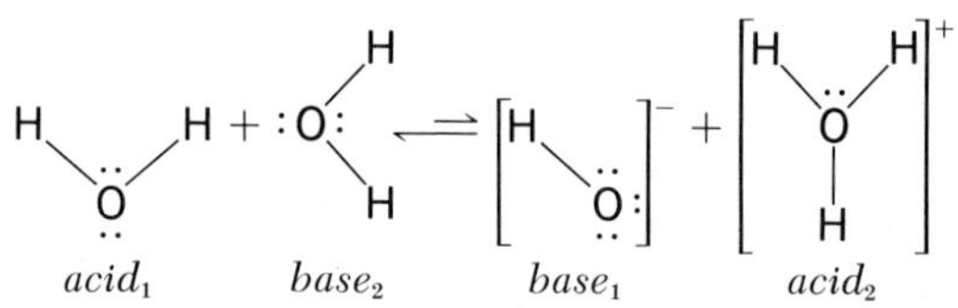

Any water solution with an equal concentration of OH^- and H_3O^+ ions is said to be **neutral.** If an excess of OH^- is present, as in a solution of sodium hydroxide, the solution is **basic** (alkaline); if an excess of H_3O^+ is present, as in a solution of hydrochloric acid, the solution is **acidic.** A substance that engenders an excess of OH^- is a *base.* A substance that produces an excess of H_3O^+ is an *acid.* A substance that has no effect on either concentration is *neutral.*

Self-ionization is exhibited by many pure liquids; for example, in pure liquid ammonia,

$$\underset{acid_1}{:NH_3} + \underset{base_2}{:NH_3} \rightleftharpoons \underset{\substack{base_1 \\ \text{(amide ion)}}}{:\ddot{N}H_2^-} + \underset{\substack{acid_2 \\ \text{(ammonium ion)}}}{[HNH_3]^+}$$

We can generalize by saying that when a liquid undergoes self-ionization, its conjugate acid and its conjugate base are formed (Equation 5, p. 316).

The equilibrium constant for the autoprotolysis of water is

$$\begin{aligned} K_w &= [H_3O^+][OH^-] \\ &= 1.0 \times 10^{-14} \text{ (at 25°C)} \\ &= 9.6 \times 10^{-14} \text{ (at 60°C)} \\ pK_w &= 14 \text{ (at 25°C)} \end{aligned}$$

17.4 ▫ TYPES OF PROTOLYTIC REACTION

Before the introduction of the Brönsted-Lowry theory, many reactions actually involving transfer of a proton were thought to be unique, and were given specific names. Several typical protolytic reactions are listed in Table 17.1 to emphasize their similarity.

(a) Neutralization The reaction between an acid and a base was called neutralization (p. 309) even before the days of Arrhenius. In dilute aqueous solutions of strong acids, the only acid present to react with the OH^- ion is H_3O^+. Hence, in terms of the Brönsted-Lowry theory, neutralization is the reaction between H_3O^+ and OH^-, as shown in reaction 4 in Table 17.1. The term neutralization has been extended to include all protolytic reactions between the conjugate acid and conjugate base of the same substance. Hence reaction 4′, Table 17.1, represents a neutralization reaction between NH_4^+, the conjugate acid of NH_3, and amide ion, NH_2^-, the conjugate base of the same molecule.

TABLE 17.1 ▪ TYPES OF ACID-BASE (PROTOLYTIC) REACTION

Type	$Acid_1$ + $Base_2$ ⟶ $Acid_2$ + $Base_1$
(1) Ionization of neutral acid	$HCN + H_2O \longrightarrow H_3O^+ + CN^-$
(2) Ionization of neutral base	$H_2O + NH_3 \longrightarrow NH_4^+ + OH^-$
(3) Autoprotolysis (self-ionization) of water	$H_2O + H_2O \longrightarrow H_3O^+ + OH^-$
(4) Neutralization in water	$H_3O^+ + OH^- \longrightarrow H_2O + H_2O$
(4′) Neutralization in liquid ammonia	$NH_4^+ + NH_2^- \longrightarrow NH_3 + NH_3$
(5) Ionization of conjugate base of a diprotic acid	$HS^- + H_2O \longrightarrow H_3O^+ + S^{2-}$
(6) Ionization of conjugate acid of a diprotic base	$H_2O + H_2NNH_3^+ \longrightarrow H_3NNH_3^{2+} + OH^-$
(7) Hydrolysis of an anion	$H_2O + CN^- \longrightarrow HCN + OH^-$
(8) Hydrolysis of a cation	$NH_4^+ + H_2O \longrightarrow H_3O^+ + NH_3$
(9) Hydrolysis of hydrated cation	$Fe(H_2O)_6^{3+} + H_2O \longrightarrow H_3O^+ + Fe(OH)(H_2O)_5^{2+}$

In general, *neutralization is the reverse of an autoprotolysis:*

$$\xleftarrow{}\ \textbf{autoprotolysis}$$

$$\underset{\textit{conjugate acid of HX}}{H_2X^+} + \underset{\textit{conjugate base of HX}}{X^-} \rightleftharpoons 2HX \qquad (5)$$

neutralization $\longrightarrow$

(b) Polyprotic acids and polyprotic bases Some acids, such as $(HO)_2SO_2$, H_2S, and $(HO)_3PO$, have more than one proton that can be donated to a base. When such **polyprotic acids,** as exemplified by the **diprotic** acid H_2S, dissolve in water, they release their protons in sequence:

$$H_2S + H_2O \rightleftharpoons H_3O^+ + HS^- \qquad K_{a_1} = 6.3 \times 10^{-8}$$
$$HS^- + H_2O \rightleftharpoons H_3O^+ + S^{2-} \qquad K_{a_2} = 10^{-14}$$

The values for the ionization constants reveal that HS^- is a much weaker acid than H_2S. This is understandable since the hydrosulfide ion, HS^-, has a negative charge and so is more likely to resist losing another proton to form the sulfide ion, S^{2-}. Conversely, the sulfide ion has a double negative charge and therefore is more likely to attract a proton than is the singly charged base, HS^-; the sulfide ion is a much stronger base than HS^-.

In oxyacids, only protons bonded to oxygen are donated to water. Thus, hypophosphorous acid is monoprotic, and phosphorous acid is diprotic:

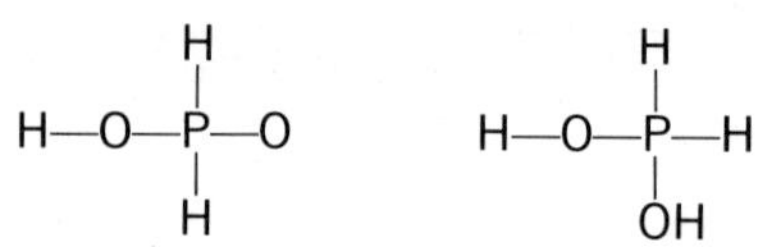

hypophosphorous acid *phosphorous acid*

Clearly, the H atoms bonded to phosphorus do not dissociate.

There are also basic compounds that can accept more than one proton in a stepwise process. For example, each nitrogen atom in hydrazine, $H_2\ddot{N}\ddot{N}H_2$, has an unshared pair of electrons available for bonding with a proton. As is typical of all such **polyprotic bases,** the ionization constant for the second step is much smaller than that for the first step:

$$H_2\ddot{N}\text{—}\ddot{N}H_2 + H_2O \rightleftharpoons \left[H_2\overset{H}{\underset{}{N}}\text{—}\ddot{N}H_2\right]^+ + OH^- \qquad pK_{b_1} = 6.04$$

$$\left[H_2\overset{H}{N}\text{—}\ddot{N}H_2\right]^+ + H_2O \rightleftharpoons \left[H_2\overset{H}{N}\text{—}\overset{H}{N}H_2\right]^{++} + OH^- \qquad pK_{b_2} = 15$$

This is to be expected, because the $H_2N\text{—}NH_3^+$ already has a positive charge and resists accepting a second proton.

(c) Reactions of anions with water (hydrolysis) An aqueous solution containing the conjugate base of a weak acid is basic, as exemplified by a solution of sodium cyanide ($Na^+ + CN^-$). Cyanide ion is sufficiently basic to remove a proton from water,

$$\underset{base_1}{CN^-} + \underset{acid_2}{H_2O} \rightleftharpoons \underset{acid_1}{HCN} + \underset{base_2}{OH^-}$$

The equilibrium constant* for this reaction with water (hydrolysis) is

$$K_b = \frac{[HCN][OH^-]}{[CN^-]} \tag{6}$$

The strength of the basic anion as measured by the magnitude of K_b is related to the strength of its conjugate acid and therefore to K_a. We illustrate the general relationship by considering HCN and CN^-. The equation for the ionization of HCN is

$$HCN + H_2O \rightleftharpoons H_3O^+ + CN^-$$

and the equilibrium constant, K_a, for this reaction is

$$K_a = \frac{[H_3O^+][CN^-]}{[HCN]} \tag{7}$$

If we multiply Equations 6 and 7, we get

$$K_a \cdot K_b = \frac{[H_3O^+][\cancel{CN^-}]}{[\cancel{HCN}]} \cdot \frac{[\cancel{HCN}][OH^-]}{[\cancel{CN^-}]} = [H_3O^+][OH^-]$$

Since $[H_3O^+][OH^-] = K_w$ (p. 327),

$$K_a \cdot K_b = K_w$$

At 25°C, $K_w = 1.0 \times 10^{-14}$; therefore, at this temperature,

$$K_a \cdot K_b = 1.0 \times 10^{-14}$$

and, in p units,

$$pK_a + pK_b = pK_w = 14$$

Although the equation $K_a \cdot K_b = K_w$ permits quantitative predictions to be made concerning the extent of reaction of anions with water, we will concern ourselves only with the following qualitative conclusions:

(1) The conjugate base of an acid, such as perchloric acid, which is much stronger than water, hydrolyzes very slightly, if at all. For example, ClO_4^-, Cl^-, SO_4^{2-}, and NO_3^- do not hydrolyze measurably. The presence of these anions leaves water neutral.

(2) The conjugate base of an acid, such as nitrous acid, $pK_a = 3.29$, which is moderately stronger than water ($pK_w = 14$), undergoes a degree of hydrolysis until a state of equilibrium is reached. Examples of

* These constants are also referred to as **"hydrolysis" constants,** K_b.

such ions are NO_2^- and $HOPO_3^{2-}$ (hydrogen phosphate). The presence of these anions leaves water slightly basic.

(3) The conjugate base of an acid, such as $HOPO_3^{2-}$, $pK_a = 12$, which is approximately equal in strength to water, hydrolyzes almost completely. Examples of such ions are PO_4^{3-} and S^{2-}. The presence of these anions makes water distinctly basic.

(4) The conjugate base of an acid much weaker than water hydrolyzes completely. Another way of stating this fourth conclusion is that any base stronger than OH^- reacts with water so that OH^- remains as the only base. This situation is another manifestation of the "leveling effect." Since ammonia, hydrogen, and methane all have pK_a values greater than 30, their conjugate bases are all converted by water to the OH^- ion:

$:\ddot{N}H_2^-$	$+ H_2O \longrightarrow$	NH_3 +	OH^-	(as from sodium amide, $NaNH_2$)
$:H^-$	$+ H_2O \longrightarrow$	H_2 +	OH^-	(as from calcium hydride, CaH_2)
$H_3C:^-$	$+ H_2O \longrightarrow$	CH_4 +	OH^-	(as from sodium methide, $NaCH_3$)
base$_1$ (*strong bases of different strength; all stronger than* OH^-)	*acid*$_2$	*acid*$_1$	*base*$_2$ (*only base present*)	

The conjugate base of a weak diprotic acid is capable of behaving either as an acid or as a base, as exemplified by HS^-,

$$\underset{base}{HS^-} + H_2O \rightleftharpoons H_2S + OH^-$$

$$\underset{acid}{HS^-} + H_2O \rightleftharpoons S^{2-} + H_3O^+$$

The HS^- ion is amphoteric.

(d) Reactions of cations with water Conjugate acids of weak bases, such as ammonium ion, NH_4^+, are strong enough to donate a proton to water,

$$NH_4^+ + H_2O \rightleftharpoons NH_3 + H_3O^+$$

and

$$K_a = \frac{[H_3O^+][NH_3]}{[NH_4^+]}$$

Since

$$K_a = \frac{K_w}{K_b} \quad \text{and} \quad K_b = 1.8 \times 10^{-5}$$

then

$$K_a = \frac{10^{-14}}{1.8 \times 10^{-5}} = 5.6 \times 10^{-10}$$

Salts of trivalent cations, such as aluminum sulfate, $Al_2(SO_4)_3$, and chromic nitrate, $Cr(NO_3)_3$, give acidic water solutions. Because of their high cationic charge density (p. 128), Al^{3+} and Cr^{3+} form covalent bonds with six water molecules,

$M^{3+} + 6H_2O \longrightarrow$ 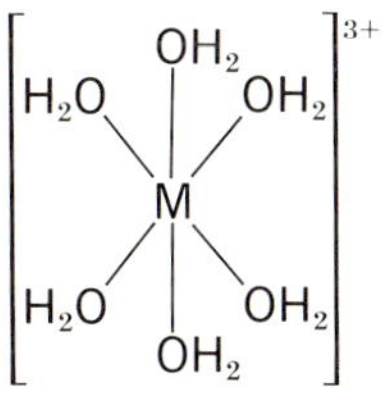 (8)

(where M = Al, Cr). These hydrated ions act as acids. $Al(H_2O)_6^{3+}$ is a triprotic acid; the end product is hydrated aluminum hydroxide, which forms as a gelatinous precipitate when an aqueous solution of aluminum sulfate is allowed to stand,

$$Al(H_2O)_6^{3+} + H_2O \rightleftharpoons H_3O^+ + Al(OH)(H_2O)_5^{2+}$$
$$Al(OH)(H_2O)_5^{2+} + H_2O \rightleftharpoons H_3O^+ + Al(OH)_2(H_2O)_4^{+}$$
$$Al(OH)_2(H_2O)_4^{+} + H_2O \rightleftharpoons H_3O^+ + Al(OH)_3(H_2O)_3$$

(hydrated aluminum hydroxide)

acid$_1$ *base*$_2$ *acid*$_2$ *base*$_1$

This precipitate has good adsorptive properties, and therefore Al(III) salts are used for water purification in the coagulation process.

Some divalent cations may also form acidic hydrated cations. For example, an aqueous solution of $Zn(NO_3)_2$ is acidic because of the presence of $Zn(H_2O)_4^{2+}$. On the other hand, an aqueous solution of $Ca(NO_3)_2$ is neutral. In general those divalent cations possessing a noble gas configuration and all univalent cations, such as Na^+ and Ag^+ are not acidic in water solution.

When a strong base such as OH^- is added to hydrated aluminum hydroxide, the precipitate dissolves. Since each of the three water molecules is capable of donating a proton, the addition of hydroxide ion, OH^- (as NaOH), converts the hydroxide to the aluminate ion, $Al(OH)_4(H_2O)_2^-$,

$$Al(OH)_3(H_2O)_3 + OH^- \longrightarrow Al(OH)_4(H_2O)_2^- + H_2O$$

Since $Al(OH)_3(H_2O)_3$ also dissolves in acid,

$$Al(OH)_3(H_2O)_3 + 3H_3O^+ \longrightarrow Al(H_2O)_6^{3+} + 3H_2O$$

it is an example of an amphiprotic or amphoteric substance. It is noteworthy that $Fe(H_2O)_6^{3+}$ is a triprotic acid; on standing in water it forms ferric hydroxide, $Fe(OH)_3(H_2O)_3$. Yet hydrated ferric hydroxide does not dissolve in alkali hydroxides. Therefore, it does not follow that all acidic hydrated cations must form amphoteric hydroxides.

17.5 □ LEWIS ACID-BASE CONCEPT

Every base has a lone pair of electrons to share with a proton. The transferred proton is electron-deficient during the process of bonding to the base. A transferred proton is not the only species to form bonds with bases; other electron-deficient substances do likewise, as exemplified by the reaction of boron trifluoride, BF_3, with NH_3.

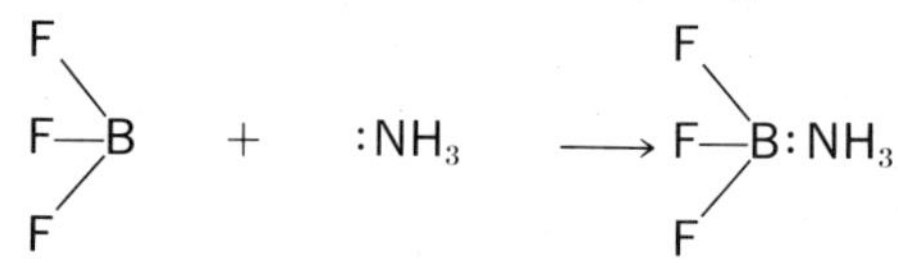

acid (*electrophile*) *base* (*nucleophile*)

Gilbert N. Lewis recognized the similarity in behavior of boron trifluoride and a transferred proton toward a base, and in 1923 enunciated a definition of acid-base reaction in terms of sharing of an electron pair—*a base donates an electron pair* in covalent bonding *and an acid accepts the pair.* The acid is called an **electrophile,** and the base is called a **nucleophile.** In the base, the atom with the lone pair of electrons is an electron-rich site, and, in the acid, the atom that accepts the pair of electrons to form a covalent bond is an electron-deficient site. The Lewis theory focuses attention on the electron pair rather than on the proton, and in so doing broadens the concept of acidity and abandons the "cult of the proton." The transferred proton of a so-called Brönsted acid is a special case of a Lewis acid.

Lewis bases are necessarily substances with lone pairs of electrons. They can be neutral molecules, such as $:NH_3$, or anions, such as cyanide ion, $:CN^-$; they are no different from the Brönsted bases we have been discussing. Lewis acids appear in many guises.

17.6 □ SOME EXAMPLES OF LEWIS ACIDS

(a) Molecules possessing an atom with fewer than an octet of electrons BF_3 falls into this category. A Lewis acid frequently used in organic chemistry is aluminum chloride, $AlCl_3$. The Al atom is electron-deficient, and can form a covalent bond with the electron-rich atom of an ether,

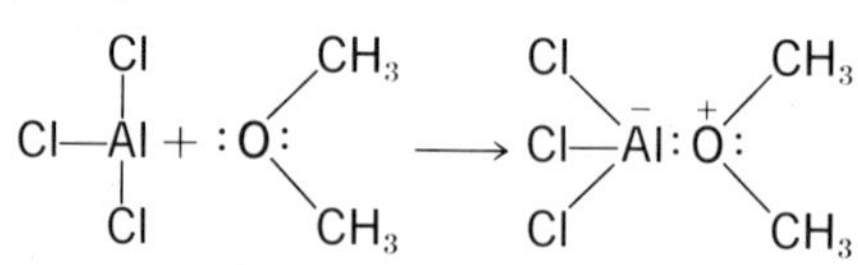

dimethyl ether

The product is shown with formal charges.

(b) Cations Theoretically, every cation is a potential Lewis acid. However, this property is negligible for the alkali metal cations such as Na^+ and K^+, and is weak for the alkaline earth cations such as Ca^{2+}, Sr^{2+}, and Ba^{2+}. Of the alkaline earth group, only Mg^{2+} and Be^{2+} show an appreciable tendency to behave as Lewis acids. Many cations form covalent bonds with Lewis bases. The reaction of the M^{3+} type of ion with water is an example of a Lewis acid-base interaction (Equation 8, p. 319). The reaction of Ag^+ and $:NH_3$ is also typical,

$$\underset{\text{Lewis acid}}{Ag^+} + 2\ \underset{\text{Lewis base}}{:NH_3} \longrightarrow [H_3N:Ag:NH_3]^+$$

17.7 ▫ INDUSTRIAL APPLICATIONS OF ACID-BASE REACTIONS

(a) Softening "temporary" hard water The hardness of water has been attributed to the presence of certain cations, notably Ca^{2+} and Mg^{2+}, that cause soap to precipitate (p. 225). Removal of the interfering cations, a process called **water-softening,** can be effected by acid-base reactions.

In certain hard waters the major anion present is bicarbonate, $HOCO_2^-$. Heating a solution containing Ca^{2+} and $HOCO_2^-$ affords a precipitate of calcium carbonate as the result of a two step sequence:

Step (1)
$$\underset{acid_1}{HOCO_2^-} + \underset{base_2}{HOCO_2^-} \xrightarrow{\text{heat}} \underset{base_1}{CO_3^{2-}} + \underset{\substack{\text{or } (HO)_2CO \\ \textit{carbonic acid} \\ acid_2}}{H_2O + CO_2}$$

Step (2)
$$CO_3^{2-} + Ca^{2+} \longrightarrow CaCO_3(c)$$

The first step represents the autoprotolysis of the bicarbonate ion. The combination of CO_2 and H_2O can be considered to be the unstable carbonic acid. Heat displaces the equilibrium to the right by decreasing the solubility of CO_2. Water containing sufficient bicarbonate ion to cause precipitation of all interfering cations when heated is called "temporary" hard water.

The formation of calcium carbonate from "temporary" hard water is frequently objectionable, especially when it occurs in hot water boilers and pipes. The deposits of calcium carbonate and other insoluble calcium salts form an adherent scale that lowers the heat conductivity of the boiler tubes and blocks the water passage in the pipes. This may ultimately lead to bursting of the boiler tubes. It is therefore necessary to soften the water before it enters the boiler. The $HOCO_2^-$ ion can be converted to CO_3^{2-} by a base such as hydroxide ion,

$$\underset{acid_1}{HOCO_2^-} + \underset{base_2}{OH^-} \longrightarrow \underset{base_1}{CO_3^{2-}} + \underset{acid_2}{H_2O}$$

For large-scale water softening, calcium hydroxide (lime), $Ca(OH)_2$, is used as a cheap source of OH^-. The "temporary" hard water is analyzed to determine the concentration of $HOCO_2^-$. One mole of $Ca(OH)_2$ is added for every 2 moles of $HOCO_2^-$ present. Any Ca^{2+} ions remaining are removed by adding sodium carbonate (*washing soda*), Na_2CO_3. For softening water in the home, household cleansers are usually reinforced with the basic substances, ammonia and sodium phosphate, Na_3PO_4.

When $HOCO_2^-$ is not the major anion present, the water is said to have "permanent" hardness. In such cases, the Ca^{2+} and Mg^{2+} ions can be converted to insoluble carbonates by adding washing soda, Na_2CO_3, or removed by ion-exchange methods (p. 323).

Chemists have developed synthetic detergents, "syndets," with the distinct advantage over soap of not forming precipitates with calcium ion (p. 224).

(b) Preparation of sodium bicarbonate; the Solvay process Another industrial application of acid-base reactions is the very efficient Solvay process for the manufacture of sodium bicarbonate, also called **baking soda,** $NaHOCO_2$. The most abundant source of Na^+ is sodium chloride, and the only source of $HOCO_2^-$ is carbon dioxide. Unfortunately, since CO_2 is a weak acid, $K_1 = 4.47 \times 10^{-7}$, it reacts with water to give a very low concentration of $HOCO_2^-$,

$$CO_2 + H_2O \rightleftharpoons H_3O^+ + HOCO_2^-$$

However, the equilibrium can be shifted to the right by reducing the concentration of H_3O^+. This displacement can be achieved by adding a base, but it must be a weak enough base so that the proton is not removed from $HOCO_2^-$. The desired results are best attained by mixing ammonia and carbon dioxide in an aqueous solution,

$$CO_2 + H_2O \rightleftharpoons HOCO_2^- + H_3O^+$$
$$NH_3 + H_3O^+ \rightleftharpoons NH_4^+ + H_2O$$

The net reaction is

$$CO_2 + H_2O + NH_3 \longrightarrow NH_4^+ + HOCO_2^- \qquad (9)$$

A mixture of NH_4^+, $HOCO_2^-$, Na^+, and Cl^- ions could result in the formation of four possible salts, NaCl, NH_4Cl, $Na(HOCO_2)$, and $NH_4(HOCO_2)$. These salts are all classified as soluble. However, under the conditions of the reaction, water at 0°C is saturated with NH_3, NaCl, and CO_2. Under these circumstances, $Na(HOCO_2)$ is the least soluble of the four salts and precipitates from the solution. Thus the equation for the overall reaction is

$$Na^+ + Cl^- + NH_3 + CO_2 + H_2O \longrightarrow NH_4^+ + Cl^- + NaHOCO_3(c) \qquad (10)$$

The CO_2 is supplied by roasting calcium carbonate: $CaCO_3 \longrightarrow CaO + CO_2(g)$. The CaO, thus formed, is used to regenerate NH_3 from the ammonium chloride, the by-product in Equation 10,

$$CaO + 2NH_4^+ + 2Cl^- \longrightarrow Ca^{2+} + 2Cl^- + 2NH_3(g) + H_2O \qquad (11)$$

To summarize, we see that the consumed raw materials are NaCl and $CaCO_3$, and the sole by-product is an aqueous solution of calcium chloride (Equation 11). The calcium chloride is used as a drying agent, and for melting ice on highways. Ammonia, the most expensive ingredient, is regenerated.

Solid sodium bicarbonate can be converted to solid sodium carbonate by heating,

$$2NaHOCO_2 \longrightarrow Na_2CO_3 + H_2O + CO_2$$

(c) Ion-exchange resins An **ion-exchange material** is a solid consisting of a molecular network structure containing electrically charged sites of one particular charge sign, and mobile ions of the other charge sign. In a **cation exchanger** the immobile molecular network is anionic; the mobile ions are cations. In an **anion exchanger** the charge signs are reversed. Figure 17.1 is a schematic representation of a cation exchanger. Ion exchange can be used, as the name implies, to substitute one ion for another in a solution. For example, if water containing Ca^{2+} ions flows through a column of cationic exchanger containing Na^+ ions, displacement occurs, and Na^+ ions appear in the eluent. This particular process thus converts **hard water** to **soft water** (see p. 225). The action is an equilibrium that may be represented as

$$[Na_x^+ \text{ resin}^{x-}] + \frac{x}{2} Ca^{2+} \rightleftharpoons [Ca_{x/2}^{2+} \text{ resin}^{x-}] + xNa^+$$

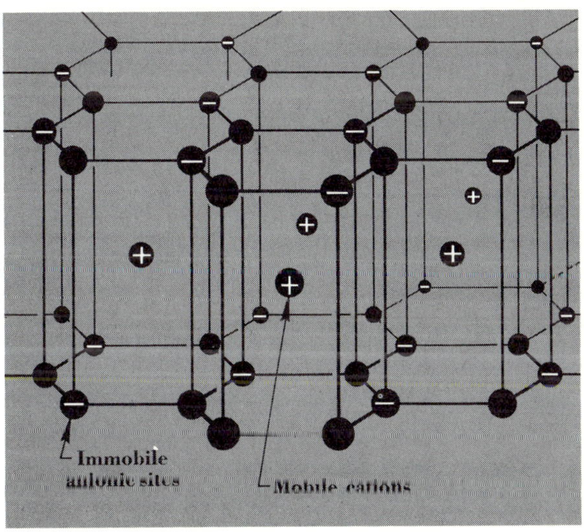

Fig. 17.1 Cationic exchanger (schematic). Some of the atoms are charged.

A cationic resin treated with acidified water becomes populated with $H^+(aq)$ ions; an anionic resin in basic solution takes up $OH^-(aq)$ ions. When water containing mineral impurities is treated with two such resins in series, the saline cations and anions are removed and are replaced by H^+ and OH^-, respectively. This combination of substitutions amounts to the replacement of salts by water, and thus constitutes a **demineralizing** action, comparable to distillation. Thus, the removal of $CaCl_2$ from water may be represented as

$$[H_x^+ \text{ resin}^{x-}] + [(OH^-)_x \text{ resin}^{x+}] + \frac{x}{2} Ca^{2+} + xCl^- \longrightarrow [Ca_{x/2}^{2+} \text{ resin}^{x-}] + [Cl_x^- \text{ resin}^{x+}] + xH_2O$$

17.8 ▫ PREPARATION OF ACIDS

(a) Direct combination of elements Only binary acids can be synthesized by direct combination of elements. The hydrogen halides and hydrogen sulfide can be prepared by this method. The applicability of this method, however, depends on the stability of the hydride, few of which can survive the temperature needed to promote the reaction. Although HF and HCl are sufficiently stable, the cost of the free halogen makes this method commercially impractical; the halogens are more expensive than their hydrides.

(b) Reaction of salts with sulfuric acid A large number of acids are made by reacting their salts with concentrated sulfuric acid. Fortunately, most common acids, such as $HONO_2$, HF, HCl, and CH_3COOH, are volatile. They are easily removed from the reaction mixture so that the equilibrium is driven in the direction of the product,

$$\underset{(concentrated)}{(HO)_2SO_2} + A^- \longrightarrow HA(g) + HOSO_3^-$$

Sulfuric acid is used because it is cheap, strong, stable, and nonvolatile (boils with decomposition at 340°C). This method is not applicable to the preparation of HBr and HI. These compounds are oxidized by concentrated $(HO)_2SO_2$,

$$8HI + (HO)_2SO_2 \longrightarrow 4I_2 + H_2S + 4H_2O$$
$$2HBr + (HO)_2SO_2 \longrightarrow Br_2 + SO_2 + 2H_2O$$

Other methods such as the reactions of nonmetallic oxides with water and nonmetallic halides with water are discussed in Chapter 20.

PROBLEMS

1. **Conjugate acid-base pairs** (a) List the conjugate acids of H_2O, Cl^-, SO_4^{2-}, NH_3, HF. (b) List the conjugate bases of H_2O, HBr, NH_4^+, $(HO)_2PO_2^-$, NH_3, OH^-.

2. Conjugate acid-base pairs Complete the following protolytic reactions. Indicate the conjugate acid-base pairs. If not shown over the arrow, the solvent is the second reactant.

(a) $NH_3 + CH_3COOH \longrightarrow$
(b) $H^- + NH_3 \longrightarrow$
(c) $O^{2-} + H_2O \longrightarrow$
(d) $NH_2^- + CH_3OH \longrightarrow$
(e) $H_3O^+ + O^{2-} \xrightarrow{H_2O}$
(f) $HCl(g) + H_2SO_4(l) \longrightarrow$

3. Amphiprotism Write two equations to illustrate amphiprotism of (a) $HOCO_2^-$; (b) $(HO)_2PO_2^-$; (c) $Zn(OH)_2(H_2O)_2$. Designate the conjugate acid-base pairs.

4. Reaction of ions with water Classify the aqueous solutions of the following salts as acidic, basic, neutral, or doubtful: (a) $NaNO_3$; (b) NaCN; (c) NH_4Cl; (d) K_3PO_4; (e) K_2HOPO_3; (f) $FeCl_3$; (g) $BiCl_3$; (h) $CaCl_2$; (i) K_2CO_3; (j) $K_2Al_2(SO_4)_4(H_2O)_{24}$ (alum); (k) $(HONH_3)^+OAc^-$ (hydroxylammonium acetate).

5. Lewis concept Categorize each of the following substances as a Lewis acid, base, or neither of these: (a) anhydrous $ZnCl_2$; (b) Cl^-; (c) anhydrous $SnCl_2$; (d) H_2; (e) $B(CH_3)_3$; (f) CH_3OCH_3. Compounds (a) and (c) are covalent.

6. Lewis concept Using structural formulas, formulate and complete the following equations. Specify the electrophile and the nucleophile.

(a) aluminum hydride + lithium hydride (Li^+H^-) $\longrightarrow$
(b) anhydrous zinc sulfate + water $\longrightarrow$
(c) boron trifluoride + ethyl ether ($C_2H_5OC_2H_5$) $\longrightarrow$
(d) ammonia + hydrogen chloride $\longrightarrow$

7. Autoprotolysis Write equations for the autoprotolysis of the following pure liquids: (a) $(HO)_2SO_2$; (b) CH_3OH; (c) $HONO_2$; (d) HCOOH (formic acid).

8. Neutralization Write an equation for a neutralization reaction occurring in each of the following pure liquids: (a) ammonia; (b) acetic acid, CH_3COOH.

9. Polyprotic acids Write equations (designate conjugate pairs) for the stepwise protolysis in water of (a) H_2S; (b) CO_2; (c) $(HO)_2\underset{\displaystyle |\atop H}{P}O$; (d) $(HO)_3PO$; (e) PO_4^{-3}; (f) ethylene diammonium ion, $(H_3NCH_2CH_2NH_3)^{2+}$.

10. Leveling effect Explain the fact that $C_2H_5OH_2^+$ is the strongest acid that can exist in ethyl alcohol, $C_2H_5\ddot{O}H$.

ADDITIONAL PROBLEMS

11. Protolysis; H-D exchange Write the equation for the equilibrium established when (*i*) CH_3COOH is dissolved in D_2O, (*ii*) CH_3COOD is dissolved in H_2O.

12. Relative acidity Write equations and account for the fact that (a) acetic acid is a strong acid in liquid NH_3; (b) nitric acid is a base in pure sulfuric acid.

13. Leveling effect (a) Elaborate on the statement, "The strongest base and acid that exist in liquid NH_3 are NH_2^- and NH_4^+, respectively." (b) What are the strongest base and acid, respectively, that can exist in pure CH_3OH? (c) Generalize about the strongest acid and base that can exist in any autoprotolytic solvent.

14. Practical protolysis Write an equation for each of the following practical protolytic reactions:

(a) $NaHOCO_2(aq) + (HO)_2SO_2$ (*"wet" fire extinguisher*)
(b) $NaHOCO_2(aq) + K_2Al_2(SO_4)_4(H_2O)_{24}(aq)$ (*baking powder reaction*)
(c) Calcium hydride + water (*filling weather balloons*)
(d) Calcium carbide (Ca^{2+} :C≡C:$^{2-}$) + water (*preparation of acetylene*)

15. Neutralization The heat of reaction at 25°C of NaOH and HCl, or KOH and HNO_3, is −13.82 kcal/mole. That for CH_3COOH and NaOH is −13.52 kcal/mole. Account for the difference in the observed data with the aid of thermochemical equations.

16. Amphoterism Mg liberates H_2 from acidic media faster than does Al, but Al liberates H_2 faster from alkaline media. Explain. *Hint:* Consider the behavior of insoluble $Mg(OH)_2$ and $Al(OH)_3$ toward excess base.

18 ■ CALCULATIONS OF IONIC EQUILIBRIUM

Because of the great abundance and importance of water, the majority of the reactions encountered by the reader occur in aqueous solutions. Because of the remarkable effectiveness of water as an ionizing solvent, most of these reactions involve ions, often along with neutral molecules or undissolved solids. Ionic reactions usually (not always) come rapidly to equilibrium, and the properties of the solution depend on the concentrations of the species present at equilibrium. For all these reasons, calculations of ionic equilibria deserve careful study.

18.1 □ THE IONIZATION OF WATER

Water undergoes self-ionization (p. 314).

$$2H_2O \rightleftharpoons H_3O^+ + OH^-$$

Thus, even perfectly pure water contains minute amounts of H_3O^+ and OH^-. Their concentrations must be equal in pure water, for two reasons: (1) In the reaction $2H_2O \rightleftharpoons H_3O^+ + OH^-$, one H_3O^+ ion is formed for every OH^- ion formed, and conversely. (2) The solution must be electrically neutral, which means that the total positive charge equals the total negative charge; when H_3O^+ and OH^- are the only ions present, this requires that $[H_3O^+] = [OH^-]$. The equilibrium concentrations of H_3O^+ and OH^- in pure water at 25°C are found by letting $x = [H_3O^+] = [OH^-]$, and substituting into the equilibrium condition,

$$[H_3O^+][OH^-] = K_w$$
$$x^2 = 1.0 \times 10^{-14}$$
$$x = 1.0 \times 10^{-7}$$

At 50°C, however, the concentrations are higher:

$$x^2 = K_w = 5.5 \times 10^{-14}$$
$$x = 2.3 \times 10^{-7}$$

If the water contains some source of H_3O^+ or OH^-, other than H_2O itself, we can no longer expect that these two concentrations will be equal. For example, we might make $[H_3O^+] = 0.10$ mole/liter by adding a suitable quantity of acid. Then, at 25°C,

$$(0.10)[OH^-] = 1.0 \times 10^{-14}$$
$$[OH^-] = 1.0 \times 10^{-13} \text{ mole/liter}$$

18.2 □ pH AND pOH

In most aqueous solutions, we are concerned with very small concentrations of H_3O^+ and OH^-. In the preceding chapter, the convenient notation $pK = -\log K$ was introduced. In the same way, we write $pH = -\log[H_3O^+]$, and $pOH = -\log[OH^-]$.* Values for pH and pOH are usually between 0 and 14, but not necessarily; if $[H_3O^+] = 2$ moles/liter, $pH = -\log 2 = -0.3$.

EXAMPLE 1 Find the pH of a solution in which $[H_3O^+] = 6.38 \times 10^{-6}$ mole/liter.

ANSWER

$$\begin{aligned} pH &= -\log(6.38 \times 10^{-6}) \\ &= -(\log 6.38 - 6 \log 10) \\ &= -\log 6.38 + 6 \log 10 \\ &= -0.805 + 6 = 5.195 \end{aligned}$$

Note the fact, surprising at first, that the pH in this case is between 5 and 6, not between 6 and 7. The reason is that 6.38×10^{-6} is between 10^{-5} and 10^{-6}, not between 10^{-6} and 10^{-7}.

EXAMPLE 2 Calculate $[H_3O^+]$ for a solution of pH 8.37.

ANSWER

$$\begin{aligned} [H_3O^+] &= 10^{-pH} = 10^{-8.37} = 10^{-9+0.63} = 10^{0.63} \times 10^{-9} \\ &= (\text{antilog } 0.63) \times 10^{-9} \\ &= 4.3 \times 10^{-9} \end{aligned}$$

Answers should never be given as base 10 raised to a fractional exponent.

EXAMPLE 3 A certain solution has pH 3.89 at 25°C. Find pOH and $[OH^-]$.

ANSWER As with other equilibrium constants, we can write $pK_w = -\log K_w$. Since

$$[H_3O^+][OH^-] = K_w$$

* Historically, pH preceded pK. The notation "pH" was introduced in 1909, as an abbreviation of "power of hydrogen."

then

$$\log[H_3O^+] + \log[OH^-] = \log K_w$$
$$pH + pOH = pK_w = 14.00$$

at 25°C.

$$pOH = pK_w - pH = 14.00 - 3.89$$
$$= 10.11$$
$$[OH^-] = 10^{-10.11} = 10^{-11+0.89}$$
$$= 7.77 \times 10^{-11} \text{ mole/liter}$$

18.3 ▫ ACIDIC, BASIC, AND NEUTRAL SOLUTIONS

An aqueous solution is said to be

acidic if $[H_3O^+]$ is greater than $[OH^-]$, or $pH < pOH$;
neutral if $[H_3O^+] = [OH^-]$, or $pH = pOH$;
basic if $[OH^-]$ is greater than $[H_3O^+]$, or $pH > pOH$.

At 25°, **pH = 7.00 for a neutral solution; when pH < 7, the solution is acidic, and when pH > 7, the solution is basic.** The number 7 is worth remembering, but one should also remember that it refers to 25°C, and that the neutral pH is somewhat different at other temperatures.

Throughout the remainder of this chapter, you may assume that the temperature is that for which the data are given, usually 25°C.

18.4 ▫ IONIZATION OF WEAK ACIDS

We may represent a general acid by HA; it may be either a neutral molecule (as acetic acid, CH_3COOH) or an ion (as ammonium ion, NH_4^+; hydrogen sulfate ion, $HOSO_3^-$). In this section, however, we shall assume that HA is uncharged. The equilibrium constant for the reaction

$$HA + H_2O \rightleftharpoons H_3O^+ + A^-$$

is the ionization constant of HA, as in Chapter 17, and is represented by K_a or K_{HA}. Table 18.1 lists K_a for a number of common acids.

Let c be the number of moles of HA added per liter of solution. The usual problem is to find $[H_3O^+]$ or pH when K and c are given.

There are two reactions that occur in the solution:

$$HA + H_2O \rightleftharpoons H_3O^+ + A^- \quad (1)$$
$$2H_2O \rightleftharpoons H_3O^+ + OH^- \quad (2)$$

Reaction (2) produces equal numbers of moles of H_3O^+ and OH^-. We have noticed that, in this acidic solution, $[OH^-] \ll [H_3O^+]$. Another way of saying the same thing is: $[H_3O^+]$ produced by reaction (2) (= $[OH^-]$) is very small in comparison to total $[H_3O^+]$. Most of the H_3O^+ in the solution is produced by the ionization of the acid HA (reaction 1), not by the ionization of water (reaction 2)—a result to be expected unless HA is an extremely weak or an extremely dilute acid. Thus, when we say that the solution is markedly acidic, we are also saying that the production of H_3O^+ by reaction (2) is negligible, and there is only one important reaction: reaction (1).

In this special, but common, case, where only one reaction need be considered, the calculations can be performed as they were in Section 14.8 (p. 255). If we start with c moles/liter of HA, the number of moles/liter of H_3O^+ and A^- must be the same (x), and the remaining number of moles/liter of HA is $c - x$.

$$\underset{c-x}{\mathrm{HA}} + \mathrm{H_2O} \rightleftharpoons \underset{x}{\mathrm{H_3O^+}} + \underset{x}{\mathrm{A^-}} \qquad \text{moles/liter at equilibrium}$$

$$K_a = \frac{[\mathrm{H_3O^+}][\mathrm{A^-}]}{[\mathrm{HA}]} = \frac{x^2}{c - x} \tag{3}$$

K_a and c are known and we can readily solve for x as shown in the following examples.

EXAMPLE 4 Referring to Table 18.1, find the pH of a 0.200 M solution of formic acid.

ANSWER Here, $c = 0.200$ mole/liter and $K_a = 1.76 \times 10^{-4}$. Then

$$\frac{x^2}{0.200 - x} = 1.76 \times 10^{-4}$$

This quadratic equation can be solved without undue labor, but a simplifying approximation is often applicable. The acid is weak, meaning that only a small fraction of it ionizes. We thus expect that x, representing the concentration of each dissociation product (H_3O^+ and A^-), will be much less than c, representing the total concentration of acid. If this approximation is valid, $0.200 - x$ can be replaced by 0.200 without serious error, and the equation becomes

$$\frac{x^2}{0.200} = 1.76 \times 10^{-4}$$

$$x^2 = 3.52 \times 10^{-6}$$

$$x = 5.93 \times 10^{-3}$$

For most purposes, we may consider 6×10^{-3} negligible in comparison with 0.200. (Accurate solution of the equation gives $x = 5.84 \times 10^{-3}$.) Then

$$\mathrm{pH} = -\log x = 3 - \log 5.93$$

$$= 2.23$$

The ionization of a weak acid is commonly described by giving the **degree of ionization** or **degree of dissociation** (represented by α), which is the fraction of the total acid present that ionizes. In terms of con-

TABLE 18.1 ▪ IONIZATION CONSTANTS OF ACIDS AND BASES IN WATER AT 25°C

Name	Formula		Constant	
	Monoprotic acids		K_a	
Acetic acid	CH_3COOH		1.75×10^{-5}	
Benzoic acid	C_6H_5COOH		6.30×10^{-5}	
Bromoacetic acid	$CH_2BrCOOH$		1.38×10^{-3}	
Chloroacetic acid	$CH_2ClCOOH$		1.40×10^{-3}	
Dichloroacetic acid	$CHCl_2COOH$		5×10^{-2}	
Formic acid	$HCOOH$		1.76×10^{-4}	
Hydrocyanic acid	HCN		7.2×10^{-10}	
Hydrofluoric acid	HF		6.7×10^{-4}	
Nitrous acid	$HONO$		4.5×10^{-4}	
Phenol (carbolic acid)	C_6H_5OH		1.3×10^{-10}	
Propionic acid	CH_3CH_2COOH		1.34×10^{-5}	
Trichloroacetic acid	CCl_3COOH		2×10^{-1}	
	Polyprotic acids	K_1	K_2	K_3
Carbonic acid[a]	$(HO)_2CO$	4.47×10^{-7}	5.62×10^{-11}	—
Hydrogen sulfide	H_2S	6.3×10^{-8}	10^{-14}	—
Malonic acid	$CH_2(COOH)_2$	1.40×10^{-3}	2.1×10^{-6}	—
Oxalic acid	$(COOH)_2$	6.5×10^{-2}	5×10^{-5}	—
Phosphoric acid	$(HO)_3PO$	7.52×10^{-3}	6.22×10^{-8}	4.8×10^{-13}
Phthalic acid	$C_6H_4(COOH)_2$	1.26×10^{-3}	3.1×10^{-6}	—
Sulfuric acid	$(HO)_2SO_2$	∞	1.0×10^{-2}	—
Sulfurous acid	$(HO)_2SO$	1.7×10^{-2}	6.24×10^{-8}	—
	Monoprotic bases		K_b	
Ammonia	NH_3		1.76×10^{-5}	
Aniline	$C_6H_5NH_2$		3.83×10^{-10}	
Dimethylamine	$(CH_3)_2NH$		5.04×10^{-4}	
Ethylamine	$CH_3CH_2NH_2$		4.71×10^{-4}	
Hydroxylamine	$HONH_2$		10^{-9}	
Methylamine	CH_3NH_2		4.40×10^{-4}	
Pyridine	C_5H_5N		1.56×10^{-9}	
Trimethylamine	$(CH_3)_3N$		6.34×10^{-5}	
	Diprotic base	K_1	K_2	
Hydrazine	H_2NNH_2	9×10^{-7}	9×10^{-16}	

[a] This constant is really for $CO_2 + 2H_2O \rightleftharpoons H_3O^+ + HCO_3^-$ (see p. 393).

centration,

$$\alpha = \frac{[H_3O^+]}{c} = \frac{x}{c}$$

For the preceding example,

$$\alpha = \frac{5.93 \times 10^{-3}}{0.200} = 2.97 \times 10^{-2} \text{ or } 2.97\%$$

EXAMPLE 5 Find the degree of ionization of acetic acid in a 0.500 M solution, and the pH of the solution.

ANSWER From Table 18.1, $K_a = 1.75 \times 10^{-5}$. Then, for the reaction

$$CH_3COOH + H_2O \rightleftharpoons H_3O^+ + CH_3COO^-$$

the equilibrium condition is

$$\frac{x^2}{0.500 - x} = 1.75 \times 10^{-5}$$

where $x = [H_3O^+] = [CH_3COO^-]$. If we assume that x is much less than 0.500,

$$\frac{x^2}{0.500} = 1.75 \times 10^{-5}$$
$$x^2 = 8.75 \times 10^{-6}$$
$$x = 2.96 \times 10^{-3} \text{ mole/liter} = [H_3O^+]$$

which is, for our purposes, negligible in comparison to 0.500. Then

$$\alpha = \frac{x}{0.500} = \frac{2.96 \times 10^{-3}}{0.500} = 5.92 \times 10^{-3}$$

$$pH = -\log[H_3O^+] = -\log 2.96 + 3 = 2.53$$

Let us now calculate $[H_3O^+]$ in a solution of an acid of moderate strength.

EXAMPLE 6 Find $[H_3O^+]$ in a 0.50 M solution of trichloroacetic acid, $K_a = 0.2$.

ANSWER

$$CCl_3COOH + H_2O \rightleftharpoons H_3O^+ + CCl_3COO^-$$

$$\frac{[H_3O^+][CCl_3COO^-]}{[CCl_3COOH]} = 0.2 \qquad (4)$$

At equilibrium $[H_3O^+] = [CCl_3COO^-] = x$ and $[CCl_3COOH] = 0.5 - x$. Equation (4) then is

$$\frac{x^2}{0.50 - x} = 0.2 \qquad (5)$$

Let us assume that $x \ll 0.50$. This assumption gives

$$\frac{x^2}{0.50} = 0.2$$
$$x = 0.32$$

Obviously, the assumption was a bad one, for 0.3 is clearly not negligible in comparison to 0.50. Substitution into Equation 5 shows that $x = 0.3$ is not a solution. We must solve Equation 5 as a quadratic equation (App. 2). Equation 5 may be rewritten as

$$x^2 + 0.2x - 0.1 = 0$$
$$x = \frac{-0.2 \pm [(0.2)^2 - 4(1)(-0.1)]^{1/2}}{2(1)}$$
$$= \frac{-0.2 \pm (0.44)^{1/2}}{2}$$
$$= +0.23 \text{ or } -0.43$$

A concentration cannot be negative, and the answer -0.43 must therefore be rejected; then

$$[H_3O^+] = x = 0.2 \text{ mole/liter}$$

(Why is only one significant figure retained?)

Evidently, the approximation that $x \ll c$ cannot be made for a moderately strong acid.

We now turn our attention to finding $[H_3O^+]$ in a dilute solution of a very weak acid; for example, a 1.0×10^{-5} M solution of HCN.

$$HCN + H_2O \rightleftharpoons H_3O^+ + CN^- \qquad (6)$$

$$\frac{[H_3O^+][CN^-]}{[HCN]} = 7.2 \times 10^{-10} \qquad (7)$$

If we assume that reaction (6) is the major source of H_3O^+, at equilibrium,

$$[H_3O^+] = [CN^-] = x$$

and

$$[HCN] = 1.0 \times 10^{-5} - x \approx 1.0 \times 10^{-5}$$

and Equation 7 is

$$\frac{x^2}{1.0 \times 10^{-5}} = 7.2 \times 10^{-10}$$
$$x^2 = 7.2 \times 10^{-15}$$
$$x = 8.5 \times 10^{-8} = [H_3O^+] = [CN^-]$$

This answer tells us that the solution must be basic since $[H_3O^+]$ is less than 10^{-7}. However, our common sense tells us that a dilute solution of even a very weak acid could never be basic. Our arithmetic is accurate; therefore our assumptions must be erroneous. The error lies in assuming that reaction 6 is the main source of H_3O^+ and that we can disregard the reaction $2H_2O \rightleftharpoons H_3O^+ + OH^-$.

To solve this problem, it is best to introduce a different approach. This approach assumes that for the ionization of a weak acid, HA, two conditions must be satisfied in addition to the equilibrium condition

$$\frac{[H_3O^+][A^-]}{[HA]} = K_a \tag{8}$$

(a) The solution must be *electrically neutral*. This means that the total number of positive charges is equal to the total number of negative charges. In the present case, there is only one kind of positive ion, H_3O^+; there are two kinds of negative ion, A^- and OH^-. In each liter of solution, then, the number of H_3O^+ ions must equal the total number of A^- and OH^- ions. Equivalently, the number of *moles* of H_3O^+ in a liter must equal the total number of moles of A^- and OH^- in a liter,

$$[H_3O^+] = [A^-] + [OH^-] \tag{9}$$

(b) Some of the HA molecules in the solution lose protons and become A^-. We do not know in advance what percentage of the molecules becomes A^- and what percentage remains HA. However, we may assume that every HA molecule that was put into the solution does one thing or the other; if it is not there as HA, it is there as A^-. We introduced c moles of HA per liter of solution. The total number of moles of HA and A^- present (per liter) at equilibrium must then be c,

$$[HA] + [A^-] = c \tag{10}$$

This equation may be described as the **conservation condition** for the "A" group of atoms.

Let us now apply this new method to finding $[H_3O^+]$ in our 1.0×10^{-5} M solution of HCN. Since we must consider both sources of H_3O^+, H_2O and HCN, we have two equilibrium conditions

$$\frac{[H_3O^+][CN^-]}{[HCN]} = 7.2 \times 10^{-10} \tag{11}$$

$$[H_3O^+][OH^-] = K_w = 1.0 \times 10^{-14} \tag{12}$$

The two other conditions are

Electrical neutrality $\quad [H_3O^+] = [CN^-] + [OH^-] \quad (13)$

Conservation $\quad [HCN] + [CN^-] = 1.0 \times 10^{-5} \quad (14)$

To obtain the correct answer in this case, we let

$$[CN^-] = x \text{ and } [OH^-] = y$$

From Equation 13,

$$[H_3O^+] = x + y$$

and from Equation 14,

$$[\mathrm{HCN}] = 1.0 \times 10^{-5} - x \approx 1.0 \times 10^{-5}$$

Substitution into Equations 11 and 12 gives two simultaneous equations, 11′ and 12′, respectively,

$$\frac{(x+y)(x)}{1.0 \times 10^{-5}} = 7.2 \times 10^{-10} \qquad (11')$$

$$(x+y)y = 10^{-14} \qquad (12')$$

or

$$x + y = \frac{10^{-14}}{y}$$

Substitution of $(x + y)$ in Equation 11′ gives

$$\frac{10^{-14} \times x}{y \times 1.0 \times 10^{-5}} = 7.2 \times 10^{-10}$$

$$x = 0.72y$$

Substitution in Equation 12′ gives

$$(0.72y + y)y = 10^{-14}$$

$$1.72y^2 = 10^{-14}$$

$$y = 7.5 \times 10^{-8}$$

and

$$x = 0.72(7.5 \times 10^{-8}) = 5.4 \times 10^{-8}$$

Therefore

$$[\mathrm{H_3O^+}] = x + y = 7.5 \times 10^{-8} + 5.4 \times 10^{-8}$$

$$= 1.3 \times 10^{-7} \text{ mole/liter}$$

This method can be used for all types of calculations of ionic equilibria. For example, if the acid is not extremely weak, the solution will be decidedly acidic, which means that $[\mathrm{OH^-}] \ll [\mathrm{H_3O^+}]$, and Equation 9 can be written

$$[\mathrm{H_3O^+}] = [\mathrm{A^-}]$$

We can now let

$$x = [\mathrm{H_3O^+}] = [\mathrm{A^-}]$$

and Equation 10 gives us

$$[\mathrm{HA}] = c - [\mathrm{A^-}] = c - x$$

Substituting into Equation 8 we obtain

$$\frac{x^2}{c - x} = K_a \qquad (15)$$

We also arrived at Equation 15 by the simpler route which was previously used (Equation 3, p. 330).

In summary, the usual approximations work only when the acid is neither too strong nor too weak. If it is too strong, we cannot assume that $[A^-] \ll c$. If it is too weak, or too dilute, or both, the ionization of water must be taken into account. The latter situation, fortunately, is not encountered very often. Useful rules of thumb are (1) that the decrease in [HA] can be disregarded only when $K/c < 10^{-4}$, and (2) that the ionization of water can be disregarded only when $cK > 10^{-12}$.

The simple method (p. 330) will be used in most of our calculations.

18.5 □ IONIZATION OF WEAK BASES

A base, like an acid, may be a neutral molecule or an ion; in this section, we shall assume that the base is a neutral molecule, and represent it by B. The ionization constant of B is the equilibrium constant of the reaction

$$B + H_2O \rightleftharpoons BH^+ + OH^-$$

for which the equilibrium condition is

$$\frac{[BH^+][OH^-]}{[B]} = K_b \tag{16}$$

As with weak acids, we can disregard the ionization of water. There is, then, only one important reaction in the solution:

$$\begin{array}{ccccccc} B & + H_2O & \rightleftharpoons & BH^+ & + OH^- & & \\ c-x & & & x & x & & \text{moles/liter at equilibrium} \end{array}$$

and substituting into Equation 16 gives

$$\frac{x^2}{c-x} = K_b$$

The only difference from the acid case is that x now represents $[OH^-]$ instead of $[H_3O^+]$.

Table 18.1 gives the ionization constants of some common bases.

EXAMPLE 7 Find the pH of a 0.100 M solution of ammonia, NH_3.

ANSWER

$$NH_3 + H_2O \rightleftharpoons NH_4^+ + OH^-$$

$$\frac{[NH_4^+][OH^-]}{[NH_3]} = K_b = 1.76 \times 10^{-5}$$

At equilibrium, let

$$x = [NH_4^+] = [OH^-] \qquad \text{and} \qquad [NH_3] = 0.100 - x$$

Then

$$\frac{x^2}{0.100 - x} = 1.76 \times 10^{-5}$$

Since $x \ll 0.100$,

$$\frac{x^2}{0.100} = 1.76 \times 10^{-5}$$
$$x^2 = 1.76 \times 10^{-6}$$
$$x = 1.33 \times 10^{-3} = [\mathrm{OH^-}]$$
$$\mathrm{pOH} = 3 - \log 1.33 = 2.88$$
$$\mathrm{pH} = 14.00 - \mathrm{pOH} = 11.12$$

18.6 □ CONJUGATE ACID-BASE PAIRS

In Chapter 17, conjugate acid-base pairs were defined: HA is known as the conjugate acid of A^-, and A^- is the conjugate base of HA. A simple relation between the acidic and basic ionization constants in a conjugate pair was established (p. 317): $K_aK_b = K_w$, or $\mathrm{p}K_a + \mathrm{p}K_b = \mathrm{p}K_w$. Thus, if we know the ionization constant for an acid or base, we can find the ionization constant for its conjugate base or acid.

EXAMPLE 8 Find K_b and $\mathrm{p}K_b$ for the acetate ion, CH_3COO^-.

ANSWER The ionization constant of CH_3COOH is $K_a = 1.75 \times 10^{-5}$; $K_w = 1.01 \times 10^{-14}$. Then,

$$K_b = \frac{K_w}{K_a}$$
$$= \frac{1.01 \times 10^{-14}}{1.75 \times 10^{-5}} = 5.77 \times 10^{-10}$$
$$\mathrm{p}K_b = 10 - \log 5.77 = 9.24$$

18.7 □ POLYPROTIC ACIDS

Let H_2A represent a diprotic acid. The equilibrium constants for the two reactions

$$H_2A + H_2O \rightleftharpoons H_3O^+ + HA^-$$
$$HA^- + H_2O \rightleftharpoons H_3O^+ + A^{2-}$$

are represented by K_1 and K_2, respectively, and are called the first and second ionization constants of H_2A.

$$\frac{[H_3O^+][HA^-]}{[H_2A]} = K_1$$
$$\frac{[H_3O^+][A^{2-}]}{[HA^-]} = K_2$$

For a triprotic acid, there is also a third constant K_3.

Calculations involving the ionization of polyprotic acids and bases present no great complications, provided that one makes a simplifying assumption. Because the second ionization constant is less than the first—usually much less—we can assume that *the concentration of*

A^{2-}, *which is produced in the second ionization, is much less than the concentration of* HA^-, *which is produced in the first ionization.* Calculations involving the first ionization can then be performed as if the acid were monoprotic, and the results of this calculation can then be used to find $[A^{2-}]$.

EXAMPLE 9 Find $[H_3O^+]$, $[HS^-]$, and $[S^{2-}]$ in a 0.1 M solution of H_2S.

For the reactions

$$H_2S + H_2O \rightleftharpoons H_3O^+ + HS^- \tag{17}$$
$$HS^- + H_2O \rightleftharpoons H_3O^+ + S^{2-} \tag{18}$$

we have the equilibrium conditions

$$\frac{[H_3O^+][HS^-]}{[H_2S]} = K_1 = 6.3 \times 10^{-8}$$
$$\frac{[H_3O^+][S^{2-}]}{[HS^-]} = K_2 = 10^{-14} \tag{19}$$

We focus attention on the first ionization and solve for $[H_3O^+]$ and $[HS^-]$ as if the acid were monoprotic. Hence, at equilibrium

$$[H_3O^+] = [HS^-] = x \qquad \text{and} \qquad [H_2S] = 0.10 - x \approx 0.10$$

Therefore

$$\frac{x^2}{0.10} = 6.3 \times 10^{-8}$$
$$x^2 = 63 \times 10^{-10}$$
$$x = 7.9 \times 10^{-5} = [H_3O^+] = [HS^-]$$

Since $[HS^-]$ has decreased only very little and $[H_3O^+]$ has increased very little as the result of the second ionization, we substitute these values into Equation 19.

$$\frac{7.9 \times 10^{-5} \times [S^{2-}]}{7.9 \times 10^{-5}} = 10^{-14}$$
$$[S^{2-}] = 10^{-14}$$

The molar concentration of A^{2-} formed by the second ionization, Equation 18, will be equal to K_2, as in Example 9, whenever HA^- is a weak acid and the solution is prepared by dissolving *only* H_2A in water.

There is a source of confusion that should be pointed out here. In a given solution, a given ion or molecule has only one concentration. The misconception sometimes arises that one concentration of H_3O^+ is produced by the reaction

$$H_2A + H_2O \rightleftharpoons H_3O^+ + HA^-$$

and another concentration of H_3O^+ is produced by the reaction

$$HA^- + H_2O \rightleftharpoons H_3O^+ + A^{2-}$$

We must remember that the solution has only *one* value of $[H_3O^+]$, to be used in all the equations in which this concentration appears. After all, an ion has no memory—it knows only what it is, not by what reaction it was produced.

18.8 □ WEAK ACID (OR BASE) IN THE PRESENCE OF STRONG ACID (OR BASE)

If a weak acid is present in a solution that also contains a strong acid, the concentration of H_3O^+ is determined almost entirely by the concentration of the strong acid. The same applies if the acid in which we are interested is present in small concentration relative to other acids in the solution. Corresponding statements are true for bases.

EXAMPLE 10 Find the degree of ionization of CH_3COOH and the concentration of CH_3COO^- in a solution 0.10 M with respect to HCl and 0.20 M with respect to CH_3COOH. K_a for acetic acid, CH_3COOH, is 1.75×10^{-5}.

ANSWER

$$CH_3COOH + H_2O \rightleftharpoons H_3O^+ + CH_3COO^-$$

$$\frac{[H_3O^+][CH_3COO^-]}{[CH_3COOH]} = 1.75 \times 10^{-5} \quad (20)$$

Let x be the number of moles of CH_3COOH ionized. Then at equilibrium

$$[H_3O^+] = 0.10 + x \approx 0.10$$

The 0.10 mole comes from the 0.10 M HCl, which is a strong acid; the x mole comes from the ionized CH_3COOH and is insignificantly small compared to 0.10.

$$[CH_3COO^-] = x$$

The only source of this ion is the ionized CH_3COOH.

$$[CH_3COOH] = 0.20 - x \approx 0.20$$

Equation 20 becomes

$$\frac{0.10x}{0.20} = 1.75 \times 10^{-5}$$

$$x = 3.5 \times 10^{-5} \text{ mole/liter} = [CH_3COO^-]$$

The degree of ionization is

$$\alpha = \frac{x}{0.20} = 1.75 \times 10^{-4}$$

It should be noted that the degree of ionization of the weak acid is much decreased by the presence of the strong acid; if the HCl were absent, α would be 9.3×10^{-3} instead of 1.75×10^{-4}. This effect is called the **common ion effect** (in allusion to the H_3O^+ ion, which is common to the two acids), and is an example of Le Chatelier's principle: the equilibrium

$$HA + H_2O \rightleftharpoons H_3O^+ + A^-$$

is shifted to the left by an increase in $[H_3O^+]$.

EXAMPLE 11 Find the degree of ionization of 0.500 M NH_3 in a solution of pH 13.0.

ANSWER

$$pOH = 14.0 - 13.0 = 1.0; \; [OH^-] = 1.0 \times 10^{-1}.$$

$$NH_3 + H_2O \rightleftharpoons NH_4^+ + OH^-$$

$$\frac{[NH_4^+][OH^-]}{[NH_3]} = 1.76 \times 10^{-5}$$

Let x be the moles of NH_3 ionized. Then at equilibrium

$$[OH^-] = 1.0 \times 10^{-1} + x \approx 1.0 \times 10^{-1}$$
$$[NH_4^+] = x$$
$$[NH_3] = 0.500 - x \approx 0.500$$

Therefore

$$\frac{1.0 \times 10^{-1}x}{0.500} = 1.76 \times 10^{-5}$$

$$x = 8.80 \times 10^{-5} = \text{mole of } NH_3 \text{ ionized}$$

$$\alpha = \frac{x}{0.500} = 1.76 \times 10^{-4}$$

EXAMPLE 12 Find the concentration of S^{2-} in a solution 0.10 M with respect to H_2S and 0.30 M with respect to HCl.

ANSWER For the reactions

$$H_2S + H_2O \rightleftharpoons H_3O^+ + HS^-$$
$$HS^- + H_2O \rightleftharpoons H_3O^+ + S^{2-}$$

we have the equilibrium conditions

$$\frac{[H_3O^+][HS^-]}{[H_2S]} = K_1 = 6.3 \times 10^{-8}$$

$$\frac{[H_3O^+][S^{2-}]}{[HS^-]} = K_2 = 10^{-14}$$

Since $[HS^-]$ does not interest us in this problem, we are free to eliminate it by multiplying the two equations together,

$$\frac{[H_3O^+]^2[S^{2-}]}{[H_2S]} = K_1K_2$$

or

$$[S^{2-}] = \frac{K_1K_2[H_2S]}{[H_3O^+]^2}$$

$$= \frac{6.3 \times 10^{-8} \times 10^{-14} \times 0.10}{(0.30)^2}$$
$$= 7 \times 10^{-22} \text{ mole/liter}$$

Note that $[S^{2-}]$ is *not* equal to K_2 in this case. That result (p. 338) applied only to the case in which the diprotic acid was the only solute. Here, the solution also contains HCl.

18.9 ▫ BUFFER SOLUTIONS

A solution containing an acid and its conjugate base, both moderately weak, is known as a **buffer solution;** such a solution has the property that the addition of acids or bases causes only a relatively small change in pH. The pH of pure water, or of a solution not a buffer solution, is very sensitive to the addition of small quantities of acid or base. Suppose that we want a solution with pH 4.76. Such a solution can be prepared by dissolving 1.74×10^{-5} mole HCl in a liter of water,

$$[H_3O^+] = 1.74 \times 10^{-5}$$
$$pH = 4.76$$

Now, perhaps by accident, 0.10 mole HCl is added per liter of this solution,

$$[H_3O^+] = 0.10 + 1.74 \times 10^{-5} = 0.10 \text{ mole/liter}$$
$$pH = 1.0$$

The pH has been changed drastically.

However, when a strong acid is added to a buffer solution, it reacts with the base (A^-) present in the solution,

$$H_3O^+ + A^- \longrightarrow HA + H_2O$$

Similarly, a strong base reacts with the acid (HA) in the solution,

$$OH^- + HA \longrightarrow A^- + H_2O$$

In each case, the acid-base ratio is changed, but with only a moderate effect on the pH of the solution. *Very little of the* H_3O^+ *or* OH^- *added will remain as such.* Of course, if the number of moles of H_3O^+ added exceeds the number of moles of A^- in the solution, or if the number of moles of OH^- added exceeds the number of moles of HA in the solution, the buffer is destroyed and there is a large change in pH. The following example illustrates the quantitative treatment of buffer action.

EXAMPLE 13 Calculate the pH of (a) a solution 0.50 *M* with respect to CH_3COOH and 0.50 *M* with respect to CH_3COONa; (b) the same solution after 0.10 mole of HCl per liter has been added to it. Assume that the volume is unchanged.

ANSWER (a)

$$CH_3COOH + H_2O \rightleftharpoons H_3O^+ + CH_3COO^-$$

$$\frac{[H_3O^+][CH_3COO^-]}{[CH_3COOH]} = 1.75 \times 10^{-5}$$

Let x be the moles of CH_3COOH ionized. Since CH_3COONa is a salt which dissociates completely, at equilibrium we find the indicated molar concentrations

$$\begin{array}{llll} CH_3COOH + H_2O \rightleftharpoons & H_3O^+ + & CH_3COO^- & \\ 0.50 - x & x & 0.50 + x & \text{moles/liter at equilibrium} \end{array}$$

Recognizing that $0.50 \pm x \approx 0.50$, we have

$$\frac{[H_3O^+](0.50)}{0.50} = 1.75 \times 10^{-5}$$

$$[H_3O^+] = 1.75 \times 10^{-5}$$

$$pH = 4.76$$

(b) The effect of adding the HCl has been to convert 0.10 mole/liter of CH_3COO^- to 0.10 mole/liter of CH_3COOH. $[CH_3COO^-]$ has gone down from 0.50 to 0.40 mole/liter; $[CH_3COOH]$ has gone up from 0.50 to 0.60 mole/liter. We now solve the problem by writing the equation

$$\begin{array}{lll} CH_3COOH \quad + H_2O \rightleftharpoons & H_3O^+ + & CH_3COO^- \\ 0.60 - x \approx 0.60 & x & 0.40 + x \approx 0.40 \end{array}$$

and, as before,

$$\frac{0.40x}{0.60} = 1.75 \times 10^{-5}$$

$$x = [H_3O^+] = \frac{0.60}{0.40} \times 1.75 \times 10^{-5}$$

$$= 2.6 \times 10^{-5}$$

$$pH = 4.58$$

If we had added NaOH, the effect would have been to convert CH_3COOH to CH_3COO^-.

The decrease in pH is only 0.18, to be contrasted with a decrease of 3.76 for the unbuffered solution. In the buffer solution, the addition of HCl has multiplied $[H_3O^+]$ by only 1.5; in the unbuffered solution, $[H_3O^+]$ has gone from 1.74×10^{-5} to 0.10, and has thus been multiplied by 5.8×10^3.

A buffer solution can also be prepared from a weak base and its conjugate acid. Thus, with a base B and its conjugate acid BH^+, the equation

$$\frac{[BH^+][OH^-]}{[B]} = K_b$$

shows that when $[BH^+] = [B]$, then $[OH^-] = K_b$, or $pOH = pK_b$. If, then, we are told to prepare a buffer solution of given pH, we should seek an acid with $pK_a \approx pH$, or a base with $pK_b \approx pOH = pK_w - pH$. If we cannot find an available acid for which $pK_a = pH$ exactly, we choose the one that is closest and adjust the acid-base ratio to make the pH come out right.

Blood is a buffered solution normally with a pH of 7.4. This pH is maintained within 0.02 pH units in spite of numerous acid-base reactions in the cells. Blood plasma is buffered mainly by dissolved CO_2 and bicarbonate ion, $HOCO_2^-$, and by proteins and their conjugate acids. Blood corpuscles are buffered mainly by $(HO)_2PO_2^-$ and its conjugate base $HOPO_3^-$. Oxyhemoglobin, the oxygenated state of hemoglobin, and its conjugate acid also buffer the corpuscles. The importance of effective buffering of blood is apparent when we realize that in human beings death usually occurs when the pH falls outside the range of 6.9–7.7.

18.10 ▫ INDICATORS AND TITRATION

An indicator is an acid (or base) having a different color from its conjugate base (or acid). Let us represent the acid form of the indicator by HIn and the base form by In^-, without implying that the acid is necessarily uncharged. An example is methyl orange, for which HIn is red and In^- is yellow. The equilibrium condition for the reaction

$$HIn + H_2O \rightleftharpoons H_3O^+ + In^-$$

is

$$\frac{[H_3O^+][In^-]}{[HIn]} = K_i$$

or

$$\frac{[HIn]}{[In^-]} = \frac{[H_3O^+]}{K_i}$$

The indicator should be used in such small quantity that $[H_3O^+]$ is determined by the other constituents of the solution. The ratio $[HIn]/[In^-]$, which determines the color, is then fixed by the pH of the solution. With methyl orange, for example, the solution will be red if $[HIn] \gg [In^-]$, yellow if $[In^-] \gg [HIn]$, and varying shades of orange when $[HIn]$ and $[In^-]$ are comparable. As a rough rule, we may assume that gradations of color are detectable when $[HIn]/[In^-]$ is between 0.1 and 10. When this ratio is greater than 10, the solution has the color of HIn, and this color does not change perceptibly if the ratio is made 1000 or 1,000,000 instead of 10 (unless, of course, the indicator accepts another proton, resulting in a second color change). Likewise, if the ratio is less than 0.1, only the In^- color is seen.

TABLE 18.2 ACID-BASE INDICATORS

Indicator	Colors		pH range
	Acid	Base	
Methyl violet	Green	Violet	0–2
Methyl orange	Red	Yellow	3–4.5
Methyl red	Red	Yellow	4–6
Bromthymol blue	Yellow	Blue	6–8
Neutral red	Red	Yellow	7–8
Litmus	Red	Blue	5–8
Phenolphthalein	Colorless	Red	8–10
Thymolphthalein	Colorless	Blue	9–10.5
Alizarin yellow	Yellow	Violet	10–12
1,3,5-Trinitrobenzene	Colorless	Red	12–14

Table 18.2 lists some common indicators, their colors, and the pH ranges in which the colors show perceptible gradations.

Quantitative analysis, or the determination of the quantities of component substances in a sample, is an important chemical study. One procedure used in quantitative analysis involves the addition of a solution of known composition to a solution of the constituent whose quantity is to be determined. The solution of known concentration is called a **standard solution. Titration** is the process of adding measured volumes of one solution to another solution until a reaction is complete. The measuring instrument used is the buret (Fig. 18.1), a long tube with volume markings and a stopcock at the bottom.

The aim of a titration is the addition of a quantity of the standard solution **chemically equivalent** *to the quantity of the unknown.* The reaction may be an acid-base reaction, an oxidation-reduction, the precipitation of a slightly soluble salt, or any other reaction which goes practically to completion, without complicating side reactions. The stage at which equivalent quantities of the standard solution and the unknown have been mixed is called the **equivalence point of the titration.** The **end point,** signaled by a sudden change in a physical or chemical property of the solution, indicates the end of the titration and the buret reading is taken.

Frequently the titration reactants do not produce a color change. It is then necessary to add an indicator to the solution being titrated to indicate the end point. A starch solution, for example, is used as the indicator in titrations with iodine solution; an excess of one drop of iodine solution imparts an unmistakable blue color to the resultant solution.

In titrations of acids and bases, when the acid and base have been mixed in equivalent quantities, they are said to have **neutralized** each other. This word is misleading, however, for the resulting solution may

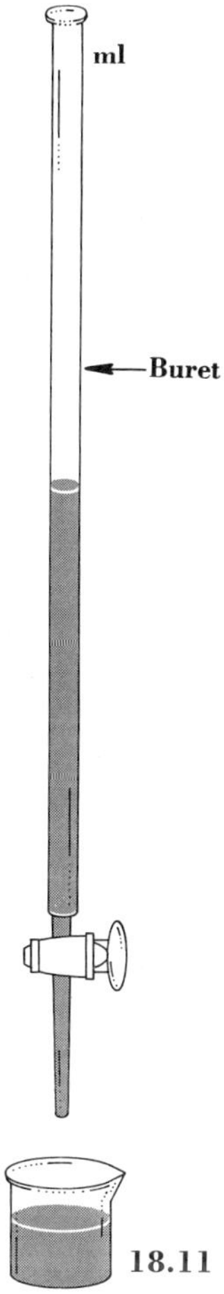

Fig. 18.1 Buret setup for a typical titration.

be neutral, acidic, or basic. Indicators possess a characteristic pH range over which they exhibit a visible color change. How, then, do we decide which indicator to use? Since the object of any titration is to mix equivalent quantities of reactants, *the choice of an indicator is determined by the pH of the solution at the equivalence point.*

Let us say that an acid solution is in a beaker or a flask; the basic solution is then in the buret. Assume that the base is an ionic hydroxide, for example, NaOH. As the base is added to the acid, the pH of the solution in the beaker increases. The dependence of pH on the quantity of base added is shown in Fig. 18.2. If the acid is strong (Fig. 18.2a), the initial effect of adding the base is merely to decrease the concentration of the acid; if the acid is weak (Fig. 18.2b), the initial effect is to form a buffer solution containing the acid and its conjugate base (p. 341). In either case, the pH rises slowly at first. *The curve becomes almost vertical when equivalent quantities of acid and base are present.* Hence, to provide the signal for the chemist to stop adding solution and to take the volume reading, the indicator must be so chosen that its color change is somewhere within the pH range spanned by the steeply rising portion of the curve. Figure 18.2a shows that either phenolphthalein or methyl red is suitable for the HCl–NaOH titration, while Fig. 18.2b shows that phenolphthalein, and not methyl red, is suitable for the acetic acid–NaOH titration. Similarly, in a titration involving a strong acid and a weak base, the indicator would have to change color on the acid side.

At the equivalence point, *the solution contains the product* of neutralization of the acid and base, but, by the definition of the equivalence point, it contains *no excess of either the acid or the base used in the titration.*

18.11 ▫ SOLUBILITY PRODUCT

Silver chloride is an ionic salt. It is only slightly soluble in water, but, like other "insoluble" salts, it does have a measurable solubility. The dissolving process may be represented by the equation

$$AgCl(c) \rightleftharpoons Ag^+ + Cl^-$$

The equilibrium condition for this reaction is

$$[Ag^+][Cl^-] = K$$

The solid is omitted from the reaction quotient (p. 252). K is called the **solubility product** or **solubility product constant** of AgCl. It is also represented by K_{sp}. When this equilibrium is established, the solution is saturated, and the solubility product thus gives the product of the Ag^+ and Cl^- concentrations in a saturated solution. The individual con-

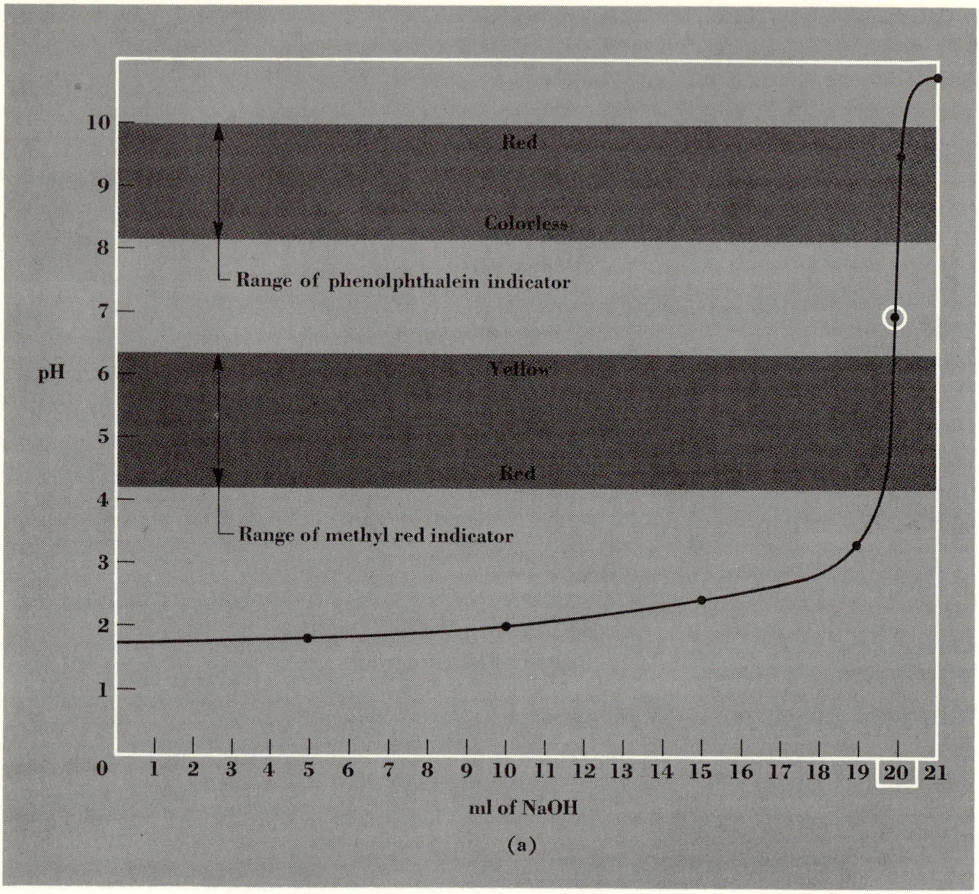

Fig. 18.2 Titration curve; pH values plotted against the corresponding volume added from a buret. (a) 20.00 ml of 0.1000 M HCl diluted to 100.0 ml and titrated with 0.1000 M NaOH; (b) 20.00 ml of 0.1000 M HOAc (acetic acid) diluted to 100.0 ml and titrated with 0.1000 M NaOH. Temperature is 25°C. Also shown is the pH range of the indicator, the range of pH values through which visual changes in color occur. The equivalence point is circled.

centrations may be changed—by adding NaCl or $AgNO_3$, for instance—but *the product, at any one temperature, remains constant.* The solubility product principle applies only to slightly soluble salts. Solubility products of some slightly soluble salts are given in Table 18.3.

Solubility and *solubility product* must not be confused with each other. The solubility of a salt is the quantity present in a unit amount of a saturated solution, expressed in moles per liter, grams per 100 ml, or other units. The solubility depends on what else is in the solution. The solubility *product*, being an equilibrium constant, depends only on temperature. There is a connection, of course, between the solubility and the solubility product; if one is known, the other can be calculated.

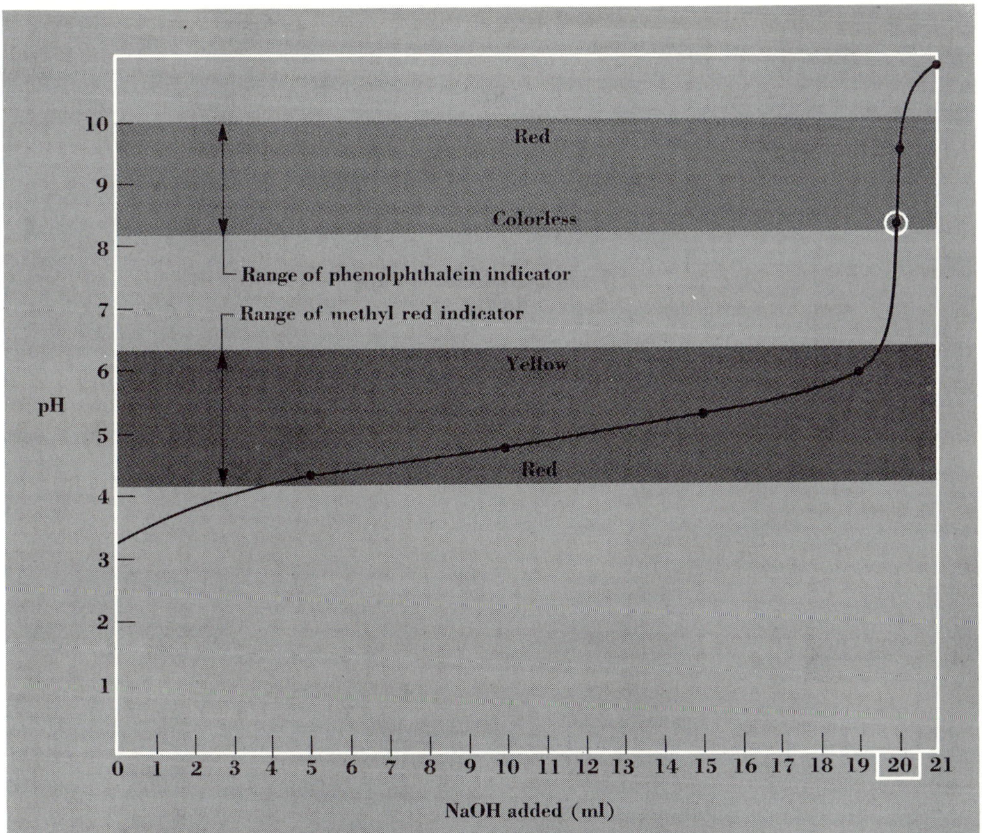

(b)

EXAMPLE 14 Find the solubility (in moles/liter) of AgCl at 25°C in (a) pure water; (b) a 0.0100 M solution of NaCl.

ANSWER (a) The solubility is equal to the concentration of Ag^+ or Cl^-; in the absence of other ions, these two concentrations must be equal. Let s be the solubility in moles/liter;

$$\underset{}{AgCl(c)} \rightleftharpoons \underset{s}{Ag^+} + \underset{s}{Cl^-}$$

$$[Ag^+][Cl^-] = K$$
$$s \cdot s = 1.73 \times 10^{-10}$$
$$s = 1.32 \times 10^{-5} \text{ mole/liter at equilibrium}$$

(b) The number of moles of AgCl that dissolve per liter (the new solubility, s') is equal to the number of moles of Ag^+ in the solution; $s' =$

TABLE 18.3 SOLUBILITY PRODUCTS IN WATER

Salt	Temperature (°C)	K
AgBr	25	7.7×10^{-13}
AgCl	25	1.73×10^{-10}
Ag_2CrO_4	25	1.15×10^{-12}
AgI	25	1.5×10^{-16}
Ag_2S	25	1.9×10^{-49}
$Al(OH)_3$	25	3.7×10^{-15}
$BaCO_3$	25	8.1×10^{-9}
$BaCrO_4$	18	1.6×10^{-10}
BaF_2	25.8	1.73×10^{-6}
$BaSO_4$	25	1.08×10^{-10}
$CaCO_3$	25	8.7×10^{-9}
CaF_2	26	3.95×10^{-11}
CdS	18	3.6×10^{-29}
CoS	18	3×10^{-26}
CuS	18	8.5×10^{-45}
FeS	25	1.3×10^{-17}
Hg_2Cl_2[a]	25	2×10^{-18}
HgS	25	2.4×10^{-52}
MgC_2O_4	18	8.57×10^{-5}
MnS	25	7.9×10^{-13}
NiS	18	1.4×10^{-24}
$PbCrO_4$	18	1.77×10^{-14}
PbI_2	25	1.39×10^{-8}
PbS	25	2.3×10^{-27}
$PbSO_4$	18	1.06×10^{-8}
ZnS	25	2.5×10^{-24}

[a] $Hg_2Cl_2(c) \rightleftharpoons Hg_2^{2+} + 2Cl^-$.

$[Ag^+]$ and $[Cl^-] = 0.0100 + s'$. Then, substituting into the equilibrium condition,

$$[Ag^+][Cl^-] = K$$
$$s'(0.0100 + s') = K$$

It is a very good approximation that $s' \ll 0.01$; practically all the Cl^- is contributed by the NaCl. Then

$$0.01s' = K = 1.73 \times 10^{-10}$$
$$s' = 1.73 \times 10^{-8} \text{ mole/liter}$$

We observe that the solubility (s') is much less in the NaCl solution than in pure water (s). This result could have been predicted from Le Chatelier's principle: the equilibrium

$$AgCl(c) \rightleftharpoons Ag^+ + Cl^-$$

is shifted to the left by addition of Cl^- or Ag^+; that is, the solubility of AgCl is decreased. This effect—like the effect of a strong acid in repressing the ionization of a weak acid—is known as the **common ion effect,** because the decrease in solubility is caused by addition of a salt having an ion in common with the slightly soluble salt.

When the salt dissolves to give unequal numbers of positive and negative ions, each concentration must be raised to a power equal to the coefficient of that ion in the equation. For example:

Salt	Reaction	Equilibrium condition
CaF_2	$CaF_2(c) \rightleftharpoons Ca^{2+} + 2F^-$	$[Ca^{2+}][F^-]^2 = K$
Hg_2Cl_2	$Hg_2Cl_2(c) \rightleftharpoons Hg_2^{2+} + 2Cl^-$ (not $2Hg^+$)	$[Hg_2^{2+}][Cl^-]^2 = K$
$Al(OH)_3$	$Al(OH)_3(c) \rightleftharpoons Al^{3+} + 3OH^-$	$[Al^{3+}][OH^-]^3 = K$
$Ca_3(PO_4)_2$	$Ca_3(PO_4)_2(c) \rightleftharpoons 3Ca^{2+} + 2PO_4^{3-}$	$[Ca^{2+}]^3[PO_4^{3-}]^2 = K$

EXAMPLE 15 A solution is saturated with CaF_2. Calculate (a) $[Ca^{2+}]$ when no salt other than CaF_2 is present; (b) $[Ca^{2+}]$ when the solution is 0.050 M with respect to MgF_2; (c) $[F^-]$ when the solution is 0.20 M with respect to $CaCl_2$.

ANSWER

$$CaF_2(c) \rightleftharpoons Ca^{2+} + 2F^- \qquad (21)$$

(a) Let $x = [Ca^{2+}]$. Then, since according to Equation 21 there are two F^- ions for each Ca^{2+} in the solution,

$$\begin{aligned}
[F^-] &= 2[Ca^{2+}] = 2x \\
[Ca^{2+}][F^-]^2 &= K \\
x(2x)^2 &= 3.95 \times 10^{-11} \\
4x^3 &= 3.95 \times 10^{-11} \\
x^3 &= 9.88 \times 10^{-12} \\
x &= 2.15 \times 10^{-4} \text{ mole/liter}
\end{aligned}$$

(b) Again, we let $x = [Ca^{2+}]$. We may assume that most of the F^- in the solution is contributed by the MgF_2, and very little by the dissolving of CaF_2. Since 0.050 M MgF_2 gives 0.10 mole/liter of F^-

$$\begin{aligned}
[F^-] &\approx 0.10 \text{ mole/liter} \\
[Ca^{2+}][F^-]^2 &= K \\
x(0.10)^2 &= 3.95 \times 10^{-11} \\
x &= 3.95 \times 10^{-9} \text{ mole/liter} = [Ca^{2+}]
\end{aligned}$$

(c) Let $y = [F^-]$. Most of the Ca^{2+} is contributed by the $CaCl_2$: $[Ca^{2+}] \approx$ 0.20 mole/liter. Then

$$[Ca^{2+}][F^-]^2 = 3.95 \times 10^{-11}$$
$$(0.20)(y)^2 = 3.95 \times 10^{-11}$$
$$y^2 = 1.98 \times 10^{-10}$$
$$y = 1.4 \times 10^{-5} \text{ mole/liter} = [F^-]$$

EXAMPLE 16 The solubility of thallium(I) iodide, TlI, in water is 6.4×10^{-3} g/100 ml at 20°C. Find its solubility product.

ANSWER In order to calculate the solubility product, we must first express the solubility in terms of molarity (moles per liter). The molecular weight of TlI is 331.3 g/mole; the molarity of the saturated solution is

$$6.4 \times 10^{-3} \frac{\text{g}}{100 \text{ ml}} \times \frac{1000 \text{ ml}}{\text{liter}} \times \frac{1 \text{ mole}}{331.3 \text{ g}} = 1.93 \times 10^{-4} \frac{\text{mole}}{\text{liter}}$$
$$= [Tl^+] = [I^-]$$

The solubility product, for the reaction $TlI(c) \rightleftharpoons Tl^+ + I^-$, is

$$K = [Tl^+][I^-]$$
$$= (1.93 \times 10^{-4})^2$$
$$= 3.72 \times 10^{-8}$$

In an unsaturated solution of, say, TlI, there is no equilibrium between solid and ions, and we therefore cannot expect to have $[Tl^+][I^-] = K$. If the solution is unsaturated, $[Tl^+][I^-] < K$. One use of the solubility product is to decide whether precipitation is possible when the ions have certain concentrations. We calculate the appropriate product of concentrations (Q, the reaction quotient) and compare it with K. If $Q > K$, the solution is supersaturated, and precipitation is possible (but it may not occur if the solution remains supersaturated); if $Q < K$, the solution is unsaturated, and precipitation is impossible.

EXAMPLE 17 A solution is prepared by mixing equal volumes of 0.010 M $MgCl_2$ and 0.020 M $Na_2C_2O_4$ solutions at 18°C. Is it possible for MgC_2O_4 to precipitate in the resulting solution?

ANSWER The reaction in question is

$$MgC_2O_4(c) \rightleftharpoons Mg^{2+} + C_2O_4^{2-}$$

We ask whether the concentrations of the two ions are large enough for equilibrium to be established in this reaction.

When the solutions are mixed, the volume is doubled (to a very good approximation, though volumes are not strictly additive on mixing), and the concentrations of Mg^{2+} and $C_2O_4^{2-}$ are half what they were originally: $[Mg^{2+}] = 0.005$, $[C_2O_4^{2-}] = 0.010$ mole/liter. Then

$$Q = [Mg^{2+}][C_2O_4^{2-}] = 5 \times 10^{-3} \times 10^{-2} = 5 \times 10^{-5}$$

which is less than $K = 8.57 \times 10^{-5}$ (Table 18.3). It is impossible for MgC_2O_4 to precipitate.

EXAMPLE 18 What is the minimum concentration of S^{2-} which can cause precipitation of CdS from a 0.10 *M* solution of Cd^{2+} at 18°C?

ANSWER The minimum concentration is that corresponding to a saturated solution (or, more precisely, to a very slightly supersaturated solution). In a saturated solution,

$$CdS(c) \rightleftharpoons Cd^{2+} + S^{2-}$$

$$[Cd^{2+}][S^{2-}] = 3.6 \times 10^{-29}$$

$$[S^{2-}] = \frac{3.6 \times 10^{-29}}{0.10} = 3.6 \times 10^{-28} \text{ mole/liter}$$

18.12 ▫ EFFECT OF pH ON SOLUBILITY

When an acid is weak, the concentration of its anion is affected by the pH of the solution. For example, the reaction

$$HS^- + H_2O \rightleftharpoons H_3O^+ + S^{2-}$$

is shifted to the left by an increase in $[H_3O^+]$. In turn, the solubility of a sulfide is affected by the concentration of S^{2-}. When acid is added to a saturated solution of ZnS, the concentration of S^{2-} is decreased, and the reaction

$$ZnS(c) \rightleftharpoons Zn^{2+} + S^{2-}$$

is shifted to the right, meaning that the solubility of ZnS increases. In general, *the solubility of a salt of a weak acid is increased by addition of a strong acid to the solution.*

The effect is especially pronounced with sulfides, which are salts of the very weak acid HS^-, and this case is of interest because of its importance in qualitative analysis. If any two of the three concentrations $[H_3O^+]$, $[S^{2-}]$, and $[H_2S]$ are given, the third can be calculated from the equation (Example 12, p. 340),

$$\frac{[H_3O^+]^2[S^{2-}]}{[H_2S]} = K_1K_2$$

$$= 6.3 \times 10^{-8} \times 10^{-14} = 6.3 \times 10^{-22}$$

In a solution of H_2S, saturated at 1 atm pressure and 25°C, $[H_2S] = 0.10$ mole/liter; under these conditions,

$$[H_3O^+]^2[S^{2-}] = 6.3 \times 10^{-23}$$

EXAMPLE 19 Find $[Zn^{2+}]$ in a saturated solution of ZnS in which $[H_3O^+] = 0.30$ and $[H_2S] = 0.10$ mole/liter at 25°C. The K for ZnS is 2.5×10^{-21}.

ANSWER First we must calculate $[S^{2-}]$:

$$[S^{2-}] = \frac{6.3 \times 10^{-23}}{[H_3O^+]^2} = \frac{6.3 \times 10^{-23}}{(0.30)^2} = 7.0 \times 10^{-22} \text{ mole/liter}$$

Then

$$[Zn^{2+}][S^{2-}] = 2.5 \times 10^{-24}$$

$$[Zn^{2+}] = \frac{2.5 \times 10^{-24}}{7.0 \times 10^{-22}} = 3.6 \times 10^{-3} \text{ mole/liter}$$

It should, perhaps, be pointed out again that a given solution has only one concentration of (for example) S^{2-}. The value of $[S^{2-}]$ must be the same in the H_2S equilibrium condition and in the ZnS equilibrium condition.

18.13 ▫ COMPLEX IONS

A **complex ion** may be defined as an ion composed of several parts, each of which has some independent existence in solution as an ion, molecule, or atom. We would not count SO_4^{2-} as a complex ion, for S^{6+} and O^{2-}, or other such pairs, do not exist in aqueous solution. A complex ion, in our sense, consists of a positive metal ion, M^{x+} (usually a transition metal), and a number of **ligands,** L, which are uncharged molecules or negative ions. The formula of the complex ion is ML_n, where n is usually 2, 4, or 6. Formation of a complex ion is a Lewis acid-base reaction (p. 320). The cation, such as Zn^{2+}, is the acid and the ligand, such as NH_3, is the base.

$$\underset{\textit{Lewis acid}}{Zn^{2+}} + \underset{\textit{Lewis base}}{4\ :NH_3} \longrightarrow \underset{\textit{complex ion}}{Zn(NH_3)_4^{2+}}$$

The ligand invariably has at least one lone pair of electrons. Some other examples of complex ions are $Ag(NH_3)_2^+$, $Ag(S_2O_3)_2^{3-}$, $Cu(NH_3)_4^{2+}$, $Fe(CN)_6^{3-}$, $Fe(CN)_6^{4-}$, $Zn(OH)_4^{2-}$.

The dissociation of the complex ion $Ag(NH_3)_2^+$ may be represented as

$$Ag(NH_3)_2^+ \rightleftharpoons Ag^+ + 2NH_3$$

The equilibrium constant for this reaction is called the **dissociation constant** or **instability constant** of $Ag(NH_3)_2^+$,

$$\frac{[Ag^+][NH_3]^2}{[Ag(NH_3)_2^+]} = K$$

EXAMPLE 20 A solution is prepared by dissolving 0.20 mole $AgNO_3$ and 0.50 mole NH_3 in H_2O to make 1.00 liter of solution. Find the concentration of Ag^+ in the solution. The instability constant for $Ag(NH_3)_2^+$ is 6.0×10^{-8}.

ANSWER The Ag^+ will be almost completely converted to 0.20 mole $Ag(NH_3)_2^+$, and will consume in the process 0.40 mole NH_3, leaving an excess of (0.50 − 0.40), or 0.10 mole of NH_3. If x is the $[Ag^+]$ at equilibrium, the process may be represented as

	Ag^+	+	$2NH_3$	$\longrightarrow$	$Ag(NH_3)_2^+$
Initial concentrations	0.20		0.50		0 mole/liter
Moles reacting per liter	$0.20 - x$		$2(0.20 - x)$		$0.20 - x \approx 0.20$ mole/liter
Final concentrations	x		$0.50 - 0.40 + 2x \approx 0.10$		

Then, for the reaction

$$Ag(NH_3)_2^+ \rightleftharpoons Ag^+ + 2NH_3$$

we have the equilibrium condition

$$\frac{[Ag^+][NH_3]^2}{[Ag(NH_3)_2^+]} = 6.0 \times 10^{-8}$$

$$\frac{x(0.10)^2}{0.20} = 6.0 \times 10^{-8}$$

$$x = 1.2 \times 10^{-6} \text{ mole/liter} = [Ag^+]$$

When a complex ion is formed, the concentration of the uncomplexed (actually hydrated) metal ion is greatly diminished. The result is that salts of this metal become more soluble, as equilibria of the type $AgCl(c) \rightleftharpoons Ag^+ + Cl^-$ are shifted to the right.

EXAMPLE 21 Calculate the quantity of NH_3 that must be added to dissolve 0.010 mole AgCl in 1.00 liter of water. K for AgCl is 1.73×10^{-10}, K for $Ag(NH_3)_2^+$ is 6.0×10^{-8}.

ANSWER In the final solution, $[Cl^-] = 0.010$ mole/liter, and $[Ag(NH_3)_2^+] = 0.010$ mole/liter very nearly. In effect, the reaction is

$$AgCl(c) + 2NH_3 \longrightarrow Ag(NH_3)_2^+ + Cl^-$$

or, for the numbers of moles in one liter of this solution,

$$0.010AgCl + 0.020NH_3 \longrightarrow 0.010Ag(NH_3)_2^+ + 0.010Cl^-$$

First, we must find the maximum permissible concentration of Ag^+:

$$[Ag^+][Cl^-] = 1.73 \times 10^{-10}$$

$$[Ag^+] = \frac{1.73 \times 10^{-10}}{0.010} = 1.73 \times 10^{-8}$$

Then the equation for the instability constant of $[Ag(NH_3)_2^+]$ may be solved for $[NH_3]$:

$$\frac{[Ag^+][NH_3]^2}{[Ag(NH_3)_2{}^+]} = 6.0 \times 10^{-8}$$

$$[NH_3]^2 = \frac{6.0 \times 10^{-8} \times 0.010}{1.73 \times 10^{-8}}$$

$$= 3.47 \times 10^{-2}$$

$$[NH_3] = 0.19 \text{ mole/liter}$$

The total quantity of NH_3 that must be added is 0.020 (to form 0.010 mole of complex ion) + 0.19 (remaining in solution) = 0.21 mole.

PROBLEMS

When the temperature is not given, take it to be the temperature for which the necessary data are tabulated, most often 25°C.

1. pH and pOH Find the pH and pOH of the following solutions and state whether the solution is acidic, basic, or neutral. (a) 0.0010 M HCl; (b) 0.0010 M NaOH; (c) 0.0050 M $Ca(OH)_2$; (d) 3.2×10^{-4} M HCl; (e) 1.0×10^{-12} M HCl; (f) 1.00 M H_3O^+; (g) 10^{10} M H_3O^+; (h) A solution made by mixing 500 ml of 2.0×10^{-2} M HCl and 500 ml of 2.0×10^{-5} M NaOH.

2. Weak acids Find (a) the degree of ionization, and (b) the pH of a 0.60 M solution of nitrous acid, HONO.

3. Weak acids An aqueous solution of a certain monoprotic acid is 0.10 M and has pH 3.00. Find K_a and pK_a for this acid.

4. Weak bases Calculate the pH and pOH of a 0.50 M solution of dimethylamine, $(CH_3)_2NH$, in H_2O.

5. Conjugate bases Find the basic ionization constant, K_b, of the phthalate ion, $C_6H_4(COO^-)_2$.

6. Mixture of acids A solution is 0.50 M with respect to HCl and 1.00 M with respect to HF. Calculate the concentration of F^- in this solution.

7. Titration At the end point of the following titrations, will the resulting solution be acidic, basic, neutral, or unpredictable? Answer without doing calculations. (a) HCl versus NaOH; (b) NH_3 versus HCl; (c) CH_3COOH versus NaOH; (d) NH_3 versus CH_3COOH; (e) HCOOH versus $CH_3CH_2NH_2$.

8. Buffer solutions (a) Find the pH of a solution 1.00 M with respect to NH_3 and 0.80 M with respect to NH_4Cl. (b) Find the pH of this solution after 0.10 mole HCl per liter has been added to it. (c) A solution of pH 9.34 is prepared by adding NaOH to pure water. Find the pH of this solution after 0.10 mole HCl per liter has been added to it.

9. Indicators Select an indicator suitable for determining the pH of solutions with pH around 7.

10. Solubility product The solubility of Li_2CO_3 in water is 1.33 g per 100 ml solution at 20°C. Calculate its solubility product.

11. Solubility from solubility product Calculate the concentration of CrO_4^{2-} in a solution saturated with Ag_2CrO_4 and containing (a) no other solute, (b) 0.50 mole $AgNO_3$ per liter.

12. Precipitation A 0.0020 M solution of $AgNO_3$ and a 0.0010 M solution of NaCl are mixed in equal volumes at 25°C. Is it possible for AgCl to precipitate?

13. Solubility and pH Find the solubility, in moles/liter, of CdS in a solution in which $[H_2S] = 0.10$ mole/liter and the pH is (a) 2.0, (b) 9.0.

14. Complex ions A solution contains 0.1 M $Cd(CN)_4^{2-}$ and 0.1 M CN^-. K for $Cd(CN)_4^{2-} = 10^{-17}$. Find $[Cd^{2+}]$.

ADDITIONAL PROBLEMS

15. Weak acids Calculate the pH and pOH of a 0.080 M solution of benzoic acid.

16. Weak acids Find the pH of a 0.50 M solution of phenol.

17. Weak acids Calculate (a) the degree of ionization and (b) the pH of chloroacetic acid in a 0.010 M solution. (Caution: Some of the usual approximations may not be valid here.)

18. Polyprotic acids Calculate (a) the pH and (b) the concentration of SO_3^{2-} in a 0.20 M solution of $(HO)_2SO$. (Be careful what approximations you make.)

19. Polyprotic acids Calculate (a) $[H_3O^+]$, (b) $[(HO)_2PO_2^-]$, (c) $[(HO)PO_3^{2-}]$, and (d) $[PO_4^{3-}]$ in a 0.20 M solution of phosphoric acid, $(HO)_3PO$ (or H_3PO_4).

20. Mixture of acids Find the degree of ionization of 0.30 M propionic acid in a solution of pH 2.50.

21. Buffer solutions Find the pH of a solution 0.50 M with respect to formic acid and 0.20 M with respect to sodium formate.

22. Buffer solutions A buffer solution of pH 5.00 is to be prepared from propionic acid and sodium propionate. The concentration of sodium propionate must be 1.00 mole/liter. What should be the concentration of the acid?

23. Indicators A certain indicator has $pK_a = 6.0$. Its acid form is yellow and its base form is blue. What will be the color of the indicator in a solution of pH (a) 4, (b) 6, (c) 6.5, (d) 9?

24. Solubility from solubility product Find the solubility of $PbSO_4$ in H_2O, in moles/liter and grams/100 ml.

25. Solubility product The solubility of CdC_2O_4 in water is 3.37×10^{-3} g/100 ml at 0°C. Calculate (a) its solubility product; (b) its solubility, in grams/100 ml, in 0.05 M $Cd(NO_3)_2$ solution.

26. Solubility from solubility product Calculate the concentration of Ag^+ in a solution saturated with AgI and 0.15 M with respect to CaI_2.

27. Precipitation 50 ml of a 0.0100 M solution of $BaCl_2$ and 20 ml of a 0.0100 M solution of NaF were mixed at 25.8°C. (a) Find $[Ba^{2+}]$ and $[F^-]$ in the mixed solution before any reaction occurs. (b) Is it possible for BaF_2 to precipitate?

28. Solubility and pH Find the minimum pH at which FeS will precipitate from a solution 0.10 M with respect to H_2S and 0.020 M with respect to Fe^{2+}.

29. Solubility and pH Find the minimum pH at which PbS will precipitate from a solution 0.10 M with respect to H_2S and 0.020 M with respect to Pb^{2+}.

30. Solubility and buffered solutions (a) Mixed to make one liter of solution are

0.1 mole Fe^{3+}, 0.1 mole NH_4^+, and 0.1 mole NH_3. Will $Fe(OH)_3$ precipitate? K_b for $NH_3 \approx 10^{-5}$, and K for $Fe(OH)_3 = 10^{-23}$. (b) In part (a), substitute 0.1 mole Mg^{2+} for 0.1 mole Fe^{3+}. K for $Mg(OH)_2 = 5.5 \times 10^{-11}$. Will $Mg(OH)_2$ precipitate?

31. Conservation and electrical neutrality Find $[H_3O^+]$ in (a) 1.00×10^{-7} M HCl; (b) 1.00×10^{-6} M phenol.

32. Complex ions and solubility What must be the concentration of NH_3 to prevent precipitation of AgBr from a solution 0.020 M with respect to Br^- and 0.040 M with respect to $Ag(NH_3)_2^+$? K for $Ag(NH_3)_2^+ = 6.0 \times 10^{-8}$.

33. Complex ions The instability constant of $HgCl_4^{2-}$ is 1.10×10^{-16}. Find the concentration of Hg^{2+} remaining in a 0.10 M solution of $Hg(NO_3)_2$ after 1.0 mole NaCl per liter has been dissolved in the solution.

ANSWERS

2. (a) 0.027; (b) 1.8.

5. 3.3×10^{-9}.

8. (a) 9.34; (b) 9.24; (c) 1.0.

11. (a) 6.6×10^{-5}; (b) 4.6×10^{-12} mole/liter.

13. (a) 5.7×10^{-11} mole/liter.

17. (a) 0.31; (b) 2.51.

18. (a) 1.30; (b) 6.24×10^{-8}.

22. 0.75 mole/liter.

25. (a) 2.83×10^{-8}; (b) 1.13×10^{-5} g/100 ml.

27. (a) $7\ 14 \times 10^{-3}$, 2.86×10^{-3} mole/liter.

28. 3.5.

30. (a) yes; (b) no.

33. 8.5×10^{-17} mole/liter.

19 ■ CHEMICAL KINETICS

19.1 □ INTRODUCTION

A chemical equation expresses the stoichiometric relations between reactants and products but tells us nothing about the rate of the reaction. The rate of a reaction is the change in concentration of a reactant per unit time or the number of moles of a reactant converted to products per unit time. Thus, although we may write

$$C(c) + O_2(g) \longrightarrow CO_2(g) \qquad \Delta H = -94.1 \text{ kcal}$$

exposing lumps of coal* or diamond to air produces no measurable reaction. At ordinary temperatures, the rate of this reaction is immeasurably slow. Explosions, on the other hand, appear to be immeasurably fast. Many reactions proceed at rates neither extremely slow nor extremely rapid.

An example of a reaction that proceeds at a conveniently measurable rate is the decomposition of dinitrogen pentoxide dissolved in carbon tetrachloride, the data for which are plotted in Fig. 19.1.

$$2N_2O_5(sol)\dagger \longrightarrow 4NO_2(sol) + O_2(g)$$

During the course of the reaction, the concentration of the reactant decreases systematically.

A balanced equation such as

$$2NO(g) + Br_2(g) \longrightarrow 2NOBr(g) \qquad (1)$$

tells us nothing about the **mechanism of the reaction,** the actual series

* Coal is composed largely of amorphous carbon and carbon compounds.

† Remainder: *sol* means *in solution.*

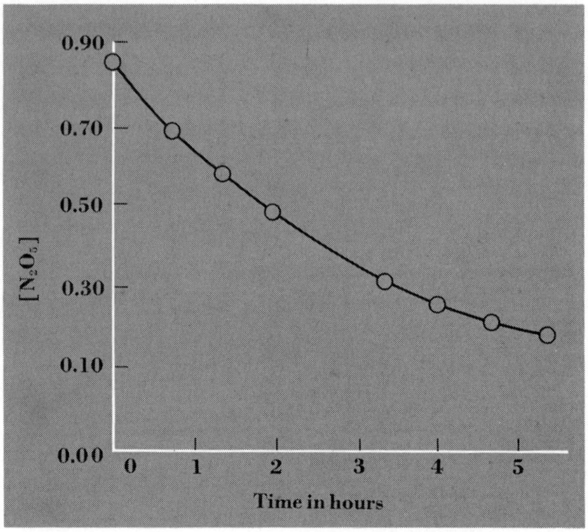

Fig. 19.1 Plot of the reaction time and the corresponding concentration of dinitrogen pentoxide.

of individual steps by which the reactants change to products. Does molecular bromine really combine simultaneously with 2 molecules of NO,

$$NO + NO + Br_2 \longrightarrow 2NOBr$$

or does one NO molecule combine with one Br_2 molecule in one step,

$$NO + Br_2 \longrightarrow NOBr_2 \qquad (2)$$

followed by a second step?

$$NOBr_2 + NO \longrightarrow 2NOBr \qquad (3)$$

Some chemical reactions do occur in a single step but, more frequently. the reaction occurs in a sequence of steps (p. 371). The addition of the successive steps 2 and 3 yields the overall reaction 1. The study of the rate and the mechanism of chemical reactions is known as **chemical kinetics.**

The combustion of coal is a typical **heterogeneous reaction,** a reaction occurring only at the interface (boundary) between two phases. The reaction between NO(*g*) and Br_2(*g*) is a typical **homogeneous reaction,** a reaction occurring in only one phase.

19.2 □ CONDITIONS AFFECTING REACTION RATES

Reactions occur at a definite characteristic rate when certain variables are fixed. These variables are (1) nature of the reactants, (2) concentration of the reactants, (3) particle size of the solid or liquid in a heterogeneous reaction, (4) temperature, and (5) presence of a catalyst.

(1) Nature of reactants Copper or silver reacts very slowly with oxygen even in a flame, while magnesium under the same conditions burns very rapidly; white phosphorus ignites spontaneously in air, whereas red phosphorus does not. When a solution containing silver ions, Ag^+, is added to a solution containing chloride ions, Cl^-, insoluble silver chloride forms very rapidly,

$$Ag^+ + Cl^- \longrightarrow AgCl(c)$$

On the other hand, the addition of magnesium ions, Mg^{2+}, to oxalate ions, $C_2O_4{}^{2-}$, produces insoluble magnesium oxalate very slowly,

$$Mg^{2+} + C_2O_4{}^{2-} \longrightarrow MgC_2O_4(c)$$

The first reaction may be completed in microseconds while the second reaction may require 24 hours.

(2) Concentration of reactants In 1864, Cato Guldberg and Peter Waage recognized that at constant temperature the rate of a homogeneous reaction is usually proportional to some power of the concentrations of the reactants. The concentration is expressed in moles of reactant per liter of the homogeneous phase. This generalization, the **law of mass action,** simply states that for the general reaction

$$a\text{A} + b\text{B} \longrightarrow c\text{C} + d\text{D}$$

the rate of formation of C or D, or the rate of disappearance of A or B, is proportional to powers of the concentrations of A and B,

$$rate \propto [\text{A}]^x[\text{B}]^y$$

in which $\propto$ denotes a proportionality. We may also write

$$rate = k[\text{A}]^x[\text{B}]^y$$

where k is a constant. The numerical value of the exponents x and y must be determined experimentally. *They cannot be deduced* from the overall reaction or from any principle or hypothesis, and they *need not necessarily be the same as a and b.* Either of these exponents could even have the value of zero. If, for example $x = 0$ the rate equation is $rate = k[\text{B}]^y$. The proportionality constant k is known as the **specific rate constant** of the reaction. The unit of k depends upon the units used for the concentration terms and the unit of time.

The sum of the powers of the concentration terms in the rate equation is known as **the order of the reaction.** For example, the reaction

$$2NO + O_2 \longrightarrow 2NO_2$$

follows the rate equation

$$rate = k[\text{NO}]^2[\text{O}_2]$$

The reaction is therefore a third-order reaction, since the sum of the exponents for the two concentration terms is 3. The reaction is said to be second order with respect to NO and first order with respect to O_2.

For this reaction the rate can be expressed in terms of disappearance of NO or O_2, or in terms of appearance of NO_2. According to the equation for the reaction, 2 moles of NO_2 are formed for each 2 moles of NO and 1 mole of O_2 consumed. The rate of consumption of NO is equal to the rate of appearance of NO_2, but is twice the rate of consumption of O_2. To summarize,

$$\text{rate of consumption of NO} = 2 \times \text{rate of consumption of } O_2 = \text{rate of production of } NO_2$$

For this reason it is necessary to specify with respect to which substance the rate is being expressed.

An example of the lack of correlation between the exponents in the rate equation and the coefficients in the chemical equation is the reaction in solution between ethyl alcohol and decaborane forming triethyl borate and hydrogen,

$$30C_2H_5OH + B_{10}H_{14} \longrightarrow 10B(OC_2H_5)_3 + 22H_2(g)$$

for which the experimentally determined rate equation is

$$rate = k[C_2H_5OH][B_{10}H_{14}]$$

The exponent for each of the two concentration terms is 1; this reaction is therefore said to be a second-order reaction, first order with respect to C_2H_5OH and first order with respect to $B_{10}H_{14}$.

EXAMPLE 1 For the decomposition of dinitrogen pentoxide, N_2O_5, dissolved in carbon tetrachloride,

$$2N_2O_5 \longrightarrow 4NO_2 + O_2$$

the following data at 30°C are given:

Reactant concentration $[N_2O_5]$	Rate of decomposition of N_2O_5
$0.170 \frac{\text{mole}}{\text{liter}}$	$0.050 \frac{\text{mole}}{\text{liter hr}}$
0.340	0.10
0.680	0.20

(a) Write the rate equation for the reaction. What is the order of the reaction?
(b) Calculate the rate constant for the reaction at 30°C.
(c) Calculate the decomposition rate at 30°C at the instant when $[N_2O_5] = 0.540$.

ANSWER (a) It is seen that doubling the concentration of the N_2O_5 doubles the rate; therefore, the rate of this reaction is proportional to the N_2O_5 concentration, or

$$rate = k[N_2O_5]$$

Since the rate is proportional to the first power of one reactant, the reaction is first order.
(b) Solving for k,

$$k = \frac{rate}{[N_2O_5]} = \frac{0.050 \dfrac{\cancel{\text{mole}}}{\cancel{\text{liter}}\ \text{hr}}}{0.170 \dfrac{\cancel{\text{mole}}}{\cancel{\text{liter}}}} = \frac{0.29}{\text{hr}}$$

and k is thus 0.29 per hour.
(c) From the previous calculation

$$rate = \frac{0.29}{\text{hr}} [N_2O_5]$$

so that when $[N_2O_5] = 0.540$

$$rate = \frac{0.29}{\text{hr}} \times 0.540 \frac{\text{mole}}{\text{liter}} = 0.16 \frac{\text{mole}}{\text{liter hr}}$$

EXAMPLE 2 The reaction

$$F_2 + 2ClO_2 \longrightarrow 2FClO_2$$

has been studied in the gaseous state and the data at 250°K may be summarized as

Rate of disappearance of F_2 (mole/liter sec)	Reactant concentration (mole/liter) $[F_2]$	$[ClO_2]$
1.2×10^{-3}	0.10	0.010
4.8×10^{-3}	0.10	0.040
2.4×10^{-3}	0.20	0.010

Calculate the rate constant for the reaction and the rate of formation of $FClO_2$ at the instant when $[F_2] = 0.010$ and $[ClO_2] = 0.020$ at 250°K.

ANSWER First, use the data to determine the rate equation, from which the rate constant is then calculated. Observe that when the concentration of F_2 is constant and the concentration of ClO_2 is quadrupled, the rate is also quadrupled; hence, the rate of the reaction is proportional to $[ClO_2]$. When the concentration of ClO_2 is constant and the concentration of F_2 is doubled, the rate is also doubled; hence the rate is proportional to $[F_2]$. The rate equation is therefore

$$rate = k[\mathrm{F_2}][\mathrm{ClO_2}]$$

from which

$$k = \frac{rate}{[\mathrm{F_2}][\mathrm{ClO_2}]} = \frac{1.2 \times 10^{-3}\ \dfrac{\cancel{\text{mole}}}{\cancel{\text{liter}}\ \text{sec}}}{0.10\ \dfrac{\cancel{\text{mole}}}{\cancel{\text{liter}}} \times 0.010\ \dfrac{\text{mole}}{\text{liter}}} = 1.2\ \frac{\text{liter}}{\text{mole sec}}$$

The rate of disappearance of F_2 is thus given by

$$rate = 1.2\ \frac{\text{liter}}{\text{mole sec}}\ [\mathrm{F_2}][\mathrm{ClO_2}]$$

so that when $[F_2] = 0.010$ and $[ClO_2] = 0.020$, the rate of disappearance of F_2 is

$$rate = 1.2\ \frac{\cancel{\text{liter}}}{\cancel{\text{mole}}\ \text{sec}} \times (0.010)\ \frac{\cancel{\text{mole}}}{\cancel{\text{liter}}} \times (0.020)\ \frac{\text{mole}}{\text{liter}}$$

$$= 2.4 \times 10^{-4}\ \frac{\text{mole}}{\text{liter sec}}$$

However, *two moles* of $FClO_2$ form for *each mole* of F_2 that disappears; hence the rate of formation of $FClO_2$ is

$$2 \times (2.4 \times 10^{-4})\ \frac{\text{mole}}{\text{liter sec}} = 4.8 \times 10^{-4}\ \frac{\text{mole}}{\text{liter sec}}$$

(3) Particle size in heterogeneous reactions Since heterogeneous reactions occur only at the surface boundary between the reacting phases, the rate of such reactions is proportional to the surface area. When a given mass is subdivided into smaller particles, the surface area is increased; hence, as the particle size is decreased, the rate of reaction increases. Lumps of coal or zinc, for example, are difficult to ignite in air, but pulverized and dispersed in air, they react explosively.

SAFETY NOTE: DISASTROUS COAL MINE EXPLOSIONS DO NOT RESULT ORDINARILY FROM IGNITION OF A GAS. BITUMINOUS ("SOFT") COAL DUST SUSPENDED IN AIR MAY BE IGNITED BY A MATCH, PRODUCING A VIOLENT EXPLOSION, THE RATE OF PRESSURE RISE REACHING AS MUCH AS 40 ATM/SEC. THE PRESSURE INCREASE RESULTS FROM THE RAPID EXPANSION OF THE GASEOUS PRODUCTS OF THE REACTION AND AIR, CAUSED BY THE HEAT EVOLVED FROM THE COMBUSTION REACTION. IN GENERAL, YOU MUST REGARD COMBUSTIBLE DUSTS SUSPENDED IN AIR AS EXPLOSION HAZARDS.

(4) Effect of temperature Temperature has a striking effect on the rate of chemical reactions. Reaction rates negligibly slow at ordinary temperatures may become appreciable and even explosive at elevated temperatures. As a very rough but useful rule, the rate constant is doubled for a rise in temperature of 10 Celsius degrees.

(5) Catalysis Life as we know it would be impossible without the phenomenon of catalysis. Almost everyone has heard of **enzymes;** these are natural substances (Sect. 26.3) that catalyze biochemically important reactions. Many industrial processes would be impossible without catalysts. A **catalyst** is a substance that increases the rate of a reaction and is recovered chemically unchanged at the end of the reaction. Although the catalyst is not consumed, it is universally agreed that the catalyst enters into the chemical reaction but is subsequently regenerated. Typical is the catalytic effect of nitrogen oxide on the rate of conversion of sulfur dioxide to sulfur trioxide. The reaction

$$2SO_2 + O_2 \xrightarrow{slow} 2SO_3$$

is slow. However, nitrogen oxide and oxygen react rapidly and the product of this reaction also reacts rapidly with SO_2,

$$2NO + O_2 \xrightarrow{fast} 2NO_2 \tag{4}$$

$$2NO_2 + 2SO_2 \xrightarrow{fast} 2NO + 2SO_3 \tag{5}$$

Addition of reactions 1 and 2 yields

$$2SO_2 + O_2 \xrightarrow{fast} 2SO_3$$

It is evident that we have substituted two fast reactions for a slow one to yield the same overall chemical reaction. In the so-called lead-chamber process for the manufacture of sulfuric acid, these oxides of nitrogen are used to catalyze the oxidation of sulfur dioxide by oxygen. The *catalyst increases the rate by changing the mechanism of the reaction.*

Substances that retard chemical reactions are known as **negative catalysts,** or **inhibitors,** or **poisons.** Physiological poisons like mercuric chloride and rattlesnake venom react with enzymes, rendering them useless for essential biochemical reactions.

19.3 □ THEORY OF REACTION RATES

(a) Collision and rate Molecules must "collide" to react. The rate of reaction is proportional to the number of colliding molecules,

$$rate \propto \text{number of colliding molecules per liter per second}$$

The requirement for collisions accounts for the dependence of the rate on a product (and not a sum) of concentration terms. Let us assume that molecule A combines directly with a molecule of B to form AB ($A + B \longrightarrow AB$). Visualize four molecules each of A and B in a box. In how many ways can the collision A + B occur? One A molecule can collide with any of four B molecules; hence it has four opportunities for collision. The same is true for each of the four A molecules, making 16 possibilities in all. Now if there were only two molecules each of A and B in the box, there would be only 2 × 2 or 4 possible collisions. Therefore, assuming the reaction rate is proportional to the number of collisions, the reaction in the box with four molecules each will be 16/4 or 4 times as rapid as in the box with two molecules each. Thus the rate is proportional to the *product* of the concentrations:

$$rate \propto [\mathrm{A}][\mathrm{B}]$$

The number of collisions calculated from the kinetic theory of gases is enormous, of the order of magnitude of 10^{32} collisions per liter per second at standard conditions; this value does not vary greatly for different gases. Hence, if "collision" were the only requirement for reaction, all gaseous reactions should proceed at practically the same explosive rate. However, for the same reactant concentration at the same temperature, 300°C, the decomposition rate of gaseous hydrogen iodide is 4.4×10^{-3} mole/liter hr, while the decomposition rate of gaseous dinitrogen pentoxide is 9.4×10^{5} moles/liter hr. Clearly, collisions between molecules cannot be the only factor involved in determining the rate of a reaction.

(b) Energy of activation Chemical reactions involve redistribution of atoms, but the redistribution of atoms requires breaking of the bonds holding the atoms together in the original molecules, and the formation of new bonds in the product molecules. For example, for the reaction $2HI \longrightarrow H_2 + I_2$ to occur, each bond holding an H atom to an I atom must be broken, and bonds must form between H atoms and I atoms. We thus postulate that molecules react only if in a collision they possess an energy equal to or greater than a certain critical value. This critical value is in excess of the energy possessed by the average reacting molecules. *The difference between the critical energy needed for reaction to occur and the average energy of the reactant molecules is called* **the energy of activation, E_{act}.**

Collisions between molecules possessing average energy at room temperature generally do not result in products; the colliding molecules bounce away from each other unchanged. Only when the collision is between molecules which together possess at least an energy equal to the average energy plus the energy of activation does reaction occur.

A molecule may have excess energy associated with several kinds of motion: translation (motion through space), rotation, vibration, and electronic excitation. The most important kind of motion for the purpose of chemical reaction is vibration. This motion may be described as the jiggling back and forth of the atoms in the molecule (or ion). For example, in a simple diatomic molecule such as H_2, the H atoms are continuously moving toward and away from each other as if they were attached to the ends of a coiled spring. The distance between

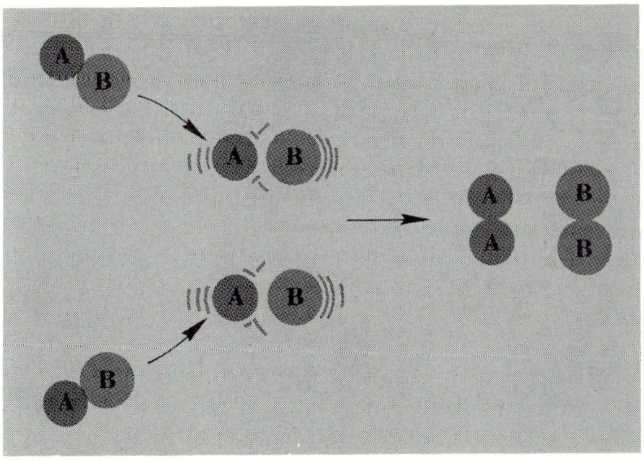

Fig. 19.2 A-B molecules acquire the energy necessary for reaction (E_{act}) upon impact; this energy converted to vibrational energy produces large vibrational amplitudes that weaken the A-B bonds.

atoms is continually changing, so that the so-called "bond distance" is actually an average value. The more the atoms move apart (the greater is the vibrational amplitude), the more easily is the bond broken.

The energy needed for reaction is obtained from the kinetic energy of colliding molecules. Upon impact, the kinetic energy of the colliding molecules can be converted in part to vibrational energy, causing the atoms in the molecule to vibrate more violently and hence the molecules to react more readily. In this way, the reacting molecules acquire the added activation energy necessary for reaction to occur (see Fig. 19.2).

(c) Entropy of activation In addition to having sufficient energy, the molecules must be properly *oriented* when colliding. By analogy, the damage in flight ("reaction") resulting from the collision of two airplanes depends not only on their speeds (energy), but also on their relative positions or orientation. Thus some molecules must be oriented in a very specific manner for reaction to occur; see Fig. 19.3. This factor, the orientation factor, is related to the **entropy of activation.** S_{act}, the entropy of activation, is defined as

$$S_{act} = \text{Entropy of molecules during effective collision} - \text{entropy of ground state.}$$

In general, then, *a chemical reaction results when reactant molecules, with sufficient energy, collide with a proper orientation.*

(d) Reaction-energy diagrams The energy changes of a chemical reaction are usually pictured by a reaction-energy diagram. The ordinate (y axis) represents energy and the abscissa (x axis) can best be thought of as representing the progression of reactants to products. The reaction is proceeding from the reactants on the left to the products on the right. Figure 19.4a is a reaction-energy diagram for a reaction for

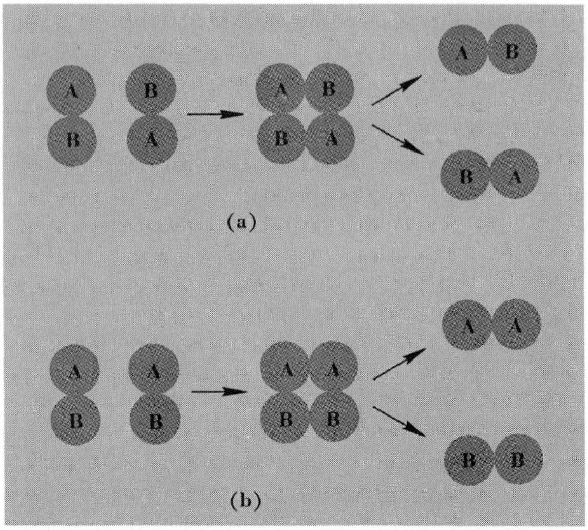

Fig. 19.3 A general representation of the possible influence of orientation of molecules on the rate of the reaction

$$2AB \longrightarrow A_2 + B_2.$$

(a) Orientation that does not lead to reaction; (b) orientation that leads to reaction.

which the products have less energy than the reactants. Of particular importance are points A, B, and C. Point A represents the average energy for the reactants. Point B, at the peak of the curve, indicates the critical energy needed by the colliding molecules in order for reaction to occur. *The difference in energy between point B and point A is the energy of activation,* E_{act}.

Point C represents the average energy, $E_{average}$, of the product molecules. The difference in energy between point C and A is the energy of the reaction, $\Delta E_{reaction}$.

Fig. 19.4 Reaction-energy diagrams for (a) a typical exothermic reaction (energy of products less than energy of reactants); (b) a typical endothermic reaction (energy of products more than energy of reactants).

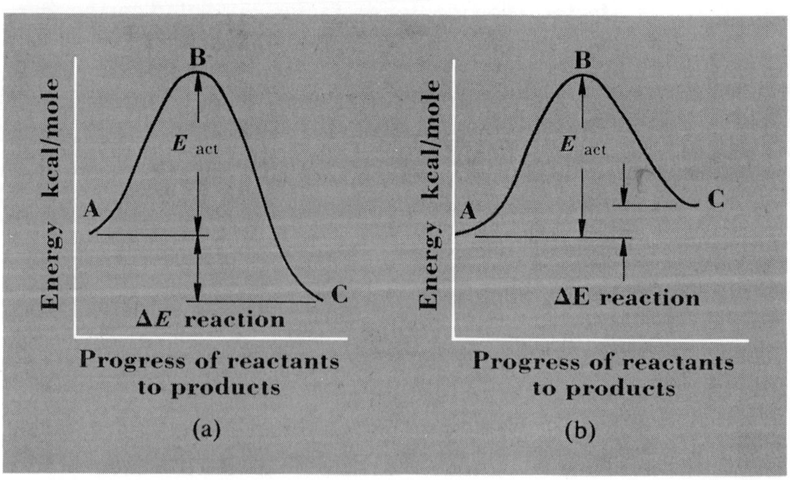

$$\Delta E_{\text{reaction}} = E_{\text{products}} - E_{\text{reactants}}$$

When the energy of the products is less than the energy of the reactants, as in Fig. 19.4a, $\Delta E_{\text{reaction}}$ is negative and the reaction is exothermic. Figure 19.4b, is a typical reaction-energy diagram for a reaction in which the average energy of the products is greater than the average energy of the reactants, so that $\Delta E_{\text{reaction}}$ is positive and the reaction is endothermic.

$\Delta E_{\text{reaction}}$ *is independent of the* E_{act}. We shall see in the next section that the rate of reaction is very dependent on E_{act}. However, the rate of reaction is not determined by $\Delta E_{\text{reaction}}$. The fact that a reaction is exothermic does not mean that it must proceed at a fast rate. There are some very exothermic reactions, for example, $2H_2 + O_2 \longrightarrow 2H_2O$, which at ordinary conditions proceed at imperceptible rates.

(e) Dependence of rate on E_{act} and S_{act} We have seen that a chemical reaction occurs when molecules with sufficient energy collide in a proper orientation. The greater the number of such effective collisions per unit time, the greater is the rate of the reaction. We have already seen how concentration affects the number of collisions and the rate. In this section, we discuss how the energy and orientation factors influence the rate.

At any given temperature not all molecules possess the same energy. Rather, because of the enormous number of collisions, the energies of molecules will vary over a wide range. Figure 19.5 illustrates an energy distribution curve. Most molecules possess an energy close to the average energy. The number of molecules with a particular energy decreases as the energy becomes smaller or greater than average. Let the line E_1 indicate the energy needed so that on collision the peak of our energy diagram, Fig. 19.4a, can be attained (reaction 1). The fraction of molecules possessing this energy is represented by the total shaded area under the curve.* $E_{1\,\text{act}}$ for this reaction is $E_1 - E_{\text{average}}$. Now let us assume that for the peak energy to be attained, the colliding molecules need possess a molecular energy represented by E_2 (reaction 2). Clearly, the fraction of molecules possessing energy E_2 is less than the fraction of molecules possessing energy E_1. In reaction 2, since fewer molecules possess the necessary energy as compared to reaction 1, there will be fewer effective collisions per unit time, and the rate of reaction 2 will be slower. $E_{2\,\text{act}}$ for reaction 2 is $E_2 - E_{\text{average}}$. $E_{2\,\text{act}}$ is greater than $E_{1\,\text{act}}$. We can now generalize the relationship between rate and E_{act}: *the greater is* E_{act}, *the slower is the rate of reaction; conversely, the smaller is* E_{act}, *the faster is the rate of reaction.*

The energy distribution curve also affords an explanation of the effect of temperature on rates of reactions. A 10 deg rise produces a negligible increase in the number of colliding molecules. On the other hand, Fig. 19.5 illustrates that at the higher temperature T_2 the fraction of the

* This is the area to the right of E_1 under curve T_1 (slanted lines).

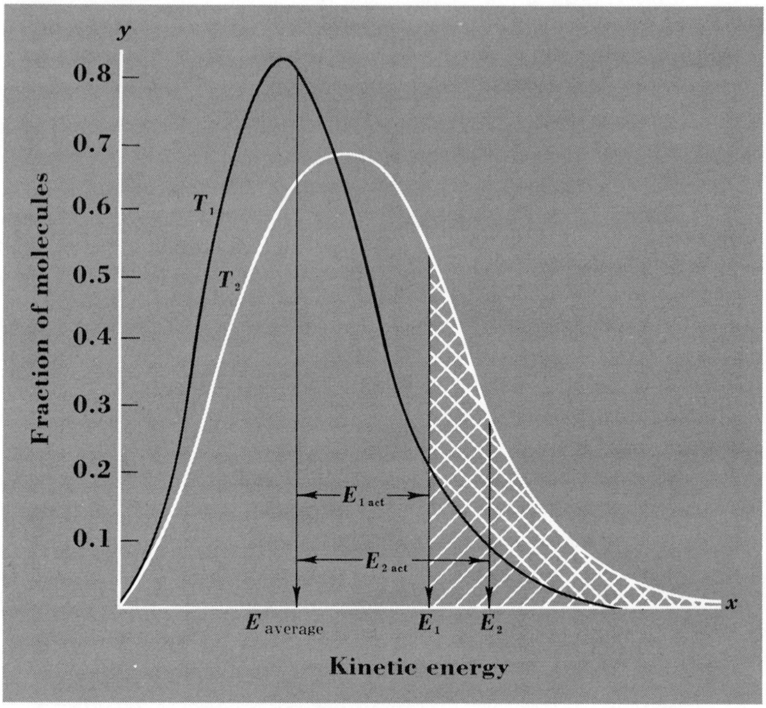

Fig. 19.5 Energy distribution curves. T_1 (black line) lower energy, T_2 (white line) higher energy.

molecules possessing E_1 or E_2 increases appreciably.* Thus the increase in reaction rates is due largely to an increase in the fraction of molecules possessing the required energy to react. At the higher temperature a greater proportion of collisions will lead to product.

Most frequently a catalyst increases the rate of a reaction by decreasing E_{act}. The catalyst provides an alternative mechanism with a lower E_{act}.

Let us now consider the orientation factor. The more specific the orientation required for a reaction to proceed the smaller is the number of effective collisions, and the slower is the rate of the reaction. Conversely, the less specific the required orientation, the greater is the number of effective collisions, and the faster is the rate of the reaction. A less specific orientation means more randomness and, therefore, greater entropy in the molecules at the moment of reaction. We say, then, that the less specific the required orientation, the *greater* is the entropy of activation, S_{act}, and the faster is the rate of reaction.

(f) The transition state theory More modern theories, developed by Henry Eyring and others, modify the concept that reacting molecules must "collide" with a certain energy. It is believed that changes in the

* Under the T_2 curve we include the crosshatched as well as the slanted line area.

arrangement of the atoms in molecules commence even when the distance between molecules is greater than the separation corresponding to a classical "collision." The chemical change is visualized as a continuous series of changes in bond distances as reactant molecules approach each other. Energy changes accompany these continuous changes in configuration or arrangement of atoms. Finally, *the reacting molecules must achieve a specific configuration* before they can form the products of the reaction. This specific configuration possessing a definite energy is known as *the transition state.*

The transition state defines the arrangement of atoms present at the peak of the energy diagram as in Fig. 19.4a (p. 366). In any transition state, at least one or more bonds are breaking, often coincidentally with the formation of new bonds. These partial bonds, in the process of breaking or forming, are shown as broken lines. Thus, the reaction $H_2 + Cl \longrightarrow H + HCl$ may be visualized as shown in Scheme I.

Scheme I

H—H + Cl	⟶	H- - -H- - -Cl	⟶	H + H—Cl
Initial reactants far apart		*The transition state*		*Products far apart*

As the reactants approach the transition state, the H—H bond begins to lengthen (and hence weaken), and the H—Cl bond begins to form. These changes continue as the transition state passes on to products in which the H—H bond is completely broken and the H—Cl bond is completely formed. *The transition state does not represent an observable substance;* in principle, it cannot be isolated. Nevertheless it is assumed to possess properties common to real molecules, such as a molecular weight, interatomic distances, ability to rotate and vibrate. Once the transition state is formed, it can do only one of two things: either it returns to the initial reactants, or it proceeds to form products. To simplify the application of the theory, it is commonly assumed that once the transition state is formed, it will decompose to the products, as shown in Scheme I.

When the reverse reaction occurs and H and H—Cl react, there is a complete reversal of the continuous changes in arrangement of atoms. The same transition state is formed. In general, *in reversible reactions, the pathway for the forward reaction must be retraced by the reverse reaction.*

Since the energy of the transition state corresponds to the peak of the reaction-energy diagram (Fig. 19.4, p. 366), E_{act} can be defined *as the difference in energy content between the transition state and the ground state.* In terms of enthalpy, we can similarly define enthalpy of activation, $H^{\ddagger}$ (same as ΔH_{act}).

$$H_{\text{transition state}} - H_{\text{ground state}} = H^{\ddagger}$$

The double dagger, ‡, is used to indicate activation, as for example, entropy of activation, $S^{\ddagger}$ (ΔS_{act}), and free energy of activation, $G^{\ddagger}$ (ΔG_{act}). The entropy of activation is similarly defined as

$$S^{\ddagger} = S_{\text{transition state}} - S_{\text{ground state}}$$

The concept of a transition state permits a scientist to make predictions, a basic aim of science. For example, for the reaction

$$\underset{\textit{triethylamine}}{(C_2H_5)_3N} + \underset{\textit{methyl iodide}}{CH_3I} \longrightarrow \underset{\textit{triethylmethylammonium iodide}}{[(C_2H_5)_3NCH_3]^+I^-}$$

we guess the product is formed by the approach of the N atom to the C atom in CH_3I, the stretching of the C—I bond, and the change in the H—C—H bond angles in the methyl iodide molecule, as shown in Scheme II.

Scheme II

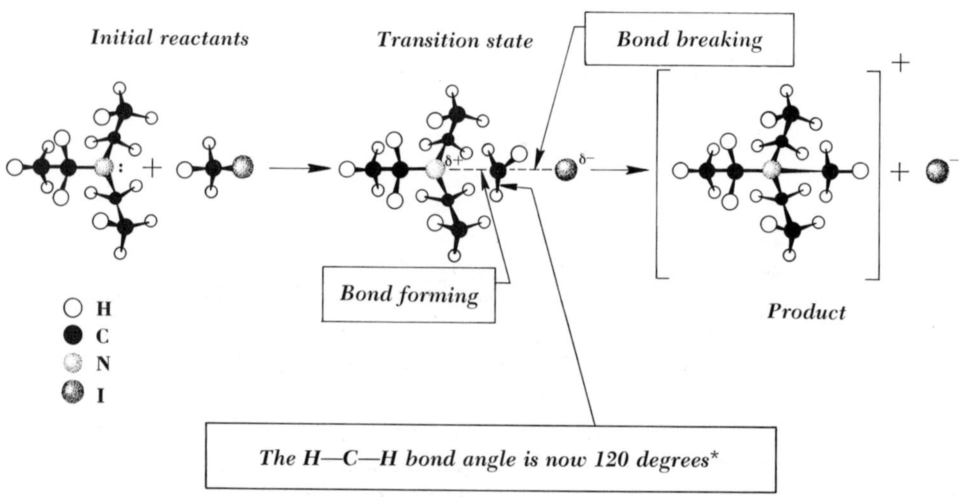

We can now predict that a change in reactants, making it more difficult for the N atom to approach the C atom, will result in an increase in the energy of activation. Thus, substitute *t*-butyl iodide, $(CH_3)_3CI$, for methyl iodide as shown.

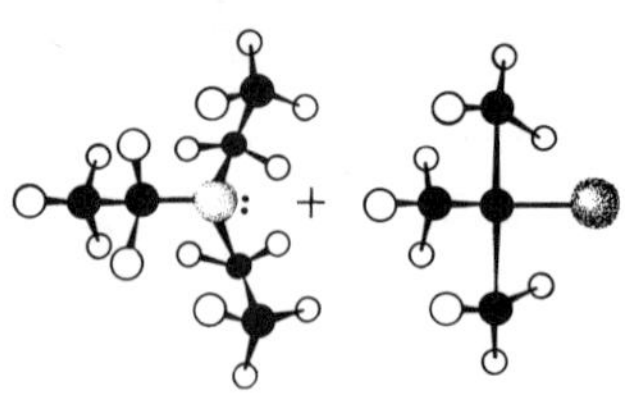

t-butyl iodide

* For this change of bond angle, the hybridization of the atomic orbitals of the C atom has changed from sp^3 in CH_3I to sp^2 in the transition state.

It should now be more difficult for the N atom to approach the C atom from which the I atom has to be detached. Consequently, the energy of activation should increase considerably, and therefore the rate of the reaction by this particular mechanism should be considerably reduced. This conclusion is in agreement with experimental results.

19.4 MECHANISM OF REACTION FROM RATE EQUATION

(a) Multistep reactions, the rate-determining step Unlike the reactions we have considered in the previous section, many reactions occur in several steps, each step passing through a transition state. The sum of the equations for all the steps gives the equation for the overall reaction. The study of the kinetics of a reaction offers the principal method of deciphering the series of processes that occur consecutively (or simultaneously) to produce the observed overall reaction. Not to underestimate the difficulty of this intellectual endeavor, let us confess immediately that many apparently simple reactions, for example, $2H_2 + O_2 \longrightarrow 2H_2O$, are devastatingly complicated, and that after a half-century of labor by many scientists the mechanisms of these reactions are not yet completely understood. Obviously, the treatment of only the simplest mechanisms will be attempted here. A study of the mechanism of the reaction $N_2 + 3H_2 \longrightarrow 2NH_3$, used for the production of ammonia, reveals that the rate of this reaction is actually controlled by the rate at which nitrogen dissociates into atomic nitrogen. Thus, *any scheme* that increases the rate of the reaction $N_2 \longrightarrow 2N$ *automatically* will increase the production rate of ammonia. This also brings us to a cardinal point in the study of reaction mechanisms: in the series of individual steps, *the slowest one determines the rate of the overall reaction and therefore determines the observed rate equation*. The slowest step is said to be *rate-determining*. These ideas may be illustrated by the following analogy.

An assembly line for the production of toasted corn puffs is set up to operate at the given rates:

Step (1) Explosion and toasting of corn kernels (125 lb/hr)
Step (2) Cooling puffed corn (130 lb/hr)
Step (3) Packaging (125 lb/hr)
Step (4) Insertion of prize in each package (15 lb/hr)
Step (5) Sealing package (130 lb/hr)

In the language of the kineticists, we would say **Step 4** is the slow step, the measured production rate will be 15 pounds per hour, and any improvement in only **Step 4** will automatically increase the production rate of toasted corn puffs.

Although the rates of reactions depend on both E_{act} (or $H^{\ddagger}$) and $S^{\ddagger}$, in most cases (but not all) E_{act} has a greater influence. For this reason the following discussion assumes that only E_{act} is decisive in determining relative rates.

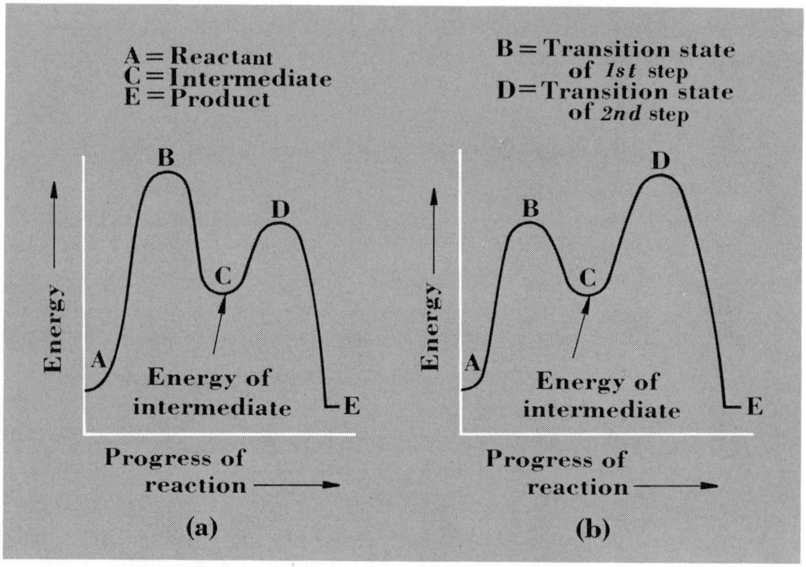

Fig. 19.6 Reaction-energy diagram for two-step reaction. (a) First step has higher transition state energy; (b) second step has higher transition state energy.

Let us consider a reaction, A ⟶ E, as proceeding in two steps:

A ⟶ C
C ⟶ E
A ⟶ E (overall reaction)

We can further elaborate the important principle that in a multistep mechanism the slowest step determines the overall rate by considering two possible reaction-energy diagrams for the mechanism of this reaction. We disregard those rare reactions in which both steps have transition states with about equal energies. In both cases (Figs. 19.6a and 19.6b) there is a dip in the energy curve at point C. This point represents the energy of an intermediate product or **intermediate.** The intermediate is usually very reactive and short-lived.* However, *unlike* transition states, intermediates exist as well-defined molecules, ions, or free radicals. Because intermediates are consumed in ensuing steps they do not appear in the equation for the overall reaction.

In Fig. 19.6a, the energy, B, of the transition state for the first step is higher than the energy, D, of the transition state for the second step. In this case, A ⟶ C is the slow step, and C ⟶ E is the fast step. The overall rate then depends on E_{act} of the first step. Let us look at the intermediate, C. It takes less energy for it to pass through the transition state D than to go back through transition state B. For this reason,

* Reactions are known in which intermediates are sufficiently stable to permit their isolation (see Problem 8c, p. 378).

practically all of the intermediate species, when formed, pass on to the second step. Very few take the reverse pathway back to A. We say, therefore, that the first step is practically irreversible. In summary, the steps are

$$\text{A} \xrightarrow{\text{slow}} \text{C}$$
$$\text{C} \xrightarrow{\text{fast}} \text{E}$$

We should bear in mind, however, that for every step in a reaction there is at least a very slight tendency for reversal. Some very energetic molecules may pass over even the highest energy barrier.

In Fig. 19.6b, the energy of the transition state, B, for the first step is lower than the energy of the transition state, D, for the second step. For this reason, the second step, C $\longrightarrow$ E, is slower and rate-determining. E_{act} for the intermediate, C, returning to starting material, A, is less than E_{act} for going to products, E. Many more molecules of the intermediate will go back to starting material than go on to products. Therefore, besides being fast, the first step is also reversible. E_{act} for the products returning to the intermediate is very large, so that the second step is practically irreversible. In summary, the steps are

$$\text{A} \underset{}{\overset{\text{fast}}{\rightleftharpoons}} \text{C}$$
$$\text{C} \xrightarrow{\text{slow}} \text{E}$$
$$\text{A} \longrightarrow \text{E (overall reaction)}$$

In some cases an intermediate reacts with one of the reactants to give the products; for example,

$$\text{A} + \text{B} \longrightarrow \text{intermediate}$$
$$\text{intermediate} + \text{A} \longrightarrow \text{products}$$
$$2\text{A} + \text{B} \longrightarrow \text{products} \qquad \text{(overall reaction)}$$

(b) Molecularity of a reaction The *number* of species (molecules, atoms, free radicals, or ions) that come together to form a transition state is called the **molecularity of the reaction.** The molecularity of a reaction must be an integral number. When one, two, or three molecules produce a transition state the reaction is said to be **unimolecular, bimolecular,** or **termolecular,** respectively. This concept is therefore theoretical, while the order of the reaction is an experimentally determined quantity (p. 359). The probability that four or more species may simultaneously combine to form a transition state is so small that reactions of molecularity higher than three are never postulated. Even termolecular reactions are rare. Most reactions are uni- or bimolecular.

(c) Use of rate equation The rate equation is one of several important experimental facts needed to elucidate a mechanism. The rate equation tells us much about the rate-determining step and any prior fast steps. If only the given reactants appear in the slow step, the rate equation tells us the kind and number of molecules of each reactant. The number of each kind of molecule is the exponent for the concentration of the particular reactant. If an intermediate is involved in the rate-controlling step, it has to be formed in a prior fast step. The rate equation then gives a clue to the identity of the given reactants involved in the slow step and in all the preceding fast steps. Any fast step(s) following the slow one may be surmised by using chemical intuition. Remember that the sum of all steps must be the same as the given stoichiometric equation.

The reaction between iodine chloride, ICl, and hydrogen illustrates the use of the rate equation to suggest a mechanism.

$$H_2 + 2ICl \longrightarrow I_2 + 2HCl \qquad rate = k[ICl][H_2] \tag{6}$$

The rate equation indicates that the slow step *might* require one molecule of ICl and one molecule of H_2:

$$ICl + H_2 \xrightarrow{\text{slow}} \text{intermediates}$$

We know this step gives intermediates because the overall reaction (Equation 6) requires a second molecule of ICl. We use our chemical intuition to suggest that the intermediate products are likely to be HI and HCl.

Step (1) $\quad ICl + H_2 \xrightarrow{\text{slow}} HI + HCl$

Another step is necessary to use a second molecule of ICl, produce a molecule of I_2 and a second molecule of HCl, and to destroy the intermediate molecule of HI. Therefore, we suggest as a plausible second step:

Step (2) $\quad ICl + HI \xrightarrow{\text{fast}} I_2 + HCl$

The addition steps (1) and (2) gives Equation 6.

The formation of nitrosyl chloride affords a more complex example.

$$2NO + Cl_2 \longrightarrow 2NOCl$$

and the rate equation is

$$rate = k_3[NO]^2[Cl_2] \tag{7}$$

The order of reaction is therefore 3, second order with respect to B and first order with respect to A.

The rate equation tells us that the slow step and any preceding fast steps require two molecules of NO and one molecule of Cl_2. Were the slow step to be

$$NO + NO + Cl_2 \longrightarrow 2NOCl$$

the reaction would be termolecular. Since termolecular reactions are rare, it is more likely that the reaction proceeds in more than one step. A reasonable two-step mechanism is the following. Since NO is a radical, it is assumed that the first step is a fast reversible formation of N_2O_2, the unstable dimer of NO. In the slow step, the dimer reacts with Cl_2.

$$\text{Step (1)} \qquad 2\ \cdot\ddot{N}{=}\ddot{O}: \ \underset{}{\overset{\text{fast}}{\rightleftharpoons}}\ :\ddot{O}{=}\ddot{N}{-}\ddot{N}{=}\ddot{O}:\ *$$
$$\text{Step (2)} \qquad :O{=}\ddot{N}{-}\ddot{N}{=}O: \ + Cl_2 \xrightarrow{\text{slow}} 2NOCl$$

Let us convince ourselves that this two-step bimolecular mechanism is consistent with the observed rate equation. The rate of the slow step (Step 2) should be

$$rate = k_2[N_2O_2][Cl_2] \tag{8}$$

The first step is fast and reversible and must reach an equilibrium state for which the equilibrium condition is

$$K = \frac{[N_2O_2]}{[NO]^2}$$

so that $[N_2O_2] = K[NO]^2$.

Substitution into Equation 8 gives

$$rate = k_2K[NO]^2[Cl_2] \tag{9}$$

This equation is identical to the observed rate equation, since k_3 in Equation 7 is equal to the product of the two constants, k_2 and K, in the derived rate equation 9.

19.5 □ CHAIN MECHANISM

When a chemical reaction proceeds by a mechanism in which a species consumed in one step is regenerated in a second step, the set of steps is referred to as a **chain mechanism.** Thus, in a chain reaction each step yields *one* species capable of producing the reaction for each *one* species consumed. For example, in the chain reaction

$$:\ddot{\underset{\cdot\cdot}{Cl}}\cdot \quad + H_2 \longrightarrow HCl + H\cdot \tag{10}$$

$$H\cdot \quad + Cl_2 \longrightarrow HCl + \quad :\ddot{\underset{\cdot\cdot}{Cl}}\cdot \tag{11}$$

each $:\ddot{\underset{\cdot\cdot}{Cl}}\cdot$ atom used to produce an HCl molecule also forms an $H\cdot$ atom, which subsequently produces an HCl molecule and regenerates the $:\ddot{\underset{\cdot\cdot}{Cl}}\cdot$ atom; the $:\ddot{\underset{\cdot\cdot}{Cl}}\cdot$ atom produced in Step (11) reacts to form another HCl molecule and $H\cdot$ atom as in Step (10), and so on.

At ordinary temperatures, hydrogen and chlorine form hydrogen chloride slowly in the dark but explosively in sunlight. The chlorine

* Although the formation of N_2O_2 is fast, the equilibrium constant for the reaction is extremely small because at the temperature of the reaction N_2O_2 is very unstable.

absorbs radiation* below 4785 Å, producing atomic chlorine,

$$Cl_2 + h\nu \rightleftharpoons 2\ :\ddot{\underset{..}{Cl}}\cdot$$

Without absorption of further light energy, the atoms of chlorine initiate the chain reaction illustrated by reactions 10 and 11.† Such chain reactions are fast but not necessarily explosive. They become explosive when the heat evolved in the overall reaction is so large that isothermal conditions (constant temperature) cannot be maintained; as the reaction proceeds, the heat evolved increases the temperature which, in turn, increases the rate of the reaction and the liberation of heat until, finally, the rate becomes explosive.

In a **branching chain reaction,** two or more species capable of producing the reaction are regenerated for every one consumed. For example, in

$$O^* + H_2 \longrightarrow H_2O^* \qquad (12)$$

$$H_2O^* + O_2 \longrightarrow H_2O + 2O^* \qquad (13)$$

for every one "excited"‡ oxygen atom consumed in reaction 12, two are regenerated in reaction 13. The overall rate then becomes infinite and, if the reaction is exothermic, an explosion results under isothermal conditions. At about 550°C, for example, the rate of the hydrogen-oxygen reaction forming water increases slowly as the concentration of either reactant is increased and then, at the same temperature, the mixture suddenly explodes, signaling the development of the branching chain reaction, illustrated by reactions 12 and 13.

PROBLEMS

1. Definitions Define and illustrate the following terms: (a) rate equation; (b) specific rate constant; (c) order of reaction; (d) energy of activation; (e) entropy of activation; (f) transition state; (g) molecularity; (h) mechanism; (i) concerted reaction; (j) rate-controlling step.

2. Rate equation For the gaseous reaction

$$2NO + 2H_2 \longrightarrow N_2 + 2H_2O$$

the following data at 1100°K are given:

* Radiation is represented by $h\nu$, the amount of energy in one photon when the frequency is ν (p. 16).

† The mechanism may be summarized as

$$Cl_2 + h\nu \rightleftharpoons 2Cl$$
$$Cl + H_2 \longrightarrow HCl + H$$
$$H + Cl_2 \longrightarrow HCl + Cl$$

but note that the overall reaction is obtained from the sum of the last *two* steps.

‡ The asterisk is usually used to denote excited species. An excited O* atom is in a higher electronic energy state than is a ground state (normal) O atom.

Rate of disappearance of NO (mole/liter sec)	Reactant concentrations (mole/liter) [NO]	[H$_2$]
2.4×10^{-5}	5.0×10^{-3}	2.0×10^{-3}
2.2×10^{-4}	15×10^{-3}	2.0×10^{-3}
4.4×10^{-4}	15×10^{-3}	4.0×10^{-3}

(a) Write the rate equation for the reaction. What is the order of the reaction?
(b) Calculate the rate constant for the reaction at 1100°K.
(c) Calculate the rate, moles per liter per second, at 1100°K at the instant when

$$[NO] = 1.1 \times 10^{-3} \qquad \text{and} \qquad [H_2] = 1.5 \times 10^{-3}$$

(d) Calculate the rate, moles per second, at 1100°K at the instant when

$$[NO] = 1.1 \times 10^{-3} \qquad \text{and} \qquad [H_2] = 1.5 \times 10^{-3}$$

and the volume of the reacting system is 2.0 liters.
(e) At the instant when NO is reacting at the rate 1.2×10^{-5} mole per liter per second, what is the rate at which H_2 is reacting and N_2 is forming?

3. Reaction order At 0°C, the rate of electron exchange between MnO_4^{2-} and MnO_4^-,

$$MnO_4^{2-} + MnO_4^- \longrightarrow MnO_4^- + MnO_4^{2-}$$

is

$$rate = 1.5 \times 10^3 \frac{\text{liter}}{\text{mole sec}} [MnO_4^-][MnO_4^{2-}]$$

Is this exchange reaction first, second, or third order?

4. Mechanism Give the rate equation for the reaction $2NO + O_2 \longrightarrow 2NO_2$, which involves two bimolecular reactions, namely,

$$NO + O_2 \underset{}{\overset{\text{fast}}{\rightleftharpoons}} NO_3$$
$$NO_3 + NO \xrightarrow{\text{slow}} 2NO_2$$

5. Mechanism Suggest a mechanism for each of the following homogeneous reactions from the given rate equations:

(a)

$$H_2 + Cl_2 \longrightarrow 2HCl$$
$$rate \text{ of disappearance of } Cl_2 = k[Cl_2]$$

(b)

$$2ClO_2 + 2OH^- \longrightarrow ClO_3^- + ClO_2^- + H_2O$$
$$rate \text{ of disappearance of } ClO_2 = k[ClO_2]^2[OH^-]$$

(c) With respect to parts (a) and (b), what is the order of the reaction and what is the molecularity of each step in your proposed mechanism?

6. Energy of activation Explain the increase in rate constants with increasing temperature.

7. Transition state theory (a) For the reaction

$$\underset{\textit{azodicarboxylate}}{N_2(COO)_2^{2-}} + H_3O^+ \longrightarrow HN_2(COO)_2^- + H_2O$$

the following data are given:

Solvent dielectric constant	$H^{\ddagger}$ (kcal/mole)	$S^{\ddagger}$ (cal/mole deg)
78.5	10.2	12.9
71.2	10.2	14.1

In which solvent will the rate be faster for the same concentration of reactants at the same temperature? (b) From the following data for gaseous decomposition:

Reaction	$H^{\ddagger}$ (kcal/mole)	$S^{\ddagger}$ (cal/mole deg)
$(CH_3CO)_2O \longrightarrow CH_3COOH + CH_2CO$	34.5	−4
$CH_3CHCl_2 \longrightarrow CH_2{=}CHCl + HCl$	49.5	−4

which reaction will be faster for the same concentration of reactants at the same temperature?

8. Reaction-energy diagrams Draw a reaction-energy diagram for (a) an exothermic one-step reaction, $A \longrightarrow B$; (b) a one-step reaction, $A \longrightarrow B$ with $\Delta H_{\text{reaction}} = 0$; (c) a two-step reaction in which an intermediate might be isolated.

9. Overall reaction The following mechanism has been proposed for the reaction of hydrogen with oxygen.

$$
\begin{aligned}
H_2 &\longrightarrow H + H \\
H + O_2 &\longrightarrow HO + O \\
O + H_2 &\longrightarrow HO + H \\
HO + H_2 &\longrightarrow H_2O + H \\
H + OH &\longrightarrow H_2O \\
H + H &\longrightarrow H_2
\end{aligned}
$$

Show how the equation for the overall reaction may be obtained from these individual steps.

10. Slow step In the following series of changes, for which the rate constants are given, (a) which step determines, for all practical purposes, the conversion of A to D?

$$A \xrightarrow{10^4/\text{year}} B \xrightarrow{10^{-9}/\text{year}} C \xrightarrow{10/\text{year}} D \xrightarrow{0/\text{year}}$$

(b) During most of the time that all these processes are going on, which substance, A, B, C, or D, will be present in (*i*) greatest abundance, (*ii*) least abundance?

(c) When all the processes have terminated which product is present?

ADDITIONAL PROBLEMS

11. Rate equation Minute tubes containing radon gas, known as "radon seeds," are used for cancer radiotherapy. Radon, Rn, changes to the element polonium, Po, kinetic data for which follow:

Rate of disappearance of Rn (mole/liter day)	Reactant concentration (mole/liter) [Rn]
5.43×10^{-3}	0.0300
1.62×10^{-2}	0.0900

The rate of this reaction is independent of temperature. (a) Write the rate equation for this reaction, and (b) calculate the rate at the instant when [Rn] = 0.0200.

12. Rate equation The rate of decomposition of gaseous acetaldehyde,

$$CH_3CHO \longrightarrow CH_4 + CO$$

is given by the rate equation,

$$rate = k[CH_3CHO]^2$$

(a) Make up some data from which this rate equation might have been obtained. (b) At a fixed temperature, the volume of a homogeneous reacting system decreases so that the concentrations of the reacting gases remain constant. Which of the following will change: the rate constant in liter/mole second, the rate in moles/liter-second, the rate in moles/second? If the volume remains constant during reaction, which will change?

13. Rate equation The overall equation for the reaction of vanadium(IV) and chromium(VI) in acidic perchlorate solution is

$$VO^{2+} + HCrO_4^- + H^+ \longrightarrow VO_2^+ + Cr^{3+} + H_2O$$

(a) Balance the equation. (b) The rate equation at constant acidity is

$$rate = \frac{k[VO^{2+}][HCrO_4^-]}{[VO_2^+]}$$

What will be the effect on the rate of disappearance of $HCrO_4^-$ if $[VO^{2+}]$, $[HCrO_4^-]$, and $[VO_2^+]$ are doubled? Can you make a quantitative statement about the effect of doubling $[H^+]$?

14. Rate constant The rate equation for the reaction

$$Xe^+ + CH_4 + e^- \longrightarrow XeCH_2 + H_2$$

is

$$rate = k[Xe^+][CH_4]$$

The rate constant is 4.5×10^{-13} ml/mole sec. Calculate the number of H_2 molecules formed per second per liter when the concentrations of Xe^+ and CH_4 are equal at 5.0×10^{-7} mole per milliliter. Calculate the number of CH_4 molecules consumed per second per liter.

15. Rate equation For the decomposition of nitrosyl chloride, $2NOCl(g) \longrightarrow 2NO(g) + Cl_2(g)$, the rate equation at 1020°K is

$$\text{rate of consumption of NOCl} = k[NOCl]^2 = 8.3 \times 10^5 \frac{\text{liter}}{\text{mole sec}} [NOCl]^2$$

The rate equation may also be expressed as

$$\text{rate of appearance of NO} = k'[\text{NOCl}]^2$$

and

$$\text{rate of appearance of Cl}_2 = k''[\text{NOCl}]^2$$

Calculate (a) k' and (b) k'' at 1020°K.

16. Types of reaction The reaction $A(g) + B(g) \longrightarrow AB(g)$ is carried out in a glass vessel. When glass beads are added at constant concentration of the gases, the rate of the reaction increases. Is the reaction homogeneous or heterogeneous?

17. Energy of activation The energy of activation for the decomposition of HI,

$$2HI(g) \longrightarrow H_2(g) + I_2(g) \qquad \Delta H = +2.6 \text{ kcal}$$

is 45 kcal/mole, while the energy of activation for the decomposition of nitrogen dioxide, NO_2,

$$2NO_2(g) \longrightarrow 2NO(g) + O_2(g) \qquad \Delta H = +27.0 \text{ kcal}$$

is 25 kcal/mole. At the same concentration and temperature which reaction will proceed faster? Explain your answer. (Assume $S^\ddagger$ is the same for the two reactions.)

18. Mechanism A mechanism proposed for the redox reaction between Sn^{2+} and Fe^{3+} is

$$SnCl_2 + Cl^- \longrightarrow SnCl_3^-$$

$$SnCl_3^- + Cl^- \longrightarrow SnCl_4^{2-}$$

$$SnCl_4^{2-} + FeCl_3 \longrightarrow SnCl_3 + FeCl_2 + 2Cl^-$$

$$SnCl_3 + FeCl_3 \longrightarrow SnCl_4 + FeCl_2$$

What should be the observed overall chemical reaction? What is the molecularity of each step? What is the oxidation number of Sn in the intermediates $SnCl_4^{2-}$ and $SnCl_3$? What species appears to serve as a catalyst?

19. Transition state theory If isopropyl iodide is substituted for methyl iodide (p. 370), should the rate of reaction with triethylamine be closer to the rate for methyl iodide or for t-butyl iodide? Explain. The configuration of isopropyl iodide is

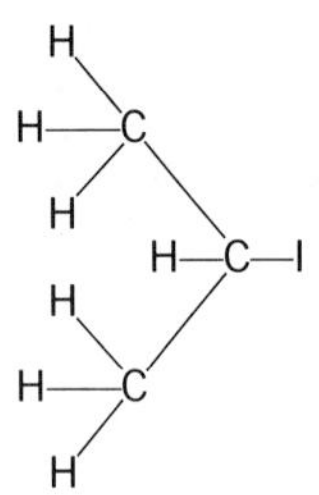

20. Transition state theory (a) Of the two bonds C—I and C—Cl, the bond energy is greater for C—Cl. When methyl chloride, CH_3Cl, is substituted for CH_3I, will the reaction rate with triethylamine (p. 370) increase, decrease, or remain

the same? Explain your answer. (b) An ethyl group in $(C_2H_5)_3N$ is replaced by a group that is of the same size but decreases the basicity of the amine. How should the rate of reaction with methyl iodide change? Justify your guess.

21. Catalysis The hydrolysis of methyl trichloroacetate to trichloroacetic acid and methyl alcohol,

$$CCl_3COOCH_3 + H_2O \longrightarrow CCl_3COOH + CH_3OH$$

is catalyzed by H^+. Should the time required to convert a given quantity of the acetate to products, in a given quantity of water at a given pH, be different in a buffered and an unbuffered solution? If so, in which solution should the time be shorter? Explain briefly.

22. Catalysis Ozone, O_3, an allotropic form of oxygen, is stable at room temperature, but when chlorine, Cl_2, is added a measurable reaction occurs. The suggested mechanism for the reaction is

$$Cl_2 + O_3 \longrightarrow ClO + ClO_2$$

$$ClO_2 + O_3 \longrightarrow ClO_3 + O_2$$

$$ClO_3 + O_3 \longrightarrow ClO_2 + 2O_2$$

$$ClO_3 + ClO_3 \longrightarrow Cl_2 + 3O_2$$

$$ClO \longrightarrow Cl + O$$

$$ClO_2 + O \longrightarrow ClO_3$$

$$Cl + \tfrac{3}{2}O_2 \longrightarrow ClO_3$$

What is the overall reaction and what is the catalyst for the reaction?

ANSWERS **2.** (a) third; (b) 4.9×10^2 liter²/mole² sec; (c) 8.9×10^{-7} mole/liter sec; (d) 1.8×10^{-6} mole/sec; (e) 1.2×10^{-5} mole/liter sec, 6.0×10^{-6} mole/liter sec.

15. (*i*) 8.3×10^5 liter/mole sec.

20 ■ THE CHEMISTRY OF THE REPRESENTATIVE ELEMENTS

20.1 □ INTRODUCTION

The representative elements are those arising from a filling up of *s* or *p* subshells (pp. 30–33). These elements constitute the A groups in our periodic table.

In this chapter, we consider particularly the relationships between the properties of the representative elements and their positions in the periodic table. Some of these trends are best illustrated by studying the hydrides, oxides, halides, and sulfides of the elements.*

20.2 □ ALLOTROPY AND PERIODICITY

Several elements exist in the form of two or more different substances. This phenomenon is called **allotropy.** When allotropic forms differ from each other in chemical bonding, they are said to be **primary modifications.** An allotropic difference in the mode of molecular aggregation, or in crystal habit, is called a **secondary modification;** such differences, which do not represent different molecular species, will not be considered here.

Primary allotropic modifications exist in C, O, P, S, As, Se, Sn, Sb, and Te. Carbon occurs as diamond (Fig. 7.2, p. 118) and as graphite (Fig. 22.1, p. 438). Both forms are very unreactive; diamond is extremely hard, while graphite is soft in consequence of the ease of layer slippage.

* Review also p. 51 to 54 for a brief introduction to the chemical properties of the elements as related to their positions in the periodic table.

At ordinary temperatures and pressures, graphite is the lower energy (more stable) form by about 0.5 kcal/g-atom. Graphite conducts heat and electricity fairly well, and has an appearance that suggests metallic luster.

The element oxygen exists as two gaseous forms: oxygen, O_2, and the much less stable ozone, O_3. Oxygen is light blue in the liquid state, is paramagnetic, and has two unpaired electrons per molecule. Ozone is dark blue in the liquid state, and spontaneously decomposes to oxygen. The bond angle in ozone is 127 degrees.

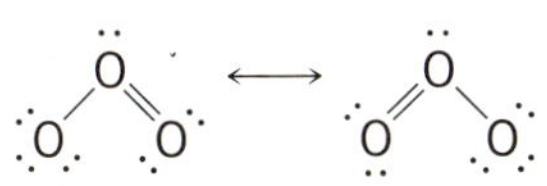

Phosphorus occurs as tetrahedral P_4 molecules (phosphorus vapor, and the "white" solid form), and as two forms consisting of large molecular networks: black phosphorus and red (or violet) phosphorus. The black form is a layer lattice that shows some metallic properties (luster, conductivity of heat and electricity); the red form lacks these attributes.

Sulfur exists as ringed molecules (S_8, in liquid sulfur below about 160°C and in the rhombic and monoclinic crystal forms), as long chains ("plastic sulfur" and the liquid form $S\mu$), and as short chains of various lengths (S_6, S_4, and S_2 in the gaseous state). Rhombic and monoclinic sulfur are secondary allotropic modifications.

Arsenic, selenium, tin, antimony, and tellurium all exist in more or less distinctly differentiated metallic and nonmetallic modifications (Table 20.1).

TABLE 20.1 ▪ ALLOTROPES OF As, Se, Sn, Sb, Te

	Allotropes	
Element	**Metallic or metalloidal**	**Nonmetallic**
As	Gray, metallic luster (density 5.73 g/cm³)	"Yellow arsenic," waxy (density 3.7 g/cm³)
Se	Gray, semiconductor of electricity	Red, amorphous or "vitreous" selenium
Sn	"White tin," metallic luster, conducts heat and electricity well, is soft and ductile	"Gray tin," diamond type lattice
Sb	Metallic luster, good conductor of heat and electricity; hard, brittle	"Yellow antimony"; also exists as a vitreous modification known as "explosive antimony"
Te	Metallic luster, poor semidonductor of electricity	Amorphous form

What generalizations can be made about the place of allotropy in the periodic classification of the elements? Oxygen is unique in exhibiting allotropy of covalently bound forms of low molecular weight. Sulfur, like oxygen, shows allotropy of covalent modifications; unlike oxygen, it can exist as an extended molecular network. Oxygen and sulfur are the most electronegative of the elements that exist in primary allotropic modifications; in consequence, neither element shows a tendency to exist in a metallic form. It is interesting to note that nitrogen, with its readiness to form multiple bonds, exists only as the stable species N_2 under ordinary conditions. In all the other cases cited, the allotropic elements occur both as distinctly nonmetallic and as more or less metallic forms. The drive toward delocalization of electrons by assumption of metallic character is evidenced even by carbon.

In summary, it may be generalized that (a) the occurrence of primary allotropy is limited to the elements in periods 2, 3, 4, and 5 of groups IVA, VA, and VIA of the periodic system, (b) the allotropy of the more electronegative elements is predominantly one of differences in covalent structure, while the allotropy of the less electronegative elements is one of difference between metallic and nonmetallic forms, and (c) the less electronegative the element, the greater is the stability of the metallic allotrope. This "allotrope region" of the periodic system corresponds roughly to the line that separates metallic from nonmetallic elements.

Notable exceptions to these generalizations are silicon and germanium, which do not exhibit primary allotropy at all. Both elements occur only in the diamond (sp^3 covalent) lattice. These elements, being beyond the second period, do not tend to form multiple bonds and so do not exist as layer lattices with extended π bonding. The structure of graphite is unique among the group IVA elements in being π-bonded. Silicon and germanium are less electronegative than carbon, however, and the promotion of their electrons to higher energy levels, even from the covalent lattice, can be readily accomplished by thermal or radiative excitation. As a result, silicon and germanium show electrical conductivity which increases with increasing temperature (the reverse of the behavior shown by true metals); they are semiconductors. These elements thus show both metallic and nonmetallic properties in a single mode of aggregation; this behavior is **semimetallic** or **metalloid** (Fig. 23.2, p. 465).

20.3 □ REACTIVITY OF HYDRIDES

The term **hydride** is used in two ways. It is the *generic* name for all binary hydrogen compounds, and the *specific* name for the binary hydrogen compounds of the group IA and IIA elements, for example, lithium hydride and calcium hydride.

Not all the representative elements form stable well-defined hydrogen compounds. It is observed, for example, that, except for groups IA and IIA, *as we proceed down the group, the stability of the hydride decreases*. Thus, we find that *no* well-defined hydrogen compounds of gallium (Ga), indium (In), and thallium (Tl) (group IIIA), lead (Pb) (group IVA), bismuth (Bi) (group VA), and polonium (Po) (group VIA) have yet been isolated in appreciable quantity.

We should expect the mode of reaction of hydrides to depend to some extent on the relative electronegativities of hydrogen and the other element. It is evident from Table 7.1 (p. 119) that hydrogen is intermediate in electronegativity. Therefore, on the basis of a comparison with hydrogen, we can divide the representative elements into three groups:

(a) Those elements that are less electronegative than hydrogen and, in their hydrides, are therefore positive relative to hydrogen. These are found in the left-hand portion of the periodic table.

(b) Those elements that are more electronegative than hydrogen, and, in their hydrides, are therefore negative. These are found in the right-hand portion of the periodic table.

(c) Those elements that have approximately the same electronegativity as hydrogen, and have little or no fractional (partial) charge in their hydrides. These are found in the center region of the periodic table.

The several ways in which hydrides react with water reveal the periodic variations in the bonding and reactivity of these compounds.

(a) No reaction The hydrides of the elements of group IVA and group VA (except for nitrogen) do not react with water. Included in this class of compounds are phosphine, PH_3, and the hydrocarbons, such as CH_4. Generally in this group of compounds the central atom has an octet of electrons, and has an electronegativity close to that of hydrogen.

(b) Basic reaction due to hydride ion The hydrides of the elements in groups IA and IIA (except for Be and possibly Mg) are essentially ionic compounds, and are called "saline" (saltlike) hydrides. These elements all have an electronegativity *much less* than that of hydrogen. The hydride ion is very basic, and reacts with water to produce H_2 and OH^-;

$$\underset{\substack{-1\\ \textit{base}_1}}{\mathrm{H}{:}^-} + \underset{\substack{+1\\ \textit{acid}_2}}{\mathrm{H}}\!-\!\ddot{\underset{\cdot\cdot}{\mathrm{O}}}\!-\!\underset{+1}{\mathrm{H}} \longrightarrow \underset{0}{\mathrm{H}}\!-\!\underset{0}{\mathrm{H}} + [{:}\ddot{\underset{\cdot\cdot}{\mathrm{O}}}\!-\!\underset{+1}{\mathrm{H}}]^- \quad (\text{oxidation number; } \textit{acid}_1,\ \textit{base}_2)$$

It is noteworthy that the above reaction is also an oxidation-reduction, as deduced from the change in oxidation numbers shown below each H atom in the equation. In this respect, H:⁻ is a reducing agent. The acceptance of a proton by hydride ion is both an acid-base and a redox reaction.

(c) Hydrolysis Several covalent hydrides react with water to give hydrogen and an oxyacid of the central atom, as typified by the reaction of diborane,

$$\overset{\delta+\ \delta-}{B_2H_6} + 6H_2O \longrightarrow 6H_2 + 2B(OH)_3$$

diborane *orthoboric acid*

Generally, during this type of hydrolysis, the more electronegative element combines with the hydrogen, and the less electronegative element combines with the OH group. In the above example, hydrogen is more electronegative and so H_2 is formed. The hydrides that hydrolyze in this manner contain a central element, less electronegative than hydrogen, that cannot exist as a cation in water because of a large cationic charge density (p. 129).

(d) Acidic reaction The hydrogen compounds of the elements in groups VIA and VIIA dissociate in water to give acidic solutions, as shown for HI and H_2Se,

$$\overset{\delta+\ \delta-}{HI} + H_2O \longrightarrow H_3O^+ + I^-$$

$$\overset{\delta+\ \delta-}{H_2Se} + H_2O \rightleftharpoons H_3O^+ + HSe^-$$

In these compounds, the bond is polarized, so that the H atom has a fractional positive charge, as shown in the above equations. However, one should not jump to the *erroneous* conclusion that the more polar the $\overset{\delta+}{H}$—$\overset{\delta-}{M}$ bond, the more acidic is the binary compound. Were this so, we could expect the more polar H-F molecule to be more acidic than the less polar H-I molecule. HI is actually a much stronger acid than HF. In general, *the acidity of binary compounds*, HX, *increases on proceeding down a group*, notwithstanding the fact that the electronegativity of the atom, X, decreases:

Compound	HF	HCl	HBr	HI
K_a (see p. 331)	7×10^{-4}	$\sim 10^{7}$	$\sim 10^{9}$	$\sim 10^{10}$

The *acidity of binary compounds increases on proceeding across a period from left to right:*

Compound	CH_4	NH_3	H_2O*	HF
K_a	10^{-50}	10^{-32}	2×10^{-16}	7×10^{-4}

We can observe this difference in acidity in a very qualitative way, by using sodium, an active metal, as a test reagent. To a rough approximation, the rate of displacement of hydrogen by sodium is an indication of the acidity of the compound. Thus, we find that methane is com-

* K_a for water $= [H^+][OH^-]/[H_2O] = 10^{-14}/55.5 = 2 \times 10^{-16}$. We must use this form to permit a fair comparison with the other constants, such as K_a for HF $= [H^+][F^-]/[HF]$.

pletely inert toward sodium, ammonia reacts slowly, water reacts violently, and it would be foolhardy to add sodium to liquid hydrogen fluoride. These findings are consistent with the indicated order of acidity.

(e) Basic reaction due to a lone pair on the central atom Ammonia is the only hydride that reacts as a base in water by virtue of a lone pair of electrons,

$$H_3N\!: + H_2O \rightleftharpoons H_3N\!:\!H^+ + OH^-$$

This property falls off sharply as we proceed down the group. Thus, the order of basicity of the hydrides of group VA elements is

$$NH_3 > PH_3 > AsH_3 > SbH_3$$

decreasing basicity →

Ammonia is more basic than water, which in turn is more basic than hydrogen fluoride. Therefore, *basicity due to one or more lone pairs of electrons on the central atom decreases* on proceeding across a period from left to right.

(f) Lewis acids The known hydrides of group IIIA possess electron-deficient atoms, and so behave as Lewis acids. Of particular interest are their reactions with ionic hydrides:

$$\underset{\text{aluminum hydride}}{AlH_3} + Li^+H^- \longrightarrow \underset{\text{lithium aluminum hydride}}{Li^+\left[\begin{array}{c} H \\ H\!:\!\ddot{\underset{..}{Al}}\!:\!H \\ H \end{array}\right]^-}$$

$$\underset{\text{diborane}}{B_2H_6} + 2Na^+H^- \longrightarrow \underset{\text{sodium borohydride}}{2Na^+\left[\begin{array}{c} H \\ H\!:\!\ddot{\underset{..}{B}}\!:\!H \\ H \end{array}\right]^-}$$

Sodium borohydride, a typical saltlike compound, is widely used as a reducing agent because it is a source of hydride ions.

20.4 ▫ THE PERIODICITY OF PROPERTIES OF THE OXIDES

(a) Basic and acid anhydrides All elements except some of the noble gases can form oxides; almost all elements form more than one. Since oxygen is the most electronegative of all elements but fluorine, the fractional charge on oxygen in all oxides except OF_2 and O_2F_2 is negative.

This fractional charge would be expected to reach its most negative values (approaching −2) in the alkali metal oxides, and its least negative values (approaching 0) in the halogen oxides. As the fractional charge becomes more negative, the structure of the oxide assumes more of the O^{2-} ionic character. As the fractional charge approaches zero, the oxide becomes more covalent. Ionic oxides react with water to give basic hydroxides; these oxides are known as **basic anhydrides,** as typified by calcium oxide, CaO,

$$\underset{\text{a basic anhydride}}{Ca^{2+}O^{2-}} + H_2O \longrightarrow \underbrace{Ca^{2+} + 2OH^-}_{\text{base}}$$

Covalent oxides react with water to give oxyacids; these oxides are known as **acid anhydrides,** as typified by sulfur trioxide, SO_3,

$$\underset{\text{an acid anhydride}}{SO_3} + H_2O \longrightarrow \underbrace{H^+ + HOSO_3^-}_{\text{acid}}$$

Each case may be considered to be a Lewis base type of attack by the *more negative oxygen* on the *more positive atom of the other compound:*

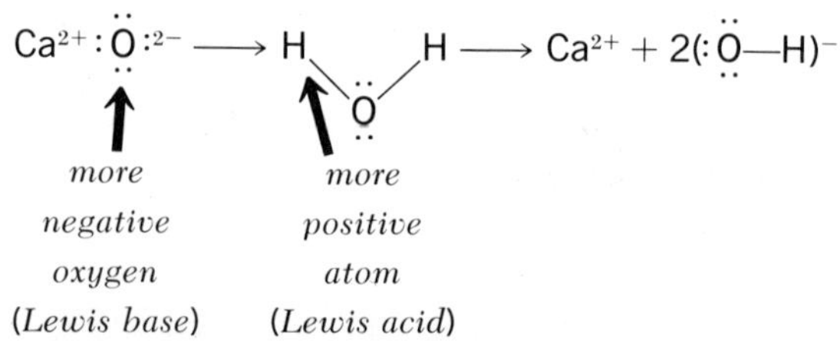

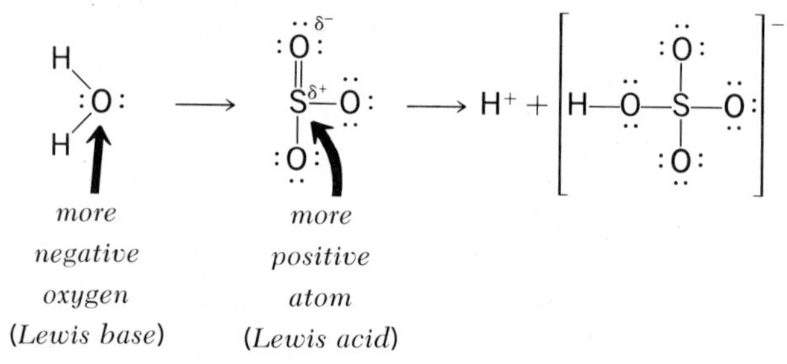

It is to be expected that oxides in which the fractional charge on oxygen is close to that of the oxygen of water will be **amphoteric,** as shown for ZnO,

$$ZnO + 2H^+ \longrightarrow Zn^{2+} + H_2O$$
$$ZnO + 2OH^- + H_2O \longrightarrow Zn(OH)_4^{2-}$$

(b) Periodicity of acidic properties Designation of oxides of different oxidation states of the elements as weakly, moderately, or strongly

TABLE 20.2 ▪ BASIC AND/OR ACIDIC PROPERTIES OF THE OXIDES OF THE ELEMENTS

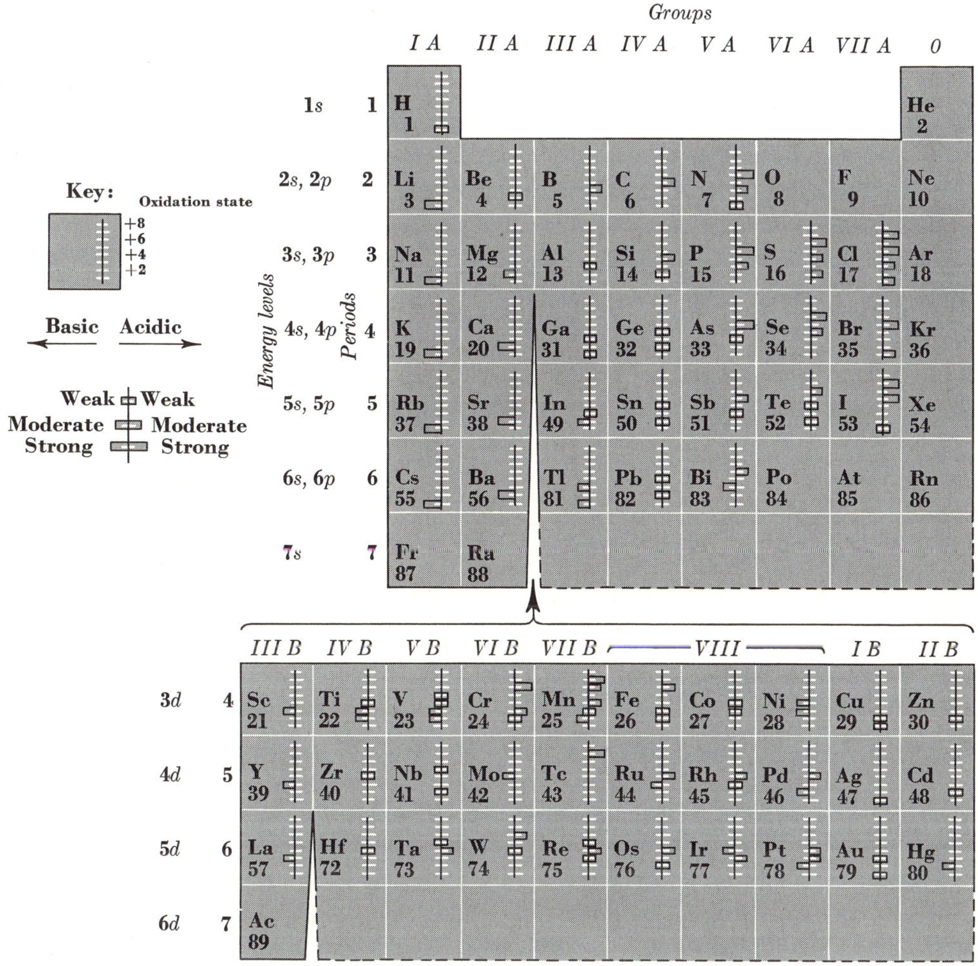

basic and (or) acidic appear in Table 20.2. The following periodic trends are observed.

(1) *The general trend within a period, from left to right, for oxidation states equal to the group number is from basic through amphoteric to acidic oxides.* The trend also holds for the hydroxy compounds as typified by the elements of the third period:

NaOH	$Mg(OH)_2$	$Al(OH)_3$	$SiO_2(H_2O)_n$	$(HO)_3PO$	$(HO)_2SO_2$	$HOClO_3$
strong base	*weak base*	*amphoteric*	*very weak acid*	*moderate acid*	*strong acid*	*strongest acid*

(2) *The general trend within a group, downward, is toward increased basicity and (or) decreased acidity of oxides and of hydroxy compounds.*

Thus, we find in group VA,

$As(OH)_3$	$Sb(OH)_3$	$Bi(OH)_3$
*(arsenious acid)** *weak acid*	*amphoteric*	*weak base*

(3) *For any element, the oxide and the hydroxy compound tend to become more acidic the higher the oxidation state of the element,* as illustrated by the chlorine oxyacids:

Acid	Formula	Oxidation number	pK_a
Hypochlorous	HOCl	+1	7.5
Chlorous	HOClO	+3	2
Chloric	$HOClO_2$	+5	−1
Perchloric	$HOClO_3$	+7	−11

(c) Acid anhydride and water; the preparation of acids The reactions of acidic anhydrides with water are used to prepare acids.

Phosphoric acid is made by reacting its acid anhydride, P_4O_{10}, with water,

$$P_4O_{10} + 6H_2O \longrightarrow 4(HO)_3PO$$

The anhydride is made by burning phosphorus in an ample supply of air.

Sulfuric acid, one of the most important industrial chemicals, is prepared by such a method. The acid anhydride of $(HO)_2SO_2$ is sulfur trioxide, SO_3. Direct oxidation of elementary sulfur affords sulfur dioxide, SO_2, which can be subsequently oxidized in the presence of a vanadium pentoxide (V_2O_5) catalyst to SO_3. The hydration reaction of SO_3 is very exothermic. The heat liberated vaporizes the water, and the steam cloud carries away much of the SO_3 in the form of a sulfuric acid mist. For more effective results, the SO_3 is dissolved in concentrated $(HO)_2SO_2$ to form pyrosulfuric acid, $H_2S_2O_7$, while water is added continuously to maintain a constant concentration of $(HO)_2SO_2$:

$$S_8 + 8O_2(g) \longrightarrow 8SO_2(g)$$
$$2SO_2(g) + O_2(g) \rightleftharpoons 2SO_3(g) \quad [\textbf{contact process}]$$
$$SO_3(g) + (HO)_2SO_2 \rightleftharpoons H_2S_2O_7$$
$$H_2S_2O_7 + H_2O \longrightarrow 2(HO)_2SO_2$$

(d) Autooxidation of nonmetallic oxides; preparation of nitric acid Nitrogen dioxide does not yield an oxyacid in which nitrogen has the same oxidation number. Instead, it is autooxidized in water to give a mixture

* The structure of arsenious acid is still in doubt. It may be hydrated arsenic(III) oxide, $As_2O_3(H_2O)_n$.

of nitric and nitrous acids,

$$\underset{+4}{2NO_2} + 2H_2O \longrightarrow H_3O^+ + \underset{+5}{NO_3^-} + \underset{+3}{HONO}$$

The autooxidation of NO_2 is the key step in the industrial preparation of nitric acid. Nitrogen dioxide is made in two steps from ammonia. Ammonia is made from nitrogen and hydrogen in the presence of an iron oxide catalyst at about 600°C and several hundred atmospheres:

$$N_2 + 3H_2 \longrightarrow 2NH_3 \qquad \text{[Haber Process]}$$

$$4NH_3 + 5O_2 \longrightarrow 4NO + 6H_2O \qquad \text{[Ostwald Process]}$$

$$2NO + O_2 \longrightarrow 2NO_2$$

$$2NO_2 + 2H_2O \longrightarrow H_3O^+ + NO_3^- + HONO$$

Autooxidation is typical of oxides of elements in groups VA and VIIA of the periodic table when the element has an even-numbered oxidation state; the autooxidation results in higher and lower odd-numbered oxidation states. Autooxidation in a basic medium results in formation of the oxyanions, as exemplified by the reactions of chlorine dioxide and dichlorine hexoxide:

$$\underset{+4}{2ClO_2} + 2OH^- \longrightarrow \underset{+3}{ClO_2^-} + \underset{+5}{ClO_3^-} + H_2O$$

chlorine dioxide

$$\underset{+6}{Cl_2O_6} + 2OH^- \longrightarrow \underset{+5}{ClO_3^-} + \underset{+7}{ClO_4^-} + H_2O$$

dichlorine hexoxide

20.5 □ STRUCTURE OF OXYANIONS

The structure of the oxyanions formed from acid anhydrides is greatly influenced by the size of the central atom. The oxyanions of the second-period elements, boron, carbon, and nitrogen,

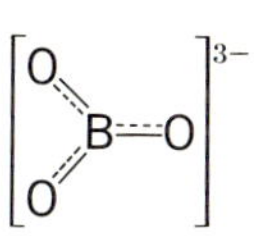

orthoborate anion

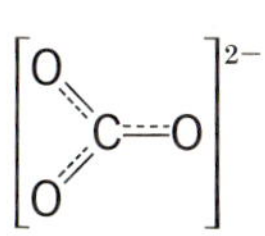

carbonate anion

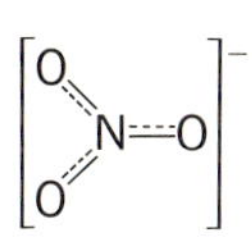

nitrate anion

have *three* oxygen atoms, each of which participates in extended p-$p\pi$ bonding with the central atom. The ions are planar and have O—M—O bond angles of 120 degrees. Each central atom uses sp^2 hybrid orbitals to form the three σ bonds. The remaining p orbital is used for the formation of the extended π bond.

The oxyanions of the third-period elements, Si, P, S, and Cl,

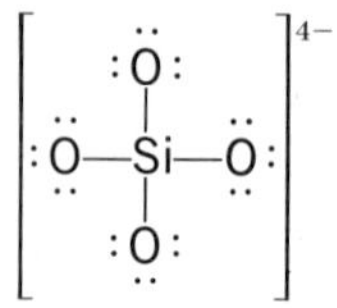

orthosilicate anion

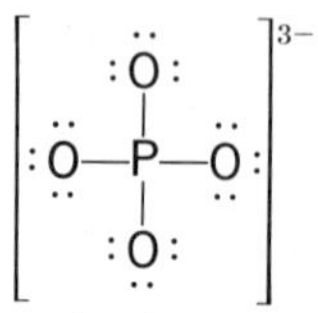

orthophosphate anion

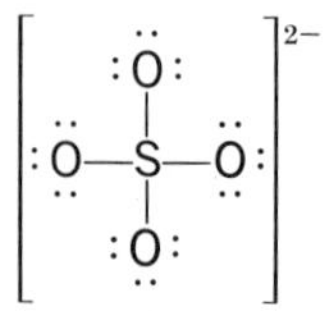

sulfate anion

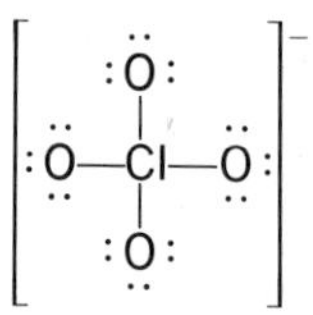

perchlorate anion

have *four* oxygen atoms bonded to the central atom. Each of these oxyanions has a tetrahedral structure, and hence the central atom utilizes sp^3 hybrid atomic orbitals to form the four σ bonds, one to each oxygen atom.

There are two main reasons for this difference in structure between oxyanions of the second- and third-period elements:

(a) The third-period elements are larger and can better surround themselves with *four* oxygen atoms.

(b) The third-period elements cannot participate in p-$p\pi$ bonding because their bonds to oxygen are too long.*

20.6 □ THE CHEMISTRY OF SOME COMMON OXIDES AND OXYANIONS

(a) Oxides of carbon Both carbon and oxygen readily form multiple bonds (p. 141), and it is therefore not surprising that all of the carbon oxides have such bonds. Four linear oxides are known:

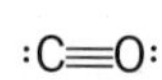

carbon monoxide

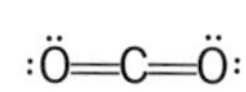

carbon dioxide

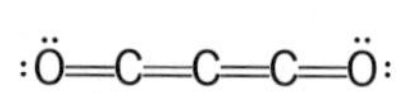

carbon suboxide (tricarbon dioxide)

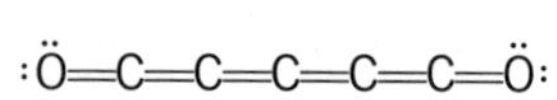

pentacarbon dioxide

Of these, CO_2 represents the most highly oxidized condition of carbon and it is therefore the ultimate oxidation product and also the major product obtained when carbonaceous matter is burned in an abundant supply of air or oxygen. When the supply of oxygen is insufficient to provide complete conversion to CO_2, the less highly oxidized product CO is formed, together with traces of the suboxides shown above.

* Sulfur is an exception (p. 405).

Carbon dioxide is a colorless, odorless, noncombustible gas with a slight acid taste. It is commercially available in steel cylinders as a liquid under the pressure of its own vapor, and also in the solid form ("dry ice") at its sublimation temperature, $-78.5°C$. Carbon dioxide is not toxic in low concentrations, but *it is toxic in high concentrations;* its asphyxiating property is greater than that produced by the exclusion of its own volume of oxygen from the air. Human beings lose consciousness when exposed for only a few minutes to air containing 10% CO_2.

Carbon dioxide is soluble in water (90 ml CO_2 gas at 1 atm in 100 ml water at 20°C) but only about 1% of it reacts to form the hydrate, carbonic acid.

$$CO_2 + H_2O \xrightleftharpoons{\sim 1\%}$$

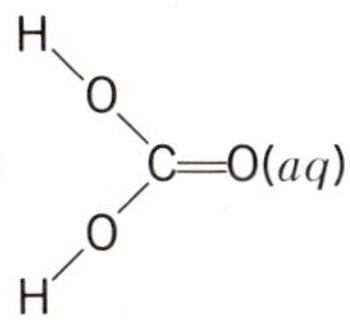

carbonic acid (unstable)

Carbonic acid itself is a moderately weak acid ($K_a \approx 5 \times 10^{-4}$, about 25 times as strong as acetic acid), but because so little of it is present when CO_2 dissolves in water the saturated solution is very weakly acidic. Carbonates and bicarbonates, which are salts of carbonic acid, are stable and well known. CO_2 gas may be prepared in the laboratory by the action of strong acids on these salts:

$$2HCl(aq) + \underset{\substack{\textit{calcium carbonate}\\ \textit{(limestone; marble)}}}{Ca^{2+}CO_3^{2-}(c)} \longrightarrow Ca^{2+}Cl_2^-(aq) + H_2O(l) + CO_2(g)$$

$$HCl(aq) + \underset{\textit{sodium carbonate}}{Na^+HCO_3^-(c)} \longrightarrow Na^+Cl^-(aq) + H_2O(l) + CO_2(g)$$

The gas is produced commercially (a) by burning coke or petroleum

$$C(amorph) + O_2(g) \rightleftharpoons CO_2(g) \qquad \Delta H = -94 \text{ kcal/mole}$$

or (b) as a byproduct of fermentation, or (c) by the "burning" of limestone:

$$\underset{\textit{limestone}}{CaCO_3(c)} \longrightarrow \underset{\textit{lime}}{CaO(c)} + CO_2(g)$$

Metals or carbon at incandescent temperatures reduce CO_2 to the element or to a lower oxide:

$$2Mg(c) + CO_2(g) \longrightarrow 2MgO(c) + 2C(amorph)$$

Carbon dioxide is a part of the respiration cycle of living organisms.

Carbon monoxide is an odorless, colorless, tasteless, flammable, highly toxic gas. Mixtures of CO in air in the concentration range between 12.5% and 74% CO by volume are explosive.

SAFETY NOTE: CO TOXICITY IS INSIDIOUS BECAUSE IT GIVES LITTLE SENSORY WARNING OF ITS PRESENCE; A CONCENTRATION OF 0.1% BY VOLUME PRODUCES UNCONSCIOUSNESS IN 1 HOUR AND DEATH IN 4 HOURS.

Therefore, the combustion products of any carbonaceous fuel should be regarded as a possible source of CO and should not be allowed to accumulate without control in occupied spaces.

Carbon monoxide may be produced in the laboratory by heating calcium carbonate with zinc dust:

$$\text{CaCO}_3(c) + \text{Zn}(c) \longrightarrow \text{CaO}(c) + \text{ZnO}(c) + \text{CO}(g)$$

or by the dehydration of formic acid, 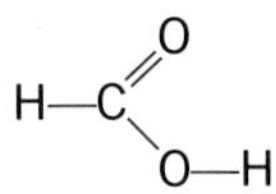, using sulfuric acid as a dehydrating agent:

$$\text{HCOOH}(l) + \text{(HO)}_2\text{SO}_2(aq) \longrightarrow \text{H}_3\text{O}^+(aq) + \text{HOSO}_3^-(aq) + \text{CO}(g)$$

It is produced on an industrial scale by the reaction of steam with hot coal or coke (water gas reaction):

$$\text{C}(amorph) + \text{H}_2\text{O}(g) \longrightarrow \underbrace{\text{H}_2(g) + \text{CO}(g)}_{\text{“}water\ gas\text{”}}$$

Carbon monoxide is an important reducing agent in metallurgical processes (see Chapter 23) where it may be produced by the incomplete oxidation of carbon or by the reduction of carbon dioxide:

$$\text{C}(amorph) + \text{CO}_2(g) \rightleftharpoons 2\text{CO}(g) \qquad \Delta H = +41 \text{ kcal/mole}$$

It is also used in organic syntheses; an example is the formation of carbonyl chloride (phosgene), using activated carbon as a catalyst:

$$\text{CO}(g) + \text{Cl}_2(g) \longrightarrow \underset{phosgene}{\text{COCl}_2(g)}$$

Carbon monoxide has the property of bonding with certain transition metals and metal ions to form **carbonyls:**

$$\text{M} + x\,\text{:C}{\equiv}\text{O:} \longrightarrow \text{M(CO)}_x$$

Examples are the formation of iron pentacarbonyl, $Fe(CO)_5$, and nickel tetracarbonyl, $Ni(CO)_4$.

Carbon monoxide can be eliminated from mixtures of gaseous pollutants by using a catalyst or flame to facilitate its conversion to CO_2; if necessary, air may be added to the gas stream to supply additional oxygen. CO can also be destroyed by disproportionation* to C and CO_2 at temperatures between 400°C and 700°C. The same result can be effected catalytically at room temperature by a mixture of manganese and copper oxides called **Hopcalite:**

$$2CO(g) \longrightarrow C(amorph) + CO_2(g)$$

(b) Oxides of nitrogen Nitrogen, like carbon and oxygen, forms multiple bonds, and the nitrogen oxides, like those of carbon, exhibit bond multiplicity. Oxides corresponding to every oxidation state of nitrogen from +1 to +6 are known.

Oxide	Oxidation number of N
N_2O, nitrous oxide, dinitrogen oxide	+1
NO, nitric oxide, nitrogen oxide	+2
N_2O_3, dinitrogen trioxide	+3
NO_2, nitrogen dioxide	+4
N_2O_4, dinitrogen tetroxide	+4
N_2O_5, dinitrogen pentoxide	+5
NO_3, nitrogen trioxide (unstable)	+6

Nitrous oxide, also called "laughing gas," N_2O (see p. 146 for structural formula) is prepared by heating ammonium nitrate† to about 200°C. This is an internal redox reaction (see Problem 8, p. 409).

$$NH_4^+NO_3^- \longrightarrow 2H_2O + N_2O(g)$$

N_2O is a colorless gas (b.p. −88°C) with a slightly sweet odor and taste, and has anesthetic properties (see p. 145). It is stable at ordinary temperatures, but will support combustion:

$$H_2(g) + N_2O(g) \longrightarrow H_2O(g) + N_2(g)$$

Nitric oxide, $:\ddot{N}{=}\dot{O}: \longleftrightarrow :\dot{N}{=}\ddot{O}:$, is prepared industrially by the high temperature oxidation of N_2 or NH_3. It may be conveniently prepared in the laboratory by reduction of sodium nitrite with ferrous sulfate:

$$NO_2^- + Fe^{2+} + 2H^+ \longrightarrow NO(g) + Fe^{3+} + H_2O$$

NO is a colorless gas. It is a paramagnetic substance (note the odd number of electrons), and condenses to a blue liquid, b.p. −152°C. It occurs as a byproduct of all combustion processes that occur in air,

* "Disproportionation" is a redox reaction within a molecule or between identical molecules.

† SAFETY NOTE: AMMONIUM NITRATE IS AN EXPLOSIVE, BUT SMALL QUANTITIES OF IT ARE NOT PARTICULARLY SENSITIVE. IN LARGE MASSES, THE BUILDUP OF INTERNAL PRESSURES ON HEATING MAY LEAD TO DISASTROUS DETONATION, SUCH AS THAT AT TEXAS CITY IN 1947.

and rapidly undergoes further oxidation to NO_2:

$$NO(g) + \tfrac{1}{2}O_2(g) \longrightarrow NO_2(g)$$

If NO loses one electron, it forms the relatively stable **nitrosonium ion,** NO^+, which is isoelectronic with the stable gases N_2 and CO. Known compounds of this ion are nitrosyl perchlorate, $NO^+ClO_4^-$, and nitrosyl hydrogen sulfate, $NO^+HSO_4^-$.

Dinitrogen trioxide, $:\ddot{O}{=}\ddot{N}{-}\underset{\cdot\cdot}{\ddot{O}}{-}\ddot{N}{=}\ddot{O}:$, is formed by cooling an equimolar mixture of NO and NO_2 to about −20°C:

$$NO(g) + NO_2(g) \longrightarrow N_2O_3\ (\textit{blue liquid})$$

Note that this reaction is the combination of two substances with unpaired electrons to form a nonradical product (see Problem 26, p. 410). N_2O_3 is the acid anhydride of nitrous acid:

$$N_2O_3(g) + H_2O(l) \longrightarrow \underset{\textit{nitrous acid (unstable)}}{HONO(aq)}$$

The ion produced when N_2O_3 or HONO reacts with a base is nitrite ion:

$$HONO + OH^- \longrightarrow H_2O + \underset{\textit{nitrite ion}}{ONO^-}$$

Nitrogen dioxide, $:\underset{\cdot\cdot}{\ddot{O}}{-}\dot{N}{=}\ddot{O}:$ (p. 121), is made industrially by oxidation of NO, and may be prepared in the laboratory by gently heating lead nitrate:

$$2Pb(NO_3)_2(c) \longrightarrow 2PbO(c) + 4NO_2(g) + O_2(g)$$

NO_2 is a reddish-brown gas. It is paramagnetic, and its free-radical character predisposes it to combine with itself (**dimerize,** see p. 496) to produce the lighter-colored dinitrogen tetroxide. The two gases exist in equilibrium at ordinary temperatures.

$$2NO_2(g) \rightleftharpoons N_2O_4(g)$$

NO_2 is a deadly poison; its great danger is often unappreciated. Initial exposure may cause inflammation of the lungs with only slight pain, but death may occur a few days later from the resulting edema. Continuous exposure to concentrations as low as 500 parts per million (0.05% by volume) may be fatal in 48 hours. NO_2 is of great significance in air pollution because it is the principal oxide of nitrogen formed in air during combustion or in secondary reactions from other oxides, and because it is involved in a series of atmospheric reactions leading to smog formation. Cigarette smoke (filtered or unfiltered) contains about 300 parts per million of NO_2; pipe tobacco smoke and cigar smoke contain higher concentrations.

NO_2 is an acidic oxide. It reacts with water to give a mixture of nitrous and nitric acids, as shown on p. 391. (See Problems 27–28, p. 411.)

If NO_2 loses an electron, it forms the relatively stable **nitronium ion,**

NO_2^+, which is isoelectronic with CO_2. Salts of this ion include nitronium perchlorate, $NO_2^+ClO_4^-$, and nitronium hydrogen sulfate, $NO_2^+HSO_4^-$.

Dinitrogen pentoxide, represented in the vapor state by the structural formula 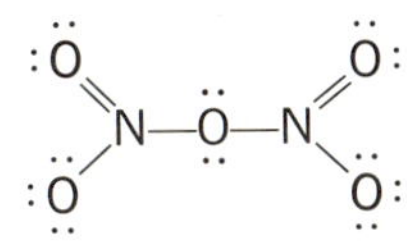, is best prepared by reaction of NO_2 with ozone.

$$2NO_2(g) + O_3(g) \longrightarrow N_2O_5(c) + O_2(g)$$

N_2O_5 is an unstable white solid (m.p. about 30°C) that decomposes readily:

$$N_2O_5(c) \longrightarrow 2NO_2(g) + O_2(g)$$

Solid N_2O_5 is an ionic salt of nitronium ion (see above), and may therefore be called nitronium nitrate, $NO_2^+NO_3^-$. N_2O_5 is the acid anhydride of nitric acid:

$$N_2O_5(c) + H_2O(l) \longrightarrow \underset{\textit{nitric acid}}{2HONO_2(aq)}$$

The corresponding anion is nitrate ion, NO_3 (see p. 146). Nitric acid is a strong acid, a strong oxidizing agent, and an effective nitrating agent (see p. 507). Examples of the oxidizing action of nitric acid are given in the following equations. Note that the state to which the HNO_3 is reduced depends both on the nature of the reducing agent and on the concentration of the acid:

$$Cu(c) + 4HONO_2(\textit{conc aq}) \longrightarrow Cu^{2+}(NO_3)_2^-(aq) + 2NO_2(g) + 2H_2O(l)$$
$$3Cu(c) + 8H^+NO_3^-(\textit{dilute aq}) \longrightarrow 3Cu^{2+}(NO_3)_2^-(aq) + 2NO(g) + 4H_2O(l)$$
$$4Zn(c) + 10H^+NO_3^-(\textit{dilute aq}) \longrightarrow 4Zn^{2+}(NO_3)_2^-(aq) + NH_4^+NO_3^-(aq) + 3H_2O(l)$$

(c) Oxides of sulfur The known oxides are

disulfur oxide	S_2O
sulfur monoxide	SO
disulfur trioxide	S_2O_3
sulfur dioxide	SO_2
sulfur trioxide	SO_3
disulfur heptoxide	S_2O_7
sulfur tetroxide	SO_4

Of these, only SO_2 and SO_3 are stable and of general interest.

Sulfur dioxide, 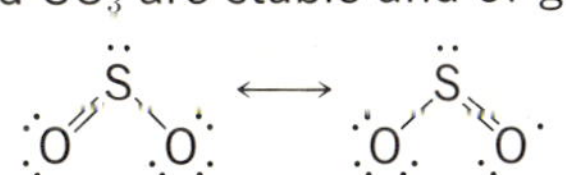, is prepared industrially by burning sulfur (p. 390), or by roasting sulfide ores (see p. 469). SO_2 is a colorless, nonflammable gas that most people can detect in air by taste in concentrations around 1 part per million by volume. In concentrations of 3 ppm or more it has a pungent, irritating odor. It is probably the most significant single air pollutant in urban atmospheres in winter, when large quantities of it are discharged to the atmosphere by burning of sulfur-containing fuels.

SO_2 is the anhydride of sulfurous acid:

$$SO_2(g) + H_2O(l) \rightleftharpoons \underset{\textit{sulfurous acid (unstable)}}{(HO)_2SO(aq)}$$

The corresponding anions are bisulfite, $HOSO_2^-$, and sulfite, SO_3^{2-}. Sulfur dioxide can serve either as a reducing agent or as an oxidizing agent, although the former is its more usual role:

$$\underset{\textit{sulfite ion as a reducing agent}}{SO_3^{2-}(aq)} + 2Fe^{3+}(aq) + H_2O \longrightarrow SO_4^{2-}(aq) + 2Fe^{2+}(aq) + 2H^+(aq)$$

$$\underset{\textit{sulfur dioxide as an oxidizing agent}}{SO_2(g)} + 2H_2S(g) \longrightarrow 2H_2O(l) + 3S(c)$$

The latter equation represents a process that can be used to remove sulfur-containing substances from natural gas by converting them to useful sulfur (see Problem 23, p. 93).

SO_2 reacts slowly with oxygen at 400°C to produce SO_3, but the process can be facilitated catalytically (see p. 390).

Sulfur trioxide, SO_3, is prepared by oxidation of SO_2. At room temperature it is a solid that may exist in any of several different crystal modifications (α-form, m.p. 16.8°C; β-form, m.p. 32.5°C; γ-form, m.p. 62.3°C). It is the anhydride of sulfuric acid (see also p. 388).

$$SO_3(g) + H_2O(l) \longrightarrow \underset{\textit{sulfuric acid}}{(HO)_2SO_2(l)}$$

The corresponding anions are bisulfate, $HOSO_3^-$, which is itself a fairly strong acid, and sulfate, SO_4^{2-}. Sulfates and bisulfates in aqueous solution are not particularly effective oxidizing agents, but hot concentrated sulfuric acid is moderately effective, as evidenced by its ability to oxidize bromides to bromine and iodides to iodine:

$$2Na^+Br^-(c) + 2H_2SO_4(l) \longrightarrow Br_2(g) + SO_2(g) + Na^+{}_2SO_4^{2-}(aq) + 2H_2O(l)$$
$$8K^+I^-(c) + 5H_2SO_4(l) \longrightarrow 4I_2(g) + H_2S(g) + 4K^+{}_2SO_4^{2-}(aq) + 4H_2O(l)$$

It will not oxidize fluorides or chlorides.

(d) Oxides of the halogens Oxygen is more electronegative than any element except fluorine. As a result, all the halogens except fluorine

TABLE 20.3 THE KNOWN HALOGEN OXIDES

Oxidation Number of the Halogen

	−1	+1	+3	+4	+5	+6	+7	+8	Mixed or unknown
F	OF_2								O_2F_2 O_5F_2 O_6F_2
Cl		Cl_2O	Cl_2O_3	ClO_2		Cl_2O_6	Cl_2O_7	Cl_2O_8	
Br		Br_2O		BrO_2 Br_2O_4					Br_3O_8
I				I_2O_4	I_2O_5				I_4O_9

acquire positive oxidation states in their oxides. The following generalizations may be made with regard to the properties of the halogen oxides (see Table 20.3):

(1) A halogen has an odd number of electrons. Therefore, any of the oxides with an odd number of halogens will also have an odd number of electrons and be paramagnetic. Such oxides will tend to dimerize (p. 496). Examples are BrO_2 and IO_2.

(2) The most stable compounds of halogens are those (like HCl and NaCl) in which the halogen is in a negative charge condition (See Section 20.8, p. 402). In their oxides, however, the halogens are in charge conditions that are considerably more positive, and these compounds are therefore powerful oxidizing agents. For example, ClO_2 has been used as a bleaching and water-purifying agent.

(3) Bond dissociation energies of oxygen-to-halogen bonds are relatively low (for example, HO—Br, 56 kcal/mole, compared with the C—O bond in ethyl alcohol, 90 kcal/mole). As a result, the halogen oxides typically decompose readily to their elements, some explosively (such as ClO_2).

The most important halogen oxyanions are those of chlorine (see p. 392). All of these oxyanions, from hypochlorite, OCl^-, to perchlorate, ClO_4^-, are oxidizing agents, although their oxidizing potentials do not vary directly with the oxidation number of the central atom:

	H^+ClO^-	$H^+ClO_2^-$	$H^+ClO_3^-$	$H^+ClO_4^-$	
$\mathscr{E}°$ for reduction of acid	1.49	1.57	1.48	1.36	volts
		strongest oxidant		*weakest oxidant*	

Perchloric acid, $HClO_4$, is a colorless, volatile, explosive liquid. It is marketed only in mixtures with water, usually containing 60–70% $HClO_4$. The acid is used as an oxidizing agent, and its magnesium salt, $Mg(ClO_4)_2$, is used as an effective dehydrating agent.

SAFETY NOTE: PERCHLORATES, WHEN MIXED WITH ANY OXIDIZABLE MATTER, FORM VERY SENSITIVE HIGH EXPLOSIVES THAT MAY DETONATE UNPREDICTABLY.

(e) Oxides of phosphorus Three oxides are known:

Phosphorus(III) oxide	P_4O_6
Phosphorus(IV) oxide	P_2O_4
Phosphorus(V) oxide (formerly written P_2O_5 and called phosphorus pentoxide)	P_4O_{10}

The III and V oxides are the more common and important ones.

Phosphorus(III) oxide is produced when white phosphorus is burned in a limited supply of air:

$$2P_4(c) + 6O_2(g) \longrightarrow 2P_4O_6(c)$$

The oxide is the anhydride of phosphorous acid,

$$P_4O_6 + 6H_2O \longrightarrow \underset{\textit{phosphorous acid}}{4(HO)_2PHO_2}$$

When phosphorus burns in an excess of oxygen, the principal product is the phosphorus(V) oxide, P_4O_{10}. This is the anhydride of phosphoric acid, $(HO)_3PO$, which is an acid of intermediate strength (see p. 331).

P_4O_{10} is a very effective drying agent.

Fig. 20.1 Structures of phosphorus oxides.

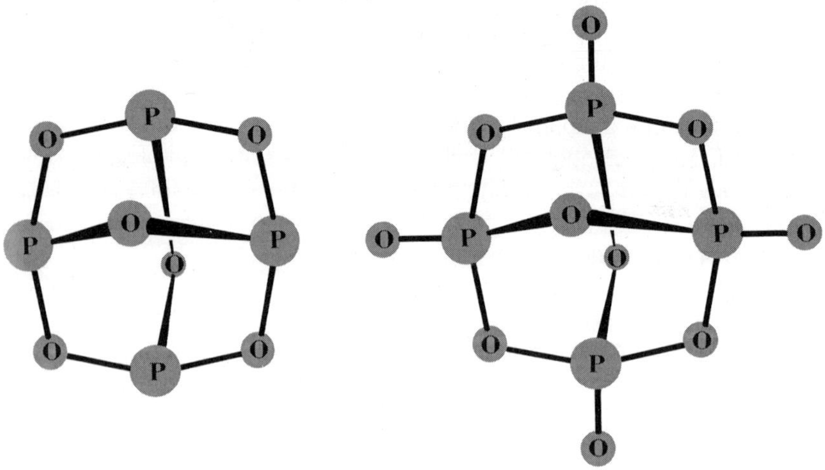

P_4O_6 P_4O_{10}

Both the P_4O_6 and P_4O_{10} oxides have molecular structures that are derived from the P_4 tetrahedron. See Fig. 20.1.

The oxyanions corresponding to P_4O_6 include $HOPHO_2^-$ and HPO_3^{2-}. The V oxyanions include phosphate, PO_4^{3-}, and the various protonated phosphate ions. The reduction of phosphate minerals by carbon is the industrial method for preparing elemental phosphorus:

$$2Ca_3^{2+}(PO_4^{3-})_2(c) + 6SiO_2(c) + 10C(amorph) \longrightarrow 6Ca^{2+}SiO_3^{2-}(c) + 10CO(g) + P_4(g)$$

The oxide P_4O_{10} is formed as an intermediate, and is then reduced by the carbon, with the overall process as depicted above.

20.7 □ PEROXIDES

In substances containing the **peroxide bond,** —Ö—Ö—, including the peroxide ion, O_2^{2-}, the oxidation number of oxygen is −1. **Hydrogen peroxide,** H_2O_2, is a very pale blue, syrupy liquid which melts at −0.89°C and boils at 152°C, if it does not explode first. It is miscible with water in all proportions; the hydrogen peroxide solution that is sometimes used as an antiseptic is about 3% H_2O_2 in water; the more concentrated form called "superoxol," sometimes used as a bleaching solution, is 30% H_2O_2. Hydrogen peroxide decomposes exothermically

$$H_2O_2(l) \longrightarrow H_2O(l) + \tfrac{1}{2}O_2(g) \qquad \Delta H = -23.5\ \frac{\text{kcal}}{\text{mole}}$$

Many substances, such as metals and metallic oxides, effectively catalyze this decomposition; therefore small quantities of impurities may encourage pure H_2O_2 or concentrated solutions to decompose explosively.

SAFETY NOTE: SOLUTIONS LIKE "SUPEROXOL" SHOULD BE KEPT CLEAN, COLD, AND VENTED TO AVOID BUILDUP OF INTERNAL PRESSURE (LEAVE THE STOPPER LOOSE).

The structure of the H_2O_2 molecule approximates two right angles in different planes.

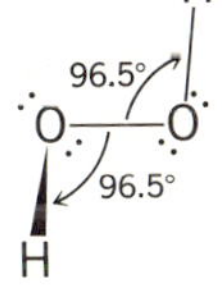

H_2O_2 typically functions as a strong oxidizing agent, although in some circumstances it behaves as a mild reducing agent.

$$\underset{\textit{oxidizing agent}}{H_2O_2(aq)} + SO_3^{2-}(aq) \longrightarrow H_2O(l) + SO_4^{2-}(aq)$$

$$\underset{\textit{reducing agent}}{5H_2O_2(aq)} + 2MnO_4^-(aq) + 6H^+(aq) \longrightarrow 5O_2(g) + 2Mn^{2+}(aq) + 8H_2O(l)$$

Hydrogen peroxide is weakly acidic in water solutions, $K_a \approx 10^{-12}$.

$$H_2O_2(aq) + H_2O(l) \longrightarrow H_3O^+(aq) + \underset{\textit{hydroperoxide ion}}{HO_2^-(aq)}$$

Metallic peroxides of the alkali metals ($M_2^+O_2^{2-}$) and of most of the alkaline earth metals ($M^{2+}O_2^{2-}$) are well known. Examples are Na_2O_2, Rb_2O_2, and BaO_2. These peroxides readily react with water or acids to form hydrogen peroxide.

$$Na_2^+O_2^{2-}(c) + 2H_2O(l) \longrightarrow 2Na^+OH^-(aq) + H_2O_2(aq)$$

Since some peroxides, notably Na_2O_2 and BaO_2, are easily formed simply by heating the metal in air, the hydrolysis of such peroxides affords an easy route for the laboratory preparation of H_2O_2. A particularly convenient procedure is the reaction of BaO_2 with sulfuric acid:

$$Ba^{2+}O_2^{2-}(c) + H_2SO_4(aq) \longrightarrow BaSO_4(c) + H_2O_2(aq)$$

The insoluble $BaSO_4$ can be filtered off and the remaining liquid phase is an aqueous solution of hydrogen peroxide.

20.8 □ PERIODICITY OF PROPERTIES OF THE HALIDES; DIFFERENCES IN OXIDATION STATES

A review of the periodicity of the chemical properties of the halides involves complexities of dealing with the four common halogens, fluorine, chlorine, bromine, and iodine, and in some cases, more than one state of the other element, as in PCl_3 and PCl_5. For simplification, we shall focus attention on the EX_G or $EX_{8\text{-}G}$ halides, where X is the halogen and E is the representative element of periodic group G. Thus for second period elements, typical halides would be LiCl, $BeCl_2$, BCl_3, and CCl_4 in the EX_G series, and NCl_3 and OF_2 in the $EX_{8\text{-}G}$ series. We shall consider mainly the chlorides as typical cases and point out sharp differences with other halides where they exist.

Each halogen is the most electronegative element in its period. Chlorine is one of the most electronegative elements, and it therefore bears a fractional negative charge in most of its compounds. As the fractional charge on chlorine becomes more negative (approaching -1) the halides become more ionic. In general, the halides of the electropositive, metallic elements such as the alkali metals (group IA) and the alkaline earths (group IIA) are ionic salts. They have high melting points and, in a fused state, conduct an electric current. The ionic character increases on proceeding down the group of the metal and up the group of halogens. The most ionic halide is therefore presumably FrF; however, we are relatively unfamiliar with the chemistry of the synthetic radioactive element francium, and we therefore consider CsF to be the most ionic halide we know.

Ionic character of the halides decreases on proceeding from left to right across a period. Thus, LiCl is an ionic compound, $BeCl_2$ has considerable covalent character, and BCl_3 is a typical covalent substance. We can generalize by saying that halides of typically metallic elements are ionic compounds and halides of typical nonmetallic elements are molecular covalent substances. The halides of the nonmetals and of some metals are frequently gases such as BCl_3 and NCl_3, volatile liquids such as CCl_4, or easily sublimed solids such as Al_2Br_6. Furthermore, as the oxidation state of the element increases, the corresponding halide behaves more like that of a nonmetal. We shall illustrate this fact below, when comparing the behavior of $SbCl_3$ and $SbCl_5$.

The halides of most nonmetallic elements hydrolyze to give the oxyacid of the element and the corresponding hydrohalogen acid. The oxidation number of the element is retained:

$$BCl_3 + 6H_2O \longrightarrow B(OH)_3 + 3H_3O^+ + 3Cl^-$$
$$PBr_3 + 6H_2O \longrightarrow (HO)_2PHO + 3H_3O^+ + 3Br^-$$
$$PCl_5 + 9H_2O \longrightarrow (HO)_3PO + 5H_3O^+ + 5Cl^-$$

The reaction with PBr_3 is used to prepare hydrobromic acid. In the above examples, the halogen is more electronegative than the central atom. In such cases the reaction usually proceeds as shown. If the halogen is less electronegative, the hydrolysis proceeds as shown for nitrogen triiodide:

$$NI_3 + 3H_2O \longrightarrow NH_3 + 3HOI$$
$$NH_3 + HOI \rightleftharpoons NH_4^+ + OI^-$$

As the metallic properties of the central atom increase, as is the trend on proceeding down a group, hydrolysis is incomplete and oxyhalides are formed, as shown for antimony(III) chloride:

$$SbCl_3 + 3H_2O \rightleftharpoons SbOCl + 2H_3O^+ + 2Cl^-$$

Antimony(V) chloride reacts with water as a typical nonmetallic halide:

$$SbCl_5 + 4H_2O \longrightarrow (HO)_3SbO + 5HCl(aq)$$

Carbon tetrachloride, CCl_4, is one of the few nonmetallic halides that resist hydrolysis. It is immiscible with water and relatively inert. An even more inert halide is sulfur hexafluoride, SF_6.

In contrast to CCl_4, silicon tetrachloride, $SiCl_4$, reacts vigorously with water,

$$SiCl_4 + 4H_2O \longrightarrow \underset{\text{"silicic acid"*}}{Si(OH)_4} + 4HCl$$

* Silicic acid, $Si(OH)_4$, is not a stable compound. It undergoes partial loss of water to give a gellike form of silica hydrated by an indefinite number of water molecules, $SiO_2(H_2O)_x$.

This sharp difference between the behavior of CCl_4 and $SiCl_4$ arises because silicon has *d* orbitals available for bonding, while carbon has none. Therefore a silicon atom can acquire more than eight "outer" electrons. Silicon tetrachloride acts as an electrophile and forms an octahedral complex (sp^3d^2) by bonding with two water molecules (ligands),

$$SiCl_4 + 2H\text{—}\ddot{\underset{..}{O}}\text{—}H \rightleftharpoons \underset{intermediate}{[SiCl_4(H_2O)_2]}$$

The silicon atom reestablishes its octet of outer electrons by ejecting two Cl^- ions. Each of the coordinated water molecules loses a proton to the solvent,

$$[SiCl_4(H_2O)_2] + 2H_2O \longrightarrow \underset{intermediate}{[SiCl_2(OH)_2]} + 2Cl^- + 2H_3O^+$$

By repetition of these two steps, the final products are obtained. The overall reaction is

$$SiCl_4 + 8H_2O \longrightarrow Si(OH)_4 + 4Cl^- + 4H_3O^+$$

The chlorides of germanium and of tin, $GeCl_4$ and $SnCl_4$, both typically covalent liquid substances, also react vigorously with water. Thus, $SnCl_4$ vapor reacts with water vapor to give fumes consisting of stannic acid, $Sn(OH)_4(H_2O)_2$, and hydrogen chloride. In the liquid state, the reaction is complicated by the fact that Sn(IV) complexes readily with Cl^- to give the chlorostannate ion, $SnCl_6^{2-}$.

Lead(IV) chloride, $PbCl_4$, is very unstable and at room temperature decomposes spontaneously to give lead(II) chloride.

$$PbCl_4 \longrightarrow PbCl_2 + Cl_2$$

Lead(IV) bromide and iodide are not known.

As we proceed down groups IIIA, IVA, and VA, we find that the stability of the higher oxidation state of the metal decreases, and the stability of the lower oxidation state increases. Specifically, in each of these groups, the trend is for the $G - 2$ oxidation state (where G is the group number) to become more stable at the expense of the G oxidation state as the atomic number of the element increases. For example, carbon(II) halides are unknown. Silicon(II) chloride, $SiCl_2$, has been prepared but readily disproportionates to form the more stable silicon(IV) chloride:

$$2SiCl_2 \longrightarrow SiCl_4 + Si$$

Tin(II) chloride, $SnCl_2$, and tin(IV) chloride, $SnCl_4$, are both stable compounds.

Among the group IIIA elements, the +3 oxidation state is the stable one for boron, aluminum, gallium and indium. However, thallium, the heaviest element in the group, exists mainly in the Tl^+ state.

The tendency for the lower oxidation state to become more stable than the higher oxidation state as the atomic number increases within a group in the periodic table is a general phenomenon. For example, we find that it is more difficult to reduce phosphoric acid to phosphorous acid than to reduce arsenic acid to arsenious acid, as indicated by the following half-cell reduction potentials.

$$\underset{\textit{phosphoric acid}}{(HO)_3P{-}O} + 2H^+ + 2e^- \longrightarrow \underset{\textit{phosphorous acid}}{(HO)_2\underset{\substack{|\\H}}{P}{-}O} + H_2O \qquad \mathscr{E}^0 = -0.28$$

$$\underset{\textit{arsenic acid}}{(HO)_3As{-}O} + 2H^+ + 2e^- \longrightarrow \underset{\textit{arsenious acid}}{HOAs{-}O} + 2H_2O \qquad \mathscr{E}^0 = 0.56$$

20.9 □ PROPERTIES OF SULFIDES

Sulfur and oxygen, both in group VI, have the same number of electrons in their highest shells ($2s^22p^6$, and $3s^23p^63d^0$, respectively), both exhibit a covalence of 2, and both can form divalent anions (O^{2-} and S^{2-}). There are three important differences, however, between the two elements, as described in the following paragraphs.

(a) Sulfur, unlike oxygen, can use *d* orbitals in its highest shell and thus accommodate more than an octet of bonding electrons. As a result, sulfur forms compounds like SF_4 and SF_6, for which no oxygen analogs exist.

(b) Sulfur atoms are larger than oxygen atoms and consequently have less tendency to form double bonds. Thus, O_2 itself is a multiply bonded molecule (p. 136), whereas sulfur exists as singly bonded S_8 cyclic molecules, and as linear chains of S atoms in liquid sulfur. Sulfur also adds to S^{2-} ions to form linear polysulfides:

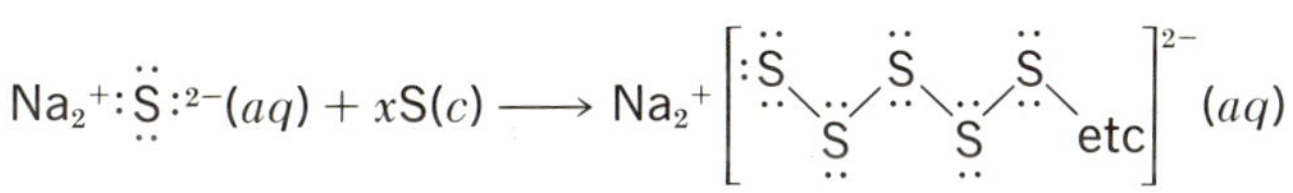

sodium polysulfide

Sulfur can participate in multiple bonding with carbon, however, as in **carbon disulfide**, $:\ddot{S}{=}C{=}\ddot{S}:$.

(c) Sulfur is less electronegative than oxygen. Therefore, the fractional charge on sulfur in sulfides is much less negative than on oxygen in oxides. Some of the values appear below:

	Fractional charge on O or S		
Na_2O	−0.81	Na_2S	−0.66
MgO	−0.50	MgS	−0.37
CO_2	−0.11	CS_2	−0.03
OF_2	+0.07	SF_6	+0.33
SO_3	−0.06 on O; +0.19 on S		

The sulfides of the metals therefore have less ionic and more covalent character than the oxides. Since ionic character contributes to the strength of a bond, the sulfides are less stable (with respect to decomposition) and lower melting than the oxides. Oxygen can therefore replace sulfur in sulfides to form the more stable oxides; thus sulfides will burn in oxygen or air (see discussion of roasting of sulfide ores, p. 469). Such replacement also occurs in aqueous solution, resulting in the hydrolysis of sulfides in which the negative charge on sulfur is sufficiently low and which are soluble enough to facilitate reaction. Such compounds are **amphoteric sulfides:**

$$Al_2S_3(c) + 6H_2O \longrightarrow 2Al(OH)_3(c) + 3H_2S(g)$$

Some sulfides, like As_2S_3, Sb_2S_3, and SnS_2, though less soluble than Al_2S_3, can be hydrolyzed in a strongly basic solution, like $KOH(aq)$. These sulfides are also considered to be amphoteric.

It is particularly important to recognize the difference between hydrogen sulfide, H_2S, and water, H_2O. The H—S—H bond angle is about 92 degrees (see p. 157), in contrast to the H—O—H bond angle which is 105 degrees. This fact implies that sulfur uses unhybridized p orbitals in its bonding with hydrogen. The H—S bond is less polar than the H—O bond, and the H bonding exhibited by water does not occur in H_2S. These concepts are consistent with the facts that H_2S is a gas at ordinary temperatures (b.p. −60°C), and that liquid H_2S, unlike water, is relatively nonpolar and is a very poor solvent for ionic substances (dielectric constant of water, 80; of H_2S, 9).

H_2S may be prepared in the laboratory by the action of a strong acid on a metallic sulfide:

$$FeS(c) + H_2SO_4(aq) \longrightarrow Fe^{2+}SO_4^{2-}(aq) + H_2S(g)$$

It can also be produced by the direct union of hydrogen and sulfur vapors at about 600°C. The gas has a characteristically offensive "rotten egg" odor, detectable by most people in air in concentrations as low as about 5 parts per billion.

SAFETY NOTE: H_2S IS EXTREMELY TOXIC; EXPOSURE TO MODERATE CONCENTRATIONS ANESTHETIZES THE SENSE OF SMELL AND THUS DESTROYS ONE'S ABILITY TO DETECT EVEN HIGHER CONCENTRATIONS. ONE OR TWO INHALATIONS OF A CONCENTRATED H_2S MIXTURE CAUSE RESPIRATORY FAILURE AND DEATH WITHIN A FEW SECONDS.

H_2S is flammable; mixtures of H_2S and air are explosive in the range of 4.3 to 46% H_2S by volume:

$$H_2S(g) + 1\tfrac{1}{2}O_2(g) \longrightarrow H_2O(g) + SO_2(g)$$

H_2S is fairly soluble in water, where it acts as a diprotic acid (p. 316).

The fact that the $[S^{2-}]$ is very sensitive to the acidity, or $[H_3O^+]$, of the solution (p. 338) is used to advantage in schemes of qualitative analysis of metal ions in aqueous solution that involve the formation of insoluble sulfides. The sulfides of some metals, such as HgS, PbS, Bi_2S_3, CuS, CdS, and others, are so insoluble that they will precipitate even from mildly acidic solutions in which $[S^{2-}]$ is as low as about 10^{-20} *M*. Separations of groups of ions may be made on the basis of the $[H_3O^+]$ at which they will or will not precipitate as the sulfide. Also helpful in qualitative cation analysis is the fact that some metallic sulfides have characteristic colors, as shown in the following tabulation.

Color	Sulfides
White	ZnS
Yellow	As_4S_6, CdS, As_2S_5, SnS_2
Orange	Sb_2S_3
Brown	Bi_2S_3
Black	CuS, HgS, PbS, CoS, NiS, FeS, Ag_2S
Pink	MnS

Some sulfides that are so insoluble that they will not react with aqueous solutions of HCl or H_2SO_4 can be dissolved in nitric acid solution. The HNO_3 oxidizes the sulfide to free sulfur; the metallic cation remains in solution. Examples of such sulfides are CuS and PbS:

$$3PbS(c) + 2NO_3^- + 8H^+ \longrightarrow 3Pb^{2+} + 3S + 2NO(g) + 4H_2O$$

20.10 ▫ COMPOUNDS OF NOBLE GASES

The apparently inert chemical behavior of the noble gases gave the original and sustaining support to the significance of the stable electronic octet. However, in 1962 even this support was weakened by the reported synthesis of compounds of the noble gases. Shortly before, it had been demonstrated that oxygen reacts with platinum(VI) hexafluoride, a deep red vapor at ordinary temperatures, to give an ionic solid, dioxygenyl hexafluoroplatinate(V), $(O_2)^+(PtF_6)^-$. Since the first ionization energy of O_2, 281.3 kcal/mole, is practically the same as that of xenon, it seemed reasonable to attempt the reaction with xenon. The PtF_6 vapor, mixed with an equimolar amount of xenon, immediately formed a yellow solid. In this reaction, xenon transfers an electron to PtF_6,

$$:\ddot{\underset{..}{Xe}}: + PtF_6 \longrightarrow [:\ddot{\underset{..}{Xe}}\cdot]^+[PtF_6]^-$$

colorless gas	*red in vapor state*	*yellow solid*

Also, xenon tetrafluoride, XeF_4, a yellow crystalline solid, was prepared by direct interaction of xenon and fluorine,

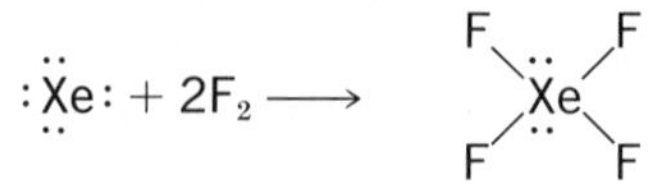

yellow crystalline solid

In XeF_4, xenon has 12 outer electrons. The dike was broken; in a short time, XeF_2, XeF_6, $XeOF_4$, KrF_2, and several others were synthesized.

PROBLEMS

1. Definitions Define and illustrate the following terms: (a) allotropy; (b) hydride; (c) basic anhydride; (d) acid anhydride; (e) oxyanion; (f) amphoteric sulfide; (g) autooxidation.

2. Allotropy Allotropy is not observed among the halogens. Offer an explanation for this fact.

3. Hydrides Complete and balance the following equations. If no reaction occurs, write "no reaction." Indicate aggregation states of the products.

(a) $Ba^{2+}H_2^-(c) + H_2O(l) \longrightarrow$

(b) $C_8H_{18}(l) + H_2O(l) \longrightarrow$

(c) $Li^+AlH_4^-(c) + H_2O(l) \longrightarrow$

(d) $HBr(g) + H_2O(l) \longrightarrow$

(e) $(H_3C)_3N\colon (g) + H_2O(l) \longrightarrow$

4. Basic and acidic anhydrides Complete and balance the following equations. If no reaction occurs, write "no reaction." Indicate aggregation states of the products.

(a) $K_2{}^+O^{2-}(c) + H_2O(l) \longrightarrow$

(b) $Al_2{}^{3+}O_3{}^{2-}(c) + HBr(aq) \longrightarrow$

(c) $Al_2{}^{3+}O_3{}^{2-}(c) + Na^+OH^-(aq) \longrightarrow$

(d) $Cl_2O_7(l) + H_2O(l) \longrightarrow$

5. Structure and acidity Arrange the following sets in order of increasing acidity in water. Account for your order. (a) H_2S, H_2Te, H_2Se. (b) SiH_4, HCl, PH_3, H_2S. (c) $HOPO_3{}^{2-}$, $(HO)_2PO_2{}^-$, $(HO)_3PO$. (d) $(HO)_2SO_2$, $(HO)_2SeO_2$, $(HO)_2SeO$. (e) $Mn(OH)_2$, $Mn(OH)_3$, $(HO)_2MnO_2$, $HOMnO_3$.

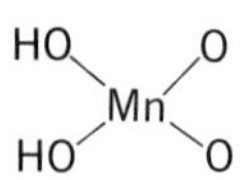

manganic acid

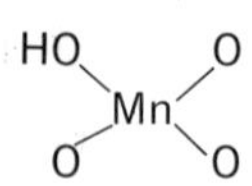

permanganic acid

6. Structure and basicity Arrange the following sets in order of increasing basicity in water. Account for your order. (a) HCl, PH_3, H_2S. (b) NH_3, PH_3, AsH_3. (c) H_2O, O^{2-}, OOH^-, OH^-. (d) NO_3^-, NO_2^-, PO_4^{3-}.

7. Carbon oxides Write the balanced equation for the complete oxidation of carbon suboxide.

8. Nitrous oxide The decomposition of $(NH_4)^+(NO_3)^-$ is an internal redox reaction. Write the partial ionic equations, identify the oxidation and the reduction, and show that the equation on p. 395 is the sum of the partial equations.

9. Nitric acid (a) Under some conditions, zinc reduces dilute nitric acid to nitrogen. Write the partial ionic equations and the overall ionic equation for this reaction. (b) Support or refute the statement, "Concentrated nitric acid is always a more powerful oxidizing agent than dilute nitric acid."

10. Sulfur dioxide Several processes have been suggested for the removal of SO_2 from gaseous atmospheric pollutants. (a) In the "Reinluft" process, SO_2 is oxidized to SO_3 on a bed of activated carbon, and the SO_3 is then converted to sulfuric acid, which is recovered. Some of the SO_3 is reduced by the carbon, however, and the SO_2 thus formed must be recycled. The carbon that is lost as CO_2 is replaced mechanically. Write the equations for each of the reactions described. Identify any substance that acts as a catalyst. (b) In the Wickert method, pulverized magnesite (magnesium carbonate) is blown into a stream of flue gas, where it catalyzes the air oxidation of SO_2. The magnesite decomposes, and the resulting metallic oxide combines with the oxidized sulfur product to form the sulfate, which is trapped in filters. Write a balanced equation for each reaction described above.

11. Halogen oxides List all the oxides shown in Table 20.3 that are likely to be paramagnetic. Is the number of such oxides large or small compared with the total number of known halogen oxides? Suggest a reason for your answer.

12. Phosphorus oxides The molecular formulas of the III and V oxides of phosphorus differ in atomic ratios from those of the corresponding nitrogen oxides. Explain these differences in terms of the relative tendencies of these two elements to participate in multiple bonding.

13. Hydrogen peroxide Write the partial ionic equations and the balanced overall equation for the oxidation of Fe^{2+} to Fe^{3+} in acidic hydrogen peroxide solution.

14. Halides The decomposition of $PbCl_4$ and the disproportionation of $SiCl_2$ are both redox reactions. For each reaction, identify all changes in oxidation numbers and state what is being oxidized and what reduced.

15. Sulfides The alkaline hydrolysis of amphoteric sulfides yields both oxy- and thioanions. (The prefix "thio" refers to sulfur in the −2 oxidation state.) Write a balanced equation for each of the following conversions: (a) As_2S_3 to AsO_2^- and AsS_2^-; (b) Sb_2S_3 to $Sb(OH)_4$ and SbS_2^-; (c) SnS_2 to $Sn(OH)_6^{2-}$ and SnS_3^{2-}. Show in each case that there are no changes in oxidation number.

ADDITIONAL PROBLEMS

16. Hydrides Metallic hydrides, such as MgH_2, give a basic reaction in water; nonmetallic hydrides like HCl are acidic in water. Does it follow that NH_3, which is basic in water, is more like a metallic hydride than is CH_4, which gives no basic reaction? Explain.

17. Hydrides LiH melts at 689°C. Would you expect liquid LiH to conduct an electric current? Write the equations for the reaction at each electrode. Write the equations for the electrode reactions in the electrolysis of the solution formed by mixing water with LiH. (Assume that the electrodes are Pt in all cases.)

18. Acid anhydrides Does the oxidation number of the central atom change when an anhydride combines with water to form an acid? Justify your answer for the anhydrides SO_2, SO_3, N_2O_3, N_2O_5, and P_4O_{10}.

19. Preparation of acids Write equations for the preparation of (a) CH_3COOH (b.p. = 118°C) from sodium acetate; (b) HI from PI_3; (c) permanganic acid, $HOMnO_3$, from its acid anhydride.

20. Properties of oxides Classify each of the following oxides as basic, acidic, or amphoteric: (a) Ga_2O_3; (b) TeO_3; (c) BaO; (d) Cs_2O; (e) As_2O_5.

21. Oxides Imagine that you wish to extinguish a fire in which magnesium is burning in air. (a) Write the equation or equations for the combustion. (b) Give your recommendations for or against each of the following agents for extinguishing the fire. Support your answer, if necessary, with chemical equations: (*i*) water; (*ii*) CO_2; (*iii*) N_2; (*iv*) Al_2O_3 powder; (*v*) Ar; (*vi*) powdered limestone.

22. Amphoterism (a) Write equations for hypoiodous acid, HOI, dissociating as (*i*) an acid, (*ii*) a base. (b) Account for the fact that HOI can do either, $pK_a = 10.4$, $pK_b = 9.5$.

23. Structure and acidity Conjecture as to why (a) the pK_a of NH_3 is 3.5, while (1) the pK_a of nitramide, H_2NNO_2, is 6.59, and (2) the pK_a of hydrazoic acid, $H{-}\ddot{N}{=}N{=}\ddot{N}{:}$, is 4.72; (b) the order of acidity of the following hydroxyl compounds is HOCl > HOOH > HOH.

24. Structure of acids Arsenious acid, H_3AsO_3, exchanges three protium atoms for deuterium atoms in D_2O. Phosphorous acid, H_3PO_3, exchanges only two protium atoms. Write equations for the exchange reactions, using structural formulas of the acids.

25. Carbon monoxide CO is produced by the action of sulfuric acid on formic acid (p. 394). In this process, one molecule of sulfuric acid protonates one molecule of formic acid, to produce an unstable intermediate, A, which is the conjugate acid of formic acid. A spontaneously loses water to form the extremely strong acid B, which is the conjugate acid of carbon monoxide. B protonates the bisulfate ion to yield the final product and regenerate the catalyst. (a) Write equations that show the mechanism of this reaction, including the structural formulas of the intermediates A and B. (b) What substance is the catalyst in this reaction?

26. Oxides of nitrogen (a) Write an equation, using Lewis (electronic) formulas, for the dimerization of NO_2. Does every atom in your formula of N_2O_4 obey the octet rule? (b) Write an equation, using Lewis formulas, for the combination of NO with NO_2 to form N_2O_3 (p. 396). Use arrows to indicate the electron shifts which you assign to obtain the correct formula for N_2O_3. (c) Write the Lewis formula for an N_2O_3 structure that would be formed if unpaired electrons on N in NO and NO_2 were to couple to produce an N—N bond. Explain why this structure would be less stable than that given in your answer to part (b). (Bond energies, kilocalories per mole, N—N, 39; N—O, 53; N—O, 145).

27. Autooxidation Write the partial ionic equations for the autooxidation of NO_2 in water to yield nitric and nitrous acids. Specify which reaction is oxidation and which is reduction.

28. Autooxidation The autooxidation of NO_2 in water (see Problem 27) may also yield a mixture of nitric acid and nitric oxide under some conditions. Write the partial ionic equations and the balanced overall equation for the reaction that gives these products.

29. Hydrogen peroxide The decomposition of H_2O_2 is an internal redox reaction. Write the partial ionic equations and state which is the oxidation and which is the reduction reaction.

30. Halides Which compound, PCl_3 or PCl_5, would you expect to be the more effective chlorinating agent? Explain.

31. Halides You are asked to select one of three chlorine compounds Li^+Cl^-, CCl_4, or ClO_2, to be used as an agent for destroying fungus growths. Which compounds would make the best choice? Justify your selection.

32. Periodate ion Potassium periodate in hot acid solution will oxidize Mn^{2+} salts to permanganates, while being reduced to the iodate, IO_3^-. Write the partial and overall ionic equations for the reaction.

33. Sulfides SF_4 is a toxic gas that attacks glass and reacts violently with water. SF_6, on the other hand, is nontoxic, odorless, and very inert. Account for the difference in reactivity between these two compounds.

34. Sulfides Write the partial ionic equations for the action of nitric acid on PbS.

35. Sulfuric acid Write the partial ionic equations for the oxidation by H_2SO_4 of (a) Br^- ion, and (b) I^- ion.

21 THE CHEMISTRY OF THE TRANSITION ELEMENTS

21.1 DEFINITION; GROUPS AND TRIADS

Transition elements are those having partly filled d or f shells. Their subdivision has been previously discussed (see p. 49). In this chapter only those with partially filled d orbitals, the so-called d-block elements are discussed. The definition is broadened to include also those elements which have partly filled d orbitals in any of their oxidation states. Hence, the coinage metals, Cu, Ag, and Au, which are $d^{10}s^1$ in their elementary states, are treated as transition metals, since Cu(II) and Ag(II) have a d^9 configuration and Au(III) has a d^8 configuration.

As was done for the representative elements, the periodicity of chemical behavior can be best discussed in terms of periodic groups, as for example IB, IIIB, IVB, VB, VIB and VIIB. Thus we talk of the chromium group, VIB, including chromium, Cr, molybdenum, Mo, and tungsten, W. However, for the nine elements in group VIII, the chemistry is similar for the three elements *across the period* rather than down a group. The three horizontal elements comprise a **triad;** (a) the iron triad, including iron, Fe, cobalt, Co, and nickel, Ni, (b) the palladium triad including ruthenium, Ru, rhodium, Rh, and palladium, Pd, and (c) the platinum triad, including osmium, Os, iridium, Ir, and platinum, Pt. The name of the most important member is used to indicate the respective triad.

21.2 METALLIC BEHAVIOR

The transition elements are metals, usually hard, high melting, and good conductors of heat and electricity. These elements are often used in forming alloys (p. 461), especially steels (p. 475). For example, molybdenum, tungsten, and chromium, even in small amounts, increase the hardness and strength of steel tremendously.

In general, the transition elements are resistant to corrosion, iron being an exception. This property makes titanium, a relatively abundant element in the earth's crust (0.6%), technologically useful. Nickel and chromium plating are used to prevent the corrosion of iron products (p. 303).

Most of the transition elements do not react with mineral acids, such as HCl and $(HO)_2SO_2$. Some do have negative standard reduction potentials, for the reaction $M^{n+} + ne^- \longrightarrow M$, and liberate hydrogen from hydrochloric acid. These include Mn, Cr, and the iron triad. Titanium, Ti, is inert in hydrochloric acid at room temperature, but is attacked in the hot acid. Silver, gold, the palladium triad, and the platinum triad, the so-called noble metals, are especially inert to acids, both to the nonoxidizing species, such as hydrochloric and hydrofluoric acids, and to the oxidizing acids, such as nitric acid.* These noble metals, however, dissolve in aqua regia, a 3 to 1 mixture by volume of concentrated hydrochloric and nitric acids, to form chloro-complex ions, as typified by Au and Pt:

$$Au + 3NO_3^- + 4Cl^- + 6H^+ \longrightarrow AuCl_4^- + 3H_2O + 3NO_2$$
$$Pt + 4NO_3^- + 6Cl^- + 8H^+ \longrightarrow PtCl_6^{2-} + 4H_2O + 4NO_2$$

Metals such as vanadium, titanium, niobium, tantalum, zirconium, hafnium, molybdenum, and tungsten are best dissolved in hydrofluoric acid, frequently mixed with nitric acid. The formation of stable fluoro-complex ions such as WF_6^- and ZrF_6^{2-} is important in abetting the dissolving process.

21.3 □ OXIDATION STATES AND BONDING

Most transition elements exhibit several oxidation states. With the notable exception of groups IB and VIII† (the three triads) the maximum oxidation number is the group number. Thus chromium and manganese have maximum oxidation numbers of +6 and +7, respectively. The group IB elements exhibit the group oxidation number of +1, but Cu(II), Ag(II), and Au(III) states are known. Table 21.1 lists the *commonly* observed oxidation states. When more than one state is shown the most important ones are in bold print, and the least important one is in italics.

* An oxidizing acid is invariably an oxyacid whose central atom undergoes a reduction in oxidation number on reacting with a metal, for example,

$$e^- + NO_3^- + 2H^+ \longrightarrow NO_2 + H_2O$$

For "nonoxidizing" acids, it is the proton which is reduced, not the anion

† Ruthenium and osmium are exceptions, for example, OsO_4 and RuO_4 are known compounds.

TABLE 21.1 COMMON OXIDATION NUMBERS OF *d*-BLOCK TRANSITION ELEMENTS

	Period IV		Period V		Period VI	
Group	**Symbol**	**Oxidation no.**	**Symbol**	**Oxidation no.**	**Symbol**	**Oxidation no.**
IIIB	Sc	3	Y	3	La	3
IVB	Ti	**4**, 3	Zr	4	Hf	4
VB	V	**5, 4**	Nb	5	Ta	5
VIB	Cr	6, **3**, 2	Mo	**6**, 4	W	6
VIIB	Mn	7, 4, **2**	Tc	**7, 4**	Re	**7**, 4
VIII	Fe	**3**, 2	Ru	8, 6, 4, **3**	Os	8, 6, **4, 3**
VIII	Co	3, **2**	Rh	3	Ir	4, **3**
VIII	Ni	2	Pd	2	Pt	4, 2
IB	Cu	**2**, 1	Ag	1	Au	**3**, 1

In groups IVB through VIIB the most stable oxidation state usually corresponds to the group number. This is especially true for the heavier elements in each of these groups. Within each of these groups, as the atomic number increases the higher oxidation states become more prevalent and the lower oxidation states less prevalent. The most common oxidation states, other than the one corresponding to the group number, are the +2 and +3 states. In these, and the rarely observed +1 state, the element most frequently forms ionic bonds. For example, manganese(II) sulfate, silver(I) nitrate, and copper(II) sulfate are salts. In the higher oxidation states, covalent bonding is observed. For example, titanium(IV) chloride is a liquid, vanadium(V) fluoride is a volatile white solid, m. p. 19.5°C, and osmium(VIII) oxide, OsO_4, is a very toxic liquid. Stable +1 oxidation states occur only among the group IB elements.

In water, elements in unstable oxidation states often undergo an autooxidation (disproportionation) (p. 395) to give two more stable oxidation states, one higher and one lower than the original. The behavior of Mn(VI) and Mn(III) is typical:

$$\underset{\substack{\textit{manganate}\\(\textit{green})}}{3MnO_4^{2-}} + 4H^+ \longrightarrow \underset{\substack{\textit{permanganate}\\(\textit{purple})}}{2MnO_4^-} + MnO_2(c) + 2H_2O \qquad (1)$$

$$2Mn^{3+} + 6H_2O \longrightarrow Mn^{2+} + MnO_2(c) + 4H_3O^+$$

The equilibrium in these reactions are dependent on pH. For example, in reaction 1, acid is needed for the autooxidation of the manganate ion. Accordingly, stabilization of the manganate ion requires very basic solution. Not all autooxidations are pH dependent, as seen by the reaction of Au(I)Cl:

$$3AuCl \longrightarrow AuCl_3 + 2Au$$

Elements in the less stable states can only exist in certain complex ions or in insoluble solids. Under these conditions the equilibria favor the normally less stable state. This behavior is illustrated by the chemistry of copper. The relative stabilities of the copper(I) and (II) states are indicated by the following reduction potentials:

$$Cu^+ + e^- \longrightarrow Cu \qquad \mathscr{E}^\circ = 0.520$$
$$Cu^{2+} + e^- \longrightarrow Cu^+ \qquad \mathscr{E}^\circ = 0.153$$

whence

$$2Cu^+ \longrightarrow Cu + Cu^{2+} \qquad \mathscr{E}^0 = 0.367$$

indicating that Cu^+ should have a tendency to disproportionate to Cu^{2+} and Cu. Nevertheless, the more stable Cu(II) state can be reduced to the less stable Cu(I) by, for example, I^- and CN^-.

$$\underset{blue}{2Cu^{2+}(aq)} + 4I^-(aq) \longrightarrow \underset{yellow\text{-}brown}{2CuI(c)} + \underset{violet}{I_2(aq)} \qquad \text{(precipitate formation)} \quad (2)$$

$$2Cu^{2+}(aq) + 10CN^-(aq) \longrightarrow 2Cu(CN)_4^{3-}(aq) + \underset{cyanogen}{N{\equiv}C{-}C{\equiv}N(g)} \qquad \text{(complex formation)}$$

Reaction 2 is used in the quantitative analysis of copper.

Rather than undergoing autooxidation, certain unstable cations undergo redox reactions with water. If the cations are oxidized, hydrogen is formed as shown,

$$2Cr^{2+} + 4OH^- + 2H_2O \longrightarrow H_2 + 2Cr(OH)_3(c)^*$$

If the cation is reduced, oxygen is evolved as shown.

$$4Co^{3+} + 2H_2O \longrightarrow 4Co^{2+} + O_2 + 4H^+$$

Cations easily oxidized by oxygen must be protected from air or preserved with pieces of the free metal. Thus the Fe(II) state is maintained by placing iron nails in the solution of its salts.

21.4 □ COLOR

Unlike the cations, anions, and oxyanions of the representative elements, all of which are colorless, the transition metal ions and their oxyanions, such as chromate, CrO_4^{2-} (yellow), dichromate, $Cr_2O_7^{2-}$ (orange), and permanganate, MnO_4^- (purple), are generally colored. The colors of the hydrated cations of the first transition series are given in Table 21.2. When no d electrons are present, as in Sc^{3+}, or when all d

* The so-called chromic hydroxide, $Cr(OH)_3$, actually is partially hydrated and is best represented as $Cr_2O_3(H_2O)x$.

TABLE 21.2 COLORS OF THE HYDRATED IONS OF THE FIRST TRANSITION SERIES[a]

Transmitted color	Ions and number of 3*d* electrons			
Colorless	Sc^{3+}(0),	Cu^{+}(10)		
Red	Co^{2+}(7),	Mn^{2+}*(5)		
Green	Fe^{2+}(6),	Ni^{2+}(8),	V^{3+}(2)	
Purple	Ti^{3+}(1)			
Violet	V^{2+}(3),	Cr^{3+}(3),	Mn^{3+}(4),	Fe^{3+}*(5)
Blue	Cr^{2+}(4),	Co^{3+}(6),	Cu^{2+}(9)	
Yellow	$FeCl^{2+}$			

[a] Symbols: * low intensity.

orbitals are filled, as in Cu^+, the ion is colorless. Only those transition metal ions that have partially filled d orbitals may exhibit color.

A study of the color, and more specifically, the spectrum in the visible region (p. 14) of transition metal ions and their complex ions fosters an understanding of the structure of these ions. This relationship of color and structure is discussed briefly later in this chapter. Before this discussion, it is necessary to understand the origin of color.

Absorption of visible or ultraviolet radiation causes the excitation of an electron from a stable to a less stable orbital. The ground state molecule (or ion) is converted to an excited state molecule (or ion) (p. 20). The difference in energy, ΔE, between the excited state, E_2, and the ground state, E_1, equals the energy of the absorbed radiation,

$$\underset{\text{energy of excited state}}{E_2} - \underset{\text{energy of ground state}}{E_1} = \underset{\text{energy of absorbed radiation}}{\Delta E}$$

The energy of the absorbed radiation is inversely proportional to the wavelength, λ; $\Delta E = h \cdot c/\lambda$ (p. 16). The longer the wavelength of the absorbed light, the lower is the energy of the electronic excitation. Colorless substances usually absorb in the ultraviolet region. Colored substances absorb in the visible region (4000–7500 Å, p. 14). Hence, it takes less energy to excite a colored substance than a colorless substance. Color appears when the energy level of the excited state is close to the ground state energy level.

If all visible light is absorbed, the substance appears black. If all but one color is absorbed, the substance will have that color. More frequently, however, only one color is absorbed, in which case the substance has the complementary* color, as shown in Table 21.3. The

* Complementary colors are two colors that, when combined, produce white or nearly white light.

TABLE 21.3 ▪ RELATIONSHIP OF ABSORBED AND OBSERVED COLOR

Absorbed wavelength (Å)	Absorbed color	Observed complementary color
4000–4500	Violet	Yellow
4500–5000	Blue	Orange
5000–5500	Green	Red
5500–6000	Yellow	Violet
6000–6500	Orange	Blue
6500–7000	Red	Green

energy of excitation, ΔE, is smaller as the visible colors vary from yellow to orange to red to violet to blue to green.

Some substances are more intensely colored than others. For example, a solution containing manganous ion, Mn^{2+}, has a pink color of weak intensity, while a solution of permanganate ion, MnO_4^-, has an intense violet color. Intensity is proportional to the number of molecules or ions being excited per unit time. Discussion of the factors determining intensity is beyond the scope of this book.

21.5 INTERSTITIAL COMPOUNDS

Many transition metals absorb hydrogen to give compounds which are not typically ionic or covalent. In most cases, the hydrides are of variable composition; they do not follow the law of definite proportions. Thus, the hydrides of titanium and zirconium are written as TiH_x and ZrH_x, respectively. Palladium metal has a remarkable ability to absorb large quantities of hydrogen gas, as much as 1000 times the volume of the piece of metal. The absorption of hydrogen is an exothermic process. On being adsorbed on the surface of the metal, hydrogen molecules undergo dissociation into atoms. These atoms then diffuse into the solid and occupy the empty spaces (interstices) between the atoms of the metal. Because of the nature of their composition such hydrides are classified as **interstitial compounds.** These interstitial hydrides are hard, nonvolatile solids.

Other small atoms such as carbon, boron, and nitrogen can also occupy the empty spaces of transition metals to form interstitial compounds. These compounds are called carbides, borides, and nitrides, respectively. Some of the heavier elements such as W and Mo form interstitial compounds with silicon. The principal chemical and physical properties of these interstitial compounds are (a) they have very high melting points, as typified by vanadium nitride, VN, m. p. 2570°C; (b) they are very hard—some borides approach diamond in hardness; (c) they are chemically very inert; and (d) they retain their metallic conductivity (p. 461). The melting points of the interstitial compounds are higher than those of the pure metals. Evidently, the presence of the foreign atoms stabilizes the crystal structure. Because of these properties, interstitial compounds are used in kilns, gas turbines, and jet engines as heat-resistant materials, and for cutting tools.

The spaces in metal crystals of Cr, Mn, Fe, Co, and Ni are just not large enough to accommodate a carbon atom. For this reason the presence of diffused carbon atoms causes a distortion of the metal lattices. As a result, the carbides of these elements are chemically more reactive than those of most other transition elements.

21.6 □ OXIDES AND OXYIONS

The oxides of the transition elements in different oxidation states are given in Table 21.4. Also shown is the relative basicity or acidity of each oxide. Oxides of the elements in the lower oxidation states (I and II) tend to be basic. Oxides of the elements in the higher oxidation states (IV to VIII) tend to be acidic; the intermediate ones tend to be either basic, acidic, or amphoteric. This trend is best illustrated by the oxides of chromium.

Oxide	Oxidation state	Chemical behavior
CrO	II	basic anhydride of $Cr(OH)_2$
Cr_2O_3	III	amphoteric—dissolves in acid to form Cr^{3+}; dissolves in base to form chromite, $Cr(OH)_6^{3-}$
CrO_3	VI	acid anhydride of very strong but unstable chromic acid, $(HO)_2CrO_2$

Because of the acidic nature of some of the oxides of transition elements, it is not surprising that several of these elements form oxyanions. Oxyanions are found in oxidation states V, VI, VII, or VIII. Thus,

TABLE 21.4* ▪ BASIC AND/OR ACIDIC PROPERTIES OF THE OXIDES OF THE d-BLOCK ELEMENTS

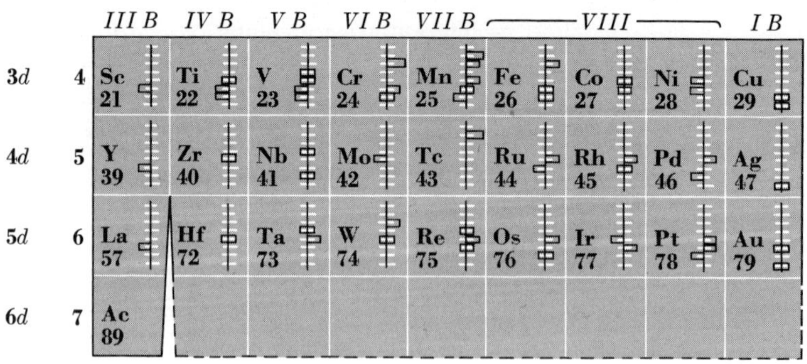

* See Table 20.2, p. 389, for key.

a glance at Table 21.1 tells us to expect to find oxyanions of ruthenium and osmium, and of the elements in groups VB, VIB and VIIB. When dealing with transition elements in oxidation states of IV or less, one should be suspicious of a name indicating the presence of an oxyanion. Thus, the mineral **ilmenite,** $FeTiO_3$, is erroneously called a "titanate." It is actually a 1:1 mixture of FeO and TiO_2.

As observed for Re, Mo, and W, when the oxidation state of the oxyanion corresponds to the most stable state of the element, the oxyanions are relatively stable and comprise the important part of the chemistry of the element. When the most stable oxidation state is lower than that of the oxyanion state, the oxyanions are strong oxidizing agents; chromate, CrO_4^{2-}, dichromate, $Cr_2O_7^{2-}$, and permanganate, MnO_4^-, are typical examples, as shown in the several half-equations below.

$$\text{(acid) } MnO_4^- + 8H^+ + 5e^- \longrightarrow Mn^{2+} + 4H_2O$$
$$\text{(basic or neutral) } MnO_4^- + 2H_2O + 3e^- \longrightarrow MnO_2(c) + 4OH^-$$
$$\text{(acid) } Cr_2O_7^{2-} + 14H^+ + 6e^- \longrightarrow 2Cr^{3+} + 7H_2O$$
$$\text{(basic) } CrO_4^{2-} + 4H_2O + 3e^- \longrightarrow Cr(OH)_3(c) + 5OH^-$$

The oxidizing power of MnO_4^- is pH dependent. In acid, MnO_4^- gains five electrons and is reduced to Mn^{2+}, whereas in base or in neutral solution it gains only three electrons to give MnO_2. Chromate in base and dichromate in acid both are reduced to the Cr(III) state.

Which ion, CrO_4^{2-} or $Cr_2O_7^{2-}$, may exist in a water solution depends on the pH. The equilibrium for their interconversion is

$$\underset{\textit{yellow}}{2CrO_4^{2-}} + 2H^+ \rightleftharpoons \underset{\textit{orange}}{Cr_2O_7^{2-}} + H_2O$$

Acid promotes the conversion of the yellow chromate ion to the orange dichromate ion; base favors the formation of CrO_4^{2-}. The reaction can be conceived as occurring in two steps through the formation of the unstable $HOCrO_3^-$ ion.

$$2CrO_4^{2-} + 2H^+ \longrightarrow 2HOCrO_3^-$$
$$2HOCrO_3^- \longrightarrow Cr_2O_7^{2-} + H_2O$$

This sequence indicates that $Cr_2O_7^{2-}$ is the bimolecular dehydrated form of $HOCrO_3^-$, and hence the name dichromate (p. 268) is assigned to it. The structure of the $Cr_2O_7^{2-}$ ion, as observed in the ammonium salt, is shown in Fig. 21.1. The Cr—O—Cr bond angle of 115 degrees indicates that the oxygen atom bridging the two Cr atoms uses essentially sp^2 hybrid atomic orbitals.

Oxyanions possessing a M—O—M bridge are called **polyanions.** The ability to form polyanions is very pronounced for the chromium group. Both molybdenum and tungsten form a very large number of complex polyanions as typified by paramolybdate, $Mo_7O_{24}^{6-}$, and paratungstate, $W_{12}O_{46}^{20-}$.

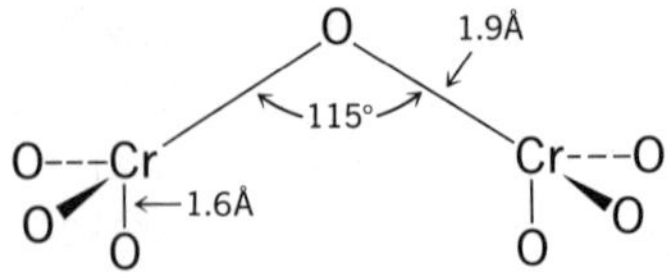

Fig. 21.1 Structural formula for dichromate ion, $Cr_2O_7^{2-}$.

Some transition elements, such as vanadium, form oxycations, for example VO^{2+}. The suffix -yl is used to indicate these states; VO^{2+} is called vanadyl. The vanadyl ion dominates the chemistry of V(IV). V(IV) forms mainly vanadyl salts such as vanadyl sulfate, $VOSO_4$, and vanadyl complex ions such as $VOCl_4^{2-}$ and VOF_4^{2-}.

21.7 □ COMPLEX FORMATION

A complex (ion or molecule) has been defined as a species composed of several parts, each of which has some independent existence in solution. It usually consists of a positive metal ion and a number of electron-rich ligands. The cation is a Lewis acid and the ligand is a Lewis base.

The ligands may be anions, such as CN^-, Cl^-, F^-, H^-, and OH^-, and (or) molecules with lone pairs of electrons, such as $H_2\ddot{O}:$, $:NH_3$, $:PH_3$, and $:C{\equiv}O:$. Some typical examples of complex compositions are

Lewis acids		*Lewis bases*		*Complexes*
Co^{3+}	+	$6NH_3$ *neutral ligand*	$\longrightarrow$	$Co(NH_3)_6^{3+}$
Fe^{2+}	+	$6CN^-$ *anionic ligand*	$\longrightarrow$	$Fe(CN)_6^{4-}$
Pt^{4+}		$+ 2NH_3 + 4Cl^-$ *mixed ligands*	$\longrightarrow$	$Pt(NH_3)_2Cl_4^0$

Notice that the complex may have a positive, negative, or zero charge. *The charge is always the sum of the charges on the individual components.* In most cases, the complexes are ions, and henceforth we shall refer to complex ions, rather than complexes.

21.8 □ COORDINATION NUMBER OR LIGANCY

The number of atoms attached to the central atom is called the **coordination number** or **ligancy.** Ligancy is one of the most puzzling problems of complex ion formation. It is undoubtedly influenced by the charge on the cation, the charge on the ligand, the relative sizes of cation and ligand, and the repulsion among ligands. However, the interplay of these factors cannot be evaluated quantitatively, and hence reliable predictions cannot be made. Fortunately, from experience, we can make a useful but not universal generalization—*a cation usually* *has a ligancy that is twice its charge.* Thus, some typical values are 2 for Ag^+, 4 for Cu^{2+}, 6 for Co^{3+}, 4 for Ni^{2+}, and 6 for Fe^{3+}. Some important exceptions are 6 for Fe(II), Co(II), Sn(II), and Sn(IV). The same cation can have more than one ligancy, as in FeF_6^{3-} and $FeCl_4^-$. However, the ligancy is rarely more than 3 times the charge of the cation.

21.9 □ WERNER'S COORDINATION THEORY

Alfred Werner in 1893 proposed a rather simple theory to explain the structure of complexes. He suggested that, in addition to the common valence, certain metals possess "auxiliary" valences. Thus, Cr(III) has a valence of three in the compound $Cr_2(SO_4)_3$, and was said to have an auxiliary valence of six in the complex $(CrCl_6)^{3-}$. The common valences are satisfied only by anions, but auxiliary valences are satisfied by either anions or molecules.

The coordination complexes that Werner studied behave in solution as stable units. The ligands lose their typical chemical behavior, as exemplified by the fact that an aqueous solution of $PtCl_4(NH_3)_2$ gives no precipitate of AgCl with silver nitrate, and does not neutralize sulfuric acid. Werner concluded that rather than being present in an uncombined state, Cl^- and NH_3 are intimately associated with Pt(IV) in a stable unit.

$$PtCl_4(NH_3)_2 \xrightarrow{Ag^+} \textit{no } AgCl, \textit{ therefore no free } Cl^- \textit{ present}$$

$$PtCl_4(NH_3)_2 \xrightarrow{(HO)_2SO_2} \textit{no } NH_4^+, \textit{ therefore no free } NH_3 \textit{ present}$$

Conclusive proof of this tight association of Pt(IV), Cl^-, and NH_3 comes from the fact that an aqueous solution of this substance does not conduct an electric current, indicating the absence of free ions. Werner bracketed the complex to indicate its unit-like behavior: $[Pt(NH_3)_2Cl_4]^0$.

Another platinum(IV)-chloride-ammonia compound, $PtCl_4(NH_3)_6$, on the other hand, gives 4 moles of AgCl per mole of complex on the addition of Ag^+, has a molar conductance indicating the presence of *five* ions, but does not neutralize sulfuric acid. Werner proposed that the formula for this compound be written $[Pt(NH_3)_6]Cl_4$, indicating the presence of the complex ion $Pt(NH_3)_6^{4+}$ and four Cl^- ions.

Most complexes, however, are not so stable as the Co(II), Pt(IV), and Cr(III) complexes that Werner studied. For example, the $Cu(NH_3)_4^{2+}$ complex is readily decomposed by acid,

$$Cu(NH_3)_4^{2+} + 4H_3O^+ \rightarrow Cu(H_2O)_4^{2+} + 4NH_4^+$$

21.10 □ COORDINATION NUMBER AND SHAPE

Werner suggested that the auxiliary valences were directed in space about the central atom, and that a definite spatial array is associated with each coordination number. We now know that this supposition is in accord with the wave-mechanical model for chemical bonding. The coordination numbers most frequently encountered are six and four. The six-coordinate complexes are all octahedral (p. 203). When the ligands are not all identical, some distortions occur because the ligands are no longer equidistant from the central atom. The four-coordinate complexes, such as $Fe(III)Cl_4^-$ and $Co(III)Br_4^-$, are usually tetrahedral, but some are square-planar. The square-planar complex has the cation at the center and the ligands each at one of the corners of the square (Fig. 21.2a). The square is the shape for Pt(II) and Au(III), as in $Pt(II)(Cl)_4^{2-}$ and $Au(III)Cl_4^-$, and occurs frequently among Ni(II) and Cu(II) complexes. Recent findings seem to indicate that square-planar complexes are really cases of distorted octahedra, in which two loosely held molecules of water occupy the two vacant positions:

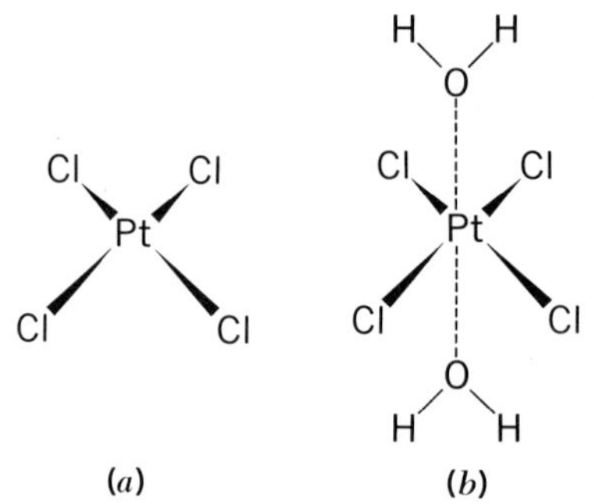

Fig. 21.2 (a) A typical square planar complex, $PtCl_4^{2-}$, (b) square-planar complex; actually a distorted octahedron (one molecule of H_2O above and one molecule below the plane of the paper).

A coordination number of 2 is observed especially for Cu, Ag, and Au, in their +1 oxidation state, for example, $Cu(I)Cl_2^-$, $Ag(I)(NH_3)_2^+$, and $Au(I)Cl_2^-$. The shape of these complexes is linear. Coordination numbers of 5 are rare and of 3 are very rare. The few well-defined complexes with coordination number 5 are trigonal bipyramids, for instance iron pentacarbonyl, $Fe(CO)_5$.

21.11 ▫ BONDING IN TRANSITION METAL COMPLEXES

Werner's coordination theory made no mention of the nature of the bonding between the central atom and ligand, yet the theory was able to account for much of the chemistry and some of the structure of complexes. However, if we wish to understand more about complexes, their color, stability, ease of formation, and magnetic properties, we must concern ourselves with the nature of the bonding. We have already seen that bonding between atoms can be viewed as two extremes: (1) electrostatic interaction between oppositely charged ions (or between an ion and a dipolar molecule); and (2) an overlap of atomic orbitals to give a covalent bond. Most often, there is a blending of the two extremes, so that many bonds have both ionic and covalent character. The same blending of bonding types prevails for the transition metal complexes.

The valence bond theory, which dates from 1935, is the contribution mainly of Linus Pauling. He introduced the concept, previously used elsewhere, that *nonequivalent atomic orbitals hybridize to give a set of equivalent hybrid atomic orbitals whose characteristic orientation determines the geometry of the complex.* Bonding may then be regarded as a result of the overlapping of a *filled* atomic orbital of the ligand with an *empty* hybrid atomic orbital of the central atom. Most of the literature on the chemistry of complex ions from 1935 to the mid 1950's used the concepts and language of this theory. Hence, this theory is a landmark in the development of chemistry. As for all theories, experiments were designed as rigorous tests. Under this pressure, inadequacies and deficiencies appeared. The theory is adequate qualitatively for transition metal complexes, but does not permit quantitative predictions of their spectra, stabilities, and magnetic behavior. The valence bond theory is still serviceable for complexes of the representative elements such as AlF_6^{3-}.

21.12 □ CRYSTAL FIELD THEORY*

This theory was developed by Hans Bethe (1929) and John Van Vleck (1932), and explored by physicists. However, it was not until 1952 that Leslie Orgel popularized its use for the inorganic chemist. It is also an oversimplification, but has several advantages over the valence bond theory. The theory assumes that the interaction between the transition metal ion and the ligand is purely electrostatic; it does not consider any orbital overlap. Essentially, the theory is concerned with the effect that the electron-rich ligand has on the degeneracy (p. 37) of the d orbitals of the central cation. The theory is applicable only to complexes of cations with partially filled d orbitals.

(a) Crystal field splitting Recall the shape and spatial orientation of the five d orbitals, as shown on p. 36. The d_{z^2} orbital is oriented along the z axis, the $d_{x^2-y^2}$ orbital is oriented along both the x and y axes, and the d_{xy}, d_{xz}, and d_{yz} orbitals are oriented between the x, y, and z axes. We now consider the formation of an octahedral complex, as an illustration of this theory. Analogous reasoning is used to arrive at the electronic structures of tetrahedral and square-planar complexes. The metal ion is at the origin of the x, y, and z axes, and the six ligands approach from both ends along each of these three axes† (Fig. 21.3).

* *Crystal field theory* was so named because its early proponents were interested in ionic crystals.

† For uniformity, the same axes are used to define the d orbitals as are used to describe the octahedral approach of the ligands.

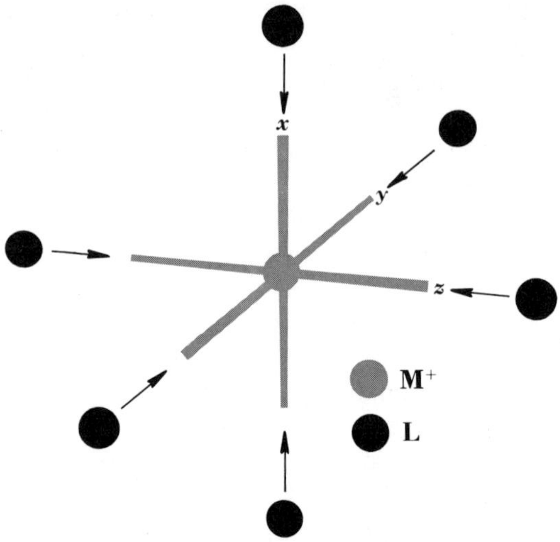

Fig. 21.3 Approach of ligands along x, y, and z axes in an octahedral complex.

This approach permits the electron-rich ligands to be at the maximum distance from each other. As the electron-rich ligands approach, they exert a repulsive force on the electrons in each of the d orbitals of the metal. The d_{z^2} and the $d_{x^2-y^2}$ orbitals are oriented along an axis, and so lie directly in the path of the approaching ligands. The other three d orbitals are oriented in space between the approaching ligands. Consequently, the electrons in the d_{z^2} and $d_{x^2-y^2}$ orbitals suffer more repulsion than do electrons in the d_{xy}, d_{xz}, and d_{yz} orbitals.

Although all five d orbitals are raised to higher energy levels, the d_{xy}, d_{xz}, and d_{yz} orbitals are raised less in energy than the d_{z^2} and $d_{x^2-y^2}$ orbitals. Thus when the ligand approaches, it is more favorable for an electron to be in the d_{xy}, d_{xz}, and d_{yz} orbitals than in the other two orbitals. Thus, the crystal field theory predicts that under the influence of the electrostatic field generated by the ligands, the five degenerate d orbitals are split into two groups of different energy (Fig. 21.4). The difference in energy between the two groups is called the **crystal field splitting energy,** and is designated by the symbol Δ:

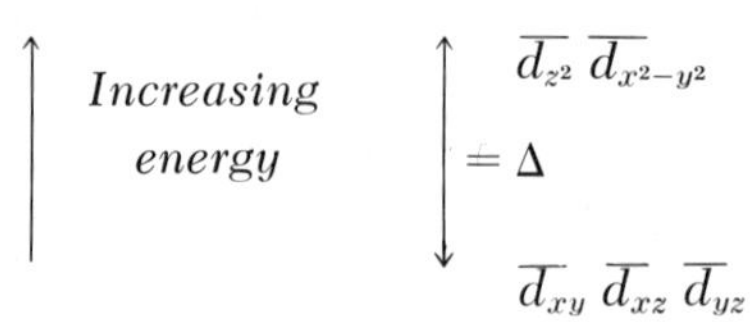

(b) Factors influencing Δ It was previously mentioned that the complex ions of transition metals with partially filled d orbitals are frequently

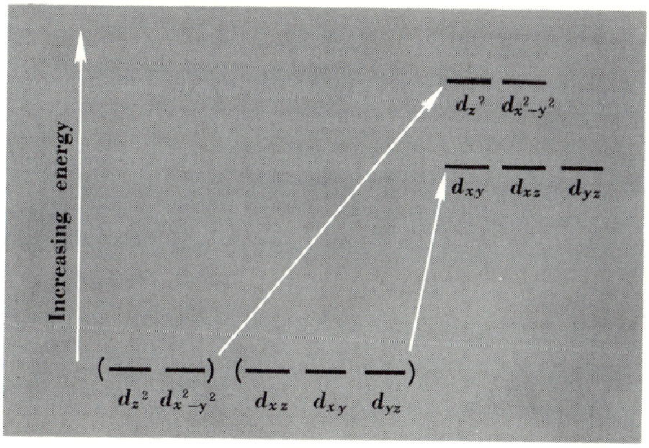

Fig. 21.4 Energy level diagram illustrating the splitting of the five degenerate *d* orbitals in an octahedral crystal field.

colored. The crystal field theory suggests that color appears when, for example, in a typical octahedral complex, $Ti(H_2O)_6^{3+}$, the 3*d* electron is excited from the lower to the higher energy level. In fact, the energy of the absorbed photon should equal the value of Δ. Studies of the absorption spectra* of complex ions show that the value of Δ depends on the nature of the ligand. The splitting power for some of the more common ligands decreases in the order: $CN^- > NO_2^- > NH_3 > H_2O > OH^- > F^- > Cl^- > Br^- > I^-$. In terms of splitting power, we talk about strong and weak ligands. Thus, CN^- is a very *strong* ligand and, at the other end of the scale, I^- is a very *weak* ligand.

The greater the crystal field splitting energy, Δ, the larger is the excitation energy, ΔE. The effect of the splitting power of ligands on Δ, and the resulting effect on the wavelength of light absorbed by the complex ion is illustrated by comparing the following three colored Co(III) complex ions:

$Co(CN)_6^{3-}$	$Co(NH_3)_6^{3+}$	$Co(H_2O)_6^{3+}$
yellow	*yellow-orange*	*blue*
Wavelength of absorbed light is shortest	Wavelength of absorbed light is intermediate	Wavelength of absorbed light is longest
↓	↓	↓
ΔE is largest	ΔE is intermediate	ΔE is smallest
↓	↓	↓
Δ is largest	Δ is intermediate	Δ is smallest
↓	↓	↓
CN^- has strongest field splitting power	NH_3 has intermediate field splitting power	H_2O has weakest field splitting power

In applying this reasoning, one must be certain that there is no change in the geometry of the complex when the ligand is altered.

The cation also plays a role in establishing the value of Δ. To deter-

* An ultraviolet spectrometer is used to measure accurately the wavelength of the absorbed radiation.

mine the nature of this role, it is imperative to compare only those cations with the same number of *d* electrons and the same geometry. Thus, the behavior toward the same ligand of Fe(II) and Co(III), which both have six *d* electrons (d^6 ions), or V(II) and Cr(III), which are both d^3 ions, can be compared. It is observed that the crystal field splitting in $V(H_2O)_6^{2+}$ is less than in $Cr(H_2O)_6^{3+}$, a fact that can be interpreted in terms of the charge on the cation. The Cr^{3+} ion exerts a greater attraction for water molecules than does the V^{2+} ion, because of the greater positive charge. Hence, the water molecules approach the Cr(III) ion more closely than they approach the V(II) ion, and so exert a stronger crystal field effect on the *d* electrons of Cr(III).

It is possible that the value of Δ is so small that radiation in the infrared region may cause excitation. In such cases, the complex ion would be colorless. Hence, lack of color in a complex ion of a frequently colored cation may be due to either a very large or very small value for the crystal field splitting energy.

(c) Distribution of electrons in the *d* orbitals; weak and strong crystal field complexes According to the crystal field theory, properties such as color, paramagnetism, stability, and geometry are greatly influenced by the manner in which the metal ion *d* electrons are distributed in the *d* orbitals. The distribution of electrons in an atom is influenced by two factors: (1) the tendency for electrons to resist pairing (Hund's rule), and (2) the tendency for electrons to occupy orbitals of lower energy before entering orbitals of higher energy. The factor that controls the electron distribution is determined by the strength of the field exerted

TABLE 21.5 ELECTRONIC DISTRIBUTION IN STRONG AND WEAK FIELD COMPLEXES

No. of *d* electrons	Typical ion	Weak field (spin-free)	Strong field (spin-paired): Lower energy	Strong field (spin-paired): Higher energy
d^1	Ti^{3+}	↑ _ _ _ _	↑ _ _	_ _
d^2	V^{3+}	↑ ↑ _ _ _	↑ ↑ _	_ _
d^3	Cr^{3+}, V^{2+}	↑ ↑ ↑ _ _	↑ ↑ ↑	_ _
d^4	Cr^{2+}	↑ ↑ ↑ ↑ _	⇅ ↑ ↑	_ _
d^5	Fe^{3+}, Mn^{2+}	↑ ↑ ↑ ↑ ↑	⇅ ⇅ ↑	_ _
d^6	Fe^{2+}, Co^{3+}	⇅ ↑ ↑ ↑ ↑	⇅ ⇅ ⇅	_ _
d^7	Co^{2+}	⇅ ⇅ ↑ ↑ ↑	⇅ ⇅ ⇅	↑ _
d^8	Ni^{2+}	⇅ ⇅ ⇅ ↑ ↑	⇅ ⇅ ⇅	↑ ↑
d^9	Cu^{2+}	⇅ ⇅ ⇅ ⇅ ↑	⇅ ⇅ ⇅	⇅ ↑

by the ligand. A strong field causes appreciable splitting of the d orbitals, and the lower energy orbitals are completely paired before electrons enter the higher energy orbitals. A complex with this type of electron distribution is called a **strong field (spin-paired) complex.** If the ligand exerts a weak crystal field effect there is little or no splitting, and the electrons enter the five d orbitals one at a time in each orbital, so as to minimize pairing of spins. The resulting aggregate is called a **weak field (spin-free) complex.** Table 21.5 shows the types of electronic arrangement for ions with differing numbers of d electrons. When there are 1, 2, 3, 8, and 9 d electrons there is no difference in the number of unpaired electrons in the weak and strong field complexes, so that magnetic behavior cannot be used to distinguish between the two kinds of complex. However, for the d^4, d^5, d^6, and d^7 electron states, there is a difference in the number of unpaired electrons and therefore in magnetic properties. An illustration of the application of magnetic properties to the elucidation of structure of complex ions follows. $Fe(CN)_6^{3-}$ shows paramagnetic behavior due to one unpaired electron, whereas FeF_6^{3-} has a magnetic moment corresponding to five unpaired electrons. Table 21.5 indicates that a weak field complex of Fe^{3+} has five unpaired electrons. Evidently, F^- acts as a weak ligand giving a weak field complex. On the other hand, a strong field complex of Fe^{3+} has only one unpaired electron. Evidently CN^- acts as a strong ligand, giving a strong field complex.

21.13 □ CARBONYL COMPLEXES

Carbon monoxide, :C≡O: , can serve as a ligand. It can react with certain transition elements in their elementary (zero oxidation) state. Iron and nickel form such complexes called **carbonyls.**

$$Fe + 5CO \longrightarrow Fe(CO)_5$$

iron pentacarbonyl, b.p. 103°C

$$Ni + 4CO \longrightarrow Ni(CO)_4$$

nickel tetracarbonyl, b.p. 43°C

The role of nickel carbonyl, a low boiling liquid, in the metallurgy of nickel, is discussed in Chapter 23 (p. 474). On forming $Fe(CO)_5$, Fe with 26 electrons gains 10 electrons from the five CO molecules, giving 36 electrons, the number possessed by the noble gas, krypton. Likewise, Ni with 28 electrons gains 8 more from the four molecules of CO, also giving the stable number of 36. It appears that carbonyl formation is predicated on the tendency of the transition element to acquire the noble gas configuration of the nearest inert gas.

Cobalt, having an odd atomic number of 27, cannot form a carbonyl molecule with only one Co atom. Instead, it forms $Co_2(CO)_8$. In this way,

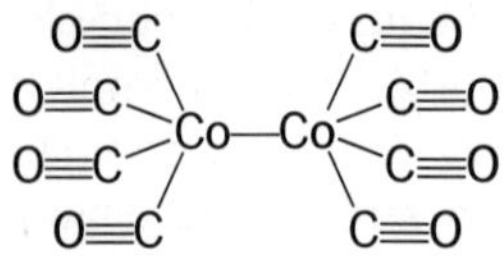

each Co atom gets the 9 electrons it needs. Co also forms a mixed carbonyl with NO and CO, $Co(CO)_3NO$. Nitric oxide is a free radical, $\cdot\ddot{N}{=}\ddot{O}$; and can be considered to be a three-electron donor to the Co atom. In this way, cobalt acquires 36 electrons [27(Co) + 3(NO) + 6 from 3CO].

Manganese, having an odd number of 25 electrons, cannot form a carbonyl with one Mn atom per molecule. However, Mn^+ reacts with 6 moles of CO (24 + 12 = 36 electrons) to give a carbonyl complex ion.

$$Mn^+ + 6CO \longrightarrow Mn(CO)_6^+$$

This tendency to acquire a noble gas electronic configuration is a convenient way to systematize information on the composition of this very limited class of compounds. It should be emphasized that this tendency is not an important factor in predicting the structure of the large host of known complexes and complex ions. The existence of $V(CO)_6$ (23 + 12 = 35 electrons) means that even for carbonyl compounds this rule is not rigorous.

Carbonyls are proving to be valuable reagents and catalysts in inorganic and organic syntheses.

21.14 ▫ GEOMETRICAL ISOMERISM

Octahedral or square-planar complexes can show geometric isomerism (p. 440); that is, there may be two or more different complexes, with different properties, containing the same central atom and the same ligands, but differing only in the spatial arrangement of the ligands. For example, the square-planar complex $Pt(NH_3)_2Cl_2$ exists as a *cis* form (Fig. 21.5a) and as a *trans* form (Fig. 21.5b), the latter having zero

Fig. 21.5 Geometric isomerism in square-planar complex, MA_2B_2.

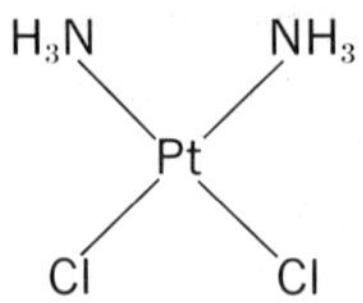

cis isomer

(*a*)

$\mu \neq 0$

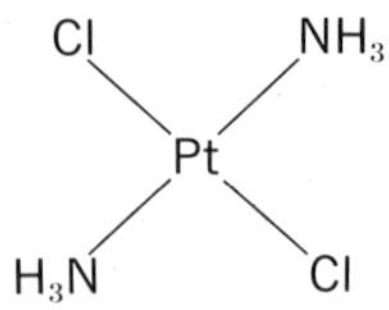

trans isomer

(*b*)

$\mu = 0$

dipole moment. We shall not consider the more complicated isomerism of the octahedral complexes.

Geometrical isomerism with respect to the central atom is *impossible* in tetrahedral complexes. The observation of geometrical isomers is therefore a way of deciding that a certain complex is square-planar rather than tetrahedral.

21.15 □ STABILITY OF COMPLEX IONS

There are two meanings of the term "stability." Thermodynamic stability relates to the position of the equilibrium between the complex ion and its components. The equilibrium constant for the decomposition of a stable complex ion is very small, for example:

$$Ni(CN)_4^{2-} \rightleftharpoons Ni^{2+} + 4CN^- \qquad K = 10^{-22} \tag{3}$$

Such equilibrium constants determine whether an insoluble compound will form from a complex ion or dissolve to form a complex ion. For example, $Cu(CN)_3^{2-}$ is more stable than $Cd(CN)_4^{2-}$ relative to the corresponding sulfides, so that only CdS is precipitated on addition of sulfide ion.

Kinetic stability ("reactivity") refers to the *rate* at which a complex ion reacts. A complex ion that reacts rapidly is called **labile;** one that reacts very slowly is **inert.** Kinetic stability and thermodynamic stability do not necessarily correlate with each other. For example, $Co(NH_3)_6^{3+}$ is thermodynamically unstable toward acid; the equilibrium constant for the reaction

$$Co(NH_3)_6^{3+} + 6H_3O^+ \rightleftharpoons Co(H_2O)_6^{3+} + 6NH_4^+$$

is about 10^{25}, yet it is inert, persisting for weeks in an acid medium. In contrast, $Ni(CN)_4^{2-}$ is thermodynamically stable (see Equation 3), but labile. Thus if cyanide ion containing the radioactive isotope C^{14} is added to the solution, the recovered complex has C^{14}-labeled cyanide ions. The exchange of $C^{14}N^-$ for $C^{12}N^-$

$$Ni(C^{12}N)_4^{2-} + 4C^{14}N^- \rightleftharpoons Ni(C^{14}N)_4^{2-} + 4C^{12}N^-$$

is rapid. Such isotopic exchange reactions are often used to measure rates of reaction.

The octahedral complexes of the first transition series are fairly labile, except for those of Cr^{3+} and Co^{3+}. The inert complexes of these two cations were among those used by Werner in his historic researches. Some Fe^{2+} complexes, such as $Fe(CN)_6^{4-}$, are also inert. Inert complexes are difficult to prepare, a fact which is important in qualitative analysis. The inert ammonia complexes of Cr^{3+} and Co^{3+} cannot be formed by the addition of NH_3 to aqueous solutions of the ions; hydroxides are formed instead.

PROBLEMS

1. **Definitions** Define and illustrate the following terms: (a) triad; (b) oxycation; (c) interstitial compound; (d) strong field complex; (e) weak field complex.

2. ***d*-Electrons** Give the number of d electrons present in (a) Rh(0); (b) Cr(III); (c) Zr(IV); (d) Os(IV); (e) Au(III). Assume that s electrons from a higher principal energy level are lost before d electrons from a lower principal energy level.

3. **Oxidation number** What is the oxidation number of vanadium in each of the substances or ions represented by the following formulas? (a) VO_2; (b) $VOCl_3$; (c) V_2O_5; (d) VO_4^{3-}; (e) $V_3O_9^{3-}$; (f) VOF_4^{2-}; (g) $V(NH_3)_6Cl_3$; (h) $K_2V_4O_9(H_2O)_7$.

4. **Oxides and oxyanions** Complete the following equations.

(a) $MnO + H_2O \longrightarrow$

(b) $Mn_2O_7 + H_2O \longrightarrow$

(c) $ReO_3 \xrightarrow{\text{heat}}$ (disproportionation to more stable oxidation states)

(d) $FeWO_4 + Al \longrightarrow$ (preparation of W)

5. **Werner theory** A compound (A) has the analysis 22.0% Co, 39.8% Cl, 31.4% N, and 6.8% H. With $AgNO_3$, 1 mole of A gives 3 moles of AgCl. NH_4Cl is not produced when hydrochloric acid is added to A. Write the structural formula and give the name for compound A.

6. **Crystal field theory** Draw the electronic configurations for the strong and weak field octahedral complexes of (a) Mn(III); (b) Ni(III).

7. **Geometrical isomerism** Give the number of geometrical isomers of the square-planar complexes of (a) $Pt(NH_3)_2ClBr$; (b) $PtNH_3ClBrF^-$. Draw the structural formulas.

8. **Crystal field splitting** $Mn(H_2O)_6^{3+}$ absorbs in the visible region at a shorter wavelength than does $Cr(H_2O)_6^{2+}$. Explain the significance of this finding in terms of the crystal field splitting energy and account for the difference in absorption.

9. **Crystal field theory** Give the electronic configuration of the central atom according to crystal field theory, for (a) $Co(NH_3)_6^{3+}$, diamagnetic; (b) CoF_6^{3-}, paramagnetic; (c) $Co(NH_3)_6^{2+}$.

ADDITIONAL PROBLEMS

10. **Oxidizing oxyanions** Ammonium salts of oxidizing anions decompose on heating. The "volcano" experiment involves heating a mound of the orange solid $(NH_4)_2Cr_2O_7$. Among the products are a large volume of the green solid Cr_2O_3 and N_2 gas. Write an equation for the reaction.

11. **Heteropolyanions** A test for PO_4^{3-} involves the formation of a yellow precipitate of ammonium molybdophosphate, $(NH_4)_3PMo_{12}O_{40}(HNO_3)_2H_2O$, on the addition of nitric acid and ammonium molybdate, $(NH_4)_6Mo_7O_{24}(H_2O)_4$. No redox occurs. Determine the oxidation number of Mo in the heteropolyanion $PMo_{12}O_{40}^{3-}$.

12. **Oxycations** Given that

$$VO^{2+} + 2H^+ + e^- \longrightarrow V^{3+} + H_2O \qquad \mathscr{E}^0 = 0.36\ V$$

Are solutions of V^{3+} subject to air oxidation?

13. Tungsten bronze In 1824, Wöhler fused $Na(HO)WO_3$ in the presence of H_2 and obtained a semimetallic substance which conducted electricity and which was chemically inert. The product, called **tungsten bronze,** is now known to have a formula Na_nWO_3, where n usually varies from 0.3 to 0.9. (a) When (*i*) $n = 0$, and (*ii*) $n = 1$, what are the oxidation states of W? (b) Suggest a structure when $n = 0.3$ to 0.9.

14. Redox with OsO_4 Alkenes (p. 440) react with OsO_4 as follows:

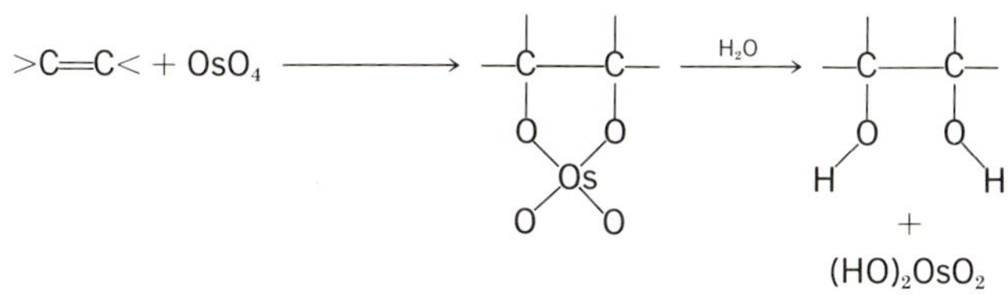

Which step(s) involves a redox reaction?

15. Peroxy species When acid dichromate solutions are treated with hydrogen peroxide, a transitory deep blue color appears. The blue species, CrO_5, is extractable by ether in which it is more stable. (a) Write a balanced ionic equation for this nonredox reaction. (b) CrO_5 possesses a peroxy bond, O—O, but is *not* paramagnetic. Suggest a reasonable structure. (The structure is still in question.)

16. Carbonyl complexes Give a reasonable structural formula for (a) $Mn_2(CO)_{10}$; and (b) $Fe_2(CO)_9$, both of which are symmetrical molecules.

17. Reactions Write balanced ionic equations to represent each of the following observations. (a) SO_2 gas turns the orange solution of acidified $Cr_2O_7^{2-}$ to a green color (b) H_2S gas decolorizes a purple solution of MnO_4^-, leaving a brown-black precipitate. (c) MnO_2 is dissolved by acidified H_2O_2 leaving a pale pink solution. (d) The addition of KNO_2 to a pink solution of Co^{2+} acidified with acetic acid, CH_3COOH, decolorizes the solution, leaving a yellow precipitate of $K_3Co(NO_2)_6$. (e) Black CoS is dissolved by a solution of NaCN. (f) $Co(OH)_2$ is precipitated on the addition of NaOH to a solution of Co^{2+}, but not on the addition of concentrated NH_3. (g) Zirconium dissolves in HF and $HONO_2$.

18. Crystal field theory Given the following standard reduction potentials:

(1) $Co(H_2O)_6^{3+} + e^- \longrightarrow Co(H_2O)_6^{2+} \qquad \mathscr{E}^\circ = +1.84\ V$

(2) $Co(NH_3)_6^{3+} + e^- \longrightarrow Co(NH_3)_6^{2+} \qquad \mathscr{E}^\circ = +0.1\ V$

(3) $Co(CN)_6^{3-} + e^- \longrightarrow Co(CN)_6^{4-} \qquad \mathscr{E}^\circ = -0.84\ V$

(a) Explain the significance of these $\mathscr{E}^\circ$ values in terms of relative stabilities of the Co(II) and Co(III) states in these three complexes. (b) Offer an explanation in terms of the crystal field theory for the differences in the three $\mathscr{E}^\circ$ values.

19. Werner theory A compound with the *empirical* formula $Pt(II)Cl_2(NH_3)_2$ (a) does not liberate NH_3 in the presence of HCl; (b) has a molar conductance corresponding to two ions; (c) gives no precipitate of AgCl with $AgNO_3$; (d) has a molecular weight of 600 g/mole. Write the formulas for two possible isomers of the compound. What type of isomerism is exhibited?

20. Ligand strength (a) Draw the molecular orbital structure for $:C{\equiv}N:^-$ and NO_2^-. (b) Do any other ligands in the series shown on p. 425 have π bonds? (c) Draw a conclusion about the crystal field effect of ligands that possess π bonds. (d) About where could CO fit into the series?

21. Practical use of complexes Explain the following phenomena in terms of complex formation, and give formula for the complex ion:

(a) In photography the fixing bath, $Na_2S_2O_3$, dissolves AgBr from the film.
(b) Gold is oxidized by O_2 only when NaCN is present.
(c) Platinum does not react with nitric acid, but reacts with aqua regia (3 parts HCl, 1 part $HONO_2$).

22. Symmetry Indicate the kind, number, and location of each symmetry element in (a) *cis*-$PtBr_2Cl_2$; (b) *trans*-$PtBr_2Cl_2$.

22 ■ ORGANIC CHEMISTRY

22.1 □ WHAT IS ORGANIC CHEMISTRY?

The adjective "organic" implies a product that originates from living things (organisms). Until about the middle of the nineteenth century, most chemists thought that the atoms in organic molecules were bound together not merely by the electrical forces typically present in inorganic molecules like NaCl, but also by some vital force that partook of the essence, and the mystery, of life itself. As life must come from previous life, so it was believed that an organic chemical must come from another organic chemical, and ultimately from a living organism, the only source of the vital force. This rationalization was used to account for the fact that chemists could not synthesize organic chemicals from the elements or from inorganic compounds. As such syntheses began to be accomplished, however, the "vital force" theory gradually fell into disrepute.*

Most chemical substances existing in or derived from living organisms contain carbon. For this reason, after the demise of the vital force theory, it was convenient to retain the term "organic chemistry" to refer to the chemistry of carbon compounds. This definition is not precise, however; organic chemists do not usually concern themselves with carbonate rocks, such as limestone, $CaCO_3$. Some definitions require that a compound have at least one C—H or C—C bond to be classified as organic, but these restrictions are not popular among

* The synthesis of urea from ammonium cyanate by Friedrich Wöhler in 1828, usually cited as the evidence that destroyed the vital force theory, was actually inconclusive at the time because his "inorganic" ammonia was prepared from animal hoofs and horns, not from nitrogen and hydrogen. Probably a more decisive turning point was the synthesis of acetic acid from the elements by Hermann Kolbe in 1845.

organic chemists who work with refrigerants (for example, Freon, CCl_2F_2), solvents (carbon tetrachloride, CCl_4, or carbon bisulfide, CS_2), toxic gases (phosgene, $COCl_2$), or fertilizers (urea, $(H_2N)_2C{=}O$). Thus, there is no universally accepted exact definition of "organic chemistry."

22.2 □ THE BONDING OF CARBON

Since the vast majority of compounds that concern organic chemists do contain carbon, it is important for us to review the bonding of this element.

Carbon ($1s^2 2s^2 2p^2$) does not usually participate in ionic bonding. The C^{4+} ion, which would be isoelectronic with helium ($1s^2$), does not exist in chemical systems, because it would require a very high ionization energy (3381 kcal/mole) to remove the four electrons, and because the charge density of C^{4+} would be intolerably high (Table 7.2, p. 129). Thus, CF_4 is covalent, not ionic. The C^{4-} ion, isoelectronic with neon ($1s^2 2s^2 2p^6$), is thought to exist only in beryllium carbide, Be_2C, and aluminum carbide, Al_4C_3, in which the attraction of the four extra electrons by the nearby Be^{2+} or Al^{3+} ions suffices to overcome their mutual repulsions.

There are molecules in which a carbon atom can participate in both ionic and covalent bonding. When methyl chloride, CH_3Cl, reacts with potassium, potassium methide, CH_3K, is formed,

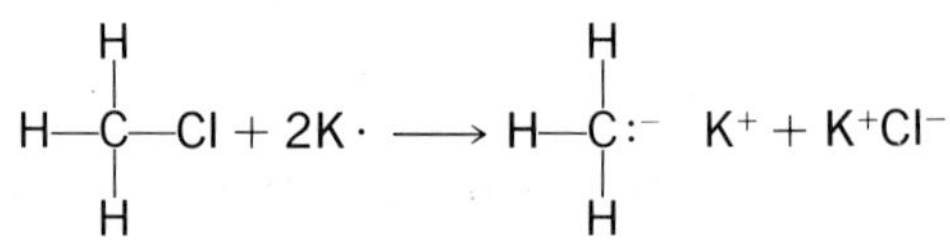

methyl chloride *potassium methide*

Potassium methide is a crystalline, ionic solid, in which there are three covalent C—H bonds, and one ionic bond between the $[:CH_3]^-$ anion and the K^+ cation. Such carbon anions, called **carbanions,** are almost invariably bonded only to the highly electropositive elements of group IA or IIA. With the less electropositive metallic elements, carbon forms a covalent bond. For example, dimethylmercury, $H_3C{-}Hg{-}CH_3$, is a typical covalent molecule (b.p. 96°C).

The cationic analog of CH_3^- is CH_3^+. Such a species, called a **carbonium ion,** has only a sextet of electrons about the carbon atom and, as expected, reacts readily in order to complete the octet. Carbonium ions are formed as very short-lived intermediates during the course of certain organic reactions (p. 453). Some carbonium ions have been characterized in solution under especially favorable conditions.

The most typical bonding of carbon is covalent, using sp^3 (tetrahedral), sp^2 (trigonal), or sp (linear) hybridized states. The particular hybridization sets a limit on the number of σ bonds (4 for sp^3, 3 for

sp^2, and 2 for sp). The deficit between the number of σ bonds and the total covalency of 4 for carbon is made up by π bonds:

Hybridization	σ bonds	π bonds	Bond multiplicity
sp^3	4	0	4 single bonds
sp^2	3	1	2 single + 1 double bond
sp	2	2	2 double bonds, or 1 single + 1 triple bond

Particularly noteworthy is the fact that carbon atoms can connect various combinations of these bonds to form continuous or branched chains or cycles of carbon atoms in a variety far exceeding that exhibited by any other element.

22.3 □ ALKANE HYDROCARBONS*; ISOMERISM AND HOMOLOGY

Hydrocarbons are compounds that contain only carbon and hydrogen.

A manufacturer of pure hydrocarbons offers for sale 18 different compounds, called **octanes,** whose molecular formula is C_8H_{18}. Imagine that we order a gallon of each. Seventeen of them look alike; they are clear colorless liquids at room temperature. The eighteenth is a white crystalline solid. Each octane has the same molecular weight, 114.23 g/mole, corresponding to the formula C_8H_{18}. We determine their boiling points, and find that they range from 99° to 126°C; some of the boiling points are so close to others that they cannot easily be distinguished from each other. Now we try a somewhat more heroic experiment. We take 18 identical new automobiles with empty gasoline tanks and add one of the octanes to each. Of course, this is a bit difficult with the solid octane, but we can liquefy it by dissolving it in a little commercial gasoline. We find that some of the octanes serve very well as gasolines; others cause some engine "knock"; some knock so badly that we cannot drive the automobile.

Clearly, what we have here are 18 different substances that are distinct in their properties despite their common formula. Such compounds are called **isomers.** The formula C_8H_{18}, assuming a covalence of 4 for carbon and 1 for hydrogen, is consistent only with single C—H and C—C bonds and with continuous or branched, but *not cyclic*, C—C linkages. This fact may be difficult to visualize, but we shall try to illustrate it by considering one of the octanes, which serves as the standard of 100-octane gasoline. Imagine that we construct a model of this molecule from atomic models by starting with methane, 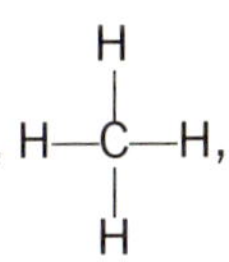,

* See Appendix 3 for a discussion of nomenclature of the alkanes and some of their simple derivatives.

and adding C and H atoms with the provision that with each addition *we maintain the proper covalences for C and H* (4 and 1, respectively). Starting with the model of methane, if we remove one H, we must add CH_3,

$$\left.\begin{array}{l} CH_4 \xrightarrow{-H} H_3C \\ H_3C \xrightarrow{+CH_3} H_3C{-}CH_3 \end{array}\right\} \quad (1)$$

The net addition is thus CH_2.

A similar second step gives

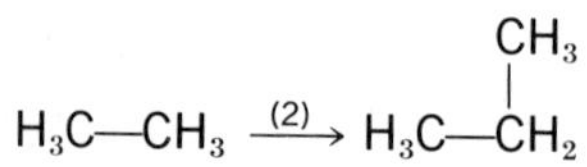

Continuing, we have

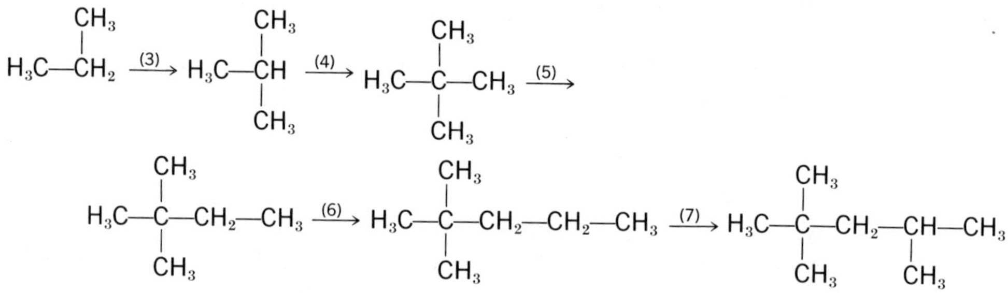

Note the following:

(a) We started with CH_4, and each successive net addition was CH_2. We added CH_2 seven times, thus $CH_4 + 7(CH_2) \longrightarrow C_8H_{18}$.

(b) We had no choice in steps (1) and (2), because all the hydrogens that we might have removed were equivalent; we cannot distinguish one from another. But in steps (3) through (7) there were other possibilities than the ones we chose to construct the model we wanted. These other choices could have led to any of 18 possible isomers. But no matter which sequence we chose, we could have added CH_2 seven times, yielding C_8H_{18} as the final formula. These compounds, which differ among each other in the *sequence of linkages of atoms*, are called **structural isomers.** We shall not illustrate all the possibilities, but the ambitious reader, with the aid of a set of molecular models, is invited to try them.

(c) If we had made a cycle, for example,

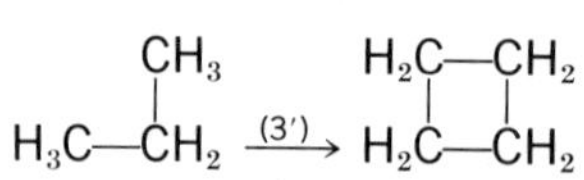

we would have added only C, and thus have had two fewer H atoms than would exist without the cycle. Therefore any cyclic arrangement would have a different molecular formula and would *not* be an isomer of noncyclic hydrocarbon with only single bonds.

If we had chosen any number of carbon atoms other than 8, we could calculate the molecular formula of a noncyclic singly bonded hydrocarbon on the basis of the addition of the required whole number of CH_2 units to the parent hydrocarbon, CH_4. This sum can be expressed by the *general formula for an alkane*, C_nH_{2n+2}, where n is a whole number. A series of compounds containing the same types of bonds, which can be described by a general formula, is called a **homologous series.** Among carbon compounds the molecular formulas of successive members of such a series differ from each other by CH_2.

22.4 ▫ THE SHAPES OF ALKANE MOLECULES. CONFORMATION

The bond angle corresponding to sp^3 hybridization is 109.5 degrees. The molecule of methane, CH_4, is thus arranged as a body-centered tetrahedron (p. 137). A chain of σ-bonded carbon atoms, which maintains the tetrahedral C—C—C angles, is therefore shaped as a series of 109.5 degree zig-zags.* But the electronic charge of the σ bond is symmetrically distributed around the bond axis (p. 135); as a result, σ-bonded atoms can rotate more or less freely about the bond. This means that a chain of singly bound carbon atoms in an alkane can be arranged in any zig-zag shape in space so long as the tetrahedral bond angles are preserved. Two of these arrangements, called conformations, are shown in Fig. 22.1 for models of the carbon skeleton of a continuous 8-carbon chain. Note that the linear structural formula represents only the projection of one particular conformation; such a formula does not mean that the chain is a straight line in space. *Conformations do not represent isomerism;* the conformers shown in Fig. 22.1 are the carbon skeletons of the *same isomer* of C_8H_{18}. As another example, the structural formulas

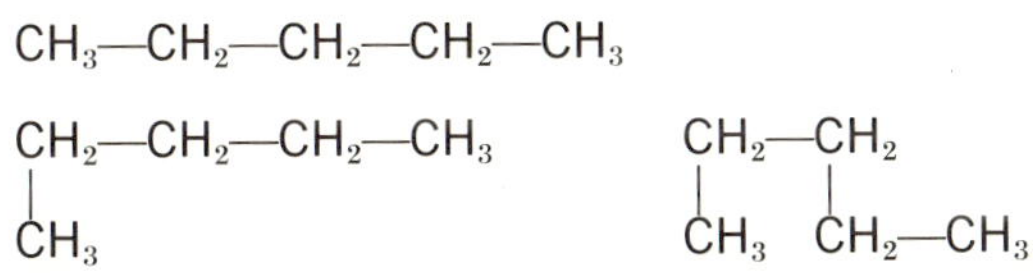

all represent the same compound, n-pentane, because all the carbon sequences are continuous, not branched.

The alkanes with continuous (unbranched) carbon chains are called "normal" alkanes (abbreviation n). Table 22.1 gives the names and properties of the first ten of these. Note that the first three, methane, ethane, and propane, are not designated n because they have no branched isomers from which they need be distinguished. As the number of carbon atoms increases, the number of possible isomers increases sharply: 9 isomers of heptane, 18 of octane, 35 of nonane, 75 of decane, and so on.

* Not always exactly. There may be slight distortions.

Fig. 22.1 Model of carbon skeleton of *n*-octane. (a) Conformation whose projection is linear.

22.5 □ CHEMICAL PROPERTIES OF THE ALKANES

Chemically, the alkanes are characterized by their inertness. For this reason they are also called **paraffins,** from the Latin *parum affinis*, meaning "of little affinity." They do not react with bases such as

TABLE 22.1 NAMES AND PHYSICAL PROPERTIES OF NORMAL ALKANES, C_nH_{2n+2}

Name	Molecular formula	Condensed formula	Boiling point °C	Melting point °C
Methane	CH_4	CH_4	−161.5	−182.5
Ethane	C_2H_6	CH_3CH_3	− 88.6	−183.3
Propane	C_3H_8	$CH_3CH_2CH_3$	− 42.1	−187.7
n-Butane	C_4H_{10}	$CH_3CH_2CH_2CH_3$	− 0.5	−138.4
n-Pentane	C_5H_{12}	$CH_3(CH_2)_3CH_3$	36.1	−129.7
n-Hexane	C_6H_{14}	$CH_3(CH_2)_4CH_3$	68.7	− 95.3
n-Heptane	C_7H_{16}	$CH_3(CH_2)_5CH_3$	98.4	− 90.6
n-Octane	C_8H_{18}	$CH_3(CH_2)_6CH_3$	125.7	− 56.8
n-Nonane	C_9H_{20}	$CH_3(CH_2)_7CH_3$	150.8	− 53.5
n-Decane	$C_{10}H_{22}$	$CH_3(CH_2)_8CH_3$	174.1	− 29.7

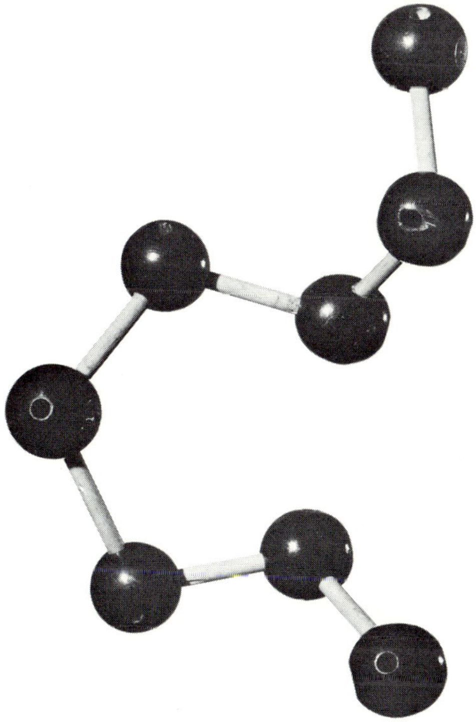

Fig. 22.1(b) Conformation that does not have a linear projection.

sodium hydroxide, or with acids; even concentrated sulfuric acid leaves them unchanged. The very active alkali metals, lithium, sodium, and potassium, are stored under liquid alkanes such as kerosene. Under ordinary conditions, the alkanes are inert toward oxygen, chlorine, and bromine. However, at elevated temperatures, alkanes burn in the presence of oxygen to form carbon dioxide and water. The reaction, typified by the combustion of propane,

$$C_3H_8 + 5O_2 \longrightarrow 3CO_2 + 4H_2O \qquad \Delta H = -483 \text{ kcal/mole}$$

is highly exothermic. For this reason, alkanes are used as fuels. The heat evolved from such reactions is known as the **heat of combustion.**

Methane is the major constituent of natural gas, widely used for heating and cooking. Propane and butane are used as liquefied petroleum gas [LPG ("bottled gases")] in areas where natural gas is not distributed by pipeline. Octane is a constituent of gasoline. Kerosene and diesel and fuel oils are also important sources of power and heat.

Although the alkanes do not normally react with chlorine and bromine, in the presence of strong sunlight or ultraviolet radiation a reaction occurs in which a hydrogen atom is displaced by a halogen atom,

$$CH_4 + Cl_2 \xrightarrow{\text{light}} CH_3Cl + HCl$$

22.6 □ ALKENES AND ALKYNES; UNSATURATED HYDROCARBONS; GEOMETRICAL ISOMERISM

Carbon atoms can form multiple bonds with each other. For example, the gaseous compound ethylene (boiling point, $-104°C$) has the molecular formula C_2H_4. The molecule C_2H_4 has 12 valence electrons. Were it to be saturated, it would need 14 valence electrons, or two more. Hence, C_2H_4 lacks two electrons of being saturated, and therefore must have a double bond. Several ways of writing the structure of ethylene are

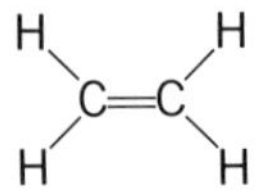

structural formula

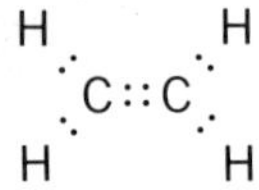

Lewis (electron-dot) formula

$H_2C{=}CH_2$

condensed formula

ethylene (common name) or *ethene* (IUPAC name)

Hydrocarbons like ethylene, that possess one carbon-to-carbon double bond (C═C), belong to the **alkene** homologous series. The suffix *-ene* indicates the presence of the C═C bond. Since the alkenes have two fewer H atoms than the corresponding alkanes, their general formula is C_nH_{2n}, where n is a whole number larger than 1.

The presence of a double bond usually increases the opportunities for isomerism. Thus, although there is only one *n*-butane, there are three isomeric *n*-butenes. It is easy to see from the condensed structural formulas that there are two different possible locations for the double bond in a continuous chain of 4 carbon atoms.

$H_2C{=}CHCH_2CH_3$ *1-butene* $\qquad$ $H_3CCH{=}CHCH_3$ *2-butene*

(*The location of the double bond is indicated by the lower-numbered doubly bound carbon atom.**) How does the third isomer arise? The fact that a π bond requires lateral or side-to-side orientation of two p orbitals (pp. 142–143) means that a double bond (which includes one π bond) fixes the two entities which it holds together in one of *two definite positions*. (Think of setting two pencils next to each other side by side. How many ways can you do this? Two—the points can be at the same end, or the points can be at opposite ends.) In one position the two methyl groups will be on the same side (*cis*) of the double bond, in the other they will be on opposite sides (*trans*) of the double bond.

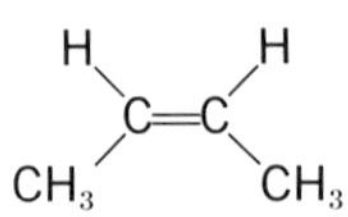

cis-2-butene

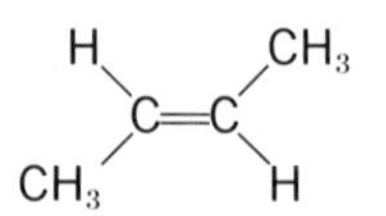

trans-2-butene

* See Appendix 3, p. 555, on numbering of carbon atoms.

This type of isomerism is called **geometrical** or **cis-trans isomerism.** In the case of 1-butene,

1-butene (*no geometrical isomerism*)

an ethyl (C_2H_5) group will be on the same side as a hydrogen atom, no matter which position the entities assume with respect to the double bond. The lack of free rotation about the double bond of 1-butene therefore has no structural consequences, and no geometrical isomerism exists for this compound. *In order for geometrical isomerism to exist, each carbon atom of the double bond must have two different atoms or groups attached to it.*

Carbon atoms can also be joined by a triple bond. The simplest example is acetylene, C_2H_2 (boiling point, −84°C),

H—C≡C—H   H:C:::C:H

It is the first member of the **alkyne** homologous series, whose general formula is C_nH_{2n-2}. The ending *-yne* indicates the presence of the —C≡C— bond. Both alkenes and alkynes have in common an important type of chemical property. Ethylene, a typical alkene, reacts with chlorine readily at room temperature (even in the dark and in an inert solvent such as carbon tetrachloride),

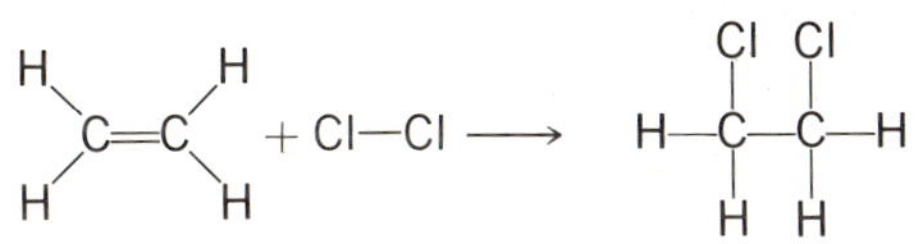

1,2-dichloroethane

As a result of this reaction, the double bond becomes a single bond, the Cl—Cl bond is broken, and two new C—Cl bonds are formed. A reaction of this type is called an **addition reaction.** Alkynes add 2 moles of halogen, as exemplified by the addition of bromine to propyne,

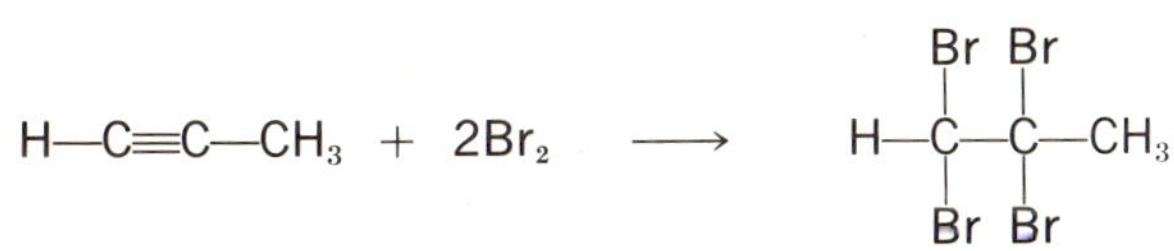

propyne (colorless)   (red-brown)   *1,1,2,2-tetrabromopropane* (colorless)

Addition reactions are typical of most molecules with multiple bonds and for this reason such molecules are said to be **unsaturated.** Saturated molecules such as the alkanes do not have multiple bonds and so do not undergo the addition reaction. The decolorization of the red brown solution of bromine (see above equation), usually in a solvent such as carbon tetrachloride, is used as a test for alkenes and alkynes.

Other reagents that add to multiple bonds include:

(a) *Hydrogen in the presence of platinum, palladium, or nickel*,

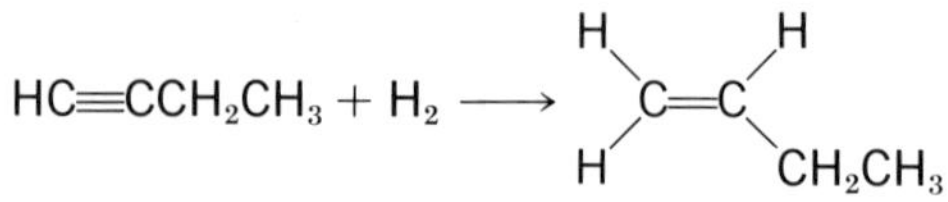

1-butyne *1-butene*

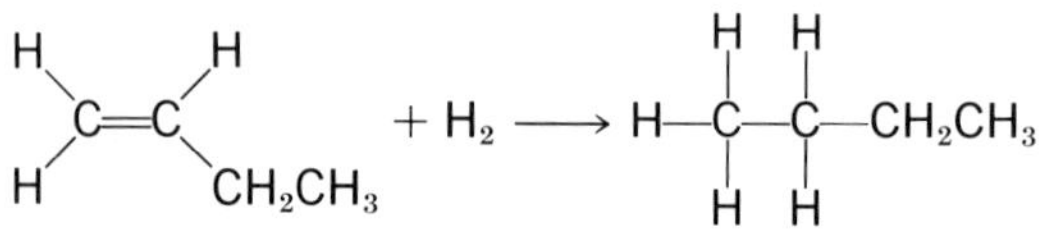

1-butene *n-butane*

(b) *Anhydrous* (lacking water) *hydrogen halides*, such as hydrogen chloride, HCl (this reagent should not be confused with aqueous HCl, hydrochloric acid),

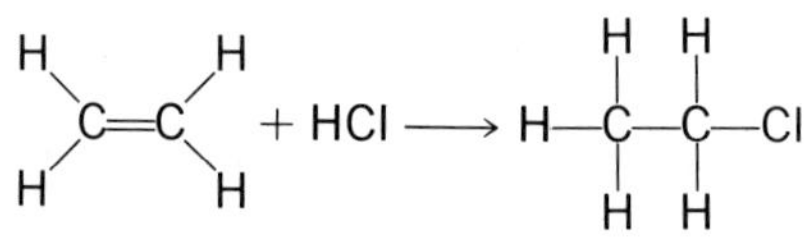

chloroethane

(c) *Water* in the presence of dilute sulfuric acid. The product is an alcohol,

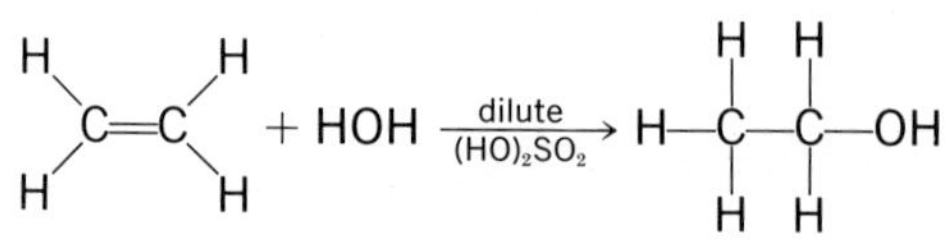

ethyl alcohol

The addition of water to ethylene can be *reversed* by heating ethyl alcohol in concentrated sulfuric acid,

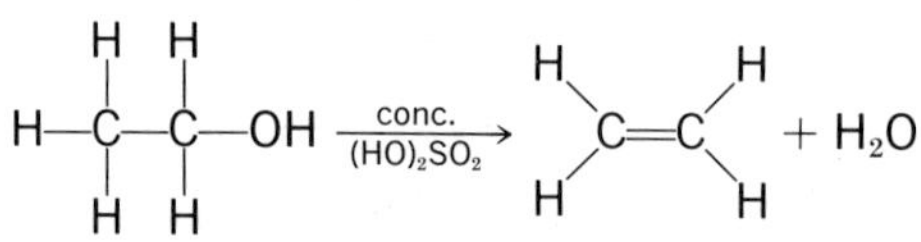

This is a useful method for making ethylene on a laboratory scale. The **dehydration** (removal of water) of alcohols is the usual method for making alkenes. This type of reaction is commonly called an **elimination.**

22.7 □ CYCLOALKANES

Carbon atoms may also be arranged in rings. Hydrocarbons containing rings of carbon atoms linked together by single bonds are known as **cycloalkanes.** Since the minimum number of carbon atoms needed to form a ring is 3, the lowest member of this series is cyclopropane,

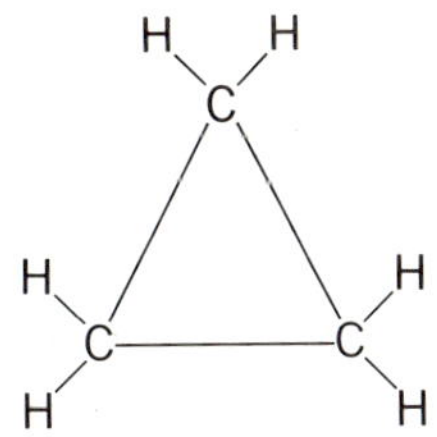

cyclopropane

To name this group of hydrocarbons, *cyclo-* is prefixed to the name of the alkane with the corresponding number of carbons. The nomenclature is indicated by the names of the next three members of the series,

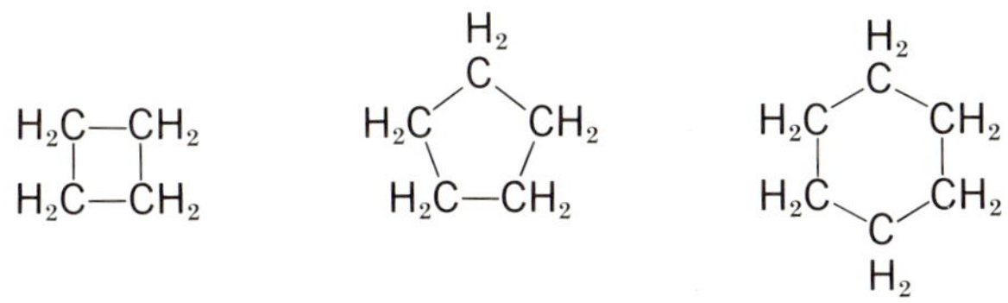

cyclobutane *cyclopentane* *cyclohexane*

The cycloalkanes have the general formula C_nH_{2n}, and therefore are isomeric with the alkenes. For example, cyclopropane (C_3H_6) and propene, $CH_3CH{=}CH_2$ (that is, C_3H_6), are isomers. These two classes of hydrocarbon are, however, chemically distinguishable. With the exception of cyclopropane, the cycloalkanes behave as saturated molecules. Thus, they do not undergo addition reactions.

22.8 ▫ DIENES; BENZENE AND AROMATIC COMPOUNDS

Molecules can have more than one double bond. Thus, hydrocarbons having two double bonds are called **dienes.** One of the more important dienes is 1,3-butadiene,

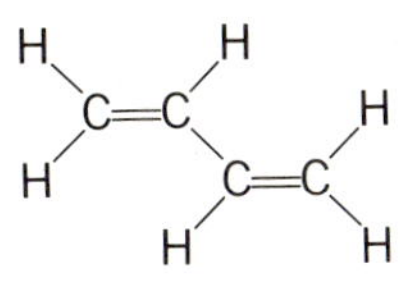

1,3-butadiene

It serves as starting material for preparing a variety of polymers (see Chapter 25).

Of particular interest is the cyclic triene called benzene. Benzene is a volatile liquid (b.p. 80°C), first isolated in 1825 by Michael Faraday by the destructive distillation* of coal. Its molecular formula is C_6H_6,

* **Destructive distillation** is a distillation in which vaporization is accompanied by decomposition.

and the structure suggested by August Kekulé in 1865 is

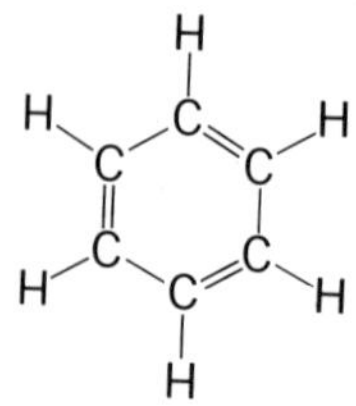

Kekulé structure for *benzene*

or more simply

This ring system is an integral unit of a very large number of hydrocarbons with the general formula C_nH_{2n-6}, where n must be larger than 5, which are referred to as benzenoid* compounds. Coal tar, a distillate from soft coal, is a source of many aromatic compounds, some of which are formulated below.

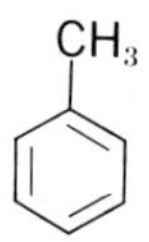

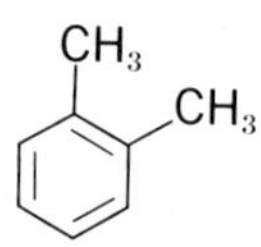

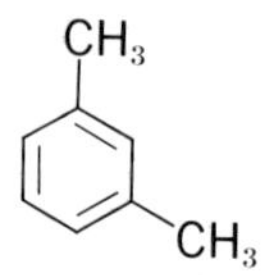

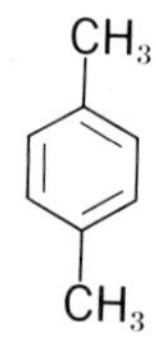

common name:	*toluene*	*o-xylene*	*m-xylene*	*p-xylene*
systematic name:	*methylbenzene*	*1,2-dimethylbenzene*	*1,3-dimethylbenzene*	*1,4-dimethylbenzene*

Note the use of the prefixes *ortho-*, *meta-*, and *para-* to indicate the 1,2-, 1,3-, and 1,4-disubstituted benzene derivatives, respectively. There are many aromatic hydrocarbons with fused benzene rings—rings with two carbon atoms in common—as in naphthalene, $C_{10}H_8$, a white solid (m.p. 80°C) used as a moth repellent,

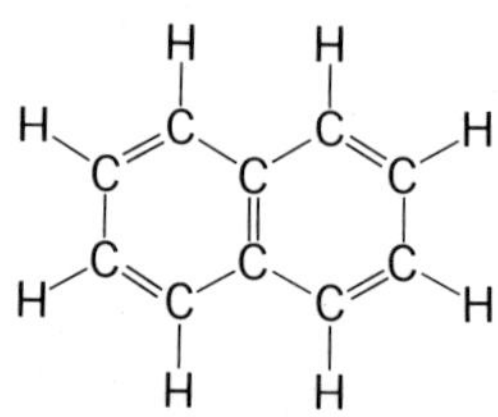

or

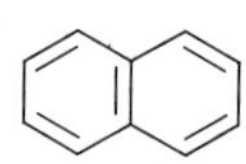

naphthalene

Graphite is an elementary form of carbon. It is a soft, black solid that conducts a current and vaporizes at 4347°C. It is a macromolecular type of substance composed of layers of *fused benzene rings*. The separation of layers is 3.35 A (Fig. 22.2a). Figure 22.2b shows a portion of the fused benzene ring layers.

Aromatic hydrocarbons are used extensively in the manufacture of

* This class of compounds is also called **aromatic** because many of their derivatives have pleasant odors. Many of them are toxic, some are carcinogenic—and therefore should not be inhaled.

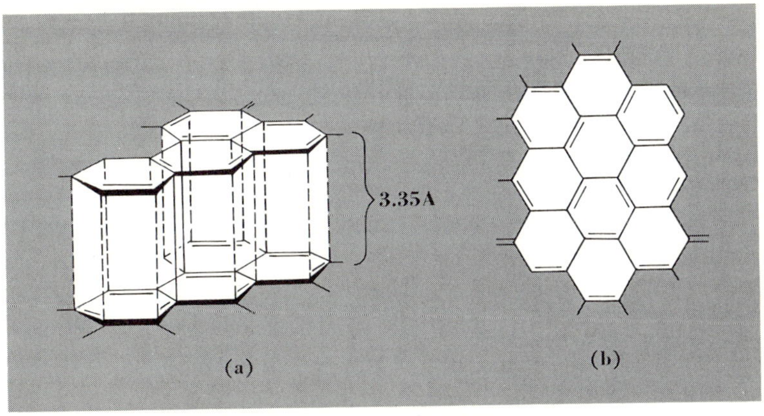

Fig. 22.2 Graphite structure. (a) Continuous layers of fused benzene-like rings. (b) A portion of one layer.

plastics, drugs, dyes, insecticides, explosives, and a host of other useful materials. A few examples of compounds derived from aromatic hydrocarbons are

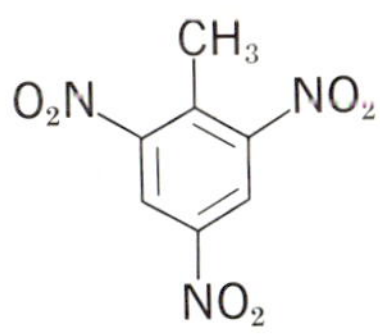

TNT (an explosive)

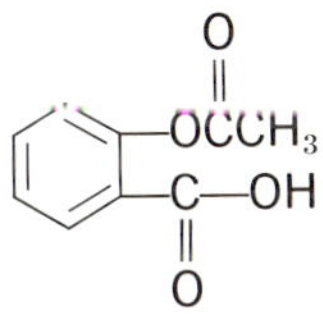

aspirin (a drug)

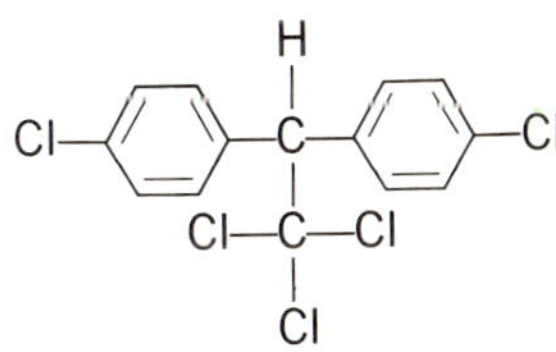

DDT (an insecticide)

The chemical reactions of benzene are not typical of an unsaturated compound. For example, it does not undergo catalytic hydrogenation at atmospheric pressure and at room temperature, as do the alkenes and alkynes. It can be converted to cyclohexane, but only under more severe conditions—for example, in the presence of Ni at 150°C, and under 10 atm pressure of H_2. This conversion with hydrogen to cyclohexane, although sluggish, is nevertheless chemical proof for the cyclic structure of benzene. Benzene also gives a negative test for unsaturation; it does not react with Br_2 in CCl_4. It does react with Br_2 in the presence of a catalyst at elevated temperatures. However, the reaction is *not* an addition. Instead, an H atom is replaced by a Br atom,

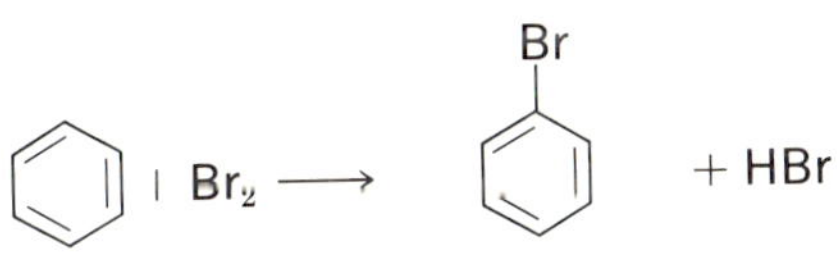

bromobenzene

Benzene evidently resists addition to its double bonds.

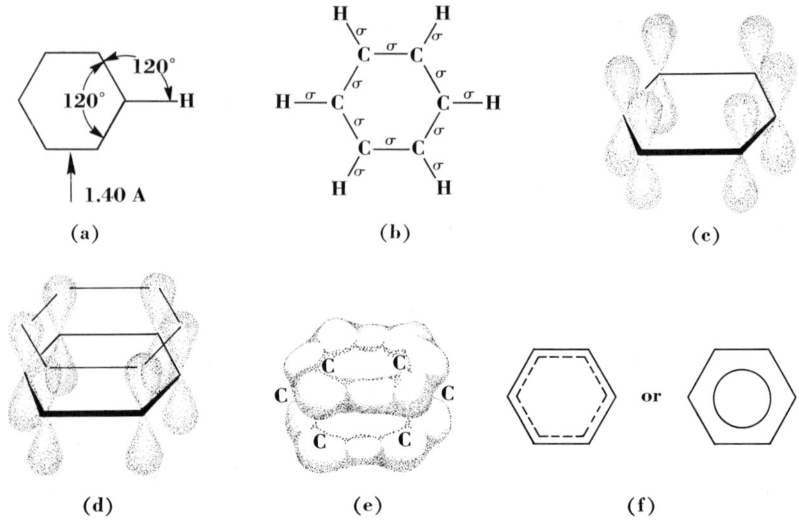

Fig. 22.3 Molecular orbital representation for benzene: (a) structure showing geometry; (b) σ-bonded skeleton; (c) p orbitals; (d) cyclic overlap of p orbitals; (e) extended π bond; (f) shorthand notations.

This behavior of benzene and other aromatic compounds has led to the idea that there is a stability inherent in this ring of six carbon atoms with three alternating double bonds. We may rationalize this behavior in terms of extended π bonding. Benzene is a planar molecule. This means that the nuclei of all 12 atoms comprising the benzene molecule lie in one plane. Each C—C—C and H—C—C bond angle is 120 degrees, and each carbon-to-carbon bond length is 1.40 Å (Fig. 22.3a). This type of geometry indicates that each C atom is using sp^2 hybridized atomic orbitals to form three localized σ bonds for the carbon-hydrogen skeleton (Fig. 22.3b). Each carbon atom utilizes three of its four valence electrons to fabricate its three σ bonds. The fourth valence electron must reside in the remaining p orbital. Each carbon atom has such a p orbital, and the six p orbitals are parallel and adjacent to each other (Fig. 22.3c). How do these p orbitals interact? We have seen (p. 135) that *a given p orbital can overlap at the same time with each parallel p orbital adjacent to it* (Fig. 22.3d). Thus, *the six electrons are delocalized in an extended, cyclic π bond encompassing all six carbons.* The electron distribution has the shape of a pair of doughnuts, one above and the other below the σ-bonded carbon skeleton. Since the electron density is somewhat greater at the carbon atoms and somewhat less between the carbons, the doughnuts are depicted with lumps at the carbon atoms (Fig. 22.3e).

This model for the bonding in benzene predicts that the bonding is uniform between any pair of carbon atoms, and hence each carbon-to-carbon bond should have the same length. The model also predicts that this length will be greater than a C═C bond length and less than a C—C bond length. This prediction agrees with experiment, as shown below.

Bond	Bond length, Å
C—C in ethane	1.54
C┅C in benzene	1.40
C═C in ethylene	1.34

Since the benzene molecule has no definite single or double bonds as written in Lewis formulas, it is represented by drawing a continuous broken line or a circle in the center of the hexagon (Fig. 22.3f). The broken line (or circle) stands for the extended π bond.

Table 22.2 summarizes the hydrocarbon homologous series referred to in the preceding sections.

22.9 ▫ FUNCTIONAL GROUP DERIVATIVES

The H atom of a hydrocarbon can be replaced by another atom or group of atoms not of the hydrocarbon variety. The resulting compounds are called hydrocarbon **derivatives,** and the atom or group is known as a **functional group.** The more common functional groups are listed in Table 22.3. The symbol **R** is used to represent an alkyl group. The chemistry of these derivatives is characteristic mainly of the functional group.

The reactions of the various functional groups constitute a large part of the subject matter of organic chemistry, and any systematic

TABLE 22.2 ▪ HYDROCARBON HOMOLOGOUS SERIES

Name	General formula	Typical example	Formula	Chemical behavior
Alk*ane*	C_nH_{2n+2}	Eth*ane*	$H_3C—CH_3$	Saturated
Alk*ene*	C_nH_{2n}	Eth*ene*	$H_2C═CH_2$	Unsaturated
Alk*yne*	C_nH_{2n-2}	Eth*yne*	$H—C≡C—H$	Unsaturated
Alka*diene*	C_nH_{2n-2}	1,3 Buta*diene*	$H_2C═CH—CH═CH_2$	Unsaturated
*Cyclo*alk*ane*	C_nH_{2n}	*Cyclo*hex*ane*		Saturated
*Cyclo*alk*ene*	C_nH_{2n-2}	*Cyclo*hex*ene*	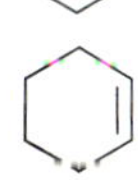	Unsaturated
Benzenoid	C_nH_{2n-6}	Benzene	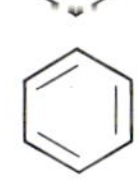	Aromatic

TABLE 22.3 ▪ COMMON FUNCTIONAL GROUPS IN ORGANIC COMPOUNDS

Group	General formula	General name	Example: Formula	Example: Name
$-Cl$	$R-Cl$	Chloride	CH_3Cl	Methyl chloride (chloromethane)
$-Br$	$R-Br$	Bromide	CH_3Br	Methyl bromide (bromomethane)
$-OH$	$R-OH$	Alcohol	CH_3CH_2OH	Ethyl alcohol (ethanol)
$-O-$	$R-O-R$	Ether	$CH_3CH_2OCH_2CH_3$	Diethyl ether
$-NH_2$	RNH_2	Amine	$CH_3CH_2CH_2NH_2$	*n*-Propylamine
$-N(=O)-O$	RNO_2	Nitro compound	$C_6H_5NO_2$	Nitrobenzene
$-C(=O)H$	$R-C(=O)H$	Aldehyde	$CH_3CH_2C(=O)H$	Propionaldehyde
$>C=O$	$R-C(=O)-R$	Ketone	$CH_3-C(=O)-C_2H_5$	Methyl ethyl ketone
$-C(=O)OH$	$R-C(=O)OH$	Carboxylic acid	$CH_3-C(=O)-OH$	Acetic acid
$-C(=O)-OR'$	$R-C(=O)-OR'$	Ester	$CH_3-C(=O)-OC_2H_5$	Ethyl acetate
$-O-O-$	$R-O-O-R'$	Peroxide	$C_2H_5-O-O-C_2H_5$	Diethyl peroxide

treatment would be too voluminous for this chapter. However, we shall cite some properties and reactions of several classes of compounds of general interest.

Alcohols are hydroxy derivatives of alkanes, ROH. They can also be considered to be alkyl derivatives of water, and indeed show some similarities to water in their chemical behavior. For example, ethanol, like water, reacts with metallic sodium with the evolution of hydrogen.

$$C_2H_5OH(l) + Na(c) \longrightarrow Na^+C_2H_5O^+\ (alcoholic) + \tfrac{1}{2}H_2(g)$$
$$HOH(l) + Na(c) \longrightarrow Na^+OH^-(aq) + \tfrac{1}{2}H_2(g)$$

Methyl alcohol, CH_3OH (also called **wood alcohol** because it can be obtained from the destructive distillation of wood), is used as a solvent and as an antifreeze. **Ethylene glycol,** $HOCH_2-CH_2OH$, is a dialcohol commonly used as an antifreeze. **Glycerol** (also called **glycerine**),

$HOCH_2—CHOH—CH_2OH$, is an important trihydroxy alcohol. Under suitable conditions, alcohols may undergo bimolecular dehydration to form **ethers.**

$$2C_2H_5OH \longrightarrow H_2O + C_2H_5—O—C_2H_5$$

Ethers, R_2O, may be regarded as dialkyl derivatives of water. Since ethers do not have any OH group, they do not react with metallic sodium. However, an ether can act as a proton acceptor, or base, in a manner analogous to the behavior of water.

$$\underset{\textit{diethyl ether}}{CH_3CH_2\ddot{\underset{\cdot\cdot}{O}}CH_2CH_3} + H:\ddot{\underset{\cdot\cdot}{Cl}}: \longrightarrow \underset{\textit{diethyloxonium ion}}{[CH_3CH_2\overset{H}{\underset{\cdot\cdot}{O}}CH_2CH_3]^+} + :\ddot{\underset{\cdot\cdot}{Cl}}:^-$$

$$H_2O + HCl \longrightarrow \underset{\textit{hydronium ion}}{H_2OH^+} + Cl^-$$

Amines are amino derivatives of alkanes, RNH_2, or alkyl derivatives of ammonia. Amines are bases which react with acids to form alkylammonium salts, in a manner analogous to the behavior of ammonia.

$$\underset{\textit{methylamine}}{CH_3NH_2} + HBr \longrightarrow \underset{\textit{methylammonium bromide}}{CH_3NH_2{}^+Br^-}$$

$$\underset{\textit{ammonia}}{NH_3} + HBr \longrightarrow \underset{\textit{ammonium bromide}}{NH_4{}^+Br^-}$$

Carboxylic or **fatty acids,** $R—C\begin{matrix}\diagup O \\ \diagdown OH\end{matrix}$, are weak acids ($K_a \approx 10^{-5}$) which react with metallic hydroxides to form salts.

$$\underset{\textit{acetic acid}}{CH_3COOH(l)} + Na^+OH^-(aq) \longrightarrow \underset{\textit{sodium acetate}}{CH_3COO^-Na^+(aq)} + H_2O(l)$$

The salt of a long chain fatty acid is called a **soap** (p. 224).

Both saturated and unsaturated fatty acids are obtained from natural sources. Some important ones are

$CH_3(CH_2)_{14}COOH$	palmitic acid
$CH_3(CH_2)_{16}COOH$	stearic acid
$CH_3(CH_2)_7CH=CHCH_2COOH$	oleic acid
$CH_3(CH_2)_4CH=CHCH_2CH=CHCH_2COOH$	linoleic acid

Carboxylic acids react with alcohols and eliminate water to form **esters.** The reaction, called **esterification,** is catalyzed by strong acids such as HCl.

$$CH_3COOH + HOC_2H_5 \xrightarrow{H^+} CH_3C(=O){-}OC_2H_5 + H_2O$$

ethyl acetate
(*an ester*)

This reaction is *not* analogous to the reaction between a strong acid and a strong base, like $ClH + HO^-Na^+ \longrightarrow Na^+Cl^- + H_2O$. Instead, the oxygen eliminated in esterification comes from the carboxylic acid, not from the hydroxy group of the alcohol. This conclusion is derived from experiments in which an alcohol "labeled" with O^{18} isotope, $RO^{18}H$, is esterified with an unlabeled acid, R′COOH. The eliminated water does *not* contain O^{18}, but the ester does, showing that the oxygen must have come from the carboxylic acid.

$$R'COOH + HO^{18}R \longrightarrow R'C(=O){-}O^{18}R + H_2O$$

The reverse of esterification, also catalyzed by strong acid, is called **ester hydrolysis.**

$$\underset{\textit{ester}}{R'COOR} + H_2O \xrightarrow{H^+} \underset{\textit{carboxylic acid}}{R'COO^-Na^+} + \underset{\textit{alcohol}}{ROH}$$

When an ester is decomposed in the presence of an alkali, the process is called **saponification,** because the carboxylate salt is a soap when R′ is a long chain of carbon atoms (p. 224).

$$\underset{\textit{ester}}{R'COOR} + Na^+OH^- \longrightarrow \underset{\textit{soap}}{R'COO^-Na^+} + \underset{\textit{alcohol}}{ROH}$$

Natural **fats** are solid esters and fatty **oils** are liquid esters of the trihydroxy alcohol **glycerol,** and are therefore called **glycerides.** Natural glycerides contain two or three different carboxylic acid components, each with an even number of carbon atoms ranging from 4 to 26. The saponification of glycerides therefore yields a mixture of soaps.

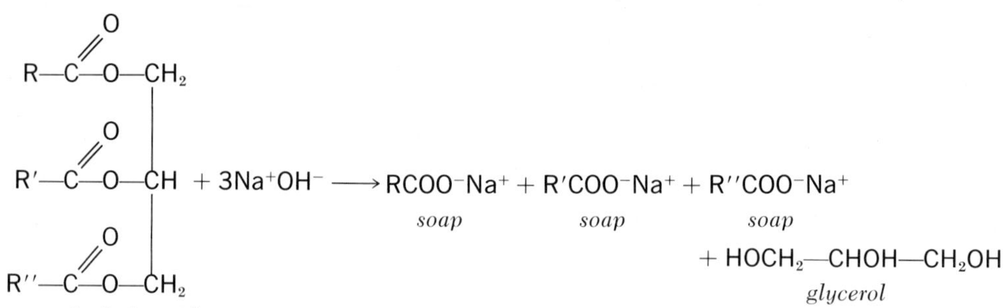

a mixed glyceride

The glycerides of saturated fatty acids (such as stearic acid) are called **saturated fats;** glycerides of unsaturated fatty acids are usually liquids at ambient temperatures and are called **unsaturated oils.** When the carboxylic acids have two or more double bonds, the glycerides are called **polyunsaturated oils.**

22.10 □ REACTIONS OF COVALENT BONDS

In the following sections we consider some general types of reactions which are typical of covalent substances. These reactions of covalent bonds may be classified as (a) decomposition, (b) displacement or substitution, (c) addition, (d) elimination, and (e) rearrangement. We have already described examples of addition and elimination reactions on pp. 441–442.

22.11 □ DECOMPOSITION

Fragmentation of a covalent molecule into two or more smaller particles without the involvement of other reactants is called a **decomposition** reaction. Heat or radiation is generally used to effect the decomposition. Exposure to an energy source increases the vibration of the atoms in the molecule, and eventually causes the rupture of the weaker bonds. Bond cleavage can occur in either of two ways

(a) *One of the atoms retains both electrons* **(heterolysis):**

$$\mathrm{A{:}\}B \longrightarrow A{:} + B}$$

Heterolysis is not typical of the decomposition of C—C bonds, because the resulting fragments would leave the carbon atoms in an ionic condition, which is unfavorable for reasons stated on p. 434. Heterolysis occurs more readily when the decomposition produces stable, uncharged entities, as illustrated by

$$\mathrm{H_3N{:}B(CH_3)_3 \overset{heat}{\rightleftharpoons} H_3N{:} + B(CH_3)_3}$$

The reader will recognize that this is the reverse of a Lewis acid-base reaction (p. 320).

(b) *Each atom takes one electron* **(homolysis):**

$$\mathrm{A{:}B \longrightarrow A\cdot + B\cdot}$$

as illustrated by the decomposition of dinitrogen tetroxide,

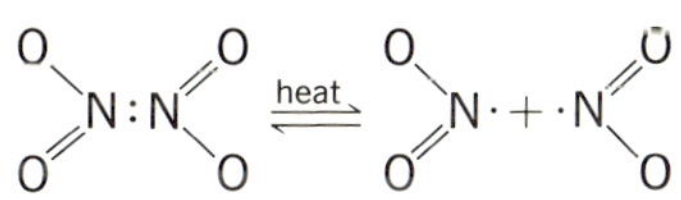

Reaction (b) leads to the formation of free radicals. Most free radicals are reactive species that rapidly undergo further reaction. Free radicals are frequently needed, especially in organic chemistry, to initiate reactions (p. 454). Compounds with weak covalent bonds, such as the bond between less active metals and carbon, or the peroxide, —O—O—

bond, are frequently used as free radical sources, as shown,

$$H_3C:Hg:CH_3 \xrightarrow{\text{heat}} \cdot Hg\cdot + 2H_3C\cdot$$

dimethylmercury

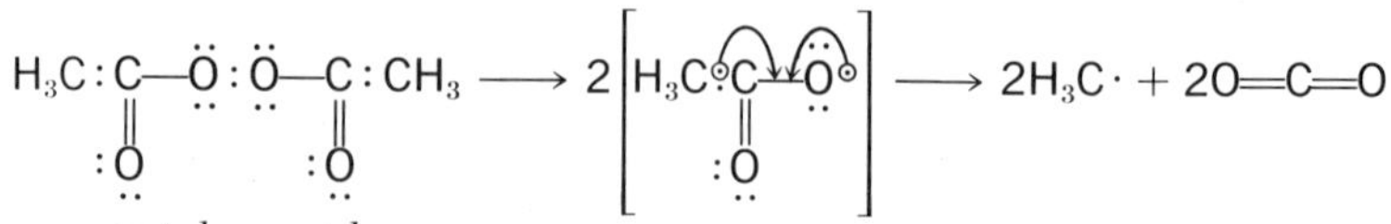

acetyl peroxide

To understand the course of these reactions, it is necessary to keep track of the assignment of electrons. To help in this "electron bookkeeping," the chemist frequently uses curved arrows to indicate the disposition of electrons, as shown above.

22.12 □ DISPLACEMENT OR SUBSTITUTION REACTIONS

In this frequently encountered type of reaction, one atom or group of atoms in the molecule is replaced by another atom or group of atoms,

$$D + A\text{—}L \longrightarrow D\text{—}A + L$$

D is the **displacing** group, L is the **leaving** group, and the displacement is said to occur on atom A. AL is frequently called the **substrate.** There are several modifications of this reaction, because of the variations in the electronic structure of D and the bonding state of A. We shall also see that there are differences in mechanism.

(a) Nucleophilic displacement on a σ-bonded carbon atom In this type of displacement reaction, D has an unshared pair of electrons and atom A participates only in σ bonding. D is a nucleophile (p. 320), and so the reaction is called a **nucleophilic displacement.** The generalized equation is

$$D: + A:L \longrightarrow D:A + :L \qquad \text{rate} = k[D][AL]$$

a nucleophilic displacement

Nucleophilic displacements on carbon are important in the synthesis of organic compounds. The thorough study of these reactions has helped to elucidate their mechanisms. The second-order reaction, as typified by

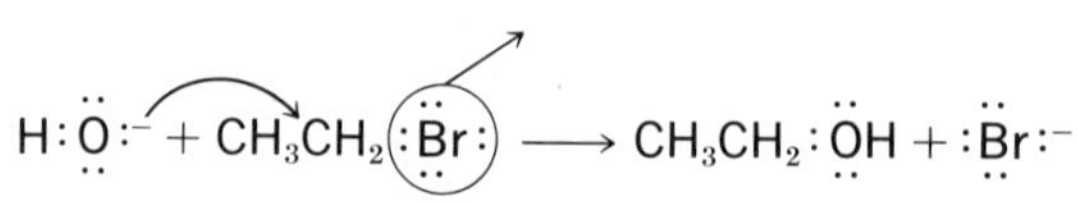

$$\text{rate} = k[OH^-][CH_3CH_2Br]$$

is termed the S_N2 reaction: S for substitution (displacement), N for nucleophilic, and 2 for bimolecular. The first-order reaction, as typified by

$$(CH_3)_3C:\ddot{Br}: + :\ddot{O}:H^- \longrightarrow (CH_3)_3C:\ddot{O}:H + :\ddot{Br}:^-$$
$$\text{rate} = k[(CH_3)_3CBr]$$

is designated the S_N1 reaction: S for substitution, N for nucleophilic, and 1 for unimolecular. The S_N1 mechanism is typical of a second type of displacement reaction; *the leaving group is ejected first, and then combination with the displacing group occurs.* The addition of Ag^+ enhances the rate of those S_N1 reactions in which halide ions are leaving groups. This observation is consistent with the proposed mechanism; the Ag^+ ion facilitates the ionization step by forming AgBr,

$$(CH_3)_3C:Br + Ag^+ \longrightarrow (CH_3)_3C^+ + Ag:Br$$
$$(CH_3)_3C^+ + 2H_2O \longrightarrow (CH_3)_3COH + H_3O^+$$

(b) Electrophilic displacement In an electrophilic displacement, an electrophile acts as the displacing group; the attacked atom must have an available pair of electrons, and the leaving group is an electrophile. The general equation for an *electrophilic displacement* is

$$E + :A:L \longrightarrow E:A: + L$$

E	:A:L		L
electrophile	*attacked atom with available electrons*		*electrophile*

The solution of aluminum chloride, $AlCl_3$, in phosgene, Cl_2CO, exemplifies the reaction. Although neither substance, alone, conducts an electric current, the solution conducts. The presence of ions is explained by the following reaction:

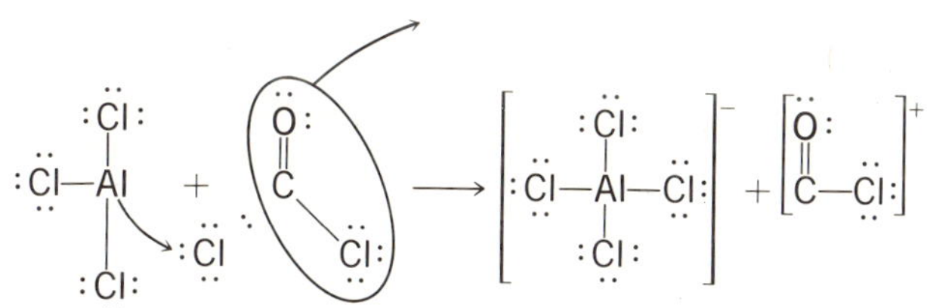

a type of carbonium ion

The electron-seeking $AlCl_3$ molecule forms a bond with the electron-rich Cl atom, and the leaving group is a carbonium ion, $[Cl{-}C{=}O]^+$.

(c) Free radical substitutions When the displacing agent is a free radical,

$$R\cdot + A:L \longrightarrow R:A + \cdot L$$

the group or atom displaced must also be a free radical. Such displacements usually occur as *intermediate* steps in reactions. This type of reaction is illustrated by the light-induced chlorination of methane,

$$CH_4 + Cl_2 \xrightarrow{\text{light}} CH_3Cl + HCl$$

In the dark, even at temperatures up to 200°C, a mixture of methane and chlorine does not react. In the presence of visible light, the reaction proceeds. We have seen (p. 451) that light energy can cause bonds to rupture. Which bond breaks, the C—H bond of methane, or the Cl—Cl bond of chlorine? The answer: the bond with the lower bond energy. The Cl—Cl bond energy is +58 kcal/mole, and the C—H bond energy is +101 kcal/mole. Hence, the Cl—Cl bond cleaves. In the first step, a molecule of chlorine absorbs a quantum of energy from the visible region and is cleaved into two chlorine atoms,

Step 1 $\quad :\ddot{\underset{..}{Cl}}:\ddot{\underset{..}{Cl}}: \xrightarrow{\text{light}} 2:\ddot{\underset{..}{Cl}}\cdot \qquad \Delta H = +58 \text{ kcal/mole}$

In a second step, a chlorine atom collides with a molecule of methane and displaces a methyl free radical from a hydrogen atom,

Step 2 $\quad :\ddot{\underset{..}{Cl}}\cdot + H:CH_3 \longrightarrow :\ddot{\underset{..}{Cl}}:H + \cdot CH_3 \qquad \Delta H = -4 \text{ kcal/mole}$

As seen from the enthalpy of reaction, step 2 is a reasonably exothermic reaction. Notice that we have broken a C—H bond ($\Delta H = +99$ kcal/mole), and formed a new Cl—H bond ($\Delta H = -103$ kcal/mole). This change accounts for the overall enthalpy of reaction of −4 kcal/mole. In a third step, the methyl free radical reacts with a molecule of Cl_2, producing the product, methyl chloride, and a Cl· atom,

Step 3 $\quad H_3C\cdot + :\ddot{\underset{..}{Cl}}:\ddot{\underset{..}{Cl}}: \longrightarrow \cdot\ddot{\underset{..}{Cl}}: + H_3C:\ddot{\underset{..}{Cl}}: \qquad \Delta H = -20 \text{ kcal/mole}$

The third step is also exothermic, since a Cl—Cl bond (+58 kcal/mole) is replaced by a C—Cl bond (−78 kcal/mole). The Cl· atom generated in step 3 can react as shown in step 2, and therefore this sequence of steps comprises a chain reaction (p. 375). Chain reactions are characteristic of free radical processes. Although free radical chain reactions vary in detail, they all have certain fundamental characteristics in common. First, we find a **chain-initiating** step (step 1), in which the absorbed energy generates the free radicals. Next, we find one or more **chain-propagating** steps (steps 2 and 3). Since, in the example cited, the propagating steps are exothermic, once sufficient energy is supplied to create the initial free radical to start the chain, the reaction should proceed "under its own head of steam."

One might suppose that the decomposition of one molecule of Cl_2 should be sufficient to supply the Cl· atoms to maintain the self-propagating chain reaction. Such is rarely the case, since the chain can be terminated when two free radicals collide. The following are examples of **chain-terminating** steps:

$$\begin{aligned} Cl\cdot + Cl\cdot &\longrightarrow Cl:Cl \\ Cl\cdot + \cdot CH_3 &\longrightarrow Cl:CH_3 \\ H_3C\cdot + \cdot CH_3 &\longrightarrow H_3C:CH_3 \\ &\qquad\ \textit{ethane} \end{aligned}$$

Whenever two radicals combine to form a nonradical species, two chains are terminated. Fortunately, the collision of two free radicals is *rare* because the concentration of free radicals in the mixture at any given instant is extremely small. Nevertheless, such collisions occasionally occur and, consequently, a continuous source of energy is necessary for the reaction. Because of chain termination steps, most free radical reactions need a continuing supply of free radicals to re-initiate the chain processes. Chain termination is another characteristic feature of free radical reactions.

We can now summarize the features of a typical free radical chain process:

$$Cl_2 \xrightarrow{\text{light}} 2\ Cl\cdot \qquad \textit{chain initiation}$$

$$\left.\begin{aligned} Cl\cdot + H:CH_3 &\longrightarrow Cl:H + \cdot CH_3 \\ H_3C\cdot + Cl:Cl &\longrightarrow H_3C:Cl + Cl\cdot \end{aligned}\right\} \quad \textit{chain propagation}$$

$$\left.\begin{aligned} H_3C\cdot + \cdot Cl &\longrightarrow H_3C:Cl \\ H_3C\cdot + \cdot CH_3 &\longrightarrow H_3C:CH_3 \\ Cl\cdot + \cdot Cl &\longrightarrow Cl:Cl \end{aligned}\right\} \quad \textit{chain termination}$$

22.13 ▫ REARRANGEMENTS

The conversion of one structural isomer into another is known as a **rearrangement reaction.** We know of many such reactions that involve organic chemicals. Some reactions require a catalyst, others proceed under the influence of heat or radiation. A simple example of a rearrangement is the conversion of 1,4-pentadiene to 1,3-pentadiene. The structure of the latter compound permits extended π bonding, and it is therefore more stable than the former (see p. 147). This enhanced stabilization is the thermodynamic drive that favors the rearrangement. The mechanism of the rearrangement involves the transfer of a H atom; this can be facilitated by either an acidic or a basic catalyst:

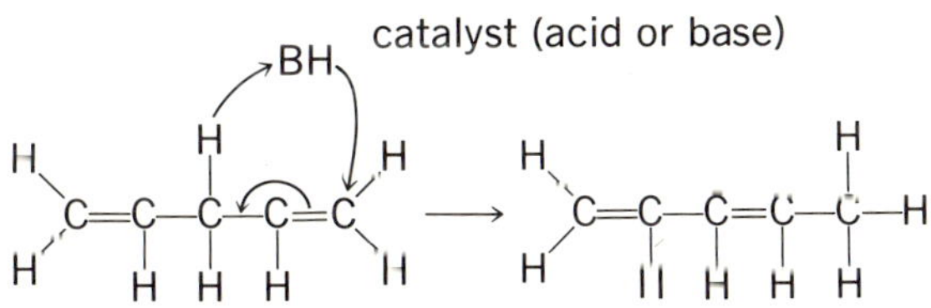

1,4-pentadiene (no extended π bonding) — *1,3-pentadiene* (extended π bonding)

PROBLEMS

1. Definitions Define and illustrate the following terms: (a) carbanion; (b) carbonium ion; (c) isomerism; (d) homology; (e) conformation; (f) derivative; (g) functional group; (h) heterolysis; (i) homolysis; (j) S_N1 reaction; (k) S_N2 reaction.

2. Bonding of carbon Problem 14 on p. 280 describes the preparation of tetraethyl lead by the electrolysis of C_2H_5MgCl, during which ethyl ions react with a lead anode. (a) Write the structural formula for C_2H_5MgCl. (b) What does this electrolytic process imply about the bonding between C and Mg? (c) Do you think that an analogous process could be based on the conductivity of diethyl mercury in ether solution? Explain.

3. Conformation How many of the following structural formulas represent different substances?

(a) 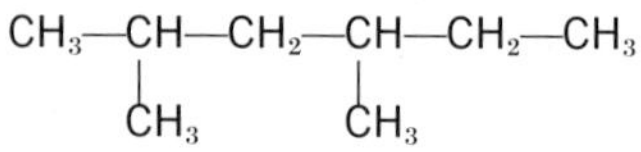

(b) 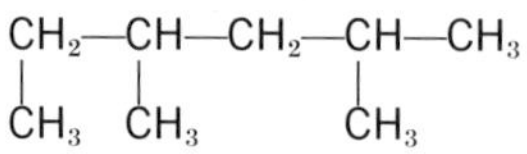

(c) 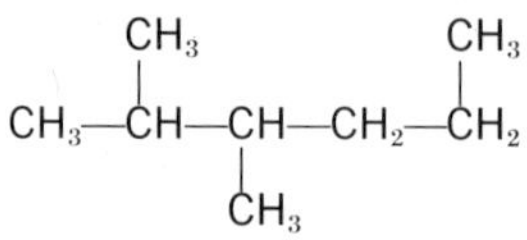

(d) 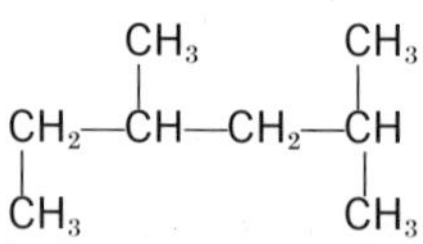

4. Homologous series Given the compounds (a) C_4H_{10}, (b) C_4H_8, and (c) C_4H_6, write a general formula for each of these hydrocarbons and name all the homologous series (when more than one is possible) to which each belongs (see Table 22.2, p. 447). Write a structural formula consistent with each series.

5. Alkane isomerism Write the structural formulas of the five isomers of C_6H_{14}.

6. Homologous series Give the structural formulas for three isomers with the formula C_3H_4, and name the homologous series to which each isomer belongs.

7. Reactions of alkanes (a) Write the reaction for the chlorination of ethane, C_2H_6, to give ethyl chloride, C_2H_5Cl. (b) Ethyl chloride can be chlorinated to give two isomers, A and B, of formula $C_2H_4Cl_2$. A can be chlorinated to give two isomers of $C_2H_3Cl_3$, while B gives only one isomer of $C_2H_3Cl_3$ on chlorination. Write the equations for the chlorination of ethyl chloride. (c) Give the structural formulas of A and B.

8. Decomposition When tetramethyl lead, $(CH_3)_4Pb$, is heated, it decomposes to yield metallic lead and a reactive substance that is capable of attacking mercury. (a) Write the equation for the decomposition of $(CH_3)_4Pb$. (b) Write the equation for the reaction between mercury and the active substance produced from tetramethyl lead. (c) In the absence of any other reagent, the active substance rapidly undergoes a reaction to produce a relatively inert gaseous product. Write the equation for this reaction.

9. Displacement Two reaction sequences are shown below for a nucleophilic displacement on a carbon atom.

Sequence (*i*), one step,

X: + CY ⟶ XC + :Y (concerted)

Sequence (*ii*), two steps,

X: + CY ⟶ XCY (slow)

XCY ⟶ XC + :Y (fast)

(a) Are both sequences consistent with the kinetics of a bimolecular (S_N2) displacement reaction? (b) Account for the fact that sequence (*ii*) cannot occur on a carbon atom that has four σ bonds. (c) Could a nucleophilic displacement of the type shown in sequence (*ii*) occur on boron in BCl_3? Explain.

10. Free radical substitution Account for the following observations: no reaction occurs when ethane and bromine are heated in the dark at 150°C; however, when a trace amount of $Pb(CH_3)_4$ is added under the same conditions, ethyl bromide and HBr are produced.

11. Addition reactions With the aid of equations show that two products are possible when HBr adds to propene. Name the products.

ADDITIONAL PROBLEMS

12. Definition of organic chemistry The compound **borazine** has the following structure:

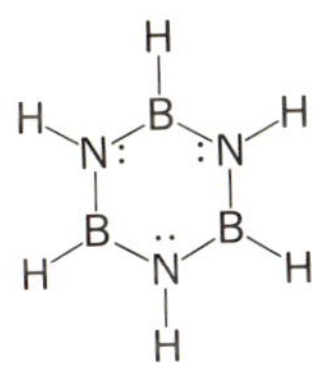

Do you think the study of this substance should be part of the subject matter of organic chemistry? Defend your answer.

13. Alkane isomerism The only octane, C_8H_{18}, that is solid at ordinary temperatures if the most highly branched of the 18 isomers. Write its structural formula.

14. Alkene isomerism Give the structural formulas for the three *structural* isomers of the alkene, C_4H_8.

15. Isomerism The *total* number of possible isomers of C_4H_8 is 6. Write all the structural formulas.

16. Geometrical isomerism State whether or not each of the following alkenes can exhibit geometrical (*cis-trans*) isomerism:

(a) $H_2C{=}C(CH_3){-}CH_3$ (b) ClFC=CHCl

(c) 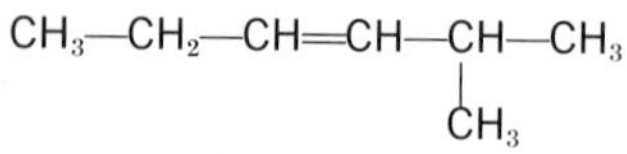

(d) 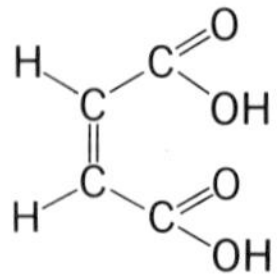 (e) 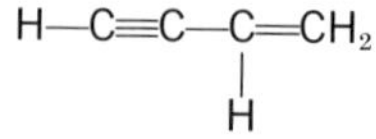

(f) 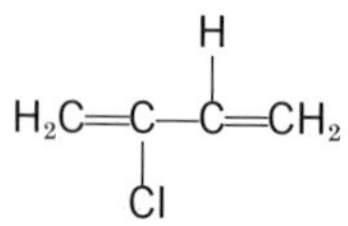

17. Unsaturation and redox (a) Write an equation for the reaction of ethylene with (*i*) hydrogen, (*ii*) bromine. (b) With the aid of oxidation numbers show that these are redox reactions. (c) In each reaction, indicate the oxidizing and reducing agents.

18. Benzene isomerism Write the structural formulas for the four aromatic isomers with the molecular formula C_7H_7Cl. When applicable, indicate which are *ortho*, *meta*, and *para*.

19. Naphthalene isomerism How many monochloro substitution products of naphthalene, $C_{10}H_7Cl$, are possible? Write their structural formulas.

20. Elimination reaction (a) How many alkenes can be formed when 2-butanol, $CH_3CHOHCH_2CH_3$, is heated with concentrated sulfuric acid? (b) Write equations for the reactions.

21. Ether A sample of diethyl ether is contaminated with traces of ethanol and water. Suggest a method of purifying the ether, giving specific laboratory directions.

22. Alcohols Pentaerythritol is a tetrahydroxy alcohol whose formula is $C_5H_8(OH)_4$. On reaction with nitric acid, $HONO_2$, it forms only one monoester, pentaerythritol mononitrate, $C_5H_8(OH)_3ONO_2$, and only one diester, pentaerythritol dinitrate, $C_5H_8(OH)_2(ONO_2)_2$. Write the structural formula for pentaerythritol.

23. Amines Write the structural formula for (a) a dialkyl derivative of ammonia; (b) a trialkyl derivative of ammonia; (c) tetramethylammonium chloride.

24. Carboxylic acids Write balanced equations for (a) the formation of potassium oleate; (b) the formation of calcium palmitate; (c) the hydrolysis of methyl acetate; (d) the saponification of glyceryl trilinoleate.

25. Decomposition (a) At 60°–100°C the following reaction occurs:

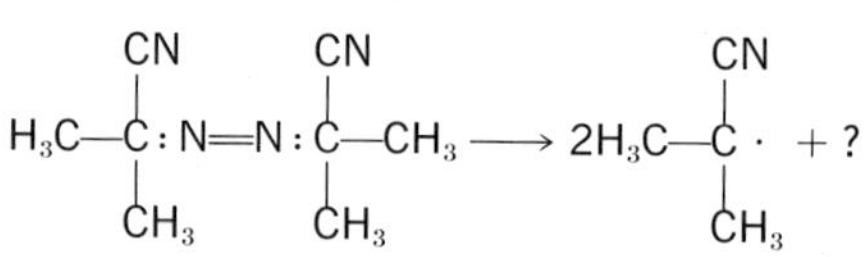

What is the formula of the missing molecule? (b) When benzoyl peroxide,

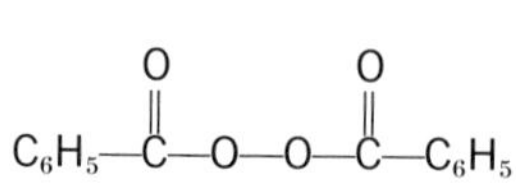

is heated, one of the products isolated is biphenyl, $C_6H_5—C_6H_5$. Suggest a mechanism for the formation of biphenyl. What gaseous substance is also produced?

26. Electrophilic substitution Account for the fact that a mixture of $(C_6H_5)_3C—Cl$ and $AlBr_3$, each of which is a nonconducting covalent compound, conducts an electric current when dissolved in a suitable solvent.

27. Rearrangement Each of the following compounds undergoes rearrangement readily. Predict the structure of the rearranged product. Explain your prediction.

(a) 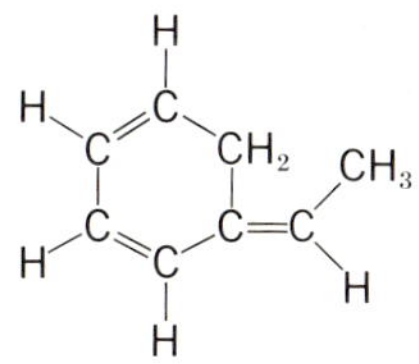

(b) 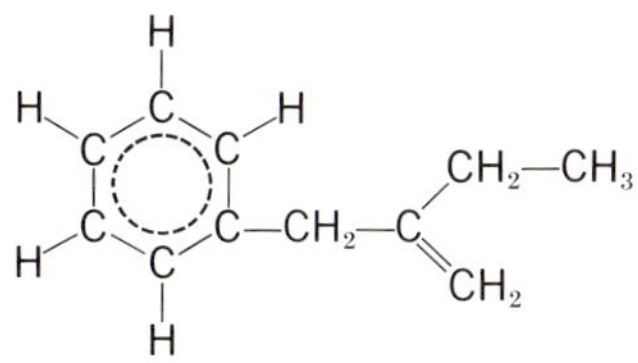

The following question is based on Appendix 3, p. 556.

28. Nomenclature Give the IUPAC name for

(a) 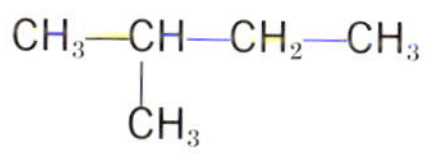

(b) 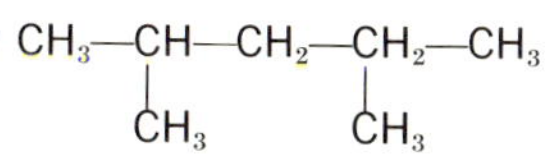

(c) 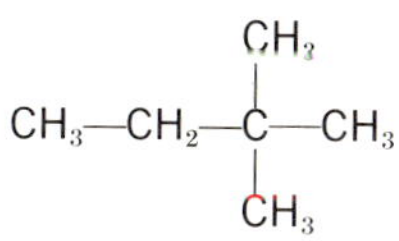

(d) 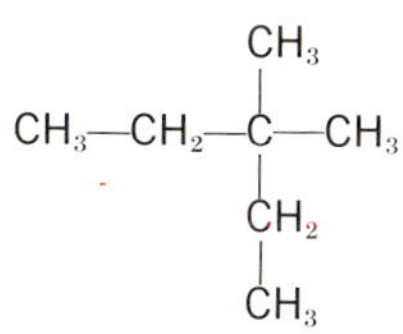

(e) 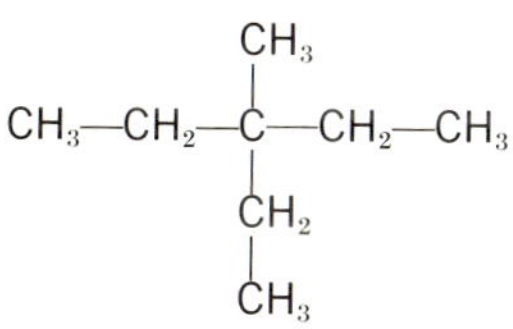

23 METALS AND METALLURGY

23.1 THE PROPERTIES OF METALS

Some 80% of the elements are metallic. Metals are widely used for decorative and structural applications, and we are sufficiently familiar with their physical properties to distinguish them with confidence from nonmetallic substances after only casual observation.

The most conspicuous attribute of metals is **luster,** the bright, highlighted appearance of a substance that reflects light well. A particularly lustrous metal is silver, a smooth surface of which throws back over 90% of the incident light; the adjective **silvery** is often used in the general sense of having high luster. Two metals, gold and copper, are colored, but nonetheless lustrous from smooth surfaces. Metals dispersed in a spongy or finely divided form appear gray to black; this results from cumulated light losses occurring in many short-distance reflections from the irregular surface before the light leaves the metal. An example is **platinum black,** the precipitate obtained on the reduction of a dissolved platinum compound (such as potassium hexachloroplatinate(IV), K_2PtCl_6). Powdered aluminum and magnesium, on the other hand, retain much of their lustrous appearance. The reflectivity of ordinary glass mirrors is, of course, due largely to their metallic backing. In fact, the word **mirror** is often used to denote any smooth metal surface, especially a deposition from a chemical reaction. For example, shortly after tetraethyl lead is heated, a **lead mirror** deposits on the adjacent cool part of the tube. Many metal surfaces are overlaid with oxide or other coatings; their natural luster reappears after mechanical or chemical removal of such deposits.

Also noteworthy are the high electrical and thermal conductivities of metals. It is significant that electrical conductivity of metals is far greater than that of electrolytes (by a factor of as much as about 10^6), that it is exhibited to a high degree in both the solid and liquid states, that it is unaccompanied by detectable chemical change, and that it increases with *decrease* in temperature. In all instances of metallic conduction, it is known that the charged particles carrying the current are electrons. Electrons play a central role, too, in the conduction of heat by metals.

The densities of metals vary over a wide range, from 0.53 g/ml for lithium at 20°C to 22.6 g/ml for iridium at the same temperature. No other kinds of substance have densities as great as those of the heavy metals. The value for granite rock, for example, is only about 3 g/ml; for concrete, 2 g/ml. Typical coordination numbers of atoms in metals are 8 or 12 (the maximum possible for spheres of identical size). It is this close packing that accounts in large measure for high metallic densities.

Metals can remain contiguous under severe conditions of deformation. A stick of light metal, such as sodium, may be extruded into a wire by a simple hand press; heavier metals may be fabricated into sheets, foils, or other forms by suitable tools and application of force. The ability of a solid to be flattened is called **malleability;** the ability to be drawn into a wire is called **ductility.** *What is unusual in metals is that crystalline structure is preserved under such stresses and strains.*

Alloys are mixtures of metals, with or without definite composition, sometimes containing a small proportion of a nonmetallic component. The physical properties of alloys are metallic, and alloys are therefore not superficially distinguishable from pure metals. This close resemblance persists even when the alloy is an intermetallic compound of definite composition (such as Cu_2Mg, Cr_3Os, Fe_5Zn_2), in sharp contradistinction to compounds of metals with nonmetals (such as NaCl), which lack metallic character. The physical nature of the intermetallic compounds implies the existence of a chemical bond quite different in kind from the ionic or covalent linkages discussed in previous chapters. The physical similarity of intermetallic compounds to the metallic elements implies that the same type of bond exists in both classes of substances.

23.2 □ THE METALLIC BOND; THE BAND THEORY OF METALS

It has long been recognized that the nature of the metallic bond is related in some way to the "free" state of electrons in metals. If these electrons are the media for the conduction of heat and electricity, if they are the vehicles that absorb and reemit energy in reflection, they must therefore be the key ingredient that imparts such unique attributes to the bonds between atoms in metals and alloys. Ideas concerning the metallic bond were proposed by Arnold Sommerfeld and

Felix Bloch in 1928, and later by Léon Brillouin and others, and have developed into what is now known as the **band theory of solids.** To understand what may constitute an electron band, and how such bands are involved in metallic bonding, it will be helpful to consider the following points.

(a) Metals typically crystallize in lattices of high coordination number. Lithium, $1s^22s^1$, for example, has only 3 electrons per atom, but exists as a body-centered cubic lattice with coordination number 8. It is clear that no assignment of valence electrons to metallic bonds can account for such a high coordination number, unless it is assumed that the valence electrons are delocalized.

(b) The combination of two like atomic orbitals forms two molecular orbitals, one more stable, the other less stable, than the original atomic orbital. In general, x atomic orbitals, when they combine, yield x molecular orbitals.

(c) In metals, where many atoms may exist together so closely packed in a crystal lattice that electron delocalization occurs, the number of molecular orbitals will depend on the number of atoms in the lattice:

Fig. 23.1 Electron bands in metallic lithium.

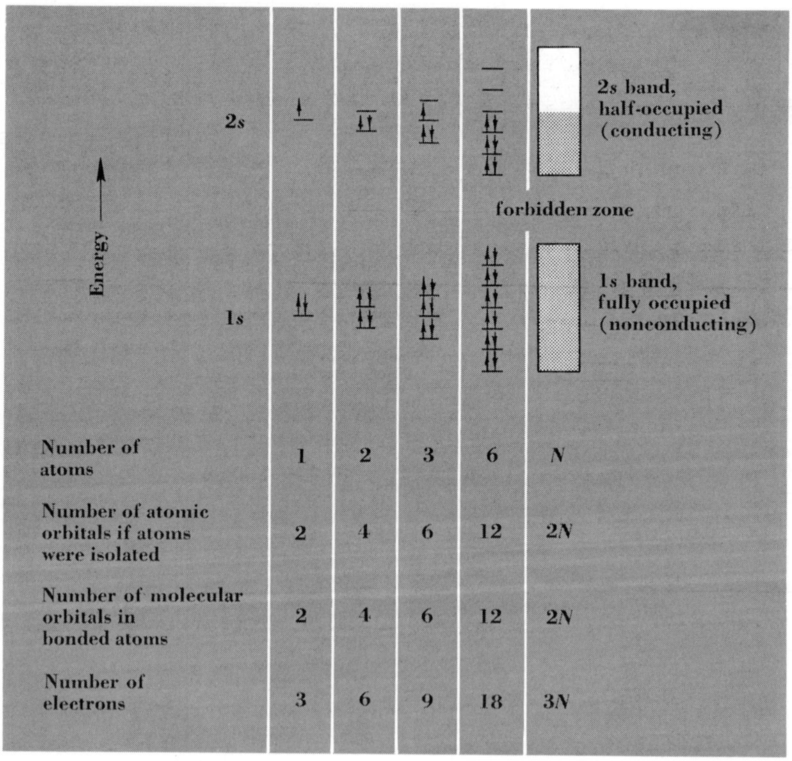

$$N \text{ atoms with } 1 \frac{\text{atomic orbital}}{\text{atom}} \text{ yield } N \text{ molecular orbitals}$$

$$N \text{ atoms with } x \frac{\text{atomic orbitals}}{\text{atom}} \text{ yield } Nx \text{ molecular orbitals}$$

Consider these relationships in the specific case of the lightest metal, lithium. Figure 23.1 shows that, as the number of atoms in a crystal of metallic lithium increases, the number of molecular orbitals increases and the energy difference between successive energy levels decreases. The smallest visible piece of lithium contains about 10^{19} atoms; for such a large number the energy difference between successive levels is so small that the levels are continuous; the group of closely associated energy levels, taken together, is called an **electron band.** In lithium, one of these bands, originating from the $1s$ atomic orbital, is fully occupied; the other, originating from the $2s$ atomic orbital, is only half-occupied. As we shall see, *metals are characterized by incompletely filled electron bands*. The energy gap between the two bands is so large that it effectively deters promotion of electrons from the lower to the higher band. Such energy gaps have come to be called **"forbidden zones."**

Let us next consider the distribution of energy bands involving both s and p orbitals. The maximum number of electrons per band will be

Energy ↑	maximum				maximum
$2p$	maximum 6 electrons per subshell	×	N atoms per crystal	=	maximum $6N$ electrons per band
$2s$	maximum 2 electrons per subshell	×	N atoms per crystal	=	maximum $2N$ electrons per band
$1s$	maximum 2 electrons per subshell	×	N atoms per crystal	=	maximum $2N$ electrons per band

Thus, for sodium (Na $1s^22s^22p^63s^1$) the electron bands are $1s^{2N}2s^{2N}$-$2p^{6N}3s^N$. The $3s$ band, having a maximum capacity of $2N$ electrons, is only half-filled; this is the conduction band of sodium.

For magnesium (Mg $1s^22s^22p^63s^2$), the electron bands are $1s^{2N}2s^{2N}$-$2p^{6N}3s^{2N}$. This completely filled band distribution appears to predict that magnesium should not be metallic. This apparent contradiction of fact is explained as follows. The relative energies of successive bands depend on internuclear distances. In metals, which are typified by close packing and relatively small internuclear distances, the adjacent bands may overlap. In the case of magnesium, the $3s$ and $3p$ bands overlap, that is, there is no "forbidden" energy zone between them, and promotion of electrons from the $3s$ to the empty $3p$ band occurs readily.

23.3 □ METALLIC PROPERTIES IN TERMS OF THE BAND THEORY

Let us now consider in more detail how an incompletely filled electron band engenders metallic character. In a completely filled band, each orbital is filled with two electrons of paired spins. When an external electric field is applied, electrons cannot flow in response to the field. A mechanical analog of this situation is a closed box completely filled with marbles. When the box is tilted, the marbles cannot roll to the low end. In an incompletely filled band, however, the effect of the applied potential is to promote the electrons flowing in the direction of the field to higher unoccupied energy levels within the band; these electrons account for a net flow of current and the substance is a metallic conductor. A rise in temperature promotes vibration of the atoms in the crystal lattice; such vibrations interfere with the movement of electrons. Electrical conductivity of metals therefore *decreases* as the temperature increases.

The labile electrons of the conduction bands can also absorb thermal energy; the ready transport of such energy by these electrons accounts for the high thermal conductivity of metals.

The absorption and reemission of photons of radiant energy by the labile electrons accounts for the high reflectivity of metals.

The destruction of the metallic lattice by mechanical stresses is rendered difficult by the fact that bonds can be readily broken and reformed by the promotion of electrons in the conduction bands to higher unoccupied levels by mechanical energy and by subsequent return of the electrons to their original levels. In the process, lattice structure may be altered, but yet remain contiguous, since the bonding electrons are not localized with respect to any particular atoms. This picture is in accord with the previously described ductility and malleability of metals.

It is possible to draw an analogy between delocalization of bonding electrons in a metallic crystal and in a small molecule, such as benzene, that exhibits extended π overlap. The aptness of the analogy is evidenced by the fact that an indefinitely extended benzene-like structure (graphite) conducts electricity and is thus partly metallic in nature (Fig. 22.2, p. 445).

23.4 □ INSULATORS AND SEMICONDUCTORS

In **insulators,** the "conduction band" is empty. The forbidden zone below the conduction band represents a large energy barrier, and promotion of electrons to the conduction band normally does not occur. The electrical conductivity of a good insulator may be as small as about 10^{-24} times that of silver.

In **semiconductors** (also called **semimetals** or **metalloids**), the conduction band is populated by electrons that are occasionally promoted to

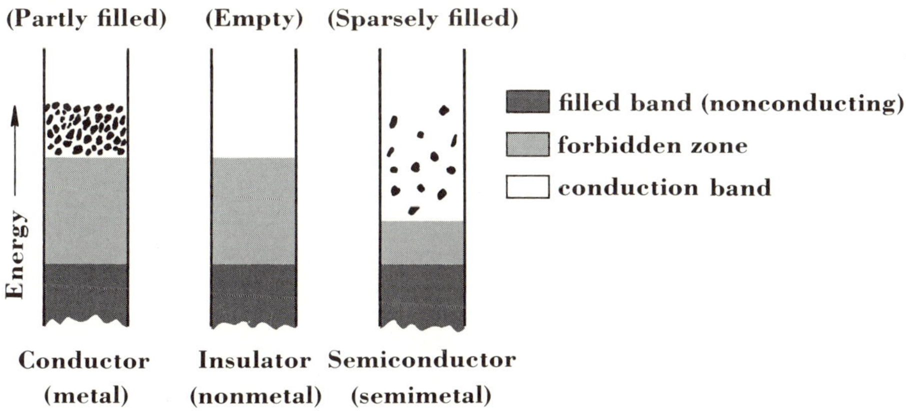

Fig. 23.2 Band theory of solids.

it by excitation from lower bands. The forbidden zone is narrower than in the case of insulators, and therefore constitutes less of a barrier to such promotion. Such substances show meager electrical conductivity. A rise in temperature favors excitations to the conduction band; electrical conductivity of semiconductors therefore *increases* as the temperature increases. Figure 23.2 illustrates, schematically, the arrangement of bands and forbidden zones in conductors (metals), insulators (nonmetals), and semiconductors. Some important semiconductors (used in transistors and other "solid-state" devices) are Ge, Si, GaAs, and InSb.

23.5 SOURCES OF METALS

The earth's crust, it is believed, was formed by the cooling of fluid matter. Under such circumstances, separation of different substances occurs first when immiscible liquids stratify and, second, when fractional crystallization (the separation of materials of different solubilities by differential crystallization from a liquid phase) occurs on cooling of the liquid. During the long geological time since the formation of the earth's crust, the various solid depositions were subjected to additional differentiating actions by atmospheric weathering, by the action of water after the earth had cooled below 100°C, and by biological processes after life began. Of all these weathering effects, the most important is the action of water. The materials now remaining in the solid state must therefore be resistant to oxidation, reduction, and biological attack, and (usually) highly insoluble. Such materials, when they are used as sources for the commercial extraction of metals, are called **ores**. Relatively pure natural solids, whether they find commercial use or not, are called **minerals.** Soluble compounds are dispersed in the oceans or concentrated in landlocked bodies of water, like Great

Salt Lake and the Dead, the Caspian, and the Aral Seas, or remain as evaporation residues, such as the Bonneville salt flats or the Chilean saltpeter deposits. The requirements for chemical stability are, specifically, resistance to oxidation by acid (H_3O^+) or reduction by water.

Metals that meet the requirement of being resistant to oxidation include bismuth, copper, mercury, silver, gold, and the platinum group.* Native (elemental) bismuth is found in Peru and Bolivia. Native copper is found in one important deposit in Michigan (so-called Lake Copper, now substantially exhausted); otherwise it is usually mined as the sulfide. The important source of mercury is also the sulfide (**cinnabar,** HgS), but the free metal is sometimes found finely dispersed in its ore. Silver is found as the free metal, the sulfide, or the chloride. In general, the metals whose reduction potential (Table 16.1, p. 294) is more positive than that of silver (+0.8 V) are mined in the native state. An example is gold (+1.4 V). Lead, on the other hand, whose reduction potential is only −0.13 V, is mined only from sulfides, PbS (**galena**), or oxygenated minerals, such as $PbCO_3$ (**cerussite**) and $PbSO_4$ (**anglesite**), and not as the native metal.

When an ore is exposed to the atmosphere or to water containing dissolved air, it must resist oxidation by oxygen as well as by acids. Of the familiar metals, only gold and the platinum metals satisfy this condition, and these metals, accordingly, are nearly always found native. Seminoble metals, such as copper and mercury, are found either native or combined, depending on the presence of oxidizing or reducing agents in the environment.

Ores that meet the requirement of being resistant to reduction are those that cannot be reduced by water. This means that the reduction potential of the ore must be *less* than that of oxygen in the half-reaction

$$O_2(g) + 4H_3O^+ + 4e^- \rightleftharpoons 6H_2O \qquad \mathscr{E}^\circ = +1.23 \text{ V}$$

This condition is satisfied, for example, by compounds of metals in the following oxidation states: Cu(II), Sn(IV), Fe(III), Mn(II), and Mn(IV). Compounds of Mn(VII) and Pb(IV), on the other hand, have reduction potentials exceeding +1.23 V and therefore cannot survive in the presence of liquid water. (See Table 16.1, p. 294.) These predictions are well borne out by the following facts. Copper occurs as CuS (**chalcocite, covellite**). The only important mineral of tin is SnO_2 (**cassiterite**). Iron is mined as Fe_2O_3 (**hematite**). Manganese is found as MnO_2 (**pyrolusite**) and as $MnCO_3$ (**rhodochrosite**), never as manganates, Mn(VI), or permanganates, Mn(VII). The most highly oxidized form of lead in ores is +2; the oxide PbO_2 does not occur in nature.

* The greater the reduction potential of a metal, the more **noble** it is said to be. Metals like gold and platinum are called **noble metals.** In chemical parlance, nobility implies inertness.

23.6 □ THE WINNING OF METALS FROM ORES: GENERAL CONSIDERATIONS

Metals occur in ores in an impure and, generally, oxidized condition. The winning (the word reflects our esteem for metallurgical success) of metals therefore involves purification processes and usually reduction. Physical or chemical treatment of materials may also be needed to prepare them for effective further processing.

The first rough purification steps are called, simply, **concentration.** The last purification, which almost always takes place after reduction, is called **refining.** Between these initial and final efforts may be interposed any of a large variety of metallurgical processes, many of which are adaptations of methods developed in antiquity. Some of these processes will be briefly described in the ensuing sections, in the order in which they are most likely to be used in commercial practice.

23.7 □ PRELIMINARY PHYSICAL TREATMENT

Crushing and grinding operations are used to subdivide large pieces of ore to a workable particle size. Excessively fine particles also pose handling problems, and must therefore be agglomerated. A common method of agglomeration, used especially in the treatment of iron ore fines,* is **sintering,** which involves incipient melting of solid material.

Fig. 23.3 Sintered agglomerates of Portland cement (*Photomicrograph courtesy Walter C. McCrone Associates, Inc.*).

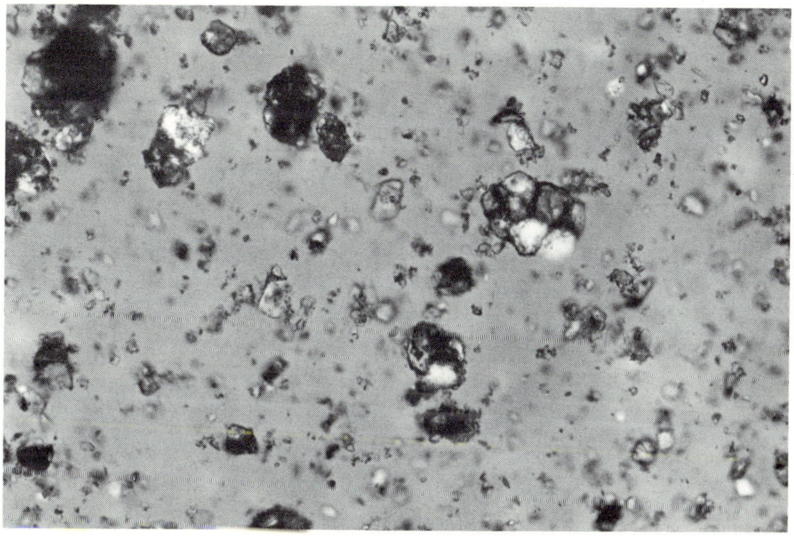

* "Fines" means, in industrial parlance, fine particles.

The edges of a solid melt before the flat surfaces, because the more highly curved edges have fewer molecules around them to help constrain them in the solid lattice. Material that is partially melted, therefore, tends to fuse together at the sharp points of contact; the aggregate thus formed is called a **sinter** (Fig. 23.3). In some processes, blast furnace flue dust, which is rich in carbon, is mixed with the ore fines to serve as a fuel. During the sintering, some sulfur is lost as SO_2; thus the physical agglomeration is accompanied by some chemical action.

23.8 □ PHYSICAL CONCENTRATION

The simplest method of concentration is the removal of lumps of impurities by sorting or sieving operations. The portion of the ore that does not contain metal is considered to be impurity. It is collectively called **gangue** and includes sand and clays. When the mineral differs greatly in density from the gangue, gravity or inertial methods (which depend on differences in masses of particles) may be used. A primitive process is the **panning** of gold, in which the lighter gravel is washed down a sluice, leaving the heavier gold behind. Spiral whirling of an air-borne suspension of ore in a conical chamber called a **cyclone separator** is an effective inertial process. Magnetic separation is useful in separating ore rich in Fe_3O_4 (**magnetite**—the lodestone of antiquity) from gangue. By using magnets of different strength it is possible to separate the ore into several different concentrations. It is sometimes possible to effect an **electrostatic separation.** By this method one material (usually the gangue) retains an electric charge better than another, and can be attracted to a charged surface from which it falls or is scraped off. **Distillation** methods are sometimes used, notably in the refining of mercury.

By far the most important method of physical concentration is **flotation.** In this process, ground or crushed ore is suspended in water by air bubbling, and a differentiation is made among various types of particle on the basis of the extent to which they are wetted by water. Particles that adhere tenaciously to water do not leave the aqueous phase. Thus, an oxide (SiO_2, for example) will be more strongly hydrogen-bonded to water than a sulfide, because the oxide is more highly polar. Particles that lack affinity for water, on the other hand, are held at the boundaries between the water and the bubbles by surface tension. These nonwetted components rise to the surface as a froth and can be skimmed off and recovered, and thus separated from the wetted particles which do not float. Older techniques employed agents like pine oil or mineral oil, which coat the ore in preference to the gangue; the ore was thus rendered nonwettable by water and was separated in the froth. Modern methods accomplish much more than merely the separation of ore materials from gangue; preferential wetting of particular ores is effected by agents of high specificity, called **collectors.** The typical collector is a molecule, one end of which has an ionic or polar group that attaches itself to the ore to be separated, and the other end of which is hydrocarbon in nature to shield against the wetting action of water. An example is potassium ethylxanthate,

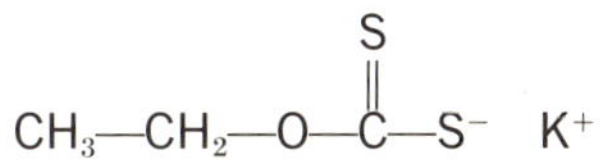

23.9 □ CHEMICAL LEACHING

Chemical leaching is a method of extraction with the use of a liquid that renders the desired component soluble by some chemical action. For such a process to be commercially valuable in metallurgy, the leached product must be so stable that a fairly high proportion of the metal is in the desired state at equilibrium. Because the cyano complex ions are very stable, and because NaCN and KCN are relatively cheap raw materials, cyanide leaching has been used for the recovery of valuable metals from low-grade ores. The overall reactions for silver and gold, respectively, are

$$4Ag + 8CN^- + O_2 + 2H_2O \rightleftharpoons 4Ag(CN)_2^- + 4OH^-$$
$$4Au + 8CN^- + O_2 + 2H_2O \rightleftharpoons 4Au(CN)_2^- + 4OH^-$$

The metals are recovered from their cyanide salts by electrolytic reduction.

23.10 □ ROASTING

Ores used in ancient metallurgy were, for the most part, surface deposits of oxides, carbonates, or native metal. The oxides and carbonates were reduced to metal by heating with wood charcoal, the gaseous products, CO and CO_2, escaping readily to the atmosphere. As mining developed in the pursuit of ores at greater depths, sulfides were increasingly encountered. Sulfides are difficult to reduce with carbon, however, and it was therefore found advantageous to convert these sulfides to oxides prior to reduction. **Roasting** is the process whereby such conversion is accomplished by heating in air. With more noble metals, careful control of the roasting process may effect reduction to the metallic state. If the temperature is fairly low (about 500°C) and the concentration of SO_2 in the gaseous environment is high, sulfates may be produced. This is an undesirable effect, because sulfates are stable and high temperatures are needed to decompose them.

Representative equations are

$$HgS + O_2 \longrightarrow Hg + SO_2$$
$$2ZnS + 3O_2 \longrightarrow 2ZnO + 2SO_2$$
$$(\textit{undesirable}) \quad 2ZnO + 2SO_2 + O_2 \longrightarrow 2ZnSO_4$$

23.11 □ REDUCTION

Reducing agents available for treatment of ores can be classified into three major categories: (1) the atoms in the mineral that have low oxidation numbers and may therefore serve as reductants, (2) added chemical reductants, and (3) electrons at the cathode of an electrolytic cell.

When a metal is produced by direct decomposition, the anionic portion of the mineral may be considered to be the reducing agent. Thus, the equation given above for the roasting of HgS may be considered to be the production of mercury by decomposition of the sulfide, followed by oxidation of the sulfur.

Among chemical reductants, the cheapest and most widely used is carbon, usually in the form of coke (the carbonaceous residue from the destructive distillation of coal). Carbon, being a solid, does not make intimate contact with ore particles. Instead, its partial oxidation product, gaseous CO, is the effective reducing agent in most instances. The source of oxygen for conversion of carbon to CO is air enriched with oxygen, or sometimes even pure oxygen.

Other reducing agents may be selected from hydrogen up to sodium (Table 16.1, p. 294); the choice depends on the reducing potential needed, and on the cost that can be justified.

The metallurgy of iron The metal produced in greatest quantity by reduction of an ore is iron. The ore used is usually an oxide mixed with gangue. For facility in handling, it is advantageous if the products obtained from metallurgical reduction are in a liquid condition. The siliceous gangue, however, does not liquefy at the temperature at which coke reduces iron oxide, but instead forms a soft vitreous mass that would make handling very difficult. A liquefying agent (**flux**) is therefore used to react with the gangue to convert it to a **slag,** a product obtained as an upper liquid layer in a metallurgical process. The slag serves the additional function of shielding the metal on which it floats from air oxidation. The flux is lime, CaO, formed by the decomposition of limestone (calcium carbonate). The carbon dioxide produced by this decomposition reacts with carbon to produce carbon monoxide. The changes are depicted by the following idealized equations:

$$\underset{\textit{limestone}}{CaCO_3(c)} \longrightarrow \underset{\textit{lime}}{CaO(c)} + CO_2(g)$$

$$\underset{\substack{\textit{lime}\\(\textit{flux})}}{CaO(c)} + \underset{\substack{\textit{silica}\\(\textit{gangue})}}{SiO_2(c)} \longrightarrow \underset{\substack{\textit{calcium silicate}\\(\textit{fusible slag})}}{CaSiO_3(l)}$$

The major sequence of reduction reactions is shown in the schematic diagram of the blast furnace (Fig. 23.4).

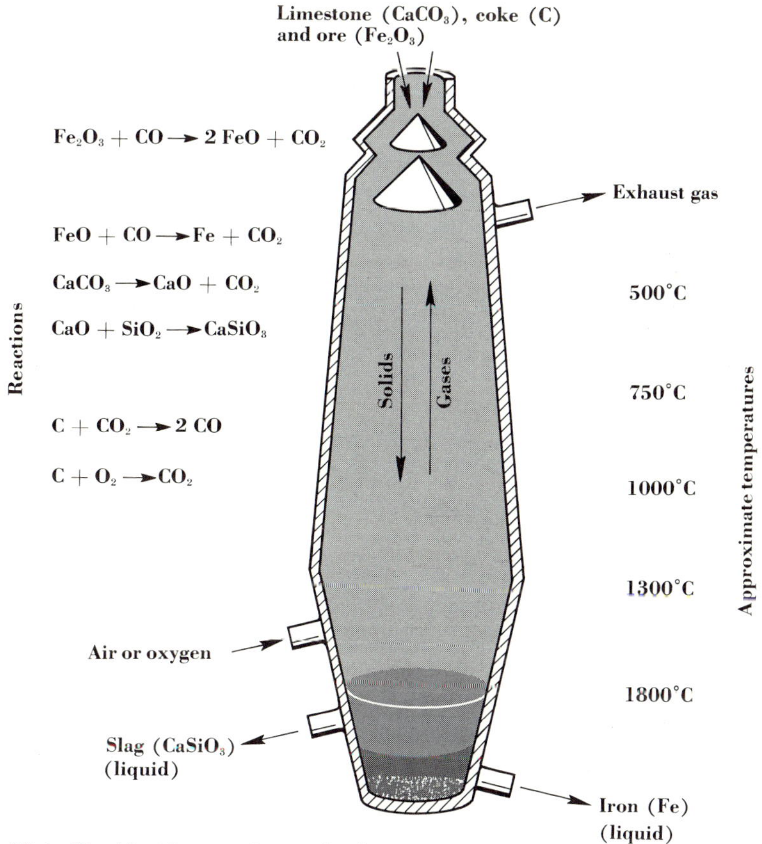

Fig. 23.4 The blast furnace for making iron (schematic).

Other chemical reduction processes Carbon dissolves in metals to some degree, often with the formation of carbides; the removal of small residues of carbon from the metal is therefore difficult. When a pure product is needed, or when a more powerful reductant must be used, a metal with high reducing potential like aluminum may serve. In the Goldschmidt or aluminothermic process, chromium may be prepared directly from its oxide,

$$Cr_2O_3(c) + 2Al(l) \longrightarrow 2Cr(l) + Al_2O_3(c)$$

The reaction is vigorously exothermic and quite rapid. The driving forces of the reaction include the reducing potential of aluminum and, as a major factor, the high lattice energy of Al_2O_3. Tantalum, niobium, and tungsten can also be produced by aluminothermic methods.

Hydrogen gas can reduce anything that carbon can, and may be thought to have the important advantage of leaving no impurity. It may leave some of itself, however, in the form of a hydride. It is costly, forms explosive mixtures with air, and is not yet widely used.

23.12 □ ELECTROMETALLUGRY

The reduction potentials of the alkali metals and the alkaline earth metals are so low that chemical reductants are not available for the conversion of ores to the pure metals. Instead, the metals are prepared by the electrolysis of their fused salts.

A noteworthy example of commercial electrolytic reduction of an ore is the metallurgy of aluminum. The important source of the metal is **bauxite,** $Al_2O_3(H_2O)_2$. Ferric oxide and silica are the usual contaminants. Since ferric oxide will be reduced under any conditions used for the preparation of aluminum, the bauxite must be purified first if the aluminum is to be free of iron. The separation is effected by extracting the amphoteric alumina (Al_2O_3) with hot aqueous NaOH, and decanting the hot solution from the nonamphoteric ferric oxide and from the siliceous residue. The purified alumina, to be used in the electrolysis, is recovered by reversal of the equilibrium on cooling,

$$Al_2O_3 + 2OH^- + 7H_2O \rightleftharpoons 2[Al(OH)_4(H_2O)_2]^-$$

The problem of reduction of Al_2O_3 stems from the low reduction potential of aluminum. Chemical reduction would require the use of a very powerful reducing agent like metallic sodium, which was in fact used in the early preparations of small quantities of aluminum. Attempts to use carbon were failures—aluminum carbide was the major product. Aqueous solutions are inappropriate media, because it is easier to reduce water (reduction potential, −0.83 V) than aluminum ion (−1.67 V) and therefore only hydrogen is produced at the cathode during such electrolyses. A suitable electrolyte would be one not undergoing electrolysis itself but serving as a solvent for the ionic dissociation of Al_2O_3.

The electrolyte now used is sodium hexafluoroaluminate (**cryolite**), Na_3AlF_6. Charles M. Hall, in 1886, found that Al_2O_3 would dissolve in liquid cryolite (m.p. 1000°C) and that the solution conducted the electric current. The cryolite itself undergoes no appreciable electrolysis, because AlF_6^{3-} is resistant to oxidation and Na^+ is resistant to reduction at the voltages used. In modern industrial practice, the aluminum accumulates as a cathodic melt which is tapped off. Carbon is used as the anodic material; it undergoes oxidative attrition with evolution of CO_2. The molten aluminum itself is the cathode. The reactions are

$$\textit{(cathode)} \qquad Al^{3+} + 3e^- \longrightarrow Al$$
$$\textit{(anode)} \qquad O^{2-} + C - 2e^- \longrightarrow CO_2$$

A schematic diagram of the cell appears in Fig. 23.5.

23.13 □ REFINING: ELECTROLYTIC METHOD

Refining refers generally to any final method of bringing a metal to its highest degree of purity. Electrolytic methods are very effective and

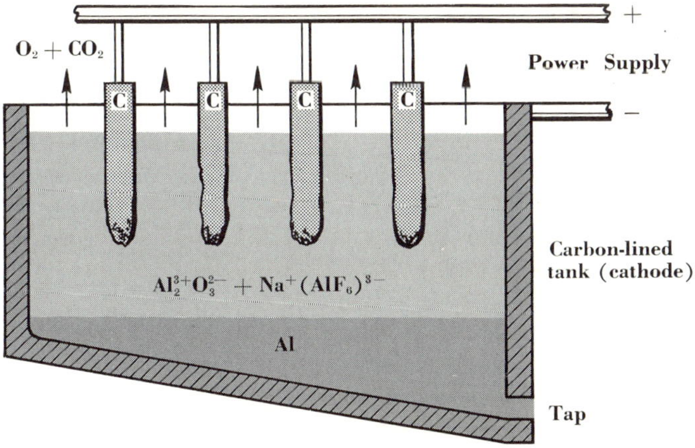

Fig. 23.5 Apparatus for electrolytic production of aluminum by the Hall process, using carbon anodes (schematic).

are used in a variety of processes. A discussion of the electrolytic refining of copper will illustrate the application.

The contaminants to be removed from impure copper may be classified, electrolytically, as (1) metals that are stronger reducing agents than copper, (2) metals that are weaker reducing agents than copper, and (3) nonmetallic substances. Let us exemplify materials in these categories by Zn, Ag, and SiO_2, respectively. The refining cell uses the impure copper as the anode, pure copper as the cathode, and aqueous $CuSO_4$ as electrolyte. In practice, the cell operates at voltages that produce only the following reactions:

Anodic actions:

$$Cu \longrightarrow Cu^{2+} + 2e^-$$
$$Zn \longrightarrow Zn^{2+} + 2e^-$$
$Ag \longrightarrow$ no reaction (because applied voltage is less than oxidation potential for Ag)

$SiO_2 \longrightarrow$ no reaction

Cathodic actions:

$$Cu^{2+} + 2e^- \longrightarrow Cu$$
$Zn^{2+} \longrightarrow$ no reaction (because applied voltage is less than deposition (reduction) potential for Zn)

The net effect is that the nonmetals and the weakly reducing ("noble") metals, such as Ag, Au, or Pt, precipitate as a residue called the **anode sludge,** which can profitably be reworked for recovery of the noble metals. The strongly reducing metals, such as Zn, remain as ions in solution from which they can be removed from time to time. The only cathodic deposition is copper; this selectivity produces copper of purity exceeding 99.95%.

23.14 ▫ REFINING: NONELECTROLYTIC METHODS

The Mond process An interesting method of refining metals is the formation of a volatile carbonyl compound, separation from nonvolatile residual impurity, and recovery of the metal by decomposition of its carbonyl. In the **Mond process** for nickel (named for the former Mond plant near Swansea, South Wales), the crude metal is produced by roasting the sulfide, NiS, and reducing the resultant oxide at 350°–400°C with "water gas," a mixture of carbon monoxide and hydrogen. Water gas is produced by the action of steam on hot carbon (p. 394). The equations are

$$\mathrm{C}(amorph) + \mathrm{H_2O}(g) \longrightarrow \mathrm{CO}(g) + \mathrm{H_2}(g)$$
$$\mathrm{NiO}(c) + \mathrm{H_2}(g) \longrightarrow \mathrm{Ni}(c) + \mathrm{H_2O}(g)$$
$$\mathrm{NiO}(c) + \mathrm{CO}(g) \longrightarrow \mathrm{Ni}(c) + \mathrm{CO_2}(g)$$

The crude nickel is then treated at 60°C with carbon monoxide, and the volatile carbonyl is formed,

$$\mathrm{Ni}(c) + 4\mathrm{CO}(g) \rightleftharpoons \mathrm{Ni(CO)_4}(g)$$

Because about 97% of the reduction of the nickel oxide is brought about by hydrogen, the gaseous effluent from the reduction is substantially carbon monoxide and steam. After removal of the water by condensation, this gas is suitable for use in carbonyl formation. After purification, the carbonyl is decomposed on nickel pellets at 180°C, the released carbon monoxide being recirculated for reuse. The nonvolatile residue contains copper, cobalt, sulfur, and some precious metals.

The Parkes process This interesting extractive method of refining is advantageously used to remove silver and gold from lead. The lead is melted and thoroughly agitated with about 2% of molten zinc, with which it is practically immiscible. Both silver and gold are much more soluble in zinc than in lead, and therefore will preferentially dissolve in the zinc. Zinc is lighter and freezes at a higher temperature than lead; the mixture can then be cooled and the solid zinc-silver-gold solution skimmed off the top of the molten lead. The zinc can be removed from the precious metals by distillation.

The refining of iron The refining of iron is essentially a selective oxidation of its impurities, especially carbon, and to a lesser extent silicon, phosphorus, and sulfur. In the "open-hearth" process, the impure iron from the blast furnace (**pig iron**) is mixed with a melt consisting of iron scrap, iron ore, and lime. The ore is the oxidant. The oxidized products that do not pass off as a gas are separated from the iron as a slag. Some of the reactions are

$$Fe_2O_3 + 3C \longrightarrow 3CO + 2Fe$$
$$2Fe_2O_3 + 3Si \longrightarrow 3SiO_2 + 4Fe$$
$$5Fe_2O_3 + 6P \longrightarrow 3P_2O_5 + 10Fe$$
$$Fe_2O_3 + 3Mn \longrightarrow 3MnO + 2Fe$$
$$S + CaO + C \longrightarrow CaS + CO$$

The oxidizing agent used may also be air (**Bessemer process**) or, as increasingly used in recent years, oxygen (**oxygen top-blowing process**).

The iron is then treated with various addition agents to modify its properties; the resulting product is called **steel.** Steelmaking is not, strictly speaking, refining, because it is not a purification process. The changes in composition, however, are strictly controlled and are monitored analytically during the manufacturing process. Addition agents include **ferromanganese** (an alloy containing Fe, Mn, and C), **ferrosilicon** (an alloy of Fe and Si), and other materials as required. Carbon contents of steels range from about 0.05% to about 1.50%. Nonferrous metallic components of steel include Mn, Ni, Cr, Mo, V, W, Co, and Cu.

PROBLEMS

1. Definitions Define and illustrate the following terms: (a) alloy; (b) forbidden zone; (c) electron band; (d) semiconductor; (e) sintering; (f) gangue; (g) flotation; (h) leaching; (i) roasting; (j) flux; (k) slag; (l) electrometallurgy.

2. Metallic properties Wax can be squeezed into a sheet; organic matter can occur as filaments like hair or fibers; fused salt conducts electricity well; cinnabar (HgS) is dense (8 g/cm^3). None of these substances is metallic. Explain in each case why the property described does not demonstrate that the substance is a metal.

3. The metallic bond Only metallic (never ionic or covalent) substances form crystals that exhibit the maximum coordination number of 12 among like-sized atoms. Account for this fact.

4. Band theory A small particle of sodium contains 1×10^{-6} mole. (a) How many atoms, (b) how many atomic orbitals, and (c) how many molecular orbitals does this particle contain? (d) Write the symbols showing the distribution of electrons in subshells for this particle of sodium.

5. Occurrence of ores You are offered the following investments: (a) purchase of stock in a brass mine; (b) purchase of stock in a mineral deposit of cobalt(III) chloride, $CoCl_3$; and (c) of potassium permanganate, $KMnO_4$. Explain your technical recommendations with regard to the likelihood that such opportunities exist. Brass is an alloy of copper and zinc.

6. Cyanide leaching Draw a diagram of an electrochemical cell in which a current is produced from the cyanide leaching of silver (p. 469). Label anode, cathode, and all reactive chemical constituents of the cell. Suggest a method of recovering silver from the cyanide complex.

7. Electrometallurgy (a) Draw a diagram of a cell which could be used for win ning sodium from NaCl. Label anode, cathode, and electrolyte. Write the equations for the half-reactions. (b) In one industrial process, magnesium oxide is electrolyzed in a fused bath of MgF_2, NaF, and BaF_2. This electrolyte is denser than liquid magnesium. Draw a diagram of a cell that could be used for such electrolysis. Label anode, cathode, and electrolyte. Write the equations for the half-reactions.

8. Reduction of ores Silicon undergoes oxidation to the dioxide,

$$Si + O_2 \longrightarrow SiO_2 \qquad \Delta H = -201 \text{ kcal/mole}$$

If elemental silicon were cheaply available, would it be useful for the industrial reduction of ores? If so, outline a practical process. If not, point out the difficulties.

ADDITIONAL PROBLEMS

9. Metallic properties Iron pyrite, FeS_2, is called "fool's gold." What are the probable reasons for this confusion? What steps would you, as an amateur prospector, take to ensure that your find was metallic?

10. Metallic properties Useful objects may be produced by a precision casting process called **powder metallurgy.** Metallic powders are first compressed to reduce void space and to increase the density, and then heated to a temperature slightly below the melting point to increase the density still further and greatly to strengthen the product. Account for these phenomena. Do you think that nonmetallic powders could be used as well? If not, what steps in the process would be unsatisfactory?

11. The metallic bond (a) When matter is heated to temperatures around 50,000°C, it enters the **plasma** state, in which electrons are extensively dissociated from atoms. What comparisons or contrasts can you draw between plasmas and metals? (b) Some substances (for example, $CaCO_3$) become metallic under extremely high pressures ($\sim 10^5$–10^6 atm), as shown by the work of Willard Libby and others. What new properties do you think might appear under such conditions to warrant describing the substance as metallic?

12. The metallic bond Is it conceivable that a substance could be found whose electrical conductivity is substantially independent of temperature in some temperature range? If so, describe the qualitative energy requirements for the electron bands of such a substance.

13. Electrometallurgy Zinc is produced at Trail, British Columbia, by electrolysis of a concentrated aqueous solution of the sulfate at high current density. It is found that the solution must be quite pure if the zinc is to be of satisfactory quality. Anodes are a silver-lead alloy; cathodes are aluminum. What types of impurity in the electrolyte would be most likely to contaminate the reduced zinc? Write the equations for the reactions that occur at each electrode.

14. Refining Although zinc can be refined electrolytically, it is sometimes refined by distillation. What factors would you wish to evaluate if you were asked to make a choice between electrolysis and distillation for refining a metal? In what way would each factor influence your choice? Explain how such evaluations might dictate the choice of electrolysis for zinc.

24 ■ NUCLEAR CHEMISTRY

In this chapter we examine more closely the nature of the bond (force) holding nucleons together in nuclei, and the properties of the nuclei such as radioactivity, artificial transmutations, fission, and fusion.

24.1 □ RADIOACTIVITY

With the discovery of x rays by Röntgen (p. 10), several investigators considered the possibility of a relationship between the phosphorescence produced by the x rays in an electron beam tube and the light emitted by natural and artificial minerals upon exposure to sunlight. Probing for such a relationship, Henri Becquerel in 1896 accidentally discovered that uranium compounds emit a radiation similar in nature to x rays. *Elements such as uranium, which spontaneously emit energy without the absorption of energy, are called* **naturally radioactive.** Artificially radioactive isotopes may be produced by adding energy to stable nuclei.

Laboratory studies using electric or magnetic fields showed that the emission is composed of three distinct types. An electric field applied to a narrow parallel beam of the emission, obtained from a naturally radioactive source, splits the beam into three beams, labeled **alpha** (α), **beta** (β), and **gamma** (γ), as shown in Fig. 24.1. The α beam is deflected to the negative plate and is therefore composed of positively charged particles; the β beam, deflected to the positive plate, must be composed of negatively charged particles. The γ beam, undeflected, must be electrically neutral.

Further experimentation has demonstrated that the α particle carries two unit positive charges, and possesses a mass equal to that of a

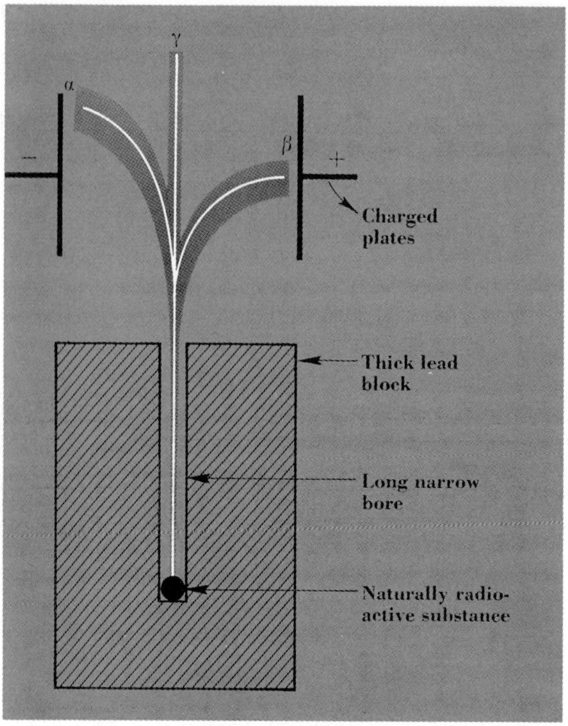

Fig. 24.1 The behavior of α, β, and γ emissions in an electric field; apparatus is enclosed in a vacuum.

helium, $^{4}_{2}He$, nucleus. Alpha particles are thus helium nuclei, $^{4}_{2}He^{2+}$, ejected at high velocities from certain radioactive substances. The electric charge and the mass of a β particle are equal to those of an electron. Beta particles are thus electrons ejected at very high velocities from certain other radioactive substances. The properties of γ rays are similar to those of x rays, except that their wavelengths are shorter. Gamma rays are therefore electromagnetic radiations or photons several million times higher in frequency than visible light; refer to Fig. 2.4 (p. 14). A **radioisotope** is an isotope showing radioactivity.

Intensive researches also demonstrated that certain radioisotopes produce other radioisotopes. Lord Rutherford, for example, found that radium continuously and spontaneously emits not only β particles but also radon, a radioactive gas.

To account for these observations, Lord Rutherford and Frederick Soddy in 1902 proposed the theory that radioactivity is the result of a spontaneous change of the atoms of one element into the atoms of another element. These nuclear reactions (transmutations, transformations, or disintegrations) involve a change in atomic number and, frequently, a change in the mass number of the radioisotope. Thus, the transmutation of radium may be represented by

$$^{226}_{88}Ra \longrightarrow {}^{4}_{2}He + {}^{222}_{86}Rn \qquad (1)$$

in which the subscript is the atomic number, and the superscript is the mass number—the number of nucleons (p. 13) in the nucleus.

In nuclear reactions a neutron can change into a proton or a proton into a neutron. Nevertheless, the total number of nucleons (the mass number) is conserved. Thus *the sum of the mass numbers of reacting nuclei and particles must equal the sum of the mass numbers of the nuclei and particles produced.* The conservation of charge requires *the sum of the atomic numbers of reacting nuclei and particles to equal the sum of the atomic numbers of the products.* The nuclear transformation represented by Equation 1 conforms to these principles. Also, in the reaction of an α particle with ${}^{27}_{13}\text{Al}$, in which a neutron is produced,

$$ {}^{27}_{13}\text{Al} + {}^{4}_{2}\text{He} \longrightarrow ({}^{31}_{15}\text{P}) \longrightarrow {}^{1}_{0}\text{n} + ? $$

the product nucleus must be ${}^{30}_{15}\text{P}$. In the reaction

$$ {}^{235}_{92}\text{U} \longrightarrow {}^{0}_{-1}\beta + ? $$

the product must be ${}^{235}_{93}\text{Np}$.

The intermediate nucleus formed, $({}^{31}_{15}\text{P})$, is known as a **"compound nucleus"** (a term coined to describe a nuclear system whose lifetime is longer than the time required for a nucleon to cross a nucleus).

A single radioactive atom emits only one particle. Usually one kind of radioactive substance emits only one kind of particle. Occasionally, however, some nuclei of an element, for example ${}^{218}_{84}\text{Po}$, emit α particles while others emit ${}_{-1}\beta$ particles. Gamma rays may be emitted along with a specific particle from a single atom.

Both α and ${}_{-1}\beta$ particles are detected from uranium ore (p. 477) because ${}^{238}_{92}\text{U}$ undergoes a succession of nuclear transformations giving intermediate radioactive atoms some of which emit α particles while others emit ${}_{-1}\beta$ particles. One of several natural radioactive series by which ${}^{238}_{92}\text{U}$ is transformed into nonradioactive ${}^{206}_{82}\text{Pb}$ is shown below. The emitted particle is indicated over the arrow.

$$ {}^{238}_{92}\text{U} \xrightarrow{\alpha} {}^{234}_{90}\text{Th} \xrightarrow{\beta} {}^{234}_{91}\text{Pa} \xrightarrow{\beta} {}^{234}_{92}\text{U} \xrightarrow{\alpha} {}^{230}_{90}\text{Th} \xrightarrow{\alpha} {}^{226}_{88}\text{Ra} \xrightarrow{\alpha} $$
$$ {}^{222}_{86}\text{Rn} \xrightarrow{\alpha} {}^{218}_{84}\text{Po} \xrightarrow{\alpha} {}^{214}_{82}\text{Pb} \xrightarrow{\beta} {}^{214}_{83}\text{Bi} \xrightarrow{\beta} {}^{214}_{84}\text{Po} \xrightarrow{\alpha} $$
$$ {}^{210}_{82}\text{Pb} \xrightarrow{\beta} {}^{210}_{83}\text{Bi} \xrightarrow{\alpha} {}^{206}_{81}\text{Tl} \xrightarrow{\beta} {}^{206}_{82}\text{Pb} $$

24.2 ▫ NUCLEAR ENERGY

The energy change of a nuclear reaction, ΔE, calculated from the difference between the masses of products and reactants in accordance with the Einstein law of the equivalence of mass and energy (p. 105), is in agreement with the measured values within the experimental errors involved in the determination of atomic masses and of the energy of the reaction. Thus, for the transmutation of radium the masses*

* Based on the mass scale ${}^{12}\text{C} = 12$ (exactly).

of the atoms involved are

$$^{226}_{88}\text{Ra} = 226.032 \qquad \begin{aligned} ^{4}_{2}\text{He} &= 4.0026 \\ ^{222}_{86}\text{Rn} &= \underline{222.024} \\ & \ 226.027 \end{aligned}$$

so that Δm, the difference in mass between the products and the initial reactants, is

$$\Delta m = 226.027 - 226.032 = -0.005 \text{ g/mole}$$

and

$$\Delta E = (\Delta m)c^2 = -5 \times 10^{-3} \text{ g/mole } (3 \times 10^{10} \text{ cm/sec})^2$$

$$= -45 \times 10^{17} \text{ ergs/mole}$$

or in kilocalories per mole,

$$\Delta E = \frac{-45 \times 10^{17} \cancel{\text{ergs}}\text{/mole}}{4.18 \times 10^{10} \cancel{\text{ergs}}\text{/kcal}} = -11 \times 10^{7} \text{ kcal/mole}$$

The measured value is $\Delta E = -11 \times 10^7$ kcal/mole. A convenient energy unit is the electron volt (eV), which is the energy gained or lost by an electron acted on by a potential of one volt. An electron volt is equal to 1.60×10^{-12} erg or 3.83×10^{-23} kcal and one electron volt per atom is equal to 23.1 kcal/mole (p. 58). Then

$$\Delta E = -11 \times 10^{7} \cancel{\text{kcal/mole}} \left\{\frac{1 \text{ eV/atom}}{23.1 \cancel{\text{kcal/mole}}}\right\} = -4.8 \times 10^{6} \text{ eV/atom}$$

There are 10^6 eV in a million electron volts (MeV), whence

$$\Delta E = \frac{-4.8 \times 10^{6} \cancel{\text{eV}}}{10^{6} \cancel{\text{eV}} \text{ atom/MeV}} = -4.8 \text{ MeV/atom}$$

Upon passage through matter, the energy of the emissions from radioactive nuclei is consumed by the ionization or excitation of atoms, molecules, or ions and by the disruption of chemical bonds.

Alpha particles usually cannot penetrate paper more than a few sheets in thickness. The walls of the ordinary glass beaker generally suffice to stop β particles. X and γ rays, however, are highly penetrating and require thick layers of lead or concrete to minimize penetration. To reduce the intensity of a γ ray of 10^{-12} cm wavelength by one half, about 1.8 cm of lead or almost 10 cm of concrete is required. Distance from the source is also important in minimizing exposure, since the intensity varies inversely with the square of the distance.

24.3 □ THE STABILITY OF NUCLEI

The **binding energy of a nucleus** is the energy evolved when the nucleons combine to form the nucleus or the energy required to separate the nucleus into its component nucleons. The formation of deuterium from protium and a neutron, ${}_0n^1 + {}_1^1H \longrightarrow {}_1^2H$, liberates 5.15×10^7 kcal/mole, calculated from the masses of particles (see Problem 3, p. 492). In comparison, the heat of the chemical reaction $H + H \longrightarrow H_2$ at 25°C is 104 kcal/mole. Evidently, the heats of nuclear reactions are about a million times larger than heats of chemical reactions. The nature of the force holding nucleons together must therefore be fundamentally different from the electrostatic force involved in chemical bonding. As in chemical bonding, gravitational and magnetic forces are not significant interaction forces. Further, the deuteron, composed of one type of charged particle (the proton) and a neutral particle, cannot be held together by electric forces. An even more striking illustration of the conclusion that nuclear forces cannot be electrical is the great stability of the nucleus of ${}_2^3He$ in which the coulombic interaction of the two protons is repulsive. In 1935, Hideki Yukawa, a theoretical physicist, suggested that *a particular particle oscillating between nucleons* with practically the velocity of light *is the force* that holds the nucleons together. He further predicted that this particle should have a mass about 275 ± 25 times the electron mass and may be electrically neutral, positive, or negative. We designate these particles, called "pi-mesons" or "pions," as π^0, π^+, π^-, the proton as p, and the neutron as n. The processes taking place in the deuteron that stabilize it are assumed to be

$$(p_1 + n_2) \longrightarrow (n_1 + \pi^+ + n_2) \longrightarrow (n_1 + p_2)$$

and the reverse

$$(p_1 + n_2) \longleftarrow (n_1 + \pi^+ + n_2) \longleftarrow (n_1 + p_2)$$

or with the π^-

$$(p_1 + n_2) \rightleftharpoons (p_1 + \pi^- + p_2) \rightleftharpoons (n_1 + p_2)$$

or with the π^0

$$(p_1 + n_2) \rightleftharpoons (p_1 + \pi^0 + n_2) \rightleftharpoons (p_1 + n_2)$$

As a result of the continuous pion transfer, protons and neutrons cannot be said to exist as such in the nucleus. Instead there are particles which are protons becoming neutrons and ones which are neutrons becoming protons. At any given instant, the nucleus has a fixed number of nucleons and an indefinite number of a variety of pions, all with a definite positive charge equal to the atomic number. At any given instant, two nucleons are sharing the same exchange particle. The basis of this theory is the new fundamental concept of an *exchange particle* (π^-, π^+, *or* π^0) *serving as a force of attraction.*

The sharing of pions by nucleons is analogous to the sharing of electrons by bonded atoms in molecules. The two resulting forces of attraction are alike in that they drop off sharply beyond a certain critical distance. Distances between bonded atoms are of the order of magnitude of 10^{-8} cm. Distances between nucleons sharing in pions are of the order of magnitude of 10^{-13} cm. This distance is called a **fermi,** named after the Italian physicist Enrico Fermi, who added much to our understanding of nuclear physics. Analogously with chemical bonding, the closer the nucleons, the stronger the nuclear bond. On the other hand, a small separation of nucleons beyond a fermi leads to instability and fragmentation of the nucleus.

This theory explains $_{-1}\beta$ decay, which is a paradoxical phenomenon, since electrons do not exist in the nucleus. Beta particles result from the decay of an *ejected* π^-. For every ejected π^-, a nuclear neutron is converted to a nuclear proton. The ejected π^- undergoes a series of transformations, the net result of which is the formation of a $_{-1}\beta$ particle and three neutrinos.*

$$\text{nuclear neutron} \longrightarrow \text{nuclear proton} + \text{ejected } \pi^-$$
$$\text{ejected } \pi^- \longrightarrow {}_{-1}^{0}\beta + 3_0\nu^0 \text{ (neutrino)}$$

24.4 □ NUCLEAR REACTIONS

(a) Radioactivity The stability of a nucleus depends on the ratio of neutrons (n) to protons (p). For the stable isotopes up to calcium, this ratio, n/p, is nearly 1, but with increasing atomic weight, it increases steadily to about 1.5. Apparently, more neutrons are needed in the heavier nuclei to dilute the mutually repulsive protons. All nuclei with atomic numbers greater than 83 are radioactive; of these only $^{238}_{92}U$ and $^{232}_{90}Th$ occur on the earth in significant amounts.

Some nuclei possess too many protons to be stable. Such nuclei alleviate this condition by emitting an α particle, a positron (a positively charged particle with the mass of an electron), or a proton, or by capturing an electron, usually from the K shell. Illustrations are

* The neutrino is a massless, chargeless particle predicted by Enrico Fermi in 1934 to maintain conservation laws in nuclear reactions.

$$^{11}_{6}C \longrightarrow {}^{7}_{4}Be + {}^{4}_{2}He \text{ (an } \alpha\text{-particle)}$$
$$^{78}_{35}Br \longrightarrow {}^{78}_{34}Se + {}^{0}_{+1}\beta \text{ (a positron)}$$
$$^{43}_{21}Sc \longrightarrow {}^{42}_{20}Ca + {}^{1}_{1}H \text{ (a proton)}$$
$$^{7}_{4}Be + e^{-} \longrightarrow {}^{7}_{3}Li \text{ (}K\text{-electron capture)}$$

$^{238}_{92}U$, and $^{232}_{90}Th$, the parents, respectively, of two natural radioactive series, have too many protons and decay by α-particle emission.

Emission of positrons and protons is observed among certain artificially radioactive atoms (see next section), not among naturally occurring radioactive elements.

When a K electron is captured, an electron from an outer shell takes its place, thereby liberating energy, usually as an x ray.

Nuclei having too many neutrons decay by $^{0}_{-1}\beta$ or $^{1}_{0}n$ emission; illustrations are

$$^{87}_{36}Kr \longrightarrow {}^{87}_{37}Rb + {}^{0}_{-1}\beta$$
$$^{87}_{36}Kr \longrightarrow {}^{86}_{36}Kr + {}^{1}_{0}n$$

(b) Artificial transmutations; transuranium elements Everyone is more or less familiar with the dreams, shattered by Daltonian chemistry, of the alchemists, who attempted to transform cheap metals into gold. With the postulation of the nuclear theory of the atom, artificial transmutation came to appear feasible. The first induced transmutation was demonstrated in 1919 by Lord Rutherford, who exposed nitrogen to α particles from a natural source and detected the production of protons. This reaction may be represented as

$$^{14}_{7}N + {}^{4}_{2}He \longrightarrow ({}^{18}_{9}F) \longrightarrow {}^{17}_{8}O + {}^{1}_{1}H$$

A nitrogen nucleus captures an α particle forming a short-lived compound nucleus ($^{18}_{9}F$), which decomposes with the emission of a proton. With the removal of the α-particle source, the reaction stops.

While treating light elements such as boron or aluminum with α particles from natural sources, Irene Curie and Frederic Joliot detected in 1934 the emission of positrons and neutrons. More important, they observed that the unexpected emission of positrons does not stop, although the emission of neutrons does stop, upon removal of the α particle source. They postulated the initial formation of an intermediate element, $^{13}_{7}N$, which has an excess of protons and decays by emitting a positron. The overall reaction occurs in two steps.

$$^{10}_{5}B + {}^{4}_{2}He \longrightarrow ({}^{14}_{7}N) \longrightarrow {}^{13}_{7}N^{*} + {}^{1}_{0}n$$
$$^{13}_{7}N^{*} \longrightarrow {}^{13}_{6}C + {}^{0}_{+1}\beta$$

in which $^{13}_{7}N^{*}$ is the first radioisotope artificially produced. An asterisk is sometimes used to indicate a radionucleus.

An α particle, proton, or deuteron, before it can be captured by a nucleus, must possess sufficient kinetic energy to overcome the repulsive force that develops as the positive particle approaches the positive nucleus. Particle accelerators, such as **cyclotrons,** were invented to increase the kinetic energy of charged particles to levels required for capture by nuclei with high atomic numbers. With the accelerators available in 1934, it was not possible to disintegrate or induce radioactivity in the elements beyond potassium. Since the neutron is electrically neutral, Enrico Fermi reasoned that a repulsive force to oppose its entry into a nucleus would probably not be set up and, therefore, that it should be possible to transform all known elements with natural sources of neutrons. The test of this hypothesis by Fermi and other investigators proved fruitful. Practically all known elements have been transformed.

Nuclear reactors ("piles"), in which controlled fission reactions (p. 486) are carried out, are excellent sources of neutrons. These neutrons are used to bombard atomic nuclei to synthesize useful isotopes. Large quantities of $^{14}_{6}C$ can be readily made in a reactor by the action of neutrons on nitrogen.

$$^{14}_{7}N + ^{1}_{0}n \longrightarrow ^{14}_{6}C^* + ^{1}_{1}H$$

A solution of nitrate salt, usually NH_4NO_3, is run through the pile. Exposure to neutrons converts the nitrogen in the NO_3^- to ^{14}C, thereby creating ^{14}C-labeled carbonate ion, $^{14}CO_3^{2-}$. The proton, $^{1}_{1}H$, formed in the reaction combines with the carbonate ion to give bicarbonate ion, $(HO)^{14}CO_2^-$. The material is recovered and sold as solid $Ba^{14}CO_3$, samples of which contain at most 5% of the ^{14}C isotope.

Some other neutron-induced isotope syntheses are the preparation of (a) tritium, (b) radioactive $^{60}_{27}Co$, used in cancer therapy, (c) radioactive $^{24}_{11}Na$, and (d) radioactive $^{38}_{19}K$.

$$\begin{aligned}
&(a)\quad ^{6}_{3}Li + ^{1}_{0}n \longrightarrow ^{3}_{1}H^* + ^{4}_{2}He\\
&(b)\quad ^{59}_{27}Co + ^{1}_{0}n \longrightarrow ^{60}_{27}Co^*\\
&(c)\quad ^{24}_{12}Mg + ^{1}_{0}n \longrightarrow ^{24}_{11}Na^* + ^{1}_{1}H\\
&(d)\quad ^{39}_{19}K + ^{1}_{0}n \longrightarrow ^{38}_{19}K + 2^{1}_{0}n
\end{aligned}$$

These reactions show the variety of processes observed as the result of neutron capture.

Until 1940, no element with an atomic number higher than 92 was known. In that year, Edwin McMillan and Philip Abelson at the Radiation Laboratory of the University of California at Berkeley exposed $^{238}_{92}U$ to a neutron source. A new radioactive isotope, $^{239}_{92}U$, was formed which decayed by $_{-1}^{0}\beta$ emission to give an element with atomic number 93. This new transuranic element in turn emitted another β particle and γ rays to give another new element of atomic number 94. These two elements were named neptunium, $^{239}_{93}Np$, and plutonium, $^{239}_{94}Pu$. The reactions are summarized below.

$$\begin{aligned}
^{238}_{92}U + ^{1}_{0}n &\longrightarrow ^{239}_{92}U + \gamma \text{ rays}\\
^{239}_{92}U &\longrightarrow ^{239}_{93}Np + _{-1}^{0}\beta\\
^{239}_{93}Np &\longrightarrow ^{239}_{94}Pu + _{-1}^{0}\beta + \gamma \text{ rays}
\end{aligned}$$

The synthesis of plutonium is now industrially important because this element is used as a source of nuclear energy.

Since these discoveries, nine more transuranic elements have been synthesized. Reactions leading to the formation of some of these elements are

$$^{239}_{94}\mathrm{Pu} + {}^{4}_{2}\mathrm{He} \longrightarrow {}^{1}_{0}\mathrm{n} + {}^{242}_{96}\mathrm{Cm}\ \text{(curium)}$$

$$^{238}_{92}\mathrm{U} + {}^{12}_{6}\mathrm{C} \longrightarrow 4{}^{1}_{0}\mathrm{n} + {}^{246}_{98}\mathrm{Cf}\ \text{(californium)}$$

$$^{252}_{98}\mathrm{Cf} + {}^{10}_{5}\mathrm{B} \longrightarrow 5{}^{1}_{0}\mathrm{n} + {}^{257}_{103}\mathrm{Lw}\ \text{(lawrencium)}$$

Note the use of stable light isotopes such as $^{12}_{6}\mathrm{C}$ and $^{10}_{5}\mathrm{B}$ in the transformations.

(c) Fission It is instructive to plot, as in Fig. 24.2, the mass number against the average energy released per nucleon, in the formation of a nucleus from nucleons. The greater the energy released per nucleon, the greater the stability of the nucleus. It is evident that the very light and the very heavy nuclei are least stable compared to the nuclei in the vicinity of $^{55}_{26}\mathrm{Fe}$. It would be natural to expect that spontaneous changes that may occur among nuclei should include the fusion of very light nuclei into heavier nuclei, and the splitting (fission) of a very heavy

Fig. 24.2 The average energy evolved per nucleon for natural nuclei.

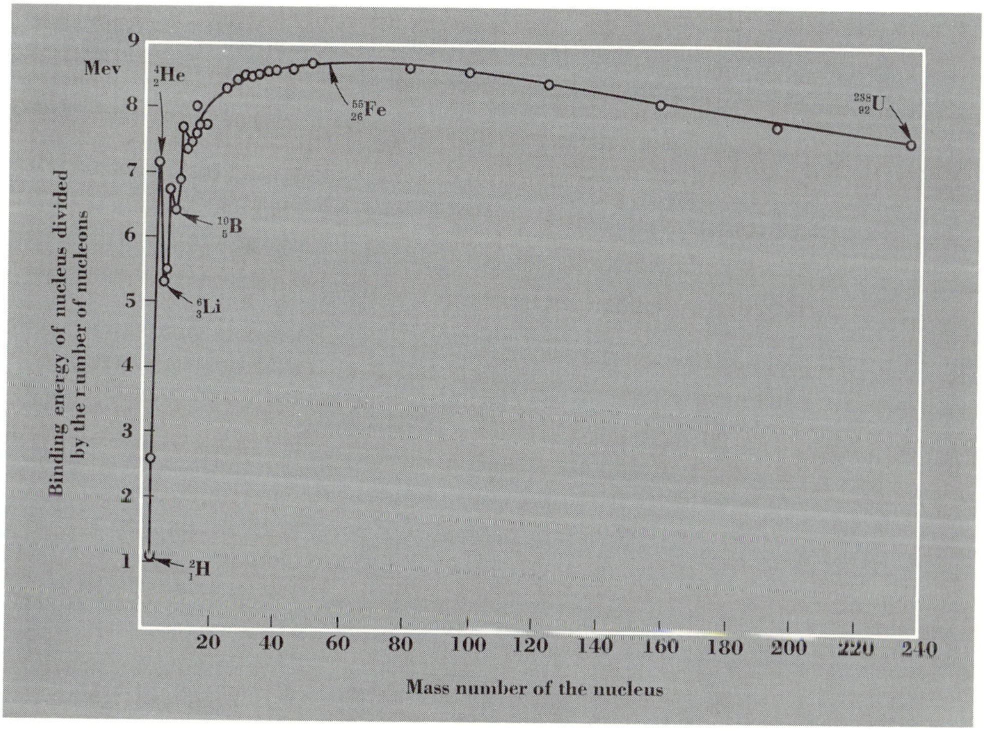

nucleus into two nuclei. It is equally evident that these reactions are rich energy sources, compared with chemical changes.

Although spontaneous fission has been detected in $_{90}$Th and heavier elements, the rate is too slow for any practical purpose because of the high energy of activation (roughly 1.4×10^8 kcal/mole) required to produce the deformation of the nucleus (a slight separation of nucleons) necessary for fission. The energy of activation is usually supplied by exposure to neutrons which form deformed unstable compound nuclei; for example,

$$^{235}_{92}U + ^{1}_{0}n \longrightarrow (^{236}_{92}U) \longrightarrow ^{137}_{53}I + ^{97}_{39}Y + 2^{1}_{0}n$$

When Fermi announced in 1934 the synthesis of transuranium elements by neutron bombardment of uranium, many investigators doubted this interpretation of the chemical analytical results. In particular, Ida Noddack intimated that the uranium nuclei were being split into smaller nuclei. Considerable interest was engendered and practically every laboratory engaged in the study of nuclear reactions undertook an analytical investigation of the possible products.

Among the products of the neutron irradiation of uranium, a radioelement was found that behaved chemically like radium. Since its radioactivity differed from that of natural radium it was assumed to be an isotope of radium. However, because of the chemical similarity of barium and radium, these results do not eliminate the possibility that *the element behaving like radium could be barium.* The chemical reactions of the compounds of barium and radium are so alike that a separation may be effected only by making use of the slight differences in solubility of the corresponding chlorides or bromides. If an unknown bromide is a radium bromide, it crystallizes along with natural radium bromide; if it is a barium bromide, it stays with the natural barium bromide. Otto Hahn and Fritz Strassman in 1939 mixed the supposed isotope of radium as the bromide with natural barium bromide and natural radium bromide and, after a long series of fractional crystallizations, found that the supposed isotope of radium could not be separated from barium. The simple, but unexpected, reason was that it was an isotope of barium. They then concluded that the uranium nucleus had divided into two fragments and it was immediately realized that a tremendous amount of energy would accompany this division, a conclusion verified by the large readings produced by this energy in an ionization chamber.* Similar readings obtained by previous investigators had been dismissed as "electronic noises." Nuclear fission was thus discovered by chemists.

* A gold-leaf electroscope serves to illustrate the principle of an ionization chamber. The gold leaves diverge when a potential difference is applied between the gold leaves (similarly charged) and the case of the electroscope; the larger the charge, the greater the leaf deflection. On exposure to radiation, ions are formed in the electroscope. The ions and electrons will collect on the oppositely charged elements of the electroscope, thus decreasing the charge on the leaves, thereby decreasing the leaf deflection. The rate of collapse of the leaves may be calibrated in terms of the charge collected.

From the principle of conservation of charge, the other nucleus formed with $_{56}Ba$ must be $_{36}Kr$; a typical fission reaction is

$$^{235}_{92}U + ^{1}_{0}n \longrightarrow (^{236}_{92}U) \longrightarrow ^{142}_{56}Ba + ^{91}_{36}Kr + 3^{1}_{0}n \qquad \Delta E = -2 \times 10^{2} \text{ MeV}$$
$$(4.6 \times 10^{9} \text{ kcal/mole})$$

Since more than one neutron is produced for each one consumed and since a ^{236}U nucleus undergoes fission in about 10^{-9} sec, it becomes possible to develop a branching chain reaction (p. 376) leading to a nuclear explosion, the release of unprecedented quantities of energy in extremely small periods of time. However, through the use of "chain breakers," materials such as cadmium that act as neutron scavengers, the number of neutrons available for the fission of $^{235}_{92}U$ may be decreased to about one for three produced. The rate of the fission reaction is then decreased, and the heat of the reaction may be utilized as a practical energy source. Also, the remaining neutrons may be used to produce desirable isotopes or to study the biological effects of neutrons.

The first controlled and sustained fission reaction was operated successfully in 1942 at the University of Chicago with comparatively impure ^{235}U and chain breakers. The first fission branching chain reaction was successfully carried out in a test explosion with pure ^{239}Pu in 1945 near Alamogordo, New Mexico.

Fission fragments are very rich in neutrons and therefore decay by emitting $_{-1}\beta$ particles or neutrons and γ radiation, ultimately forming stable elements.

(d) Fusion Tremendous amounts of energy can be released by the fusion of smaller atoms into larger ones. The largest energy release occurs with the lightest elements. Two of the more important reactions are

$$^{2}_{1}H + ^{3}_{1}H \longrightarrow ^{4}_{2}He + ^{1}_{0}n + 18 \text{ MeV}$$
$$^{2}_{1}H + ^{2}_{1}H \longrightarrow ^{3}_{2}He + ^{1}_{0}n + 3 \text{ MeV}$$

Energy from the fusion of one pair of elements induces the fusion of other pairs. Such a sustained chain reaction results in an explosion, specifically named **thermonuclear.**

However, a tank of hydrogen does not explode with the formation of deuterium because the rate of the reaction is infinitesimally small. To increase the rate, the temperature of the reactants must be increased to 10^{6}–10^{7} °C. At these temperatures, light elements are completely stripped of orbital electrons forming a completely ionized **plasma,** a gaseous system containing nuclei and electrons.

The hydrogen bomb makes use of nuclear fusion. The high temperatures needed to activate the reacting atoms is supplied by an atomic (fission) bomb. As yet, ways of controlling nuclear fusion for the benefit of man have not been devised. One of the major stumbling blocks is the inability to handle the plasma. Hopefully, research to harness nuclear fusion will be successful. The burgeoning consumption of energy will require, in the future, that nuclear fission and fusion supplement the supply of fossil fuels on earth.

(e) Stellar energy The mean temperature of the sun, $4 \times 10^6\,°K$, suffices for fusions as shown:

$$
\begin{aligned}
{}^1_1H + {}^1_1H &\longrightarrow {}^2_1H + {}_{+1}^{0}\beta \\
{}^2_1H + {}^1_1H &\longrightarrow {}^3_2He \\
{}^3_2He + {}^3_2He &\longrightarrow {}^4_2He + 2{}^1_1H
\end{aligned}
$$

Once the sun has synthesized ^3He and ^4He, further nuclear reactions involving ^3He and ^4He nuclei may lead to the production of ^{12}C, which then starts a carbon-nitrogen-oxygen cycle in which 4 protons are fused into a ^4He nucleus:

$$
\begin{aligned}
{}^{12}_6C + {}^1_1H &\longrightarrow {}^{13}_7N \longrightarrow {}^{13}_6C + {}_{+1}^{0}\beta \\
{}^{13}_6C + {}^1_1H &\longrightarrow {}^{14}_7N \\
{}^{14}_7N + {}^1_1H &\longrightarrow {}^{15}_8O \longrightarrow {}^{15}_7N + {}_{+1}^{0}\beta \\
{}^{15}_7N + {}^1_1H &\longrightarrow {}^{12}_6C + {}^4_2He
\end{aligned}
$$

The overall reaction is

$$
4{}^1_1H \longrightarrow {}^4_2He + 2{}_{+1}^{0}\beta \qquad \begin{aligned}\Delta E &= -4 \times 10^{-5} \text{ ergs per } \alpha \text{ particle} \\ &= -6 \times 10^8 \text{ kcal/mole}\end{aligned}
$$

The proton-proton chain reaction and the C—N—O cycle for the burning of hydrogen to helium were first proposed in 1938–1939 by Hans Bethe and Carl von Weizsäcker.

24.5 □ RATE OF RADIOACTIVE DECAY PROCESSES; HALF-LIFE

It is impossible to predict when, if ever, a radioactive atom will disintegrate. However, the chemist is almost never concerned with the behavior of a single particle. Even in a small weight of radioactive material there are many trillions of atoms. For example, in 1 mg of ^{238}U, there are 2.5×10^{18} atoms. With such large numbers of atoms, it is statistically possible to make precise predictions about the rate of disintegration. Experiment shows that the decay rate is directly proportional to the number of atoms present. For example, if 10^{20} atoms have a decay rate of 10^3 atoms per minute, then 4×10^{20} atoms will have a decay rate of 4×10^3 atoms per minute. Radioactive decay is a typical first-order reaction.

$$
\text{rate} = k \times N, \qquad \text{where } N = \text{number of atoms}
$$

Rates of disintegration measure the stability of nuclei. Since the rate varies with the number of atoms, absolute decay rates do not indicate the stabilities of different atoms unless the same number of atoms is used in each case. However, the time it takes for one-half

TABLE 24.1 HALF-LIVES OF SOME RADIOISOTOPES

Isotope	Decay reactions	$t_{1/2}$
${}^{3}_{1}H$	${}^{3}_{1}H \longrightarrow {}^{3}_{2}He + {}^{0}_{-1}\beta$	12.5 years
${}^{15}_{8}O$	${}^{15}_{8}O \longrightarrow {}^{15}_{7}N + {}^{0}_{+1}\beta$	2.0 min
${}^{32}_{15}P$	${}^{32}_{15}P \longrightarrow {}^{32}_{16}S + {}^{0}_{-1}\beta$	14.3 days
${}^{59}_{26}Fe$	${}^{59}_{26}Fe \longrightarrow {}^{59}_{27}Co + {}^{0}_{-1}\beta$	45.1 days
${}^{60}_{27}Co$	${}^{60}_{27}Co \longrightarrow {}^{60}_{28}Ni + {}^{0}_{-1}\beta$	5.2 years
${}^{90}_{38}Sr$	${}^{90}_{38}Sr \longrightarrow {}^{90}_{39}Y + {}^{0}_{-1}\beta$	27.7 years
${}^{131}_{53}I$	${}^{131}_{53}I \longrightarrow {}^{131}_{54}Xe + {}^{0}_{-1}\beta$	8.1 days
${}^{14}_{6}C$	${}^{14}_{6}C \longrightarrow {}^{14}_{7}N + {}^{0}_{-1}\beta$	5.73×10^{3} years
${}^{238}_{92}U$	${}^{238}_{92}U \longrightarrow {}^{234}_{90}Th + {}^{4}_{2}He$	4.5×10^{9} years

the atoms to disintegrate, the so-called **half-life,** is independent of the number of atoms. The half-life, an experimentally determined quantity, is a characteristic of each radioactive isotope, and therefore is used to indicate the relative stabilities. The shorter the half-life, the faster is the rate of decay. The half-lives of some frequently used radioisotopes are given in Table 24.1.

The half-life of ${}^{15}_{8}O$ is 2.0 minutes. Therefore, given 8 mg of this isotope to start with, 4 mg would remain after 2.0 minutes, 2 mg would remain after 4.0 minutes, and 1 mg would remain after 6.0 minutes.

EXAMPLE 1 Given 16 g of tritium, ${}^{3}_{1}H$. How much tritium will remain after 50 years?

ANSWER Since the half-life is 12.5 years (Table 24.1), 50 years represents 50 years/12.5 years or 4 half-lives. For each half-life, one-half of the remaining quantity of radioactive material decays. Therefore, in 4 half-lives

$$16\text{ g} \times \tfrac{1}{2} \times \tfrac{1}{2} \times \tfrac{1}{2} \times \tfrac{1}{2} = 1\text{ g of tritium remaining}$$

The answer to Example 1 could have been expressed as $16\text{ g} \times (\frac{1}{2})^4 = 1$ g, where 4 is the number of half-lives. Thus we see that a general equation for relating q_t, the quantity at some time t, and q_0, the initial quantity, is

$$q_t = q_0(\tfrac{1}{2})^{t/t_{1/2}} \tag{2}$$

where $t_{1/2}$ is the half-life, and $t/t_{1/2}$ is the number of half-lives. (See Example 2 below.)

24.6 RADIOCHEMISTRY

(a) Tracer elements With very few exceptions, the disintegration rate of a given radioisotope is, unlike chemical changes, independent of the chemical state and physical conditions. Hence, radioisotopes may be "traced" through chemical and physical changes, yielding valuable information about the mechanism of such changes. The fraction of ingested cholesterol, for example, that terminates in the blood stream may be determined by feeding cholesterol, enriched with ${}^{14}_{6}C^*$ of known **activity,*** and then measuring the ${}^{14}_{6}C^*$ activity of cholesterol per unit volume of blood.

(b) Radiochemical dating Dating with ${}^{14}C$ as conceived by Willard Libby, is based on the following reasoning. Natural carbon contains a small fraction of ${}^{14}_{6}C^*$. The activity of ${}^{14}_{6}C^*$ as found in living plants and animals and in the air is approximately constant at 14 disintegrations per minute per gram (d/min-g) of carbon, the activity lost through disintegration being replenished by the production of ${}^{14}_{6}C^*$ through the interaction of cosmic radiation with ${}^{14}_{7}N$ in the atmosphere. It is assumed that this value is also true for ancient times. Then, so long as a plant or an animal is alive, the activity of ${}^{14}_{6}C^*$ per gram of carbon remains constant, but upon death the ${}^{14}_{6}C^*$ continues to decay without replenishment and the activity decreases. From the known half-life, 5.73×10^3 years, for ${}^{14}_{6}C^* \longrightarrow {}_{-1}\beta^0 + {}^{14}_{7}N$, the age of archeological samples may then be determined. Radiochemical dating has made important contributions to paleoanthropology and geology. The full significance of the discoveries of hominid† fossils at Olduvai Gorge, Tanzania, in 1960, for example, could not be realized until the ages of the beds at the Gorge were accurately determined.

Uranium (Problem 20) and potassium have been used in dating. However, carbon-dating is more useful because carbon is so much more widespread.

EXAMPLE 2 The activity of the hair of an Egyptian woman is 7.50 d/min-g carbon. What is the historical age of the hair?

ANSWER The number of disintegrations per minute per gram of carbon is directly related to the quantities q_t and q_0, so that

$$q_0 = 14 \frac{\text{d}}{\text{min g}} \qquad \text{and} \qquad q_t = 7.50 \frac{\text{d}}{\text{min g}}$$

whence, using Equation 2,

$$7.50 \frac{\cancel{\text{d}}}{\cancel{\text{min g}}} = 14 \frac{\cancel{\text{d}}}{\cancel{\text{min g}}} (0.5)^{t/5730 \text{ yr}}$$

$$\log 7.50 = \log 14 + \frac{t}{5730} \log 0.5$$

from which $t = 5.10 \times 10^3$ years; the age of the hair is therefore 5.10×10^3 years.

* Activity = the number of nuclei disintegrating per unit time per gram of radioisotope.
† Pertains to very remote ancestors of man.

(c) Chemonuclear synthesis Gamma rays are of sufficient energy to cause excitation of electrons in σ bonds and thus cause fragmentation of such bonds. For this reason γ rays arising from radioactive sources are used to initiate chemical reactions. Solid methane at 77°K is polymerized rapidly and efficiently to heavy hydrocarbons by ^{60}Co gamma rays. The product is a viscous oil consisting mainly of saturated hydrocarbons containing an average of about 20 carbon atoms per molecule.

Nitrogen, mainly incorporated into proteins (p. 516), is essential to all living organisms. The cells of most plants cannot assimilate the nitrogen of the air. However, the roots of some plants, notably the leguminous variety, carry **nitrogen-fixing bacteria** capable of converting atmospheric nitrogen into nitrogen compounds that are then assimilated by the plant for protein synthesis. Lightning also fixes atmospheric nitrogen into compounds. However, artificial fertilizers, made through the fixation of nitrogen, must supplement natural sources. Under investigation is the fixation of nitrogen with oxygen to nitrogen oxide,

$$N_2 + O_2 \longrightarrow 2NO \qquad \Delta H = +42 \text{ kcal}$$

by the passage of air through a nuclear reactor serving as a source of ionizing radiation.

PROBLEMS

1. Alpha, beta, and gamma particles How is the number of protons in a nucleus changed when (a) a γ ray, (b) a β particle, or (c) an α particle, is emitted from the nucleus?

2. Conservation of nucleons and charge (a) Balance the following nuclear reactions which involve the formation of a compound nucleus [indicate symbol, mass number, and atomic number for ?]:

(*i*) ${}^{93}_{41}Nb + {}^{12}_{6}C \longrightarrow (?) \longrightarrow\ ? + 2\,{}^{1}_{0}n$
(*ii*) ${}^{51}_{23}V + {}^{16}_{8}O \longrightarrow (?) \longrightarrow {}^{65}_{31}Ga +\ ?$
(*iii*) ${}^{12}_{6}C + {}^{12}_{6}C \longrightarrow (?) \longrightarrow {}^{22}_{11}Na +\ ? +\ ?$

Not all nuclear processes involve the formation of a compound nucleus. When, for example, very high-energy particles are used, the energy is not shared among all the nucleons in the target nucleus. (b) Balance the following nuclear reactions which do not involve compound nuclei:

(*i*) ${}^{239}_{94}Pu + {}^{4}_{2}He \longrightarrow\ ? + {}^{1}_{0}n$
(*ii*) ${}^{246}_{96}Cm +\ ? \longrightarrow {}^{254}_{102}No +$ neutrons (mass number of ? is indeterminate)
(*iii*) ${}^{252}_{98}Cf + {}^{11}_{5}B \longrightarrow\ ? + 4\,{}^{1}_{0}n$
(*iv*) ${}^{252}_{98}Cf + {}^{13}_{6}C \longrightarrow\ ? + 4\,{}^{1}_{0}n$

(c) Balance the following nuclear reactions:

(*i*) ${}^{14}_{6}C +\ ? \longrightarrow {}^{14}_{6}C + 2\,{}^{1}_{0}n + {}^{1}H$
(*ii*) $? + 15\,{}^{1}_{0}n \longrightarrow {}^{254}_{94}Pu \longrightarrow\ ? + 6\,{}_{-1}\beta^{0}$

(*iii*) $^{238}_{92}U + ? \longrightarrow ^{247}_{99}Es + 5^1_0n$

(*iv*) $^{75}_{33}As + ^4_2He \longrightarrow ^{38}_{17}Cl + ? + 6^1_1H + 11^1_0n$

3. Nuclear energy Calculate the energy released in kilocalories per mole on the formation of deuterium, ^2_1H, from a proton and a neutron. $^1_0n = 1.00866$, $^1_1H = 1.00783$, $^2_1H = 2.01410$.

4. Radioactive series Thorium, $^{232}_{90}Th$, decays to the stable isotope of lead, $^{208}_{82}Pb$, by a sequence of emissions as shown: α, β, β, α, α, α, α, β, β, and α. Indicate the radioactive element arising from each step.

5. Nuclear charge How are the atomic number and atomic weight of a nucleus changed when (a) a proton, (b) a positron, (c) a neutron is emitted, and (d) an electron is captured?

6. Beta decay (a) How would you account for the production of electrons that do not exist in nuclei? (b) Positrons do not exist as particles in nuclei. Explain $_{+1}\beta$ emission (by analogy to $_{-1}\beta$ emission).

7. Half-life By beta emission, 1.000 μg of ^{107}Cd decays to 0.125 μg in 20.1 hours. What is the half-life of ^{107}Cd?

8. Dating Two wood samples of well-established age (4900 $\pm$ 200 years, by historical evidence) from Egyptian tombs possess activity of 7.8 d/min per gram of carbon. Calculate the age of these wood samples from radiochemical evidence.

9. Tracer chemistry An animal is fed ethyl alcohol, $*C_2H_5OH$, activity a counts/min. Subsequently, acetic acid, $*CH_3COOH$, activity b counts/min per 10 g liver, is isolated from the liver; total weight of liver is c g. Exhaled $*CO_2$, recovered as $Ca*CO_3$, has an activity of d counts/min per liter of air; total air volume is e liters. *C activity of blood stream is zero. What qualitative conclusions regarding the metabolism of ethyl alcohol would you draw from these results?

10. Yukawa theory In terms of the Yukawa theory, what is the possible number of "nuclear bonds" in ^2_1H, ^3_2He, and ^4_2He?

ADDITIONAL PROBLEMS

11. Electron capture Dependence of the half-life on the chemical state and physical conditions is found in radioactive processes in which orbital electrons enter directly. For example, ^7_4Be decays by electron capture; the rate is 0.14% faster for BeF_2 than for BeO. Balance the following:

$$^7_4BeF_2 + e^- \longrightarrow ?$$
$$^7_4BeO + e^- \longrightarrow ?$$

12. Protection Alpha particles may be stopped by paper or by the outer skin layers. Does it follow that an α emitter, internally located, presents no hazard to an animal?

13. ΔE Calculate ΔE for the reaction

$$^7_3Li + ^1_1H \longrightarrow 2^4_2He$$

Atomic masses: ^{7}Li = 7.016, ^{1}H = 1.008, ^{4}He = 4.003. The measured value is 39×10^{7} kcal/mole or 17 MeV.

14. Fission Is the fission reaction

$$^{63}_{29}\mathrm{Cu} + ^{1}_{1}\mathrm{H} \longrightarrow ^{24}_{11}\mathrm{Na} + ? + ^{1}_{0}\mathrm{n}$$

endothermic or exothermic? Atomic masses: ^{1}H = 1.008, ^{1}n = 1.009, ^{24}Na = 23.991, ^{39}K = 38.964, ^{63}Cu = 62.929.

15. Stellar energy The thermonuclear chain reactions (p. 488) evolve 4×10^{-5} erg per α particle. The sun radiates 8×10^{21} kcal/sec from its surface. Calculate the mass of hydrogen in tons (9×10^{5} g per ton) and the number of H atoms destroyed per second.

$$\text{rate} = k[\mathrm{H}]^2$$

16. Nuclear energy Matter is converted to energy through chemical or nuclear changes. Calculate the heat in kilocalories available at constant volume per mole of lithium for the following reactions:

(a) $^{6}_{3}\mathrm{Li} + ^{1}_{1}\mathrm{H} \longrightarrow {}^{6}\mathrm{Li}^{1}\mathrm{H}$

in which Δm is estimated to be -3×10^{-9} g/mole;

(b) $^{6}_{3}\mathrm{Li} + ^{1}_{1}\mathrm{H} \longrightarrow ^{3}_{2}\mathrm{He} + ^{4}_{2}\mathrm{He}$

for which the atomic masses are ^{1}H = 1.0078, ^{3}He = 3.0160, ^{4}He = 4.0026, ^{6}Li = 6.0151.

17. Half-life The overall equation for the decomposition of $^{238}\mathrm{U}(t_{1/2} = 4.50 \times 10^{9}$ years) is

$$^{238}_{92}\mathrm{U} \longrightarrow ^{206}_{82}\mathrm{Pb} + 8\alpha + 6\,_{-1}^{0}\beta$$

The average amount of ^{238}U in the earth's crust is 4 g per metric ton. What is the maximum number of atoms and volume of helium produced at 27°C and 1 atm per metric ton in 2.25×10^{9} years?

18. Half-life The most stable isotope of berkelium or californium has a half-life of 7×10^{3} years. Accepting 4×10^{9} years as the age of the earth and assuming berkelium was part of the primordial stock of elements, calculate the fraction of the original amount remaining. Is this quantity detectable?

19. Dating Will the production of ^{14}C* in the atmosphere by other than natural means affect radiocarbon dating (a) of existing relics; (b) of matter now living that may become future relics? If so, will the error in the age be positive or negative?

20. Dating The overall equation for the decomposition of $^{238}\mathrm{U}$ ($t_{1/2} = 4.50 \times 10^{9}$ years) is

$$^{238}_{92}\mathrm{U} \longrightarrow ^{206}_{82}\mathrm{Pb} + 8\alpha + 6\,_{-1}^{0}\beta$$

A uranium mineral mined from a stratum corresponding to the Precambrian era contains 2.4 g ^{206}Pb to 22.8 g ^{238}U. (a) Assuming no lead was leached from or added to the mineral, calculate the weight of ^{238}U that decomposed. (b) Assuming further that no uranium was leached or added to the mineral, calculate the age of the mineral. (c) Does the leakage of radon from uranium minerals introduce a positive or a negative error in the age of the mineral? (d) The half-life of $^{222}_{86}Rn$ is 3.83 days, while that of $^{219}_{86}Rn$, formed in the ^{235}U natural radioactive series, with ^{207}Pb as the end product, is 3.92 seconds. Which should yield more accurate age values, ^{206}Pb:^{238}U or ^{207}Pb:^{235}U weight ratios? Justify your answer.

21. Tracer chemistry (a) Describe an experiment that would use (*i*) $^{18}_{6}O$, (*ii*) $^{3}_{1}H^*$, to demonstrate that H_2O liquid $\rightleftharpoons$ H_2O vapor is a dynamic equilibrium. (b) Outline how you would use $^{59}Fe^*$ to demonstrate the efficiency of an iron compound for the production of hemoglobin. (c) CrO_4^{2-} does not become radioactive in the presence of $^*Cr^{3+}$ in solution. If $^*Cr^{3+}$ is added to a chromic acid plating bath, the Cr deposit is not radioactive. Using $^*CrO_4^{2-}$, the deposit is radioactive. What can you say about the mechanism of the deposition of Cr from a chromic acid bath?

ANSWERS

8. 48×10^2 years.

15. (a) 4×10^6 tons/sec; 2×10^{35} atoms/sec.

16. (b) 9.3×10^7 kcal/mole.

18. 1 part in $10^{2\times10^5}$.

20. (a) 2.8 g; (b) 7.5×10^8 years.

25 ■ POLYMERS

25.1 □ BACKGROUND AND DEFINITIONS

In Chapter 1 "molecules" were described as electrically neutral individual particles of ordinary matter whose atoms are linked by chemical bonds. The characterization of matter as an aggregation of molecules became conceptually quite useful in the study of gases during the eighteenth and nineteenth centuries. Gases that lend themselves to experimental study rarely consist of molecules larger than about 10 Å in diameter or greater than several hundred in molecular weight. Substances whose molecules are within this size range generally can be characterized by measurements of their vapor pressures or of their melting, boiling, or sublimation temperatures. The molecular weights of such substances can be determined with acceptable precision from their vapor densities or from the colligative properties of their solutions. Moreover, such measurements can be made with laboratory equipment that has been available at least since the beginning of modern chemistry. As a result of this concordance of concept and method, the first 200 years of modern chemistry were largely directed to the study of materials of low molecular weight. During the course of their investigations, however, chemists sometimes encountered materials that did not respond to studies of these kinds, materials that decomposed instead of melting or vaporizing, materials that were gummy, gluey, waxy, or resinous, materials that were insoluble in most solvents, and that, when they did dissolve, yielded solutions whose colligative properties were too minuscule to measure. Chemists found these materials so intractable that they usually threw them into the trash can. It gradually came to be recognized, however, that some naturally occurring substances of unquestioned importance to man, like cellulose, rubber, and protein, resembled the sticky residues that plagued experimenters. During the latter part of the nineteenth century, chemists began to realize that what these classes of substances had in common was the fact that their molecules were very large. It was evident that entirely new methods of investigation would have to be brought to bear to study such materials.

Substances of high molecular weight can often be characterized by their decomposition products. Thus, if natural rubber is heated, the hydrocarbon isoprene, C_5H_8, distills off. If starch is chewed, the sweet taste of glucose is detected. If egg albumin (a protein, see p. 516) is boiled in dilute sulfuric acid, the amino acids leucine, alanine, serine, glutamic acid, methionine, and some 13 others are produced (see Table 26.1, p. 517). It is reasonable to assume that these fragments detected in the decomposition of large molecules are indeed their units of construction. Thus, starch is called a **polysaccharide** (many sugar units) and, in general, materials whose molecules are made up of repetitions of individual units are called **polymers** (Greek, *many parts*). The individual unit substance of a polymer is called a **monomer.** Molecules composed of at least two different kinds of monomeric unit are called **copolymers.** When the number of units in a polymer is small and specifically known, it is often specified by use of the appropriate Greek prefix (**di**mer, **tri**mer, or so on). When the number of units is large it is not usually uniform for all the molecules in the sample, and specific numerical prefixes rapidly become inappropriate after about 4 or 5. When polymerization is so extensive that molecular weights reach the range of tens of thousands or hundreds of thousands, the substance is called a **high polymer,** and its constituent particles are called **macromolecules.**

The molecular weights of high polymers have been determined from measurements of osmotic pressure, from the ratios of sedimentation in high speed centrifuges (**ultracentrifuge**), from viscosity measurements, and by optical methods such as light scattering (Tyndall effect, p. 213).

25.2 ▫ THE DECOMPOSITION OF NATURAL HIGH POLYMERS; MONOMERIC UNITS (MERS)

The structure of isoprene (2-methyl-1,3-butadiene) produced by the destructive distillation of natural rubber is

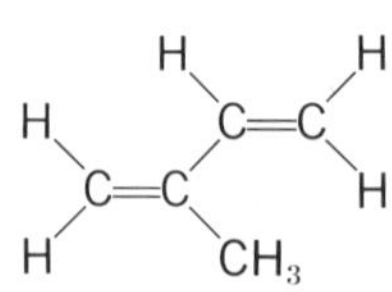

isoprene

Natural rubber itself is found to be a long chain-like structure of

the type

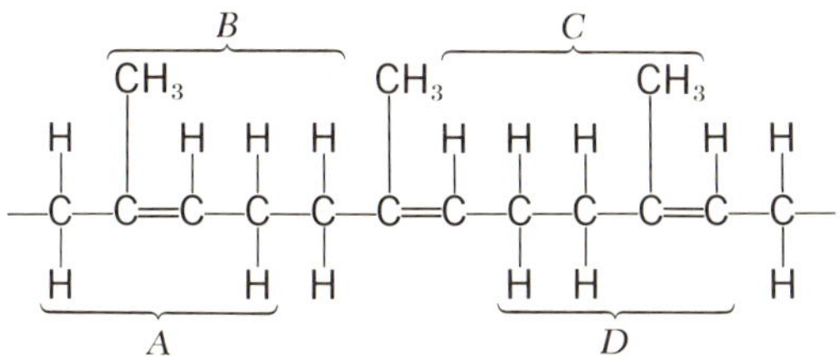

(For a more appropriate rendition of bond angles, see Table 25.1.) The formula of this macromolecule can be written by repetition of any one of the portions A, B, C, or D. However, only unit A has the same atomic skeleton as isoprene, the observed decomposition product. Furthermore, the treatment of isoprene with appropriate reagents yields a rubbery solid. The repeating monomeric unit (**mer**) of natural rubber is therefore unit A, not the other geometric possibilities. In general, a mer is the repeating unit that is known by study of decomposition or synthesis to be the building block of a macromolecule.

Table 25.1 shows the formulas of some natural organic high polymers and of their decomposition products. The identification of the decomposition product, however, is far from a full description of the polymer, for the following reasons:

(a) The mode of linkage of the monomer will influence the properties of the macromolecule. Note (Table 25.1) that in natural rubber the isoprene units are linked in the *cis* form; in gutta-percha the linkage is *trans*.

(b) The sequence and extent of monomeric linkages also are important determinants of polymer structure. Cellulose and starch are both glucose polymers, but the rigid cellulose consists of linear macromolecules of about 3500 glucose units each, whereas the more pliable starch is a lower polymer (about 500 glucose units per linear chain of cornstarch) with different structural linkage of the monomeric groups and more extensive chain branching.

(c) The shape of the macromolecules and their orientation in space with respect to each other are also critical factors affecting polymer structure. Raw and cooked egg white (albumin) differ in the shape of their constituent protein molecules.

Examination of Table 25.1 shows two kinds of relationship between monomer and polymer. In rubber and gutta-percha, the repeating unit of the macromolecule is identical in atomic composition and skeletal structure with the decomposition product. If the reaction could be reversed, a simple **addition** of isoprene units to each other would form the polymer.

In cellulose, chitin, and silk, however, decomposition is accompanied

TABLE 25.1 SOME NATURAL HIGH POLYMERS AND THEIR DECOMPOSITION PRODUCTS (THE SHADED PORTION IS THE MER)

Polymer and source	Formula	Decomposition product(s)
Cellulose (primary structural material of plants)	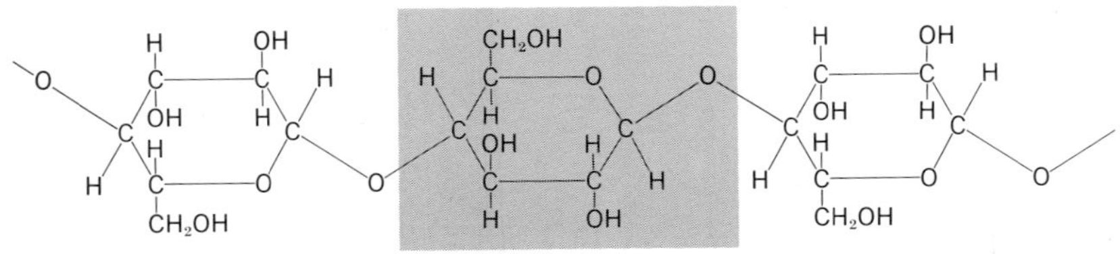	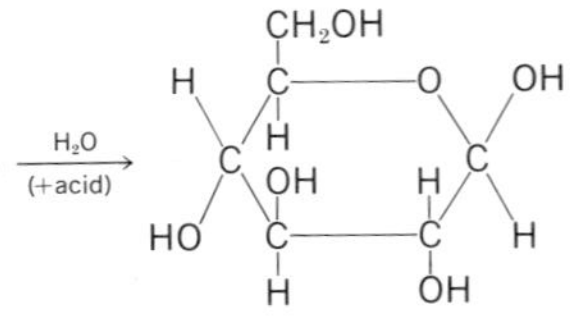 *Glucose*
Chitin (protective shell of insects, crustaceans, etc)	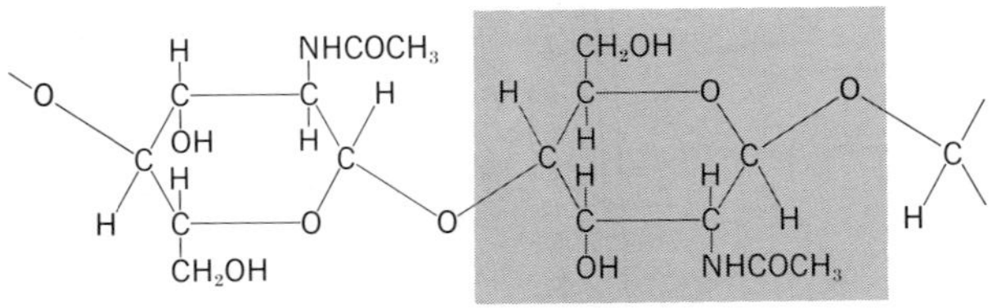	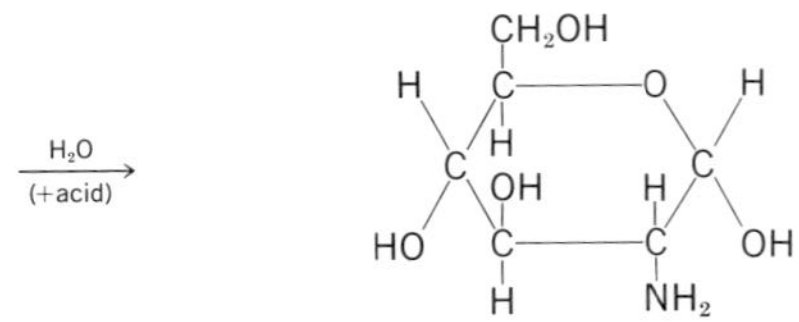 *Aminoglucose*
Natural rubber	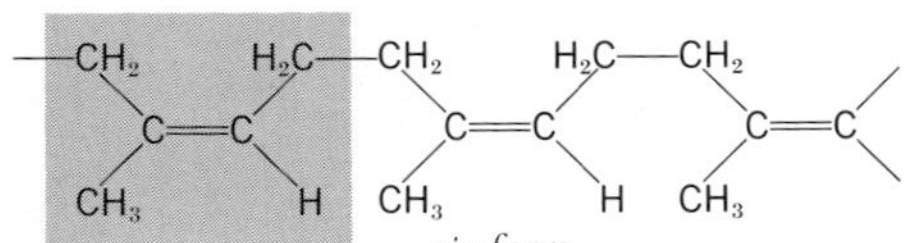 *cis-form*	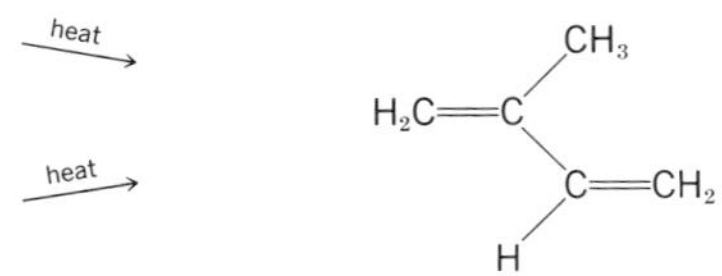 *Isoprene*
Gutta-percha	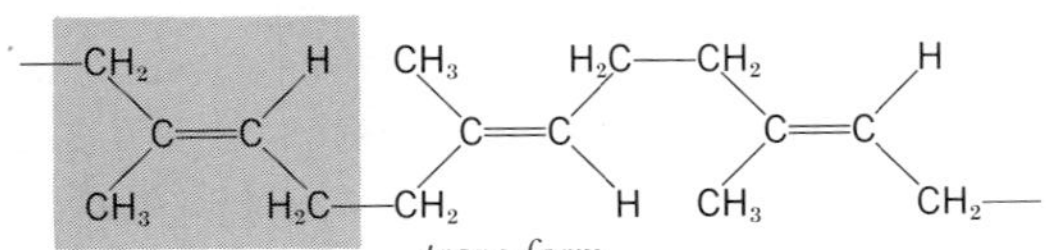 *trans-form*	
Silk (a protein)	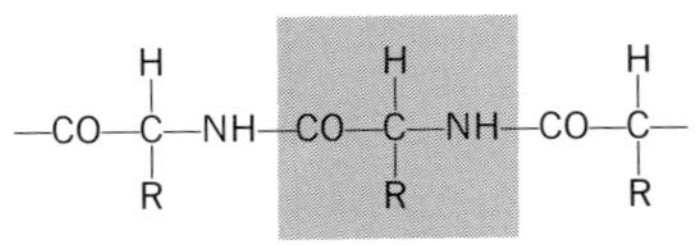 R═H, CH_3, CH_2OH, or any of about 12 other groupings.	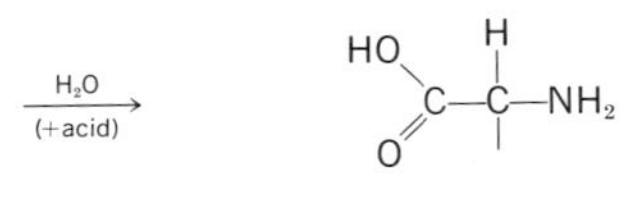 *Amino acid*

by the uptake of water. A reversal of the process, if it could be effected, would require the **condensation*** of monomers with elimination of water.

Polymers of the first type are called **addition polymers** (A-polymers); those of the second type are called **condensation polymers** (C-polymers).

25.3 ▫ THE PRODUCTION OF SYNTHETIC HIGH POLYMERS; C-POLYMERIZATION

The controlled synthesis of high polymers by condensation methods reached a high degree of industrial success as early as the 1930's, largely as a result of the pioneer work of Wallace H. Carothers. A C-polymer must be made from monomers that contain more than one functional group so as to enable intermolecular reactions to proceed continuously. Carothers succeeded in producing C-polymers by heating the salts of diamines with dicarboxylic acids at 200°–250°C, with elimination of water:

$$\underset{\textit{diamine}}{H_2N{-}(CH_2)_n{-}NH_2} + \underset{\textit{dicarboxylic acid}}{HOOC{-}(CH_2)_m{-}COOH} \xrightarrow[\text{via salt intermediates}]{\text{reaction occurs}}$$

$$\underset{\textit{dimer}}{H_2N{-}(CH_2)_n{-}NH{-}CO{-}(CH_2)_m{-}COOH} + H_2O \xrightarrow{\text{many more condensations}}$$

$$\cdots{-}CO{-}\boxed{NH{-}(CH_2)_n{-}NH{-}CO{-}(CH_2)_m{-}CO}{-}NH{-}\cdots + \text{more } H_2O$$

polymer (*Nylon*)
(shaded portion is the mer)

Nylon polymers are often identified by the numbers of carbon atoms in the monomeric diamine and diacid (n and $m + 2$ in the formulas above). The popular hosiery polymer is Nylon-66, the copolymer of hexamethylene diamine and adipic acid, each containing 6 carbon atoms per molecule ($n = 6$, $m = 4$). A Nylon can also be made by the simple polymerization of an amino acid, such as 6-aminohexanoic acid, which is polymerized to the material called Nylon-6:

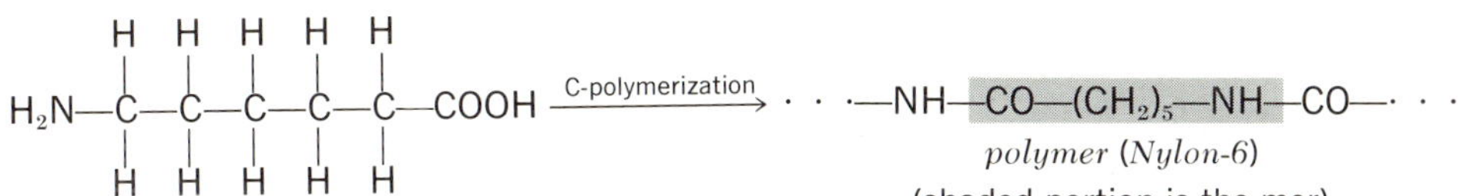

6-*aminohexanoic acid*; *polymer* (*Nylon-6*)
(shaded portion is the mer)

The single numeral 6 indicates that there is but one monomer and that it has 6 carbon atoms.

25.4 ▫ A-POLYMERIZATION

In the case of addition polymerization it is necessary for the **adduct** (addition product) of two monomers to be capable of undergoing further

* Condensation is the union of two or more molecules with elimination of a smaller unit like H_2O or NH_3.

addition, and for this process to continue to macromolecule formation. Otherwise, of course, the addition stops after a single step. Such termination occurs, for example, after the chlorination of ethylene; the product is saturated and therefore lacks the capacity for undergoing further addition or chain lengthening (p. 441). It is found that A-polymerizations proceed by a path involving the initial formation of some reactive species, such as free radicals or ions, and by the addition of the reactive species to another molecule with the *regeneration of the reactive feature*. In this way the addition reactions may proceed continuously. Pioneer studies in this field were carried out by Hermann Staudinger in the 1920's. An example of free radical A-polymerization (the curved arrows represent electron shifts that occur during reaction) is

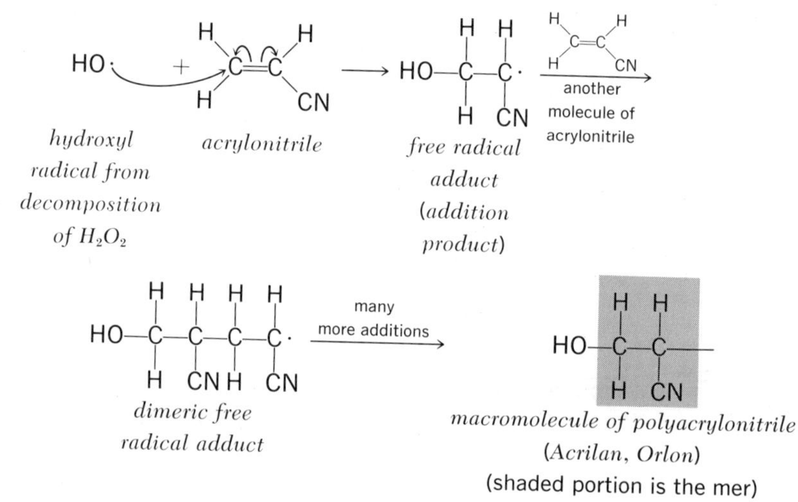

An example of ionic A-polymerization is the use of sodium amide in the production of polyvinylidene nitrile:

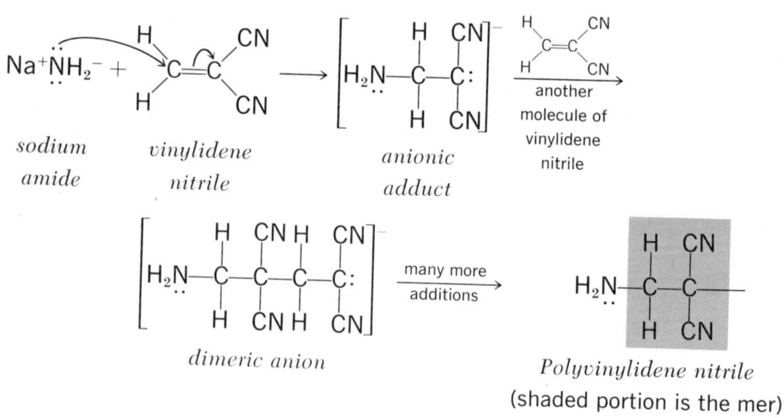

A-polymerization must deal with initiation and propagation reactions, as well as with terminations of the addition reactions by the union of two active species. Referring to the free-radical addition, we may represent these three types of reaction as

(a) Initiation of A-polymerization:

$$\underset{\textit{starter, initiator, or "catalyst"}}{S} \xrightarrow{\text{decomposition}} \underset{\textit{radical}}{2R\cdot}$$

$$R\cdot + \underset{\textit{monomer}}{M} \xrightarrow{\text{addition}} \underset{\textit{monomer radical}}{RM\cdot}$$

(b) Propagation:

$$RM\cdot + M \xrightarrow{\text{propagation}} \underset{\textit{dimer radical}}{RMM\cdot}$$

$$RMM\cdot + M \longrightarrow \underset{\textit{trimer radical}}{RMMM\cdot}$$

$$\text{etc.} \longrightarrow \underset{\textit{polymer radical}}{RM_x\cdot}$$

(c) Termination:

$$RM_x\cdot + R\cdot \xrightarrow{\text{termination}} RM_xR$$

$$RM_x\cdot + RM_y\cdot \longrightarrow \underset{\textit{nonradical polymers}}{RM_{x+y}R}$$

If the rate of termination of the reaction exceeds the rate of initiation, the reaction slows down and polymerization eventually ceases. If the initiation rate is the greater of the two, the overall reaction rate will accelerate and the polymerization may progress too far or yield undesirable products. For best control of polymerization rate, initiator (catalyst) is added just fast enough to maintain a constant concentration of radicals by producing them as fast as they are destroyed.

25.5 □ THE ORIENTATION OF MONOMERIC UNITS IN MACROMOLECULES

In a polymer made up of long, chainlike molecules, molecular motion is curtailed and rotation even around single bonds is restricted. As a result, side groups may be oriented in either an orderly or disorderly

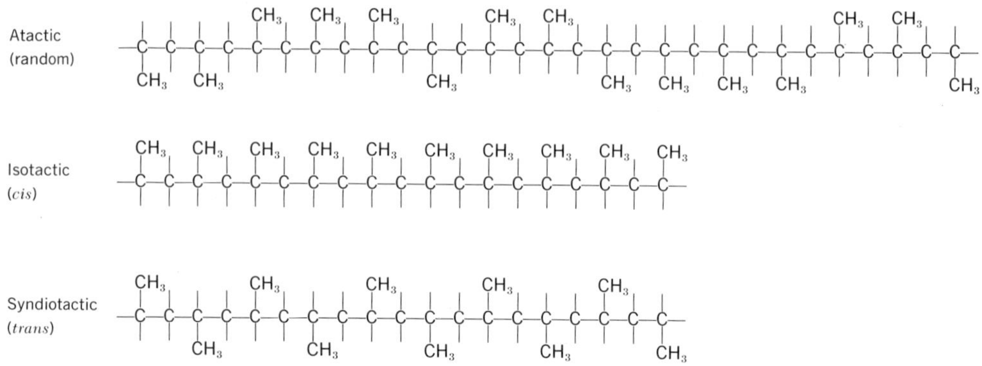

Fig. 25.1 Polypropylene chain arrangements: atactic (random); isotactic (ordered and cis); syndiotactic (ordered and trans).

pattern with respect to the chain. The name **atactic** (Greek, "without order") is used for the random arrangement, **isotactic** for a *cis* arrangement in which all the side groups lie on the same side of the chain, and **syndiotactic** for a *trans* or alternating arrangement of side groups. These are illustrated for polypropylene in Fig. 25.1. Another example of such orientations is afforded by the difference between natural rubber (*isotactic* or *cis*) and gutta-percha (*syndiotactic* or *trans*).

We have seen that molecules must provide two sites for reaction to produce a chain polymer. If more than two sites per molecule are available, then the chain may branch or a polymeric network structure may be formed. Network polymers generally extend in space in all directions and are therefore often called **three-dimensional polymers.** Chains connected to each other by occasional bridges are called **cross-linked polymers.** Short branching groups (as in the rubber or polypropylene chains) are not considered to be chain branches in the sense described above.

An example of a branched modification of a linear polymer is amylopectin, which is a branched starch polymer. The difference between the two is shown in Fig. 25.2. Note that the monomeric units from which the branches originate have formed linkages at three molecular sites.

A typical cross-linking bridge in protein is the diamino diacid cystine

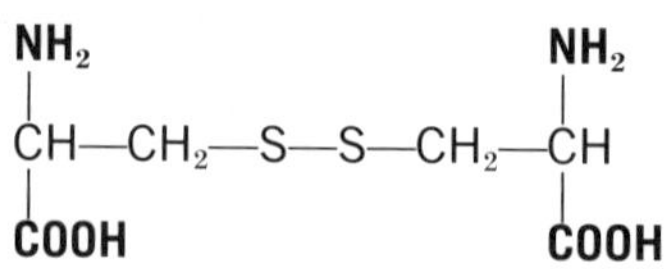

cystine

which has four sites for reaction (shown in bold face) and can serve as a ladder rung between protein chains.

A good example of difference in molecular requirements for linear

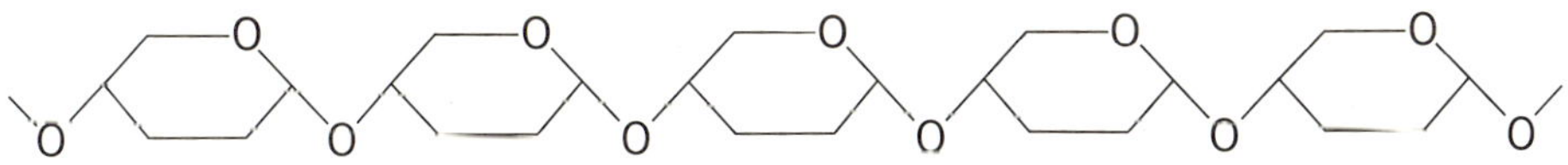

(*a*)

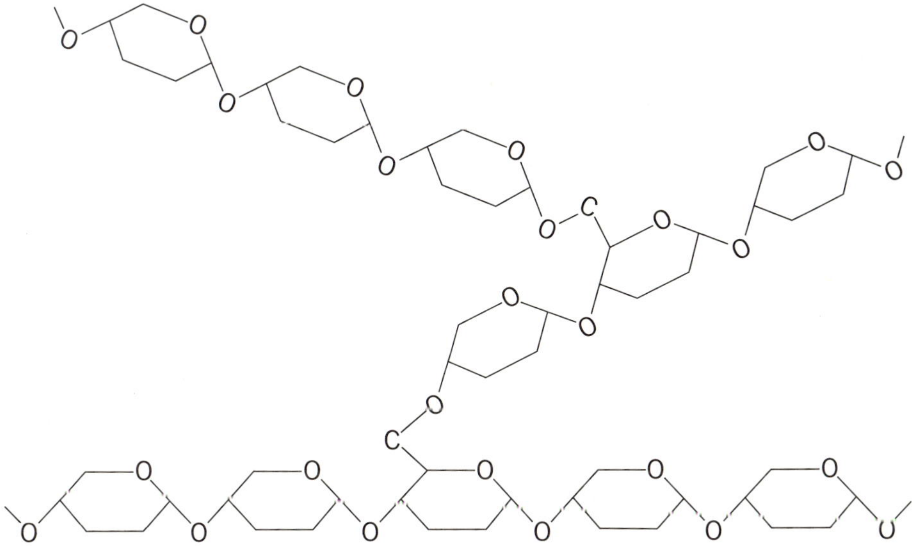

(*b*)

Fig. 25.2 "Straight" (amylose) and branched-chain (amylopectin) starch molecules. (a) Straight-chain amylose starch (structural skeleton only). (b) Branched-chain amylopectin starch (structural skeleton only).

and network polymers is the formation of polyesters from dihydroxy or trihydroxy alcohols. As was shown in Section 25.3 for diamines, reaction of a dialcohol with a diacid gives a linear copolymer. If a trihydroxy alcohol is used instead, a network copolymer is formed. An example is **Glyptal resin** (Fig. 25.3), produced from glycerol and phthalic acid, and used as an ingredient of paints and varnishes.

It is possible, by a suitable choice of catalysts, to control the orientation of monomeric units in A-polymerization that proceeds via ionic intermediates. An example of this control is the improvements in the synthesis of polyethylene ($—CH_2—CH_2—)_n$, one of the most widely used plastics. This material was formerly made by free-radical polymerization at high temperatures and pressures (up to 190°C and 1500 atm). These extreme conditions produced dissociation of atoms from the macromolecular chain with resulting branching from the newly created active sites. In the 1950's Karl Ziegler described the use of hydro- and organoaluminum catalysts like AlH_3, $LiAlH_4$, and AlR_3 (where R is an organic radical) that made it possible to manufacture polyethylene under very mild conditions (atmospheric pressure and 30°–40°C). It is thus possible to produce unbranched macromolecules regularly arranged with respect to each other.

Even more spectacular is the series of stereospecific (spatially controlled) syntheses described by Giulio Natta during the same decade. Natta's catalysts included transition metal halides (for example, $TiCl_4$) and organometallic compounds supported on a solid carrier having a layer crystalline structure. A-polymerization occurs at the surface of this solid ionic catalyst; the macromolecule may be envisaged as "growing out" of the catalyst surface. The monomer, propylene,

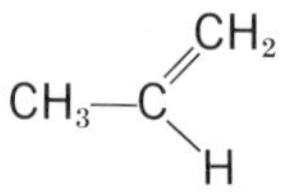

propylene

is not symmetrical with respect to the double bond, because there is a methyl group on only one side. There is therefore a sterically preferred mode of entrance of the propylene molecules to the points on the catalyst from which the macromolecules are emerging. When the catalyst is firmly fixed on the carrier and the temperature is low, this preference produces a regular, long, isotactic molecule. Under less rigidly controlled conditions, the chain alters its pattern from time to time and some degree of irregularity is introduced. When the catalyst carrier is amorphous, lacking entirely the ordered pattern of the

Fig. 25.3 Network copolymer of glycerol and phthalic acid.

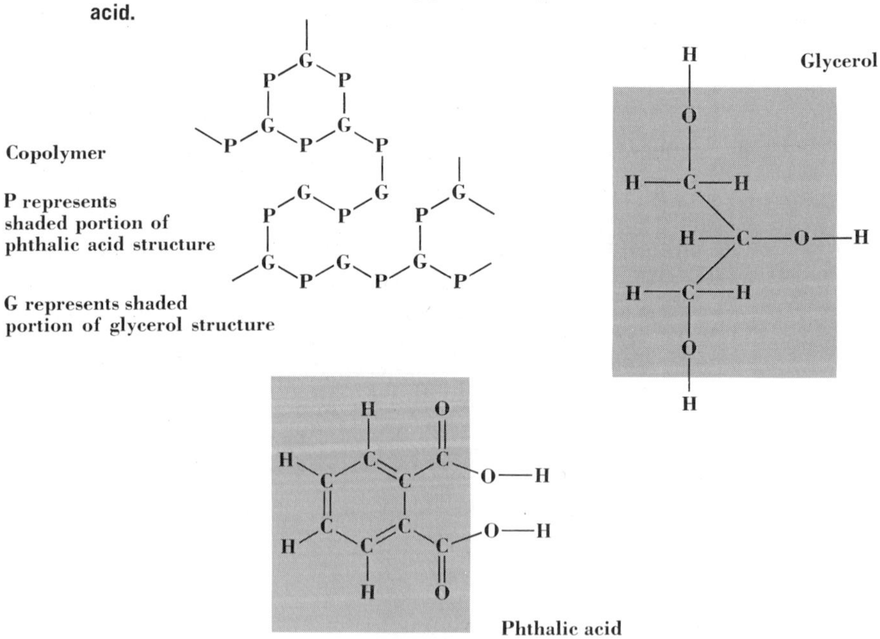

layered crystal structure, disordered atactic macromolecules are formed.

Stereospecific polymerizations have solved some problems of long standing, like the synthetic duplication of natural rubber, and promise much for the future.

25.6 □ CRYSTALLINITY OF POLYMERS

The degree to which molecules of a substance are arranged in orderly pattern with respect to each other is, of course, a measure of crystallinity. So it is also with macromolecules. Longer molecules have more geometrical opportunities for partially crystalline arrangements, however, than shorter ones have. Consider as examples (a) a pile of neatly stacked lengths of flexible rods, and (b) a similarly stacked pile of short rods of the same material. If each pile is disturbed at the same rate by

Fig. 25.4 Crystallinity in chain polymers: (a) unoriented crystallites; (b) oriented crystallites; (c) parallel crystalline areas.

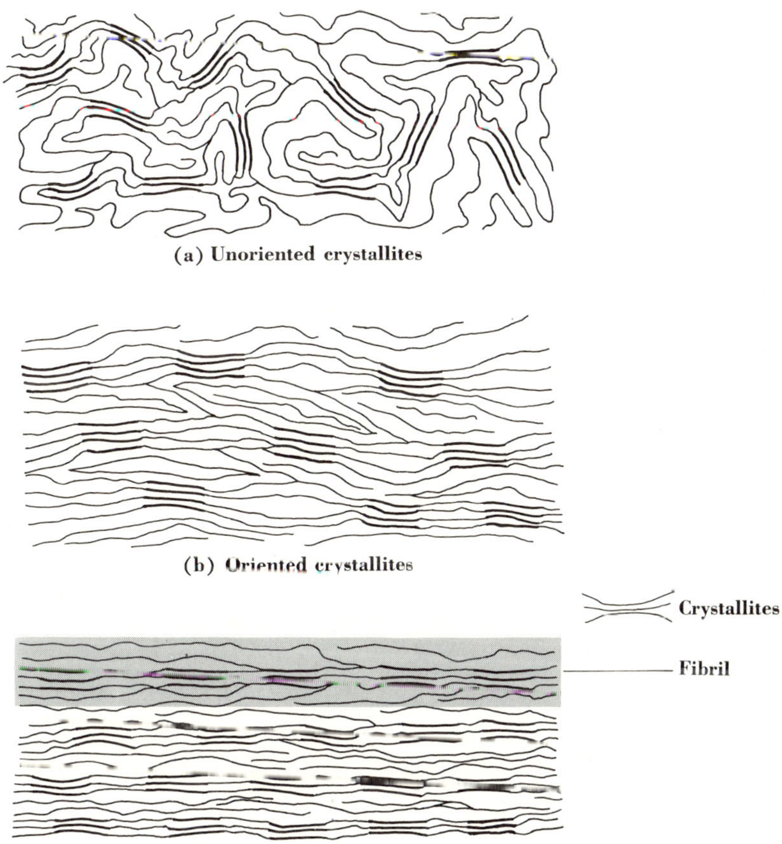

(a) Unoriented crystallites

(b) Oriented crystallites

(c) Parallel crystalline areas

the same amount of random displacement in space, the pile of long rods will become disordered (lose "crystallinity") much more slowly than the pile of short ones. This example is given to illustrate that the distinction between "crystalline" and "noncrystalline" is much less sharp for macromolecular material than for substances of low molecular weights. A small region of a macromolecular material in which portions of large molecules are linked to each other in some regular way is called a **crystallite.** It is the orderly orientation of crystallites with respect to each other that builds up crystallinity in a polymer. In fibrous polymeric structures, the small aggregates of crystallites just large enough to appear as a small threadlike unit under a high-power microscope are called **fibrils.** Figure 25.4 shows sketches of (a) unoriented crystallites in a polymer that is otherwise amorphous; (b) oriented crystallites; and (c) long parallel crystalline areas separated by amorphous areas.

25.7 ▫ IONIC POLYMERS

A number of inorganic oxyanions and oxycations form polymers in which oxygen atoms alternate with atoms of another element. The ligancy of oxygen is generally 2; that of the other element may be 2 or more. Cross-linking may lead to two- or three-dimensional network structures.

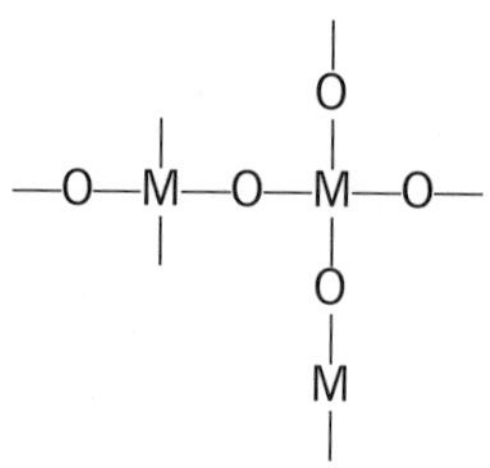

An example of a group of oxyanion polymers is the class of complex

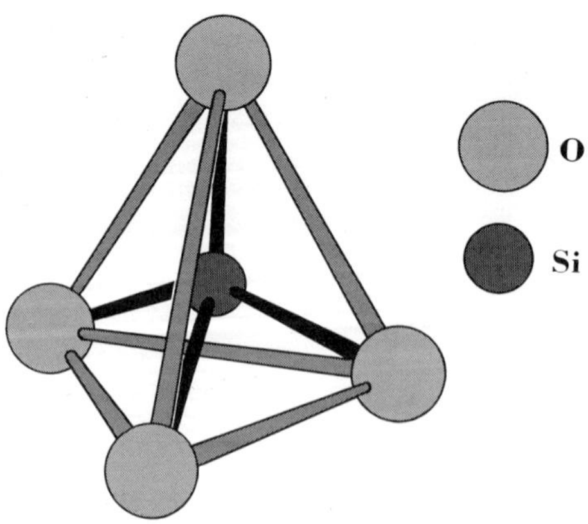

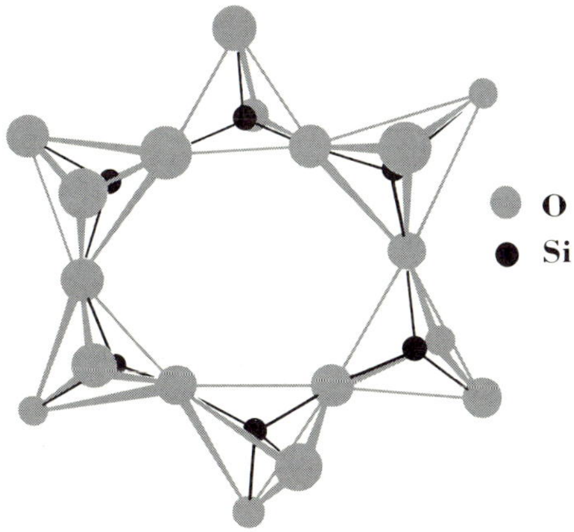

$Si_6O_{18}^{12-}$

silicates. The monomeric unit is orthosilicate ion, SiO_4^{4-}, which has a tetrahedral structure. An example of a cyclic silicate condensation polymer derived from six monomeric units is the $Si_6O_{18}^{12-}$ ion, which occurs in the mineral **beryl,** $Be_2Al_2Si_6O_{18}$. Beryl crystals are semiprecious stones; **emerald** and **aquamarine** are varieties of beryl to which traces of impurities impart characteristic tints.

Polyanions derived from chromates, phosphates, molybdates, sulfates, and tungstates are also known. (See Chapter 21.)

Inorganic polycations are generally condensation polymers of the monomeric hydrated cations. Such a monomer is hydrated ferric ion, $Fe(H_2O)_6^{3+}$, whose systematic name is hexaaquoferrate(III) ion, and whose shape is octahedral. This ion dimerizes with elimination of H^+ to form the

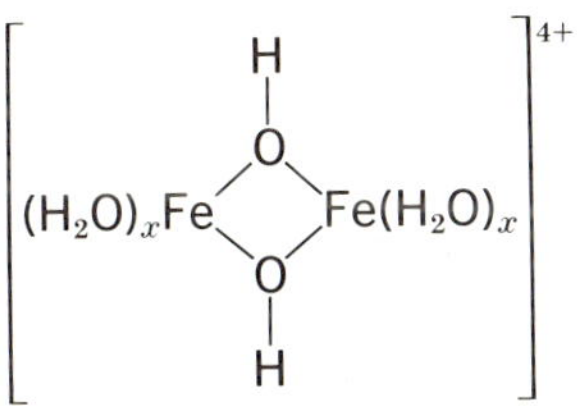

ion. More extensive condensation produces the polymer known as **hydrated ferric oxide,** $Fe_2O_3(H_2O)_x$.

Monomeric ions in an ionic polymer are relatively mobile. This feature is used to advantage in **ion-exchange resins,** as discussed on p. 323 and illustrated in Fig. 17.1.

25.8 ▫ PROPERTIES OF POLYMERS

Polymeric substances display wide varieties of physical and chemical properties; this is to be expected from the many possible kinds of macromolecular composition and arrangement. The broad range of fibrous, adhesive, plastic, filmy, foamy, rubbery materials so readily available to us as useful products attests to this macromolecular versatility. What generalizations can be made regarding the relationships between properties and makeup? Although the problems are complicated, considerable advances have been made in recent years.

Quite obviously, the chemical nature of the monomeric unit itself is a prime determinant of properties. Differences in thermal stability and mechanical strength of different polymers are related in part to differences in bonding and structure of the monomer. The chemical reactivity of a polymer is, in large measure, the reactivity of its molecular components. Natural rubber, for example, undergoes deterioration when ozone attacks the double bonds of the polymer chain; a saturated hydrocarbon chain, like polyethylene, resists such attack. Cellulosic polymers offer their free hydroxyl groups to a variety of reagents and thus make it possible to introduce useful modifications of properties. Reaction with nitric acid produces **nitrocellulose,** from which propellant (guncotton) and plastic (Celluloid) products are formed. Reaction with a derivative of acetic acid produces **cellulose acetate,** which can be fabricated into films, sheets, and other useful forms. Both reactions alter the free hydroxyl group of cellulose:

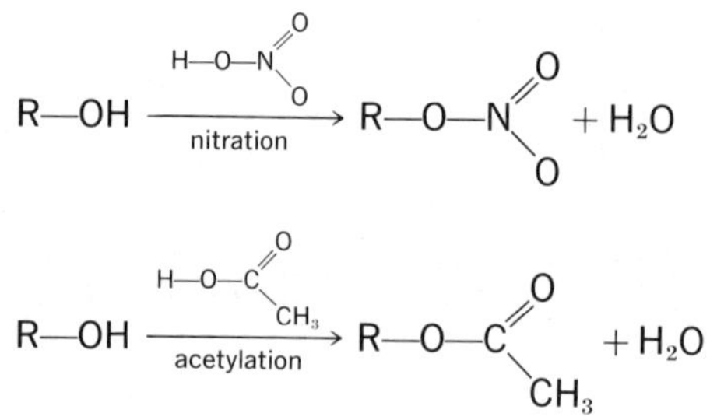

The possibilities for chemical modification and fabrication into products are very extensive.

More drastic chemical differences yield wider differences in chemical properties. Especially notable is the family of silicone polymers, in which the macromolecular chains contain —Si—O— linkages (Fig. 25.5). The great thermal stability of the O—Si bond, due in part to the p-$d\pi$ bonding* of O to Si, makes it possible to use silicone products at

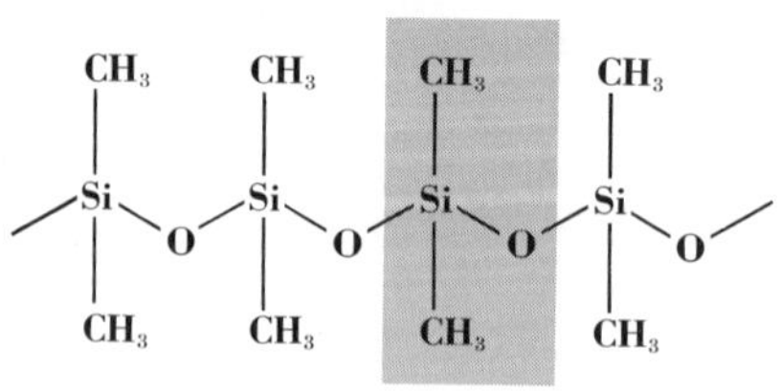

Fig. 25.5 Methylsilicone polymer. The shaded portion is the mer.

* This refers to π-bonding between p and d orbitals.

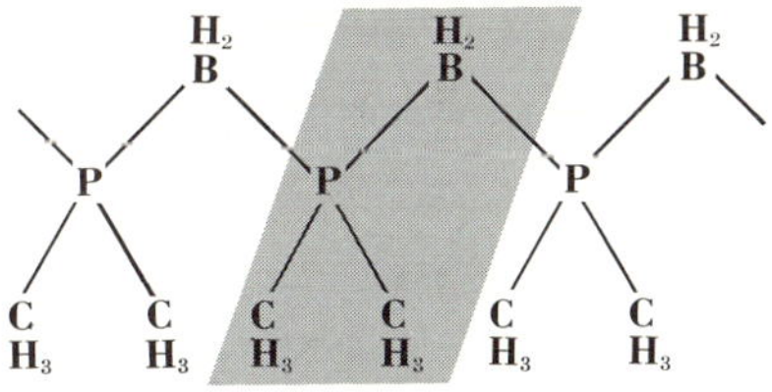

Borophane

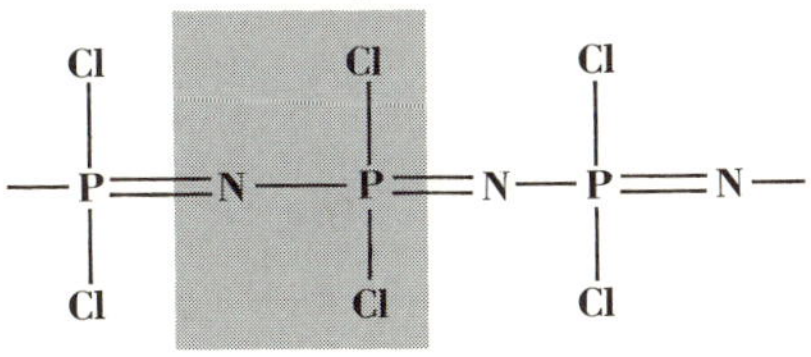

Polydichlorophosphonitrile

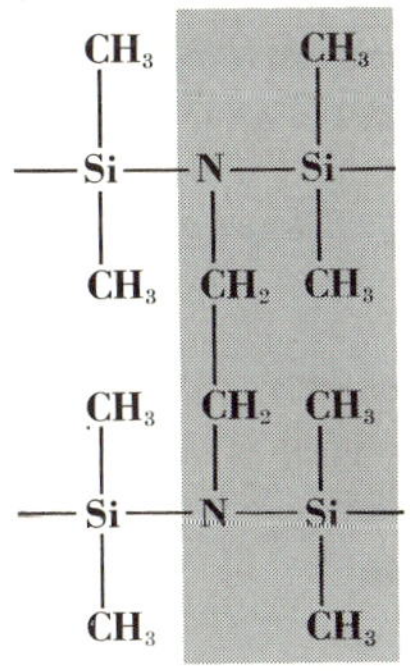

Silazane ladder polymer

Fig. 25.6 Some inorganic polymers: borophane; polydichlorophosphonitrile; silazane ladder polymer; two-dimensional boron nitride polymer. *The shaded portions are the mers.*

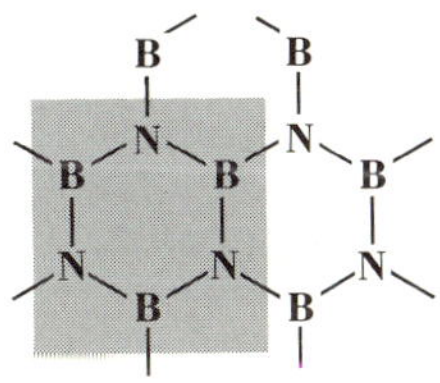

Two dimensional boron nitride polymer (The three-dimensional form, Borazon, has a diamond-like structure)

high temperatures. The hydrocarbon side chains contribute oily or lubricating properties. There have been, of course, attempts to extend the range of "inorganic" polymers by using other varieties of linkage for the chain backbones. Potential candidates include boron-carbon, boron-oxygen, arsenic-oxygen, and beryllium-oxygen. The problems are difficult, in part because our understanding of inorganic linkages, which may involve d-orbital π bonding and other less well-recognized interactions, lags behind our knowledge of the simpler σ bonds and σ-π multiple bonds typical of simple organic molecules. Another difficulty is the annoying (to the polymer chemist) tendency of inorganic systems to cyclize in units of relatively low molecular weight. Diminution of the tendency to bond multiplicity with increasing atomic number impedes the use of the valuable A-polymerization techniques used with organic monomers like ethylene. Despite such difficulties, however, many inorganic polymers have been prepared (see Fig. 25.6), and this field of endeavor seems particularly promising.

The properties of a polymer are also determined by the form and arrangement of its macromolecules. The critical factors are the molecular weight (which is a function of the degree of polymerization), the extent of branching, cross-linking, or network structuring, the steric disposition of the monomeric units, and the degree and kind of crystallinity of the macromolecules. Certainly there is room enough for variation even without alteration of chemical functional groups. Some of these relationships are approximated in Table 25.2.

TABLE 25.2 PROPERTIES AND MOLECULAR MAKEUP OF POLYMERS

Physical nature of polymer	Molecular requirements
Hard	High molecular weight High crystallinity Cross-linking or network structure *Example:* phenol-formaldehyde copolymer (Bakelite)
Strong (cannot easily be pulled apart)	High molecular weight Cross-linking or network structure Partly crystalline—crystallites embedded in amorphous matrix that acts like a cement *Example:* polymethylmethacrylate (Lucite, Plexiglas) reinforced with glass fiber
Fibrous	High molecular weight Linear macromolecules Long parallel arrangements of crystalline and amorphous regions *Example:* polyacrylonitrile (Acrilan, Orlon)
Leathery	High molecular weight Linear macromolecules with slight degree of cross-linking; fragments of the chains are free to move under stress Low crystallinity *Example:* vinyl chloride–vinyl acetate copolymer (vinyl floor covering)
Rubbery	Linear macromolecules with little cross-linking; entire chains are free to move under stress High molecular weight Low crystallinity (but crystallinity increases with elongation) *Example:* polybutadiene (Buna rubber)
Soft, waxy	Low molecular weight ($<10{,}000$) Low crystallinity *Example:* polyvinyl acetate chewing gum
Thermoplastic (able to soften and assume new shapes by application of heat and pressure)	Linear macromolecules Little cross-linking Relatively low molecular weight *Example:* cellulose acetate
Thermosetting (not able to melt or flow without decomposition)	High molecular weight Highly cross-linked or network structure *Example:* urea-formaldehyde resin

Stereoregularity (orderly arrangement in space) facilitates crystallinity and enhances those properties for which crystallinity is required. Thus, for example, stereoregularity strengthens polyethylene.

Shifts in properties also accompany the changes in molecular arrangements produced by a rise in temperature. A polymer of low crystallinity may thus progress, on heating, from a rigid glassy state, through a partly flexible leathery condition, to a rubbery condition, and finally to a flowing viscous liquid. Mechanical deformation also changes the properties of a polymer. Perhaps the most striking instance is the stretching of rubber. In the unstretched form rubber molecules undergo random motion; on stretching this motion is restricted, the entropy is reduced, and the molecules assume a linear crystalline arrangement. The consequent release of energy is familiar to anyone who has stretched a wide rubber band, touched it immediately to his lips, and felt its sudden warmth.

PROBLEMS

1. Definitions Define and illustrate the following terms: (a) monomer; (b) polymer; (c) copolymer; (d) tetramer; (e) A-polymer; (f) C-polymer.

2. Monomeric units Neoprene, a synthetic rubber introduced commercially in 1931, is able to withstand ozone and oils more effectively than natural rubber. It is made from "chloroprene,"

$$CH_2{=}CCl{-}CH{=}CH_2$$

and is known to have the following structure:

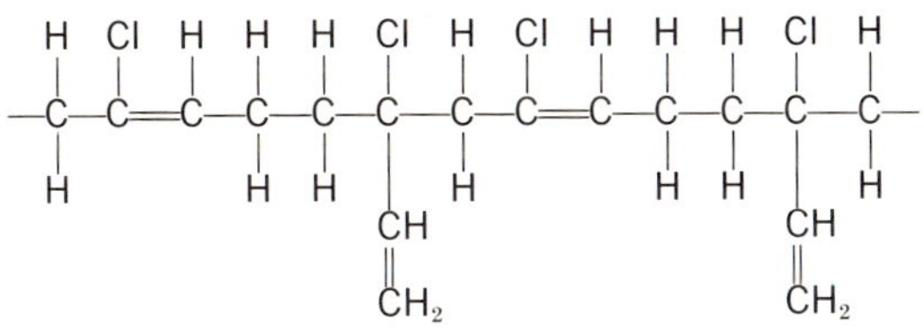

(a) Write the structures of all the geometrically possible recurring units. (b) Write the structure of the actual monomeric unit (mer) of the rubber.

3. Molecular weight Using electrical resistance thermometry, it is possible to detect temperature differences as small as about 10^{-5} Celsius degrees. Assume you have a 1% aqueous solution of a polymer of molecular weight about 100,000. Could you detect the effect of this solute on the freezing point of the solution? Support your answer by calculations (see p. 236).

4. Orientation of monomeric units Classify the monomer sequences in natural rubber and gutta-percha as atactic, isotactic, or syndiotactic. Explain your answer.

5. Polymerization Early attempts at polymerization of isoprene failed to produce a product with the properties of natural rubber. Write the structural formula of an isoprene polymer that might be produced by atactic polymerization.

6. Orientation of monomeric units Referring to Table 25.1 and Fig. 25.2, do starch and cellulose differ from each other in the steric configuration of their monomeric units, in the orientation of monomeric units with respect to each other, or both? Explain your answer.

7. Molecular sieves The Si—O bond distance is about 1.8 Å. (a) Estimate the diameter of the "hole" in the $Si_6O_{18}^{12-}$ ion. (b) Do you think it might be possible to prepare silicate polymers that could separate different substances on the basis of the sizes of their molecules? Defend your answer.

8. Network structure Draw a section of the glycerol-phthalic acid network copolymer of Fig. 25.3, showing all the atoms in each monomeric unit in place of the symbols P and G.

9. Properties of polymers In nitration and acetylation of cellulose the free hydroxyl groups are converted to —ONO_2 and —$OCOCH_3$ groups, respectively. Draw the structures of the monomeric units of cellulose nitrate and cellulose acetate.

10. Properties of polymers When polyethylene is subjected to intense ionizing radiation from a nuclear pile, it becomes much stronger, more difficult to melt, and less soluble. Account for these effects.

ADDITIONAL PROBLEMS

11. Polybutadiene The structural formula of butadiene is

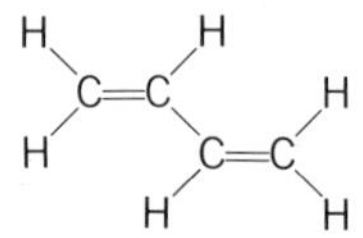

(a) Draw the structural formula of *cis*-polybutadiene and of *trans*-polybutadiene, assuming in each case that the carbon skeleton is unbranched. Show which portion of each formula is the mer. (b) Under the influence of ultraviolet irradiation, an added "sensitizer" such as phenyl disulfide

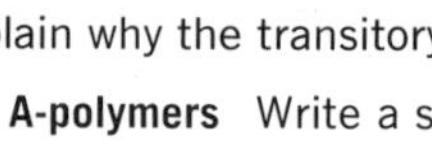

is converted into radicals which can add to the double bonds in the polymer to form transitory radical structures that are freely rotating. When the sensitizer radicals subsequently become detached, the double bonds reform. Under these conditions, irradiation of *cis*-polybutadiene converts it to a mixture of the *cis* and *trans* forms. Show these transformations with the aid of structural formulas. Explain why the transitory radical derivative of polybutadiene is freely rotating.

12. A-polymers Write a structural formula for an A-polymer formed from each of the following monomeric units. Indicate which portion of each formula is the mer.

(a) 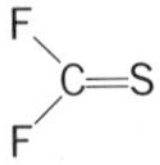

(b) 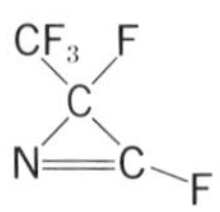

(c) Copolymer of 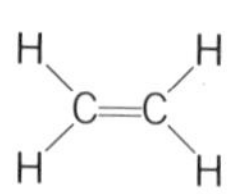 and 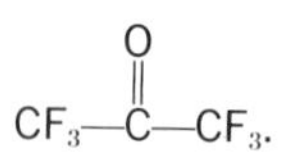

13. C-polymers Write a structural formula for a C-polymer formed from each of the following monomeric units. Indicate which portion of each formula is the mer.

(a) Copolymer of 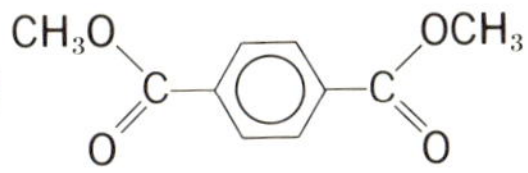 (*methyl terephthalate*)

with H—O—CH_2—CH_2—OH (*ethylene glycol*)

with elimination of methyl alcohol, CH_3OH, to form the polyester **Dacron.**

(b) Copolymer of

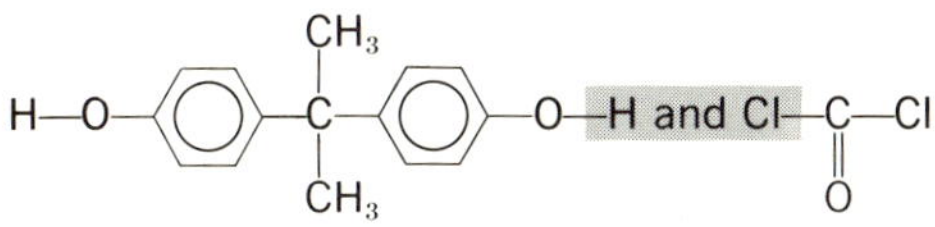

a phenol derivative *phosgene*

with elimination of HCl as shown to form **polycarbonate plastic.**

26 ■ BIOCHEMISTRY

26.1 □ INTRODUCTION

Biochemistry is the chemistry of organisms. An **organism** is anything that is alive (a "living organism"), or if not, was once alive (a "dead organism"). What, then, is the condition we call **life?** We cannot offer a rigid, precise definition, but we do know that living things are characterized by metabolism, growth, and reproduction. **Metabolism** is the process by which a body introduces into itself ("ingests") various energy-rich materials from its environment ("food"), and transforms these materials, with the release of energy, into other substances, some of which are retained by the body ("growth" or "repair") and some eliminated. **Reproduction** is the process by which one body produces another that is like itself in properties, structure, composition, and function, including metabolism and reproduction.

The distinction between an organism and a material is not always clear. A **virus** (Fig. 26.1) consists of particles several hundred Ångstrom units in length or diameter; these particles can reproduce themselves in a suitable environment but they do not ingest food, or grow, or carry on any other metabolic processes. Are viruses, then, living organisms, or are they chemical materials that consist of large molecules capable of replicating themselves under suitable conditions? To include viruses among the living the definition of life must be modified. Most broadly, we may consider anything living if it can bring order out of disorder at the expense of energy and has the capability to preserve accidental variations (called **mutations**) that may occur in the process.

Fig. 26.1 Crystalline tobacco necrosis virus, magnification 160,000 times (*J. Ultrastructure Research*, Vol. 2, page 8, 1958; courtesy of Dr. R. W. G. Wyckoff).

26.2 □ THE ROLE OF THE CELL

When a chemist studies a chemical process, he usually confines his reacting system in a container. The container and its contents, taken together, are sometimes called a **reactor.** Containers include test tubes, beakers, flasks, stainless steel kettles, and the like. The container influences the reaction by restricting the reactants to a confined space and thus maintaining some specified concentration; orifices in the container control the flow of materials into or out of the reacting system; the container material may act as a catalyst; the container walls may provide surface at which the reactants are adsorbed and concentrated or at which the progress of a reaction chain may be terminated. Of course, some reactions, such as those occurring in the atmosphere, may be considered to be unconfined, although many chemists broaden the concept of "reactor" to include an entity like the Los Angeles basin during an atmospheric stagnation.

In an organism, the structure called the **cell** may be considered to be a biochemical reactor. An organism consists of one or more cells, and the various groups of cells in a multicellular organism may be sharply differentiated as to biochemical function. The reactions in the cell are said to occur *in vivo* (Latin, "in the living organism"); the corresponding reactions outside of the cell are said to occur *in vitro* (Latin, "in glass").

The living cell is not merely a tiny membranous beaker with homogeneous contents. It is, rather, an entity of great complexity, not yet completely understood as to structure and function. There are specific sites within the cell at which specific reacting systems, metabolic or reproductive, operate. The biochemist seeks to identify these sites, and to illuminate the course and mechanism of the reactions that occur there. Sometimes he tries to remove a chemically reacting system from its cellular environment and duplicate it *in vitro*. He does this because reactions are usually easier to study under the more controllable conditions of laboratory reactors than they are *in vivo*.

26.3 □ BIOCHEMICALS

Before we examine some of the various functions of the cell, it will be helpful to describe some classes of substances of biochemical importance. This section is therefore a sort of biochemical microhandbook.

(a) **Proteins** (Greek, *proteos*, "primary") are condensation polymers of amino acids, and have molecular weights in the range of 6000 to 500,000.

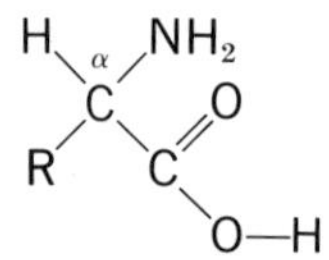

General formula for an **α-amino acid.** R may be hydrogen, an alkyl group, or some other organic group. The α-carbon atom refers to the one adjacent to the COOH group. When all four groups attached to the α carbon are different, the molecule lacks a point or plane of symmetry and is therefore dissymmetric. See Table 26.1.

The C-polymerization of two or more molecules of an amino acid involves the formation of **peptide linkages** in which one molecule of water is eliminated between one NH_2 and one COOH group. (See also Table 25.1, p. 498, showing the degradation of silk.)

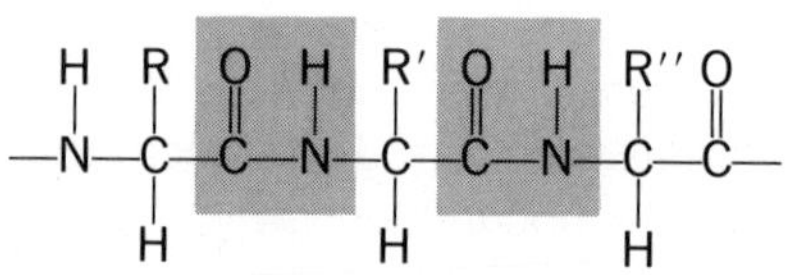

Shaded portions are peptide linkages. The C—N bond in the peptide linkage is called a **peptide bond.**

Practically all of the dissymmetric amino acid portions of protein in living organisms on earth, so far as we know, have L-configurations. (This means that their configurations are "left-handed" by comparison with an arbitrary reference standard, D-glyceraldehyde, which is "right-handed.") We may rationalize, on the basis of what we know about biochemical structures and reactions, that it is advantageous for the amino acids of organisms to be all of the same configuration, just as it would be convenient for manufacturers of scissors, baseball gloves, and other dissymmetric objects if all people had the same handedness. We do not know, however, why our amino acids are L rather than D, nor what amino acid configurations (if any) may exist in extraterrestrial organisms.

A protein may be characterized according to its *primary* and *sec-*

ondary structures. The "primary structure" of a protein is that which can be depicted by the classical structural formula of the organic chemist. Such a formula shows the sequence of atomic linkages in the molecule, without regard to conformation (p. 437) or to relatively weak linkages such as hydrogen bonding. To elucidate the primary structure of any macromolecule is a task that, until about the 1940's, was considered by most chemists to be hopelessly difficult. The first formulation of this kind was reported in 1955 by Frederick Sanger for the protein **insulin;** he was awarded the Nobel Prize in 1958. This accomplishment was facilitated by (a) highly improved methods for the rapid analysis of individual amino acids, and (b) an ingenious scheme for determining the sequence of amino acid linkages by breaking down

TABLE 26.1 ▪ COMMON AMINO ACIDS FROM PROTEINS

Name	Symbol	Formula
Glycine	Gly	H_2NCH_2COOH
Alanine	Ala	$CH_3{-}CH(NH_2)COOH$
Valine	Val	$(CH_3)_2CHCH(NH_2)COOH$
Leucine	Leu	$(CH_3)_2CHCH_2CH(NH_2)COOH$
Isoleucine	Ileu	$CH_3CH_2CH(CH_3){-}CH(NH_2)COOH$
Proline	Pro	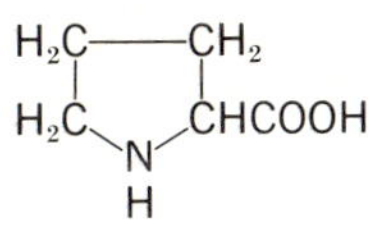
Phenylalanine	Phe	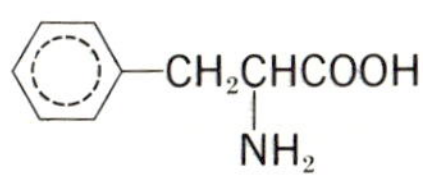
Tyrosine	Tyr	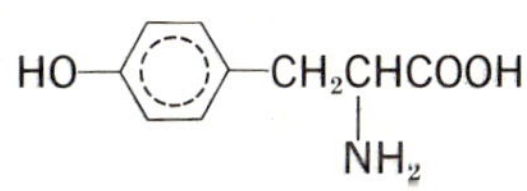
Cysteine	CySH	$HSCH_2CH(NH_2)COOH$
Cystine	CyS—SCy	$HOOCCH(NH_2)CH_2SSCH_2CH(NH_2)COOH$
Serine	Ser	$HOCH_2CH(NH_2)COOH$

TABLE 26.1 (Continued)

Name	Symbol	Formula
Threonine	Thr	$CH_3CH(OH)CH(NH_2)COOH$
Histidine	His	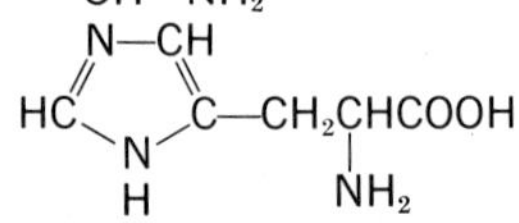
Aspartic acid	Asp	$HOOCCH_2CH(NH_2)COOH$
Glutamic acid	Glu	$HOOCCH_2CH_2CH(NH_2)COOH$
Lysine	Lys	$H_2NCH_2CH_2CH_2CH_2CH(NH_2)COOH$
Arginine	Arg	$H_2NC(=NH)NHCH_2CH_2CH_2CH(NH_2)COOH$
Tryptophan	Try	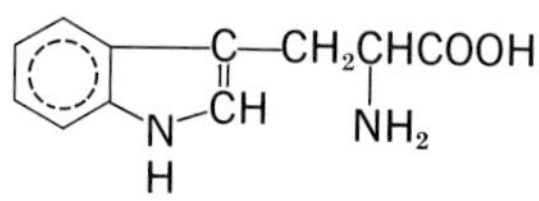
Methionine	Met	$CH_3SCH_2CH_2CH(NH_2)COOH$

the protein into fragments containing several amino acid units, repeating this a number of times so as to obtain different fragmentations, and calculating what one sequence of amino acids in the original protein could account completely for the observed breakdown products. Figure 26.2 depicts the primary structure of insulin.

The secondary structure of a protein is determined by the spatial arrangement of the amino acid chain. Evidence obtained mainly from x-ray diffraction patterns (Linus Pauling, 1951, and others) has shown that the chain is typically wound into a helix containing about 3.6 amino acid portions per turn of the helix (Fig. 26.3a). Each amino acid portion is linked to an amino acid portion in each of the adjacent turns of the helix by a hydrogen bond. This bonding takes the form shown in

Fig. 26.2 Primary structure of insulin.

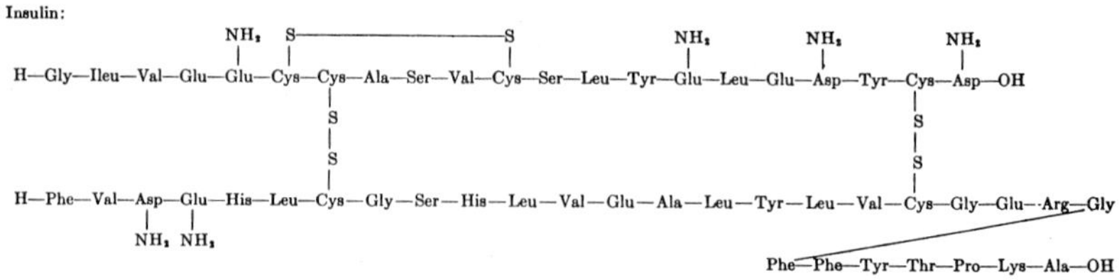

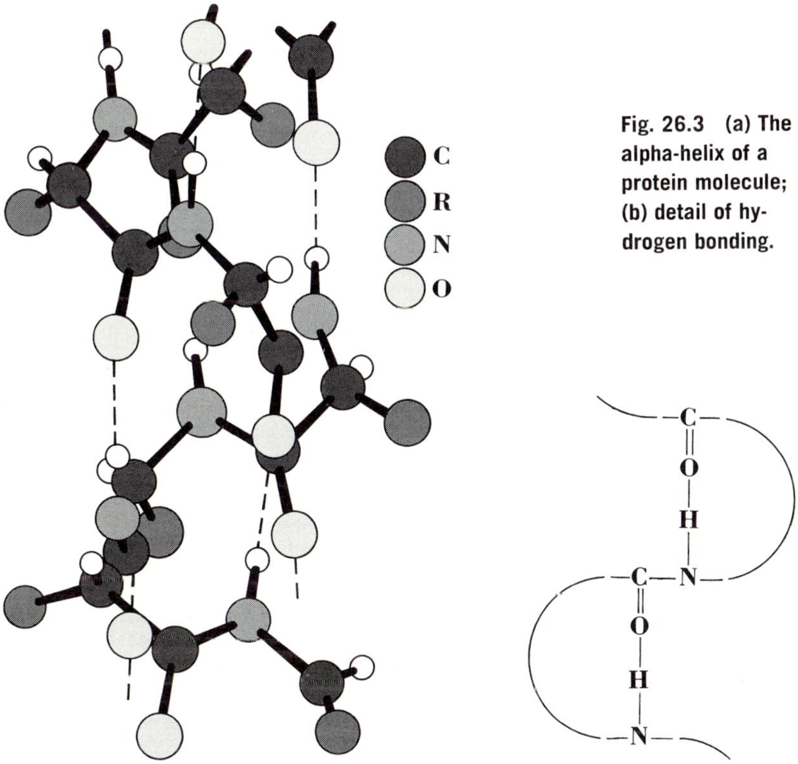

Fig. 26.3 (a) The alpha-helix of a protein molecule; (b) detail of hydrogen bonding.

Fig. 26.4 Mode of enzyme action (Schematic).

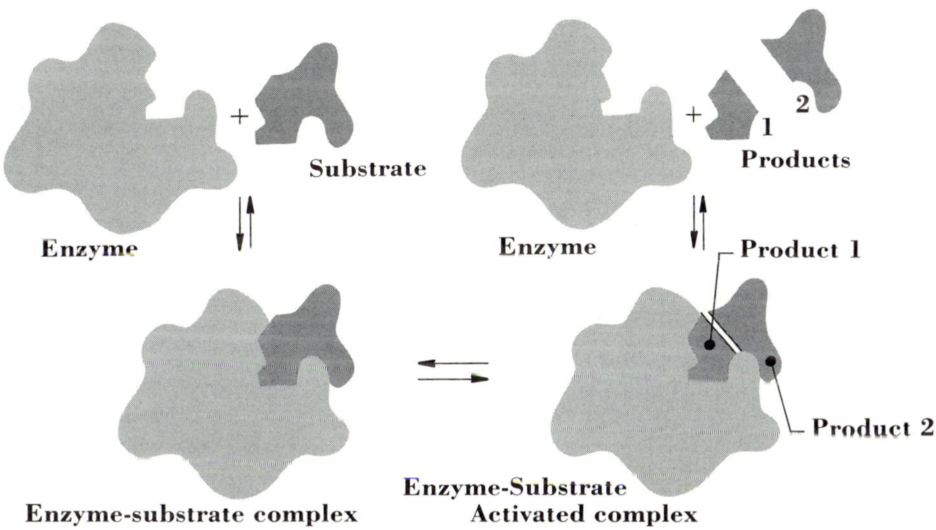

Fig. 26.3b. The entire structure is called the **alpha-helix.*** Other secondary structures of proteins include pleated sheets and random coils.

Proteins serve both structural and catalytic functions in biochemical systems. Proteins that are catalysts in biochemical reactions are called **enzymes.** They are characteristically very highly specific, each enzyme being capable of catalyzing only a particular reaction of a particular substance, called the **substrate.** This high specificity implies that steric (spatial) effects, in addition to interactions of functional groups, are critical in enzyme action, because there are many more possible shapes of molecules than there are different types of chemical bonds. Figure 26.4 shows a model of an enzyme-catalyzed decomposition of a substrate. In this sequence, the enzyme shape just accommodates the substrate molecule, and forms a complex in which chemical bonds within the substrate are weakened and decomposition occurs more rapidly (or at much lower temperature) than it would without the enzyme.

(b) **Nucleic acids** are substances which occur in the nuclei of cells

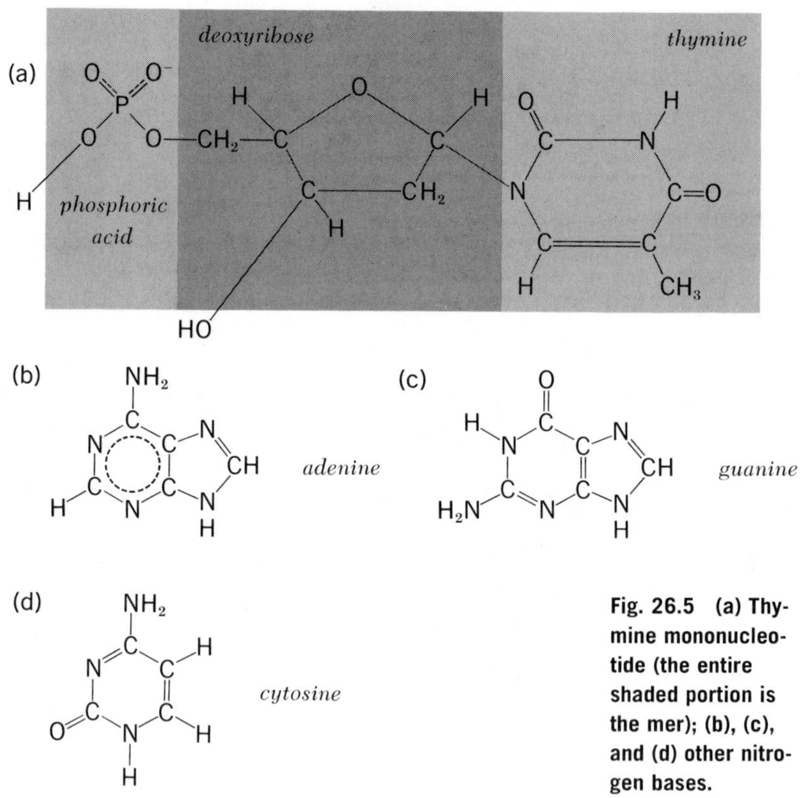

Fig. 26.5 (a) Thymine mononucleotide (the entire shaded portion is the mer); (b), (c), and (d) other nitrogen bases.

* The "alpha" refers to the alpha carbon atom, to which are attached the groups that are linked by H bonds.

and are the determinants of protein synthesis and reproduction. They also occur in bacterial cells (which have no nuclei) and in viruses (which have no cells), and are thus the genetic determinants of all living organisms. **Deoxyribonucleic acid (DNA)** is a polymer containing 2500–15,000 monomeric units called **mononucleotides.** Each mononucleotide contains a phosphoric acid portion, a carbohydrate-like portion (deoxyribose), and one of four possible organic bases. The four bases, all nitrogen compounds, are **adenine, cytosine, guanine,** and **thymine.** Formulas appear in Fig. 26.5.

The C-polymerization of mononucleotides involves elimination of water to form the bond.

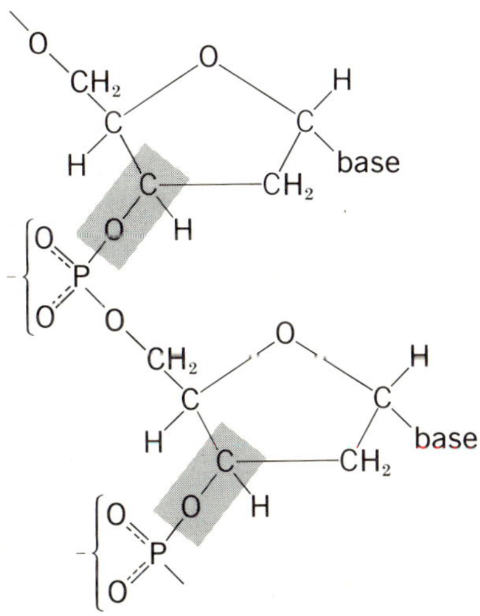

Shaded portions are the linkages formed by elimination of H_2O

The secondary structure of the DNA macromolecule is that of a double-stranded helix which is held in this configuration partially by hydrogen bonds between pairs of nitrogen bases. This model, proposed in 1953 by James D. Watson and Francis H. C. Crick, is illustrated in Fig. 26.6.

Another nucleic acid is **ribonucleic acid (RNA),** which is similar to DNA except that the carbohydrate portion is ribose instead of deoxyribose. The difference between the two is an atom of oxygen.

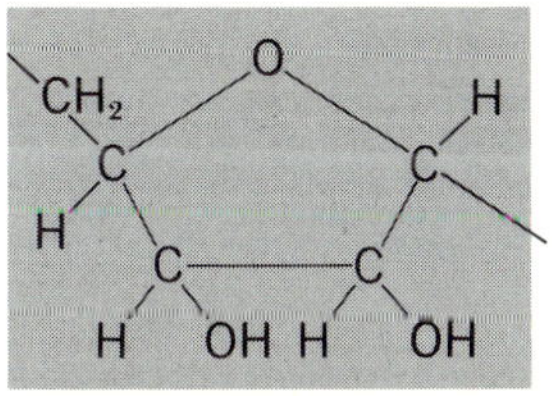

ribose portion of RNA

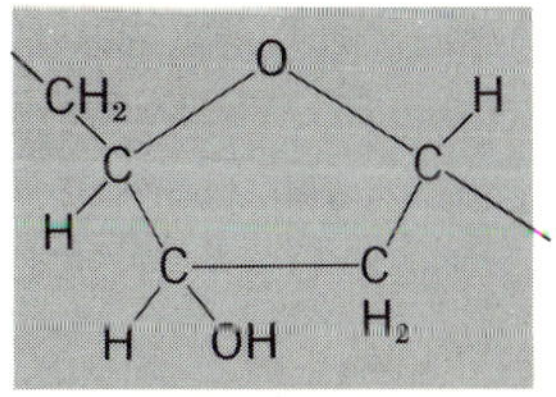

deoxyribose portion of DNA

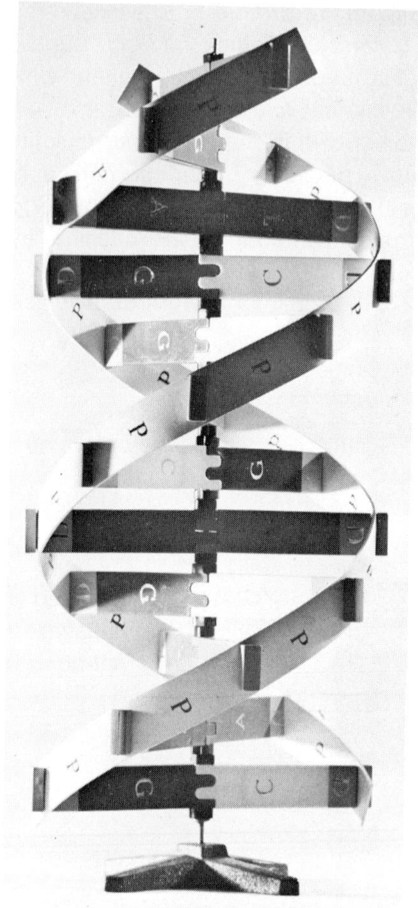

Fig. 26.6 DNA helix (model). Key: P = phosphate; D = deoxyribose; A = adenine; G = guanine; T = thymine; C = cytosine. (Photo by Fundamental Photographs, New York.)

(c) **Porphyrins** are particularly important biochemical structures in which four pyrrole rings

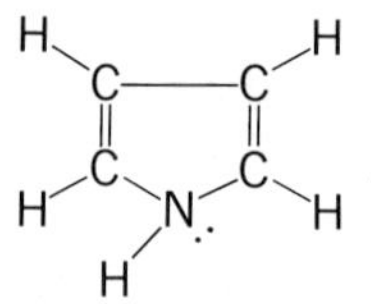

pyrrole

are linked together through 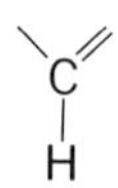 bridges to form a planar ring system with extended π bonding. The center of the system can accommodate a metal atom or ion that is linked to the four nitrogen atoms in a square-planar complex. The porphyrin structure with a central metal atom, M, is shown in Fig. 26.7; the R groups may be H, alkyl, or some other

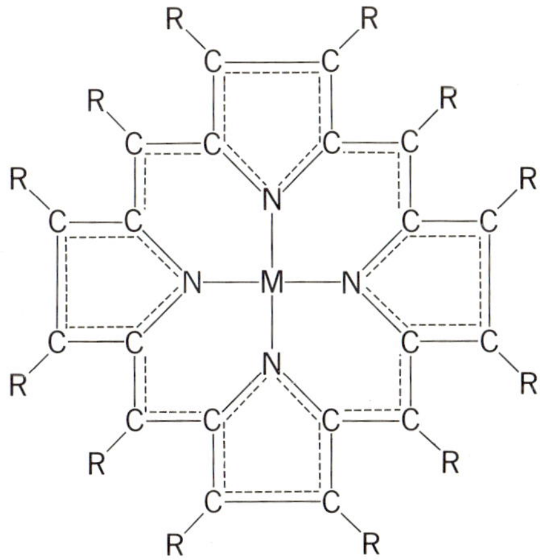

Fig. 26.7 Porphyrin structure.

derivative. The metal, M, is the site for important reversible biochemical redox reactions. **Chlorophyll** consists of porphyrin molecules in which the central metal atom is magnesium. The series of redox reactions in which it is involved is **photosynthesis** (p. 525). The hemoglobin of our blood consists of protein molecules linked to **heme** groups; heme is a porphyrin in which the central atom is iron. The sequence of redox reactions in which it plays an essential role is **respiration.**

(d) **Adenosine triphosphate, ATP,** is of great biochemical importance because it undergoes a reversible reaction which serves to store and transfer energy in many biochemical processes.

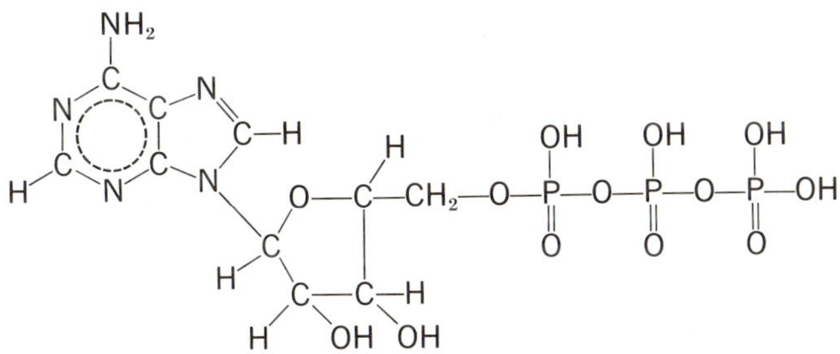

adenosine triphosphate (ATP)

The reaction is an exothermic decomposition to adenosine diphosphate (ADP) and phosphoric acid.

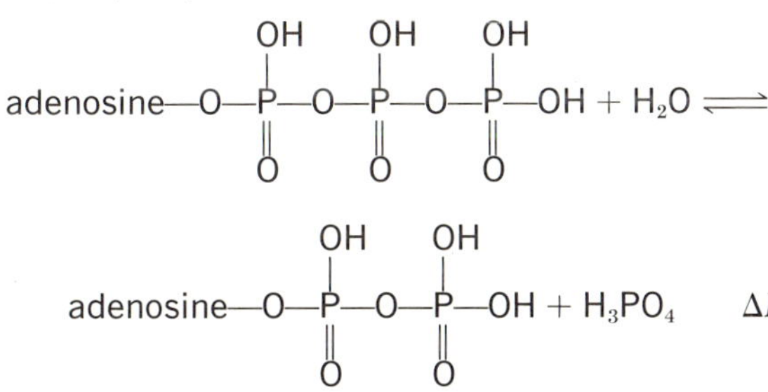

$\Delta H = -8$ kcal/mole

26.4 □ BIOCHEMICAL PROCESSES IN THE CELL

We return now to a more detailed consideration of the cell and of some of the reaction sequences which occur at various sites within it. Figure 26.8 is a general representation of a cell of a multicellular organism. Several of the indicated anatomical features are so small that they can be revealed only with the aid of an electron microscope. Some of these fine structures of the cell are nonessential inclusions, like blobs of fat, or particles of starch. Others, called **organelles,** perform essential functions and are reproduced when the cell divides. Some of these functions are well known; others still elude us.

Fig. 26.8 A typical cell revealed by an electron microscope. (Courtesy *Scientific American*, September, 1961, p. 55.)

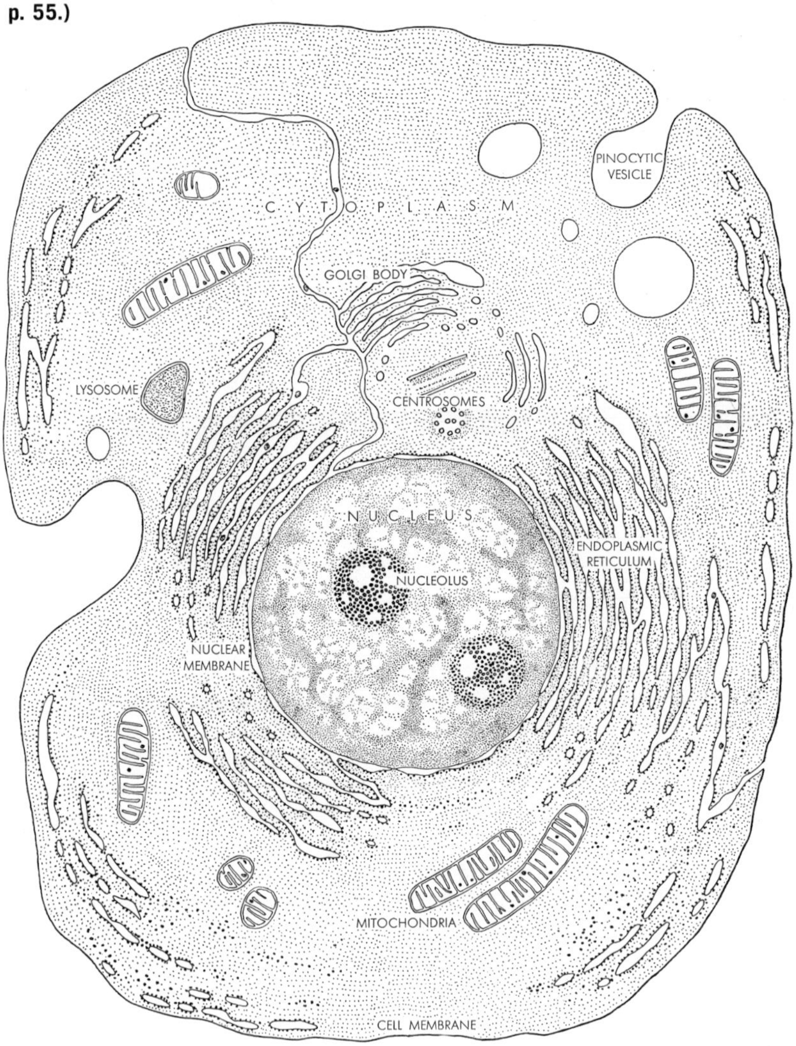

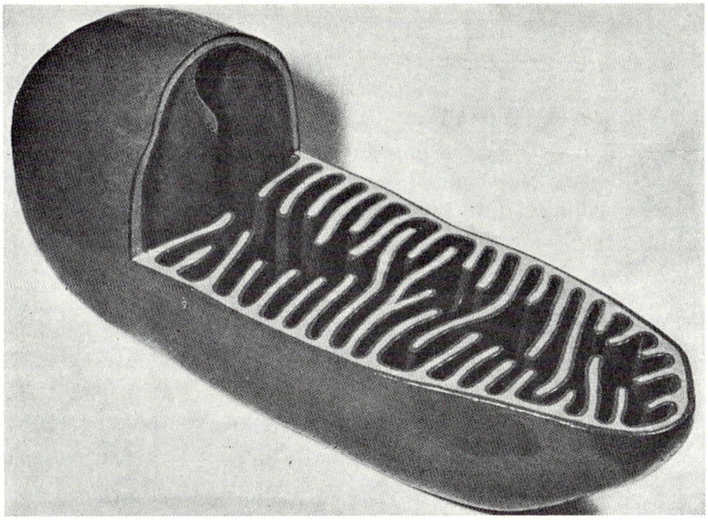

Fig. 26.9 Model of a mitochondrion, reconstructed according to electron optical photographs.

(a) The **mitochondria** (Fig. 26.9) are organelles shaped like elongated slippers; their cross-sectional diameters are about 1 micron. The highly differentiated structure of a mitochondrion contains some 40 enzymes, which control a complex series of redox reactions, including the conversion of diverse organic substances into ATP. The energy reservoir that is thus stored up is available for biomechanical work such as muscle contraction, for electrical work like the action of nerve impulses, and for the activation of other biochemical reactions. Because of these functions, the mitochondria have been called, by an analogy that not all mechanical engineers would accept, the "furnace of the cell."

(b) **Chloroplasts** are organelles that occur in plant cells and that contain the green pigment chlorophyll. Chlorophyll is the catalyst for the endothermic process of **photosynthesis,** in which glucose is synthesized from carbon dioxide

$$6CO_2 + 6H_2O \xrightarrow{\text{light quanta}} \underset{\textit{glucose}}{C_6H_{12}O_6} + 6O_2$$

$$\Delta H = 672 \text{ kcal/mole of glucose}$$

The equation above represents the sum of a sequence of steps that starts with the absorption of light quanta by chlorophyll molecules. This absorption promotes some of the electrons of the chlorophyll molecules to excited states, a condition which enables the chlorophyll to initiate a long series of redox reactions. The net effect of one set of this series is the production of oxygen from water,

$$2H_2O \longrightarrow O_2 + 4H$$

The hydrogen is not a released product of the reaction; it, in turn, participates in a series of steps whereby oxidized chlorophyll is reduced to its original condition.

(c) The **nucleus** is a well-defined structure which contains the genetic material of the cell; the nucleus thus is the site of the reproductive function.

Each time a cell divides, it reconstitutes itself. This ability of self-duplication is retained by the new cells and is transmitted repeatedly through successive generations of cells. The reliability of this transmittal accounts for the continuity of species. By what means can the structure of one cell be replicated in an indefinite number of its counterparts? Some sort of durable pattern must be preserved or reconstructed each time a cell duplicates itself. A "pattern" is a physical entity whose size, shape, and makeup carry information, such as a dress pattern, or a perforated metal plate that establishes the positions of bolt holes. In the nucleus, the fundamental determinant of replication is the DNA molecule. The sequence of base cross-links embodies the pattern, or code, for the synthesis of proteins. Protein synthesis does not occur by direct interaction of amino acids with DNA, but rather by a complex transfer mechanism in which the DNA code (the sequence of bases) is carried by a single-stranded RNA molecule, called **messenger RNA,** to structures outside the cell nucleus called **ribosomes.** The amino acids, which constitute the monomeric units of the protein, are first activated by coupling with ATP. These activated species are then transferred by an RNA called **transfer RNA** to a sequence that is determined by the sequence of bases in the messenger RNA chain. In other words, the RNA is the template on which the arrangement of amino acids in protein is fixed; DNA stamps the plate.

(d) **Other organelles.** The **lysosomes** are organelles that house enzymes which cleave protein and nucleic acid molecules to their component mers—a sort of cellular digestion. Other organelles are structurally well defined, but obscure as to biochemical function. Probably there are organelles that have not yet been discovered. The size range larger than that of the group of "small molecules" of the classical chemist (up to about 10 Å), but smaller than the macromolecules of the polymer chemist and the viruses of the biochemist (50 Å or more), is relatively unexplored, and may well be promising territory for future discoveries.

26.5 □ EPILOGUE

The subject of biochemistry is so extensive that much of it has not even been outlined or mentioned in this chapter. It was helpful to focus our attention on the cell for an introduction to the subject, but no single approach can be sufficiently inclusive. Biochemical subjects like nutrition and vitamins, the special functions of vital organs, the metabolism of minerals, and the roles of vitamins and hormones have not even been touched.

If there is any general statement that should be made about this subject, it is that biochemists necessarily assume as their working hypothesis that the laws of physical systems apply equally well to living organisms. This fact has been taken by some to prove that living beings can be interpreted, at least in principle, by the laws of physics and chemistry. Some scientists disagree with this proposition. For example, Michael Polanyi argues that the description of a living system necessarily transcends the physical and chemical laws that govern its atomic constituents. The subject is a matter of serious current debate among philosophers of science.

PROBLEMS

1. Definitions Define and illustrate the following terms: (a) metabolism; (b) reproduction; (c) protein; (d) enzyme; (e) substrate; (f) photosynthesis.

2. Peptides A **dipeptide** molecule is formed from two amino acid molecules. Write structural formulas for the four possible dipeptides formed from alanine and glycine.

3. Proteins (a) Nylon-66 (p. 499) is a copolymer of a diacid and a diamine that is characterized by amide linkages. Argue for or against the proposition that Nylon-66 is a synthetic protein. (b) Nylon-6 (p. 499) is an amino acid polymer. Argue for or against the proposition that Nylon-6 is a synthetic protein.

4. Insulin Using the information in Fig. 26.2 (p. 518), write the portion of the structural formula for insulin that includes the first five amino acids starting with glycine. Do not use the amino acid abbreviations; write your formula completely using a symbol for each atom.

5. Proteins Groups of protein molecules can orient themselves to form parallel threadlike units called fibrils (see Fig. 25.4, p. 505). Fibrils, in turn, can aggregate themselves to produce fibers large enough to be visible under the magnification of the ordinary light microscope. Fibers have been called polymers of fibrils, and fibrils polymers of proteins. Are these concepts of "higher-order polymers" acceptable extensions of our ideas of macromolecules? Defend your answer.

6. Enzymes The action of enzymes has been compared with that of a lock-and-key combination. Do you think that this is an apt analogy? Discuss this question from the point of view of (a) stereospecificity; (b) the catalytic effect of lowering the energy barrier to a process; (c) reversibility.

7. Nucleotides Write the structural formulas of (a) cytosine deoxymononucleotide; (b) guanine-thymine dinucleotide.

8. Porphyrins Iron exhibits a maximum coordination number of 6. (a) How many additional ligands could be bound to the iron atom in a heme molecule? (b) What would be the geometrical arrangement of the ligands with respect to the iron atom under these circumstances?

9. ATP For the reaction C(graphite) + $O_2(g) \longrightarrow CO_2(g)$, $\Delta H = -94.1$ kcal/mole. Compare quantitatively ATP decomposition with combustion of carbon as a source of energy. List some attributes other than the magnitude of ΔH that you think are advantageous for energy sources in biochemical systems.

10. Photosynthesis (a) Which process stores the greater amount of chemical energy, (*i*) the decomposition of one mole of CO_2 into its elements or (*ii*) the production of $\frac{1}{6}$ mole of glucose from one mole of CO_2 by photosynthesis? (b) What do you think of the possibility of developing a synthetic catalyst which would make it possible to produce one mole of glucose from CO_2 and water in a single step?

11. Genetic continuity Imagine that you purchase a dress pattern. Is it reasonable to say that this item contains *information?* Now if you use this pattern to make a dress, could you say that the information has been transmitted from the pattern to the dress material? Is it reasonable to call the transfer of information *communication?* Which is a more highly entropic process, the cutting of a fabric to match a dress pattern, or the cutting of a fabric by a child playing randomly with scissors? Would you agree that the increase of information in a system is antientropic? Is it therefore reasonable to consider that a molecule of DNA communicates with other molecules, thereby transferring information, and thus preserves the antientropic character of genetic continuity?

APPENDIX 1 ■ DEFINITIONS AND REVIEW OF PHYSICAL CONCEPTS

MATTER

The tendency to maintain a constant velocity is called **inertia.** Thus, unless acted on by an unbalanced force, a body at rest will remain at rest; a body in motion will remain in motion with uniform velocity.

Matter is anything that exhibits inertia; the quantity of matter is its **mass.** Mass is difficult to define without falling into the circle of defining mass in terms of force and force in terms of mass. However, there is no practical difficulty in deciding when two masses are equal, for their weights (see below) are then equal; an instrument which verifies this condition is a **balance.** The mass of a body is determined by choosing a combination of known smaller masses equal to the unknown mass; this process is called **weighing.** The definition of "matter" leads to a dual classification:

(a) *Matter that has mass both at rest and in motion* (*ordinary matter*). "An object at rest tends to remain at rest; an object in motion tends to remain in motion." These forms of inertia are exhibited by ordinary matter of the type that occupies space; water, air, smoke, living matter, radioactive fallout dust, electrons, and so on.

(b) *Matter that does not have rest mass*. Radiation, such as visible light, involves motion and can be deflected by gravitation; the degree of deflection depends on the force of the field. In this sense radiation exhibits inertia and has mass. In the sense that it does not exhibit mass at rest, it is not ordinary matter.

The **center of mass** of an object is that point around which all of its mass can be considered to be balanced. If the object is fixed at its center of mass, but is free to rotate around this point, it will have no tendency to do so in a gravitational field, regardless of the orientation to which it is turned.

MOTION

Motion is the change of position or location in space. Motions of objects may be classified as follows:

Translational motion occurs when the center of mass of an object changes its location. *Example:* an arrow in flight.
Rotational motion occurs when some points of a moving object remain stationary. *Examples:* a spinning top (the points along its axis are stationary); a rotating molecule (the center of mass remains stationary).
Distortion is change of shape, or the motion of the points of an object relative to one another. *Example:* a sagging metal bar.
Vibration is periodic distortion and recovery of original shape (the center of mass remains stationary). *Examples:* a struck tuning fork; a vibrating molecule.

FORCE AND WEIGHT

Force is that which changes the velocity (that is, the state of rest or motion) of a body,

$$\text{force} = \text{mass} \times \text{acceleration}$$

One **dyne** is the force needed to change the velocity of a mass of 1 g by 1 cm/sec in a time of 1 sec. This is an acceleration of (1 cm/sec) per sec, or 1 cm/sec^2, and a force of 1 dyne is 1 g cm/sec^2.

The weight, w, of a body is the force exerted on it by the earth's gravity,

$$w = m \times g$$

The weights of all bodies, when determined under the same force of gravity (same g), are proportional to their masses.

ENERGY AND WORK

The activity required to overcome inertia is called **work:**

$$\underbrace{\text{work} = \text{mass} \times \begin{pmatrix}\text{rate at which the velocity} \\ \text{is changed (}\textbf{accelera-}\\ \textbf{tion}\text{)}\end{pmatrix}}_{\text{force}} \times \begin{pmatrix}\text{distance through} \\ \text{which the mass} \\ \text{moves}\end{pmatrix} \quad (1)$$

Energy is the capacity to do work. There is energy in a mule, a moving train, a compressed spring, a stick of dynamite, a pound of uranium or of hydrogen or of chalk. Any moving object can do work by virtue of its motion because, by collision, it can overcome the inertia of another object; this capacity is called **kinetic energy.** Energy of motion may be either translational, rotational, or vibrational. The energy that a moving body loses in a *perfectly elastic* collision will be transmitted to another object, and

$$\left(\begin{array}{l}\text{energy gained by object that}\\ \quad\text{is set in motion}\end{array}\right) = \left(\begin{array}{l}\text{energy lost by object that stops}\\ \quad\text{or slows down}\end{array}\right)$$

This is a particular instance of the law of conservation of energy. The magnitude of this kinetic energy, for a body of mass m and velocity u, is derived directly from Equation 1.

Consider the energy given to a stationary body by accelerating it at a uniform rate to velocity u. Assume the energy of the stationary body is zero.

The change in velocity from 0 to $u - 0 = u$.

The acceleration, or change in velocity per unit time, is u/t.

The average velocity (distance/time, or d/t) from an initial value of 0 to a final value of u is $\frac{1}{2}(0 + u) = \frac{1}{2}u$. Then

$$d/t = \tfrac{1}{2}u$$

The kinetic energy is given by

$$\begin{aligned}\text{kinetic energy} &= \text{mass} \times \text{acceleration} \times \text{distance}\\ &= m \times \frac{u}{t} \times d\\ &= m \times u \times \frac{d}{t}\\ &= m \times u \times \tfrac{1}{2}u\\ &= \tfrac{1}{2}mu^2\end{aligned}$$

The energy of a body may stem from some attribute other than motion. A compressed spring, for example, may do work when it expands, a stick of dynamite when it explodes. In every case in which we attribute a quantity of energy to an object, we imagine that an event could occur in which this energy may be transmitted in some way by the object to another body. When we can no longer imagine such an event, we may say the object has no energy. Of course, the kinds of event that we imagine change with the times, hence the energy that we assign to objects changes. We attribute much more energy to a pound of steam when we think of its use in an atomic fusion reactor than when we think of it in a reciprocating piston engine.

Energy is frequently named according to the kind of event we imagine: for example, we say that a substance has **chemical energy** when its energy loss would be accompanied by a change in composition, **electrical energy** by generation of an electric current, and so on. The term **potential energy** is sometimes used in contradistinction to kinetic energy.

Sometimes an energy-transmitting event cannot occur until some energy is absorbed first: we can think that a rock in the crater of a volcano can lose energy by rolling down to the valley, but not before it is lifted over the rim. Such a preliminary expenditure of energy that is required to bring a system to a condition in which energy is spontaneously released is called **energy of activation.**

The fundamental unit of energy is the **erg.** It is defined in terms of force and distance.

One **erg** is the work done when a force of a dyne acts over a distance of 1 cm,

$$\text{erg} = \text{dyne cm} = \frac{\text{g cm}^2}{\text{sec}^2}$$

One **joule** (J) is 10^7 ergs.

Work may also be expressed in terms of pressure and volume.

$$\text{Work} = \text{pressure} \times \text{volume through which pressure is exerted}$$

(See p. 534 for a discussion of pressure.) To illustrate the equivalence of the above expression with that given earlier in terms of force and distance, consider a force of 1 dyne displacing a piston 1 cm² in area through a distance of 1 cm. Then,

$$\begin{aligned} \text{Work} &= \text{force} \times \text{distance} &&= \text{pressure} \times \text{volume change} \\ &= 1 \text{ dyne} \times 1 \text{ cm} &&= 1\,\frac{\text{dyne}}{\text{cm}^2} \times 1 \text{ cm}^3 \\ &= 1 \text{ dyne cm} &&= 1 \text{ dyne cm} = 1 \text{ erg} \end{aligned}$$

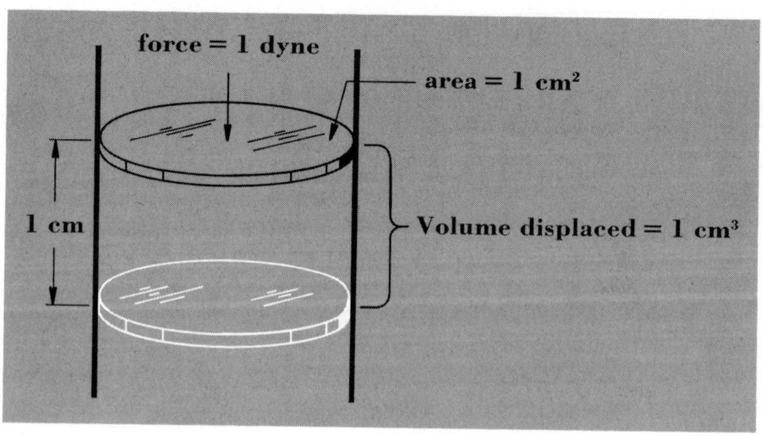

MASS

Standard International Prototype Kilogram (kg): a mass of platinum-iridium at the International Bureau of Weights and Measures, Sèvres, France.

Conversions

1 kg $\approx$ 2.205 pounds, avoirdupois
1 pound, avoirdupois $\approx$ 453.6 g

Fractions

1 gram (g)	$= 10^{-3}$ kg
1 centigram (cg)	$= 10^{-2}$ g
1 milligram (mg)	$= 10^{-3}$ g
1 microgram (μg)	$= 10^{-6}$ g
1 nanogram (ng)	$= 10^{-9}$ g
1 picogram (pg)	$= 10^{-12}$ g $=$ 1 micromicrogram ($\mu\mu$g)

LENGTH

Standard The meter (m), 1,650,763.73 wavelengths of the orange-red radiation of krypton-86. This standard was adopted in 1960 by the General Conference on Weights and Measures, replacing the International Prototype Meter, the distance between two lines at 0°C on a platinum-iridium bar at the International Bureau of Weights and Measures, Sèvres, France.

Conversion

1 inch $=$ 2.54 cm (exactly)

Multiples and Fractions

1 terameter (Tm)	$= 10^{12}$ m
1 gigameter (Gm)	$= 10^{9}$ m
1 megameter (Mm)	$= 10^{6}$ m
1 kilometer (km)	$= 10^{3}$ m
1 decimeter (dm)	$= 10^{-1}$ m $=$ 10 cm
1 centimeter (cm)	$= 10^{-2}$ m
1 millimeter (mm)	$= 10^{-3}$ m
1 micron (μ)	$= 10^{-6}$ m $= 10^{-4}$ cm
1 millimicron (mμ)	$= 10^{-9}$ m $= 10^{-7}$ cm
1 angstrom unit (Å)	$= 10^{-8}$ cm
1 fermi	$= 10^{-13}$ cm

VOLUME

Standards 1 liter = 1 dm^3 (exactly). The liter was formerly defined as the volume of 1 kg of pure water at the temperature of its maximum density (4°C) at one standard atmosphere.

Conversions

1 liter = 1000 cubic centimeters (exactly)
≈ 1.057 U.S. liquid quarts

Multiples and fractions

1 milliliter (ml) = 10^{-3} liter = 1 cm^3
1 microliter (μl) = 10^{-6} liter = 10^{-3} ml
1 nanoliter (nl) = 10^{-9} liter = 10^{-6} ml

DENSITY

Density is mass per unit volume; the densities of substances depend on temperature. When the density of a standard substance is used as a reference, the **specific gravity** is defined as follows:

$$\text{specific gravity} = \frac{\text{density of a given substance}}{\text{density of a standard substance}}$$

The temperatures of both substances must be specified; they are frequently, but not always, the same. For solids and liquids the standard substance is usually water; for gases it is usually air, or sometimes hydrogen. Thus, the specific gravity value for carbon tetrachloride, $1.594^{20°/4°}$, means that the density of this liquid at 20°C is 1.594 times as great as the density of water at 4°C. By the definition of the liter, the specific gravity$^{x°/4°}$ is numerically equal to the density in grams per milliliter at x°.

PRESSURE

Pressure is force per unit area. If the pressure on any point of matter is greater in one direction than in another, the matter will move in the direction of the lower pressure. Conversely, a stationary point of matter is subject to equal pressures in all directions. Consider an infinitely thin horizontal disk submerged in a liquid, as shown in Fig. A1. The pressure from above is exerted by the combined mass of the liquid and the atmosphere above the disk. The pressure from below is exerted by the compressed liquid and by the container which is strained by the pressure of its contents. If all the liquid above the dotted line were instantaneously removed, the pressure from below would throw the disk up. Thus, the pressure of a liquid or a gas depends only on the depth (or height), and is exerted equally in all directions. At sea level,

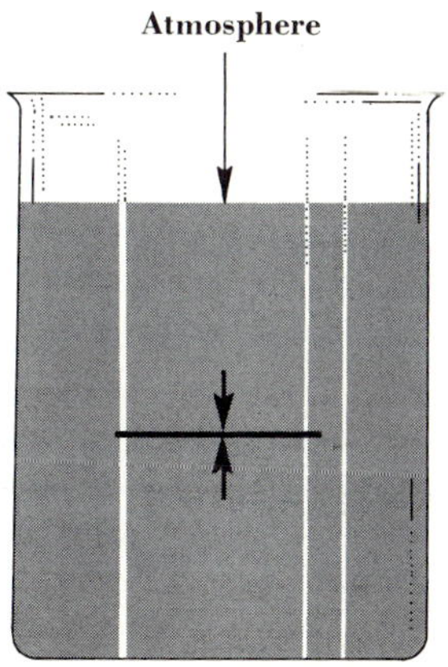

Fig. A1 Pressure exerted by a liquid.

the pressure exerted by the earth's atmosphere varies around 14.7 lb/in.^{2}. This value is equivalent to the pressure exerted by a column of mercury about 76 cm high, or of water about 34 feet high.

Standard One standard atmosphere = the pressure exerted by exactly 76 cm (= exactly 760 mm) of mercury at 0°C (density 13.5951 g/cm^3) and at standard gravity, 980.665 cm/sec^2.

It follows that

$$\begin{aligned} 1 \text{ standard atm} &= 13.5951 \text{ g/cm}^3 \times 76 \text{ cm } (\textit{exactly}) \times 980.665 \text{ cm/sec}^2 \\ &= 1.01325 \times 10^6 \text{ dynes/cm}^2 \end{aligned}$$

Fraction

$$\begin{aligned} 1 \text{ torr*} &= \frac{\text{standard atm}}{760\ (\textit{exactly})} \\ &= \text{pressure exerted by 1 mm of mercury at 0°C at standard gravity} \end{aligned}$$

A **barometer** is a device that measures atmospheric pressure. The original and simplest form was invented by Torricelli in 1643. It is made by inverting a tube longer than 76 cm filled with mercury into a dish of mercury (Fig. A2). The atmosphere will support only that height of mercury which exerts an equivalent pressure; any excess mercury will fall into the reservoir and leave a space in the upper part of the tube with zero air pressure. Note that the height in millimeters, P_b, reads in torr units only if the mercury is at 0°C and standard gravity. Otherwise, appropriate corrections must be made.

HEAT AND TEMPERATURE

One may do work on an object and yet fail to convert such work into equivalent energy of motion, electricity, magnetism, radiation, or

* Named after Evangelista Torricelli.

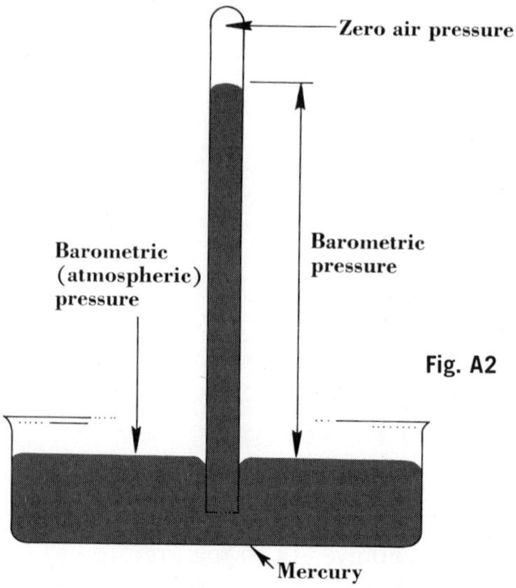

Fig. A2 Mercury barometer (schematic).

chemical or physical change. For example, one may bend an iron bar back and forth several times, or stir a liquid or a gas in a confined space, or force an electric current through a copper wire and, after all such expenditures of energy, observe that the object retains substantially its original form and position in space. We assume that this energy has been conserved, and exists as an undifferentiated internal energy called **heat,** which is retained by the object in question. It is believed that heat is the energy of random motion of all the elementary particles of a body.

If two bodies, A and B, are in contact, and there is a spontaneous transfer of heat from A to B, then A is said to be *hotter* or at a higher temperature than B. When, despite an available path, there is no net heat transfer between two bodies, they are said to be at the same temperature. **Temperature** is thus the property of a body that predisposes it to lose heat: the higher the temperature, the greater the tendency for heat to flow away from the body.

Many properties of substances change with changes in temperature. Examples are density, color, ability to conduct electricity, and ability to stimulate nerve impulses. A set of values in which temperature is related to some measured property is called a **temperature scale.** A device used to obtain such measurements is a thermometer. Additional discussion of temperature scales appears in Chapter 11.

STANDARDS

The triple point The temperature at which ice, water, and water vapor coexist at equilibrium. This temperature is 273.16°K (exactly) or 0.01°C; °K refers to the Kelvin scale, °C to the Celsius (previous name, centigrade) scale.

Absolute zero This is 0°K (exactly), or −273.15°C.

The ice point The temperature at which ice and water saturated with air exist together at standard atmospheric pressure. This temperature is 0°C (exactly).

The steam point The temperature at which steam and water exist together at standard atmospheric pressure. This temperature is 100°C.

Other points frequently used are (all taken at one standard atmosphere): boiling oxygen, −182.96°C; freezing mercury, −38.87°C; boiling sulfur, +444.60°C.

1 calorie* (cal), sometimes called the thermochemical calorie, or the defined calorie, $=4.184 \times 10^7$ ergs = exactly 4.184 joules (J).

1 kilocalorie (kcal)
1 kilogram-calorie
1 Calorie (capital C; used in expressing food energies)
} = 1000 cal = 4184 J

Conversions

0°C = 32° Fahrenheit (°F)
100°C = 212° F

There are thus fewer Celsius degrees than Fahrenheit degrees in the temperature interval between the ice point and the steam point. The ratio is (100 − 0)/(212 − 32) or 5/9. The conversion equation is

$$°C = \frac{5}{9}(°F - 32)$$

The **specific heat** of a substance is the heat required to warm 1 gram of the substance 1 Celsius degree (units: calories per gram per degree). The relationship is

$$\text{specific heat} = \frac{\text{heat absorbed or released (cal)}}{\text{mass (g)} \times \text{temp. rise or fall (C deg)}}$$

Specific heats vary markedly from substance to substance, and to some extent with temperature for any given substance. Some approximate values under ordinary conditions are: air (standard pressure), 0.25 cal/g deg; water, 1.00; ice, 0.5; alcohol, 0.58; copper, 0.09.

ELECTRICITY

There are several kinds of interaction among matter in the universe. Gravitation is a relatively weak attractive force seemingly exerted by all matter. Much stronger forces are exerted in **electrostatic interactions,** which may be either attractive or repulsive. Matter that exhibits such

* The "small calorie" or "gram-calorie" is the quantity of heat required to warm 1 g of water from 3.5°C to 4.5°C. The "normal calorie" involves the temperature change from 14.5°C to 15.5°C, and the "mean calorie" is 1/100 the heat needed to warm 1 g of water from 0°C to 100°C. All of these units are nearly the same.

behavior is said to possess **electric charge.** To account for observed phenomena, it is postulated that two kinds of electric charge exist; they are called **positive** and **negative.** The interaction among combinations of charged (+, −) or neutral (0) points of matter are

Charge of matter		Electrostatic interaction
−	+	Attraction
−	−	Repulsion
+	+	Repulsion
−	0	None
+	0	None
0	0	None

The magnitude of electrostatic force is given by Coulomb's law,

$$F = \frac{1}{D}\frac{q_1 q_2}{d^2}$$

where F is the force of attraction or repulsion (in dynes), q_1 and q_2 are the magnitudes (in esu) of two electric charges separated by distance d (in centimeters), and the **dielectric constant,** D, is a dimensionless number having the value of 1 (exact number) in a vacuum, and more than 1 in matter. Examples of values of D are 1.000590 for air at 1 standard atmosphere and 0°C, 24.3 for alcohol at 25°C, and 80.37 for water at 20°C.

A neutral body may have both + and − charges of equal total magnitude, but they need not be distributed homogeneously. We may arbitrarily assign a **center of positive charge** and a **center of negative charge** to each body, corresponding to the points around which the + and − charges, respectively, can be considered to be balanced. When these centers do not coincide, the body is said to have **electrical polarity,** or to be **polar,** and is called a **dipole.** The **dipole moment** is defined as the magnitude of the + or the − charge times the distance between the charge centers. Dipoles will be attracted or repelled to or from charged bodies or other dipoles, depending on the relative orientations of the bodies. But attractive forces tend to aggrandize themselves (by increasing the polarization of attracting bodies and by bringing about favorable orientations), whereas repulsive forces tend to diminish themselves. Hence attractive forces are statistically favored, and dipoles are attracted to charged bodies and to each other. Likewise, nonpolar bodies may be polarized by and attracted to charged bodies.

Standard

1 electrostatic unit (esu) = charge which will exert a force of 1 dyne (1 g cm per second per second) at a distance of 1 cm on a charge of equal magnitude in a vacuum or 1 dyne = esu^2/cm^2.

Conversion

1 coulomb (C) = 2.998×10^9 esu

If two points differ in electric charge, energy must be gained or lost in transferring a given charge from one point to the other. This energy is called the potential difference between the two points. When 10^7 ergs are gained or lost during the transfer of 1 C of charge, the potential difference is said to be 1 volt (V). Then

10^7 ergs = 1 V C

and since

10^7 ergs = 1 J

then

1 J = 1 V C

A movement of charged bodies is called an **electric current.** The relationship is

$$I = C/t$$

where I is the current (amperes A), and C is the quantity of charge (coulombs) which passes a given point in time t (seconds).

The magnitude of the current which can pass through a given section of matter depends on the potential difference (E, in volts) between the two points of reference and on the resistance imposed by the matter (R, in ohms, Ω). The expression is **Ohm's law,**

$$I = E/R$$

The reciprocal of the resistance, $1/R$, is called the **conductance,** K (units: reciprocal ohms, or mhos, Ω^{-1}).

MEASUREMENT; UNCERTAINTY

Measurement is the transfer of information in terms of standard units. Information may be transferred to human beings (through sensory routes) or to machines. Any measurement culminates in a sensation; the sensing element may be the human eye or other organ, or an inanimate instrument. In any event, the sensor must be stimulated by a quantity of energy if it is to receive the measurement information. As a result, the process of measurement disturbs to some degree the matter being measured. For measurements of ordinary bodies, the magnitude of this disturbance is too small to be of consequence, but for measurements of very small bodies like individual electrons the inherent uncertainty introduced is relatively great.

Significance of measurement figures Gross uncertainties in measurement are introduced through factors such as human error, malfunction of measuring devices, and experimental bias (error in one direction, as by a ruler whose "inches" are too short). Information is useful only to the extent that one can be confident of its validity. To insure such utility, each figure or digit in the numerical expression of a measurement should be significant. A **significant figure** may be defined as a number that we believe to be correct within some specified or implied limit of error. Thus, *if* the height of a man, expressed in significant figures, is written as 5.78 ft, it is assumed that only the *last figure* may be in error. Clearly, any uncertainty in the first or second figure would remove all significance from the last figure. (Not being certain of the number of feet, it is idle to speak of inches.) If we have reason to believe that the last figure will be in doubt by a *specified* amount, we may so indicate by expressions such as 5.78 ± 0.01 ft.

To count the significant figures in a number, read the number from left to right and count all the digits starting with the first digit that is not zero. The position of the decimal point should be ignored, because it is determined by the particular units employed. Thus, the measurements 12.2 cm and 122 mm are identical and therefore all the corresponding figures are equally significant.

The following examples will be illustrative:

Value	Number of significant figures
0.0301 lb	3
4.290 hours	4
1.030 mg	4
0.001030 g	4
5000 miles	Ambiguous—the zeros may have measurement significance or may be merely spacers to indicate the magnitude of the first digit
5.00×10^3 miles	3—the zeros clearly have measurement significance; they are not needed to indicate magnitude
6 pencils	If this is used in the sense of a tally ("Tom has 6 pencils"), it is an exact number, and the concept of significant figures has no meaning; if it is used as a measurement ("The average student uses 6 pencils per semester"), it has 1 significant figure
1 ft = 12 in.	These are exact numbers which express a definition, not measurements

We must guard against introduction of uncertainty by arithmetical procedures. The following rules will be helpful.

Rule 1 In addition or subtraction, any figure in the answer is significant only if *each number* in the problem contributes a significant figure at that decimal level. Therefore the value which terminates at the highest decimal level (that is, the level of greatest magnitude) will determine how far the significant figures should be carried in the answer:

$$\begin{array}{r} 308.7812 \\ 0.00034 \\ 10.31 \\ \hline 319.09 \end{array}$$

Rule 2 When a number is "rounded off" (nonsignificant figures discarded), the last significant figure is unchanged if the next figure is less than 5, and is increased by 1 if the next figure is 5 or more.

Rule 3 In multiplication and division, the number of significant figures in the answer is the same as that in the quantity with the fewest significant figures:

$$\frac{3.0 \times 4297}{0.0721} = 1.8 \times 10^{-5}$$

Rule 4 In a multistep computation, it will be convenient first to determine the number of significant figures in the answer by Rules 1, 2, and 3 above, and to round off each number that contains excess significant figures to one more significant figure than necessary. Then round off the answer to the correct number of significant figures. This procedure will preserve significance with minimum labor.

EXAMPLE 1 Evaluate the expression

$$4.3 \times \frac{311.8}{273.1} \times \frac{760}{784 - 2}$$

ANSWER Number of significant figures in final answer: 2. Round off according to Rules 1 and 2, to one extra significant figure. Note that the presence of only one significant figure in 2 does not mean that there is only one significant figure in the answer, because $784 - 2 = 782$, which has three significant figures. Thus,

$$= 4.3 \times \frac{312}{273} \times \frac{760}{782}$$

Solve and round off to two significant figures,

$$= 4.8$$

Measurement information, to be complete, must specify the units of measurement, which are sometimes called **dimensional units.** It is valid to handle such units as algebraic terms, and one will find that such practice speeds up calculations and provides an automatic check against dimensional errors. When numbers do not imply measurement units, they are said to be **dimensionless;** common examples are percentages and other ratios. In handling units algebraically, learn how to manipulate identities so that the correct answer will emerge expressed in the desired units. Thus, the identity

$$1 \text{ ft} = 12 \text{ in.}$$

means that the fraction

$$\frac{12 \text{ in.}}{1 \text{ ft}}$$

or its reciprocal

$$\frac{1 \text{ ft}}{12 \text{ in.}}$$

has the value of unity, and can be inserted as a term in the numerator or denominator of any fraction, and be raised to any power, at will.

Dimensional units, like other algebraic terms, may have positive or negative exponents. Thus, the unit of area, square feet, is expressed as ft^2. A unit of speed may be written $miles/hr$ or $miles\ hr^{-1}$. A unit of frequency, the number of times some event occurs *per second*, is sec^{-1}.

EXAMPLE 2 A fan blows 95.0 ft^2 of air per minute uniformly through a duct whose cross-sectional area is 72 in.^{2}. What is the linear speed of the air in miles per hour?

ANSWER

$$\text{Air speed} = \frac{95.0 \text{ ft}^3/\text{min}}{72 \text{ in.}^2} \times \frac{\left(12 \frac{\text{in.}}{\text{ft}}\right)^2}{5280 \text{ ft/mile}} \times 60 \text{ min/hr}$$
$$= 2.2 \text{ miles/hr}$$

APPENDIX 2 ■ REVIEW OF SOME MATHEMATICAL OPERATIONS

EXPONENTS

In the expression x^n, x is called the **base,** and n the **exponent.** When n is an integer, the expression has the value of x multiplied by itself $n - 1$ times. Thus

$$2^4 = 2 \times 2 \times 2 \times 2 = 16$$

In the expression $x^{1/n}$, $1/n$ is the exponent, and n is the index of the **root,**

$$x^{1/n} = \sqrt[n]{x}$$

Thus

$$4^{1/2} = \sqrt{4} = 2$$
$$16^{1/4} = \sqrt[4]{16} = 2.$$

In the expression $x^{m/n}$, m/n is the exponent,

$$x^{m/n} = \sqrt[n]{x^m}$$

Thus

$$4^{3/2} = \sqrt{4^3} = \sqrt{64} = 8$$

The expression x^{-n} is the reciprocal of x^n,

$$x^{-n} = \frac{1}{x^n}$$

Thus

$$4^{-1} = \tfrac{1}{4}$$

$$2^{-3/2} = \frac{1}{2^{3/2}} = \frac{1}{\sqrt{2^3}} = \frac{1}{\sqrt{8}} = \frac{1}{2\sqrt{2}} = \frac{\sqrt{2}}{4}$$

Any number to the 0th power is 1:

$$(\tfrac{1}{2})^0 = 1^0 = 8^0 = 1000^0 = (10^{23})^0 = x^0 = 1$$

When exponential expressions having the same base are multiplied, the exponents are added; when such expressions are divided, the exponents are subtracted:

$$x^m \times x^n = x^{m+n}$$

Thus

$$10^2 \times 10^3 = 10^5 \qquad \text{and} \qquad 10^6 \times 10^{-4} = 10^2$$

and

$$\frac{x^m}{x^n} = x^{m-n}$$

Thus

$$\frac{10^5}{10^2} = 10^3 \qquad \frac{10^5}{10^{11}} = 10^{-6}$$

When a power is raised to a power, the exponents are multiplied:

$$(x^m)^n = x^{mn}$$

Thus

$$(10^8)^3 = 10^{24} \qquad \text{and} \qquad (10^{-1})^2 = 10^{-2}$$

For calculations involving multiple exponential terms and dimensional units, the following procedure is recommended:

Procedure	Example
(1) Write the expression	$\dfrac{52.02\ \text{g} \times (10^8\ \text{Å/cm})^3 \times 2\ \text{atoms/unit cell}}{7.20\ \text{g/cm}^3 \times \dfrac{(2.88\ \text{Å})^3}{\text{unit cell}}}$
(2) Simplify any compound fractions; multiply exponents; evaluate dimensional units	$\dfrac{52.02\ \cancel{\text{g}} \times 10^{24}\ \cancel{\text{Å}^3}/\cancel{\text{cm}^3} \times 2\ \text{atoms}/\cancel{\text{unit cell}}}{7.20\ \cancel{\text{g}}/\cancel{\text{cm}^3} \times 2.88^3\ \cancel{\text{Å}^3}/\cancel{\text{unit cell}}}$
	(final answer is in atoms)
(3) Rewrite numbers only; check that none is omitted	$\dfrac{52.02 \times 10^{24} \times 2}{7.20 \times 2.88^3}$
(4) Decide on significant figures; avoid unnecessary computations	3 significant figures
	("2" is an exact number and does not limit significance)
(5) Perform computation; gather exponential terms	0.604×10^{24}
(6) Write answer in final form with units	6.04×10^{23} atoms

LOGARITHMS

A logarithm is an exponent:

$$N = a^x \qquad (a \text{ is the } \textbf{base}; \ x \text{ is the } \textbf{exponent})$$
$$\log_a N = x \qquad (a \text{ is the } \textbf{base}; \ x \text{ is the } \textbf{logarithm})$$

The base of logarithms used as an aid to ordinary computations is 10. Then

$$10{,}000 = 10^4$$

and

$$\log_{10} 10{,}000 = \log_{10} 10^4 = 4$$

When the base is not specified, 10 is understood,

$$\log 0.001 = \log 10^{-3} = -3$$

Because logarithms are exponents, logarithms of products are added and logarithms of quotients are subtracted:

$$\log(a \times b) = \log a + \log b$$
$$\log\left(\frac{a}{b}\right) = \log a - \log b$$

Logarithms that cannot be expressed as integral exponents are found in tables or on slide rules. Logarithmic tables present logarithms of numbers between 1 and 10; a decimal point is assumed to precede the logarithm values given in the tables. Then

$$\log 1.21 = 0.0828$$

A number that is not between 1 and 10 may be written in proper exponential form, and the logarithm obtained as follows:

$$\begin{aligned}\log 7040 &= \log(7.040 \times 10^3)\\ &= \log 7.040 + \log 10^3\\ &= 0.8476 + 3 = 3.8476\end{aligned}$$

$$\begin{aligned}\log 0.000625 &= \log(6.25 \times 10^{-4})\\ &= \log 6.25 + \log 10^{-4}\\ &= 0.7959 + (-4) = -3.2041\end{aligned}$$

The procedure can be reversed. The number, N, whose logarithm has a given value, x, is the **antilogarithm** of that value:

$$\left.\begin{aligned}10^x &= N\\ \log N &= x\\ \text{antilog } x &= N\end{aligned}\right\} (x \text{ is the } \textbf{log}; \ N \text{ is the } \textbf{antilog})$$

Rule for significant figures in antilogarithms The number of significant figures in the antilogarithm is equal to the number of decimal places (digits after the decimal point) in the logarithm. Some examples follow:

(*i*) $\log 100 = \log 10^2 = 2$
$\text{antilog } 2 = 10^2 = 100$

We assume here that these are exact numbers.

(*ii*)
$$\begin{aligned}\text{antilog } 6.3909 &= \text{antilog}(6 + 0.3909)\\ &= \text{antilog } 6 \times \text{antilog } 0.3909\\ &= 10^6 \times 2.460 \text{ (from log table)}\\ &= 2.460 \times 10^6\end{aligned}$$

Note that only four significant figures are permitted.

(*iii*)
$$\begin{aligned}-0.0079 &= 0.9921 - 1\\ \text{antilog}(-0.0079) &= \text{antilog}(0.9921 - 1)\\ &= \text{antilog } 0.9921 \times \text{antilog}(-1)\\ &= 9.820 \times 10^{-1}\\ &= 0.9820\end{aligned}$$

Note that four significant figures are permitted because there are four figures after the decimal point in −0.0079.

(*iv*)
$$\begin{aligned}-9.42 &= 0.58 - 10\\ \text{antilog}(-9.42) &= \text{antilog } 0.58 \times \text{antilog}(-10)\\ &= 3.8 \times 10^{-10}\end{aligned}$$

Note that only two significant figures are permitted.

Logarithms to the base e are **natural logarithms** and are given the symbol ln. Thus

$$\begin{aligned}\ln e^x &= x\\ \ln e^{0.2} &= 0.2\end{aligned}$$

The relation between the logarithms of the two bases, e and 10, is

$$\ln x = 2.303 \log x$$

To solve for the value of exponential functions of e, for example $e^{-3.0}$, the following procedure is recommended:

(a) Take the natural logarithm of $e^{-3.0}$, $\ln e^{-3.0} = -3.0$;
(b) Convert to the base 10 by dividing by 2.303,

$$\frac{-3.0}{2.3} = -1.3$$

(c) Take the antilogarithm,

$$\begin{aligned}-1.3 &= 0.7 - 2\\ \text{antilog } -1.3 &= \text{antilog } 0.7 \times \text{antilog}(-2)\\ &= 5 \times 10^{-2}\end{aligned}$$

Therefore

$$e^{-3.0} = 5 \times 10^{-2}$$

The notation $\exp(x)$ is often used to represent e^x.

Values of exponential functions and natural logarithms are given in tables in standard reference books. However, these tables are usually less adequate than the tables of common logarithms.

QUADRATIC EQUATIONS

A quadratic equation is one in which the highest exponent to which a variable is raised is 2. Any quadratic equation may be written as

$$ax^2 + bx + c = 0$$

The equation has two solutions, given by

$$x = \frac{-b \pm \sqrt{b^2 - 4ac}}{2a} \tag{1}$$

When a quadratic equation is applied to a problem dealing with physical reality, an impossible solution (such as a negative mass, or a quantity that violates a conservation law) may be dropped. Let us say that x is the molar concentration of hydrogen ion in a solution, as given in the following equation:

$$x^2 + 0.20x - 0.10 = 0 \qquad (x \text{ is the variable})$$

$a = 1$ (understood) $\quad b = 0.20 \quad c = -0.10$

Then, applying Equation 1, the answers are

$$x = +0.23 \text{ or } -0.43$$

The negative answer is rejected because the variable in question (concentration) cannot be negative. Then

$$x = 0.23$$

GRAPHS AND PROPORTIONALITY

A straight line is represented algebraically by the equation

$$y = mx + b$$

where x and y are variables; b is the value of y when $x = 0$ ("the y intercept"); m is the slope of the line, or the change in y per unit change in x. The plot is as shown in Fig. A3.

A nonlinear equation may be represented by a straight line if the variables are suitably expressed (Boyle's law, p. 176):

$$V = k\left(\frac{1}{P}\right)$$

If the variables are considered to be V and P, the plot is not linear. But a plot of V versus $(1/P)$ corresponds to the equation for a straight line,

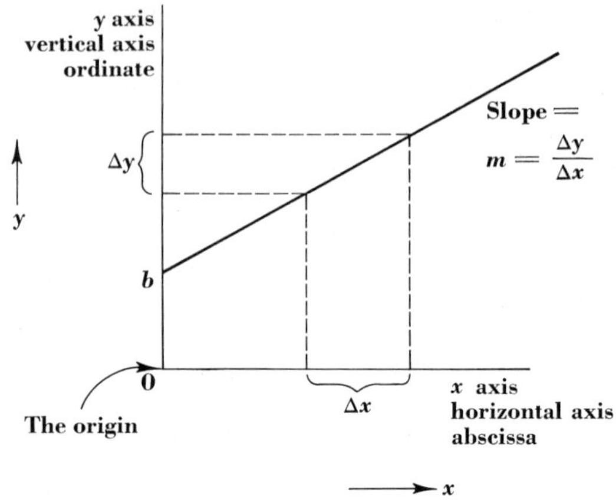

will be linear, and will pass through the origin:

$$y = mx + b$$

$$V = k\left(\frac{1}{P}\right) \quad \text{zero}$$

A **direct proportionality** is a linear relationship that passes through the origin, such as $x = ky$. Any change in one of the variables produces a proportionate (of equal portion, or equal percentage) change in the other. The symbol $\propto$ means "is proportional to." Then, $x \propto y$ is the same as $x = ky$.

In an inverse proportionality (see Boyle's law above and on p. 176).

$$x \propto \frac{1}{y}$$

or

$$x = \frac{k}{y}$$

When any one variable changes by a given factor, the other variable changes by the reciprocal of that factor. Thus when x is doubled, y is halved; when x is quadrupled, y is quartered.

Of course, other proportionalities are possible. According to Graham's law (p. 194) the rate (u) of effusion of a gas is inversely proportional to the square root of its density (d),

$$u \propto \frac{1}{\sqrt{d}}$$

or

$$u = \frac{k}{\sqrt{d}} = kd^{-1/2}$$

It is important to recognize that for a direct proportionality, the curve must pass through the origin (both variables must be zero at the same time). A physical scale that has a *natural* beginning at zero, that is, the real value of the measurement is zero when the number 0 is assigned to it, is called an **absolute scale.** Therefore, measurements may have to be converted to absolute scales before proportionalities can be established. An example is the conversion of the Celsius (centigrade) scale to the Kelvin scale (p. 178).

APPENDIX 3 ■ NOMENCLATURE

INORGANIC NOMENCLATURE

See p. 42 for naming of binary compounds and some ternary compounds containing oxygen. Further information needed for such nomenclature is given in Table A-1, which includes common valences of various elements and groups.

NOMENCLATURE OF ACIDS

There are two main classes of acids, binary acids and oxyacids.

Binary acids are compounds in which a hydrogen atom combines with one other atom, HX or H_2X. They are named by bracketing the root, derived from the name of the element X, by the prefix **hydro-** and suffix **-ic** followed by the word **acid.** The common binary acids are restricted mainly to the elements of the halogen and oxygen families (p. 386), since the water solutions of the hydrogen compounds of other elements are not sufficiently acidic to warrant the use of the name acid. The corresponding negative ion is named by adding the suffix **-ide** to the root as shown in the tabulation:

HF	*Hydro*fluor*ic acid*	F^-	Fluor*ide* ion
HCl	*Hydro*chlor*ic acid*	Cl^-	Chlor*ide* ion
HBr	*Hydro*brom*ic acid*	Br^-	Brom*ide* ion
HI	*Hydro*iod*ic acid*	I^-	Iod*ide* ion
H_2S	*Hydro*sulfur*ic acid*	S^{2-}	Sulf*ide* ion
H_2Se	*Hydro*selen*ic acid*	Se^{2-}	Selen*ide* ion
H_2Te	*Hydro*tellur*ic acid*	Te^{2-}	Tellur*ide* ion
HCN	*Hydro*cyan*ic acid*	CN^-	Cyan*ide* ion

TABLE A-1 NAMES AND COMMON VALENCES OF SOME ELEMENTS AND GROUPS

Name	Valence	Illustrative compound: Ionic	Illustrative compound: Nonionic
Metals and cations			
Ammonium	1	$NH_4^+Cl^-$	—
Copper			
Cuprous	1	—	Cu_2O
Cupric	2	$Cu^{2+}(Cl^-)_2$	—
Hydrogen	1	—	H_2O
Mercury			
Mercurous	1	—	Hg_2S
Mercuric	2	$Hg^{2+}(F^-)_2$	HgICl
Potassium	1	$(K^+)_2SO_4^{2-}$	—
Silver	1	$Ag^+NO_3^-$	AgI
Sodium	1	Na^+Cl^-	—
Barium	2	$Ba^{2+}CO_3^{2-}$	—
Cadmium	2	$Cd^{2+}(F^-)_2$	—
Calcium	2	$Ca^{2+}O^{2-}$	—
Cobalt	2	$Co^{2+}(NO_2^-)_2$	$CoCl_2$
Iron			
Ferrous	2	$Fe^{2+}SO_4^{2-}$	$FeBr_2$
Ferric	3	$(Fe^{3+})_2(SO_4^{2-})_3$	$FeCl_3$
Lead	2	$Pb^{2+}(F^-)_2$	$Pb(CH_3COO)_2$
Magnesium	2	$Mg^{2+}SO_4^{2-}$	—
Nickel	2	$Ni^{2+}O^{2-}$	$NiCl_2$
Strontium	2	$Sr^{2+}(Cl^-)_2$	—
Tin			
Stannous	2	$Sn^{2+}SO_4^{2-}$	$SnCl_2$
Stannic	4	$Sn^{4+}(O^{2-})_2$	$SnCl_4$
Zinc	2	$Zn^{2+}(F^-)_2$	$ZnBr_2$
Aluminum	3	$(Al^{3+})_2(O^{2-})_3$	AlI_3
Antimony	3	—	SbI_3
Bismuth	3	—	$BiCl_3$
Chromium	3	$(Cr^{3+})_2(SO_4^{2-})_3$	$CrCl_3$
Nonmetals and anions			
Acetate	1	$Na^+CH_3COO^-$	CH_3COOH
Arsenite	1	$K^+AsO_2^-$	—
Bicarbonate	1	$Na^+HCO_3^-$	—
Bromide	1	Na^+Br^-	HBr
Bromate	1	$Na^+BrO_3^-$	$HBrO_3$
Chloride	1	Na^+Cl^-	HCl
Chlorate	1	$K^+ClO_3^-$	$HClO_3$

TABLE A-1 (Continued)

Name	Valence	Illustrative compound	
		Ionic	Nonionic
Cyanide	1	K^+CN^-	HCN
Dihydrogen phosphate	1	$Na^+H_2PO_4^-$	—
Fluoride	1	K^+F^-	$(HF)_x$
Hydride	1	Li^+H^-	B_2H_6
Hydroxide	1	Na^+OH^-	HOH
Iodide	1	K^+I^-	HI
Nitrate	1	$Na^+NO_3^-$	HNO_3
Nitrite	1	$Na^+NO_2^-$	HNO_2
Permanganate	1	$K^+MnO_4^-$	$HMnO_4$
Carbonate	2	$Ca^{2+}CO_3^{2-}$	—
Chromate	2	$(K^+)_2CrO_4^{2-}$	—
Dichromate	2	$(K^+)_2Cr_2O_7^{2-}$	—
Hydrogen phosphate	2	$(Na^+)_2HPO_4^{2-}$	—
Oxygen			
Oxide	2	$(Na^+)_2O^{2-}$	HgO
Peroxide	2	$(Na^+)_2O_2^{2-}$	H_2O_2
Sulfate	2	$(Na^+)_2SO_4^{2-}$	H_2SO_4
Sulfide	2	$(K^+)_2S^{2-}$	CdS
Sulfite	2	$(Na^+)_2SO_3^{2-}$	—
Arsenate	3	$(K^+)_3AsO_4^{3-}$	H_3AsO_4
Phosphate	3	$(Na^+)_3PO_4^{3-}$	H_3PO_4

Included in the list is hydrocyanic acid, HCN, which is named as if it were a binary acid.

Oxyacids have the general formula $(HO)_mXO_n$. They possess at least *one* OH group attached to a central atom but need not always have a lone oxygen atom.

The naming of the oxyacids is complicated by the fact that the central element may exist in different oxidation states and there may be an acid corresponding to each state. When only one oxidation state exists, the suffix **-ic** is added to the root name of the element followed by the word **acid.** This method is illustrated by the name carbon**ic acid,** $(HO)_2CO_2$, given to the water solution of carbon dioxide. When two oxyacids are known for a given central element, the one with the higher oxidation number (p. 122) has the suffix **-ic,** the other the suffix **-ous:**

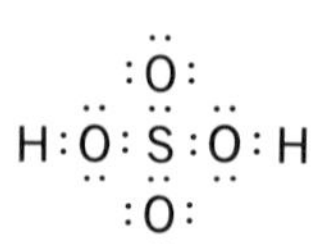

oxidation number of S = +6, sulfur**ic acid**

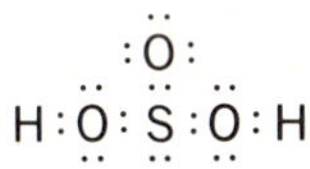

oxidation number of S = +4, sulfur**ous acid**

If an element has three oxidation states, the acid corresponding to the lowest state is designated by bracketing the root with the prefix **hypo-** and the suffix **-ous** (the H atom is assumed to have an oxidation number of +1):

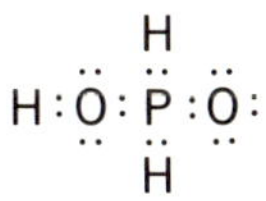

oxidation number of P = +1, **hypo**phosphor**ous acid**

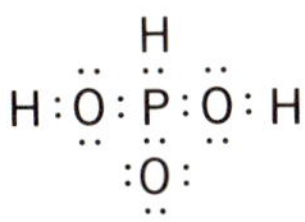

oxidation number of P = +3, phosphor**ous acid**

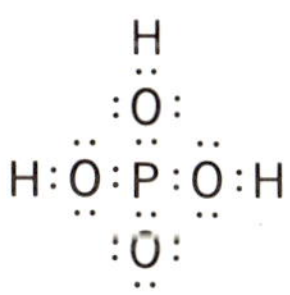

oxidation number of P = +5, phosphor**ic acid**

When four oxyacids are known, each with the central atom in a different oxidation state, the acid with the atom in the highest state is designated by the prefix **per-** and suffix **-ic.** This method is illustrated with the oxyacids of chlorine:

oxidation number of Cl = +1, **hypo**chlor**ous acid**

oxidation number of Cl = +3, chlor**ous acid**

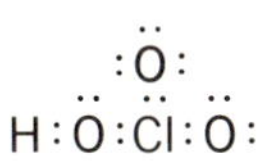

oxidation number of Cl = +5, chlor**ic acid**

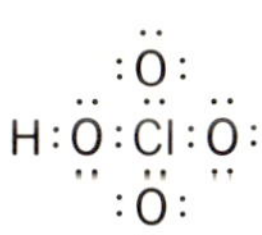

oxidation number of Cl = +7, **per**chlor**ic acid**

Oxyacids in the same oxidation state may differ in water content, and the prefixes **ortho, pyro,** and **meta** are used in order of decreasing hydration.

Prefix	Meaning	Examples	Names
Ortho	Contains most water possible	H_3PO_4, $(HO)_3PO$	Orthophosphoric acid
		H_3BO_3, $B(OH)_3$	Orthoboric acid
Pyro	Intermediate water content; an acid that appears to be formed by the removal of 1 H_2O molecule from 2 acid molecules	$H_2S_2O_7$, 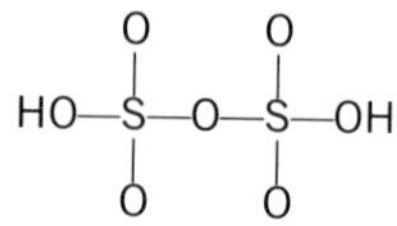	Pyrosulfuric acid
		$H_4P_2O_7$, 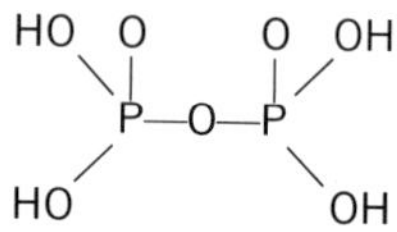	Pyrophosphoric acid
Meta	Contains least water possible	HPO_3, $HOPO_2$	Metaphosphoric acid acid
		HBO_2, HOBO	Metaboric acid

These relationships are further illustrated by the following equations for the phosphoric acids:

$$\underset{ortho}{2H_3PO_4} \longrightarrow H_2O + \underset{pyro}{H_4P_2O_7}$$

$$\underset{ortho}{H_3PO_4} \longrightarrow H_2O + \underset{meta}{HPO_3}$$

The anions of oxyacids are named by changing the suffixes **-ous** and **-ic** to **-ite** and **-ate,** respectively. More than one anion may be derived from an acid $(HO)_mXO_n$ when m is greater than 1, because it is possible for the molecule to lose more than one hydrogen atom. These anions are differentiated by indicating the number of hydrogen atoms. For example, $(HO)_2PO_2^-$ is named **di**hydrogen phosphate and $HOPO_3^{2-}$ is named **mono**hydrogen phosphate, although in practice the prefix **mono-** is frequently omitted. When no more than one H atom can be present in the anion, the prefix **bi-** is often used; $HOSO_3^-$ and $HOCO_2^-$ are commonly called **bi**sulfate and **bi**carbonate, respectively.

NOMENCLATURE OF COMPLEX COMPOUNDS

The rules presented in this section apply only to the naming of the simple coordination compounds—those with no multidentate* ligands and with only one central atom.

(1) Naming the ligand Anionic ligands are usually named by adding

* A multidentate ligand, exemplified by ethylenediamine, $H_2NCH_2CH_2NH_2$, has more than one bonding site.

the letter *o* to the stem name, for example, Cl^- (*chloro*), CN^- (*cyano*), SO_4^{2-} (*sulfato*), NH_2^- (*amido*), $—NO_2^-$ (*nitro*), —O—NO^- (*nitrito*), and $S_2O_3^{2-}$ (*thiosulfato*). Molecular ligands are given special names, for example, H_2O (*aquo*), NH_3 (*ammine*), CO (*carbonyl*), NO (*nitrosyl*), and PH_3 (*phosphine*).

(2) Naming compounds with a complex ion Four rules are observed:

(a) If the compound is ionic, the cation is mentioned first, whether it is the complex ion or not.

(b) In naming the complex, the order is negative ligand, neutral ligand, and then central atom, followed by a Roman numeral in parentheses to indicate the oxidation number.

(c) If the complex is an anion, the suffix *-ate* is appended to the name of the central atom. The oxidation number then follows this suffix.

(d) The number of each kind of ligand is specified, using the Greek prefixes *di-*, *tri-*, *tetra-*, *penta-*, and *hexa-*.

The examples below illustrate these rules:

$[Co(NH_3)_6]Cl_3$	Hexamminecobalt(III) chloride
$K_3[Fe(CN)_6]$	Potassium hexacyanoferrate(III)
$[Cr(H_2O)_4Cl_2]Cl$	Dichlorotetraaquochromium(III) chloride
$Na_3[Ag(S_2O_3)_2]$	Sodium dithiosulfatoargentate(I)

ORGANIC NOMENCLATURE

Many organic compounds are given **common** or **trivial** names. However, as the number of carbon atoms in a molecule increases, the possibilities for isomerism increase. For example, there are 9 isomers for heptane, and 18 for octane. It would be obviously foolhardy to attempt to give each isomer an unambiguous trivial name. Chemists have developed a method of nomenclature that assigns a systematic, unique name to each possible compound. The system presently used, called the IUPAC system, an abbreviation for the International Union of Pure and Applied Chemistry, will be applied to the naming of

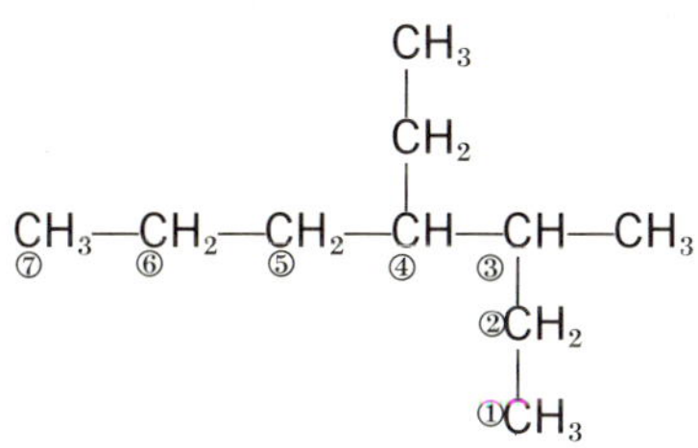

Rule 1: Selecting and naming the longest chain Name the compound as a **derivative** of the alkane represented by the longest chain or con-

tinuous sequence of carbon atoms. In the above compound, the longest chain contains 7 carbons as numbered. The compound is therefore named as a heptane, and is considered to be a *derivative* of this parent compound. The name must end in the suffix *-ane* to indicate that the compound is a member of the alk*ane* series. Except for the first four members of the alkane series, the prefix of the name is derived from the Greek word for the number of carbon atoms in the chain. Thus, for example, *hept* indicates 7 carbons.

Rule 2: Numbering the longest carbon chain Number the carbon atoms in the longest chain, as illustrated, so that the branching groups are on the lower- rather than the higher-numbered carbon atoms.

Rule 3: Locating the branching groups The location of the branching hydrocarbon groups is indicated by the number of the carbon atom of the parent chain to which they are attached. Thus, in our illustrative compound, we find a CH_3 group on C-3 and a CH_3CH_2 group on C-4.

Rule 4: Naming the branched groups Each branched group is named according to the number of carbons in its longest chain. To name the group, replace the suffix *-ane* by *-yl*. Thus, CH_3 is called meth*yl* since it stems from meth*ane*, and CH_3CH_2 is named eth*yl* from eth*ane*.

Some prefixes other than those designating alkyl groups are

fluoro	F—
chloro	Cl—
bromo	Br—
iodo	I—
hydroxy	HO—
thio	HS—
amino	H_2N—
nitro	O_2N—
phenyl	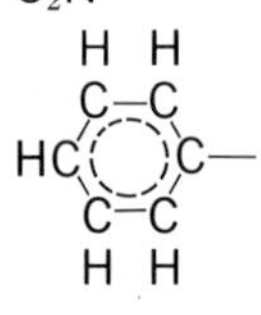

Rule 5: Arranging the names of the branched groups The names of these groups are prefixed alphabetically with the number of the carbon atom to which each group is attached. Thus, the name of the illustrative compound is 4-*ethyl*-3-*methylheptane*.

The rules of punctuation are: (*i*) words are run together; (*ii*) a dash is used to separate a number from a word; and (*iii*) a comma is used to separate numbers.

The following examples further illustrate the method:

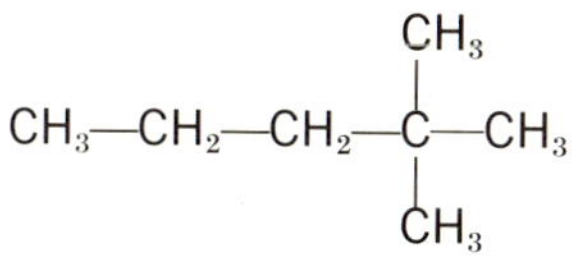

2,2-dimethylpentane

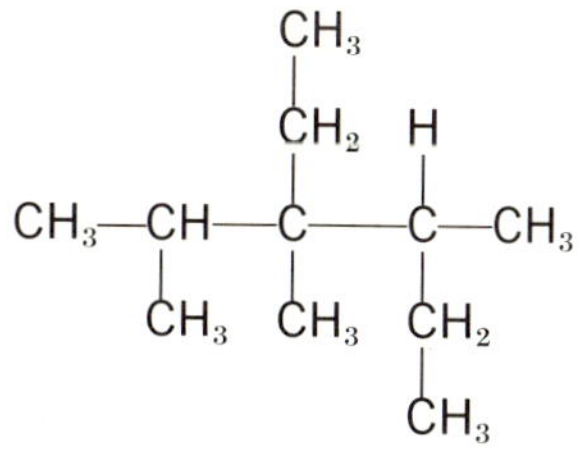

3-ethyl-2,3,4-trimethylhexane

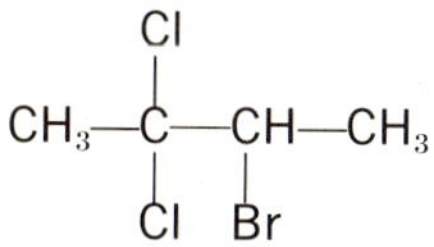

2,2-dichloro-3-bromobutane

Note the repetition of numbers when a group appears twice on the same carbon, and the use of *di-* and *tri-* to indicate the number of times the group appears as a branch.

APPENDIX 4 FUNDAMENTAL CONSTANTS*

c	Speed of light *in vacuo*	2.997925×10^{10} cm/sec
e	Charge on the electron	4.80298×10^{-10} esu 1.60210×10^{-19} C
$\mathscr{F}$	Faraday's constant	9.64870×10^{4} C/eq
h	Planck's constant	6.6256×10^{-27} erg sec/particle
k	Boltzmann's constant	1.38054×10^{-16} erg/°K molecule
$\mathscr{N}$	Avogadro's number	6.02252×10^{23} molecules/mole
R	Ideal gas constant	8.2056×10^{-2} liter atm/°K mole 8.3143 J/°K mole
T_0	Ice point	273.15°K = 0°C
V_0	Molar gas volume	22.4136 liters/mole

* Values recommended by the Committee on Fundamental Constants of the National Academy of Sciences–National Research Council, 1963 (United States).

APPENDIX 5 ■ VAPOR PRESSURE OF WATER

Temperature °C	Pressure torr	Temperature °C	Pressure torr
0	4.6	24	22.4
5	6.5	25	23.8
10	9.2	26	25.2
11	9.8	27	26.7
12	10.5	28	28.3
13	11.2	29	30.0
14	12.0	30	31.8
15	12.8	35	42.2
16	13.6	40	55.3
17	14.5	60	149.4
18	15.5	80	355.1
19	16.5	100	760.0
20	17.5	110	1075
21	18.7		
22	19.8		
23	21.1		

APPENDIX 6 ABBREVIATIONS

ampere, A
angström, Å
atmosphere, atm
atomic mass unit, amu
calorie, cal
coulomb, C
cubic centimeter, cm^3
cubic foot, ft^3
Debye unit, D
degree Celsius (Centigrade), °C
degree Fahrenheit, °F
degree Kelvin, °K
electronvolt, eV
electrostatic unit, esu
foot, ft
gauss, G
gram, g
gram atom, g-atom
hour, hr
inch, in.
joule, J
kilocalorie, kcal
kilogram, kg
liter, l
meter, m
microliter, μl
micron, μ
milligram, mg
milliliter, ml
millimeter, mm
millimicron, mμ
million electron volts, MeV
minute, min
mole, mol
molal (concentration), m
molar (concentration), M
normal (concentration), N
ohm, Ω
pound, lb
second, sec
volt, V
watt, W

APPENDIX 7 ■ LOGARITHMS

Logarithms to the Base e

N	.0	.1	.2	.3	.4	.5	.6	.7	.8	.9
1	0.000	0.095	0.182	0.262	0.336	0.405	0.470	0.531	0.588	0.642
2	0.693	0.742	0.788	0.833	0.875	0.916	0.956	0.993	1.030	1.065
3	1.099	1.131	1.163	1.194	1.224	1.253	1.281	1.308	1.335	1.361
4	1.386	1.411	1.435	1.459	1.482	1.504	1.526	1.548	1.569	1.589
5	1.609	1.629	1.649	1.668	1.686	1.705	1.723	1.740	1.758	1.775
6	1.792	1.808	1.825	1.841	1.856	1.872	1.887	1.902	1.917	1.932
7	1.946	1.960	1.974	1.988	2.001	2.015	2.028	2.041	2.054	2.067
8	2.079	2.092	2.104	2.116	2.128	2.140	2.152	2.163	2.175	2.186
9	2.197	2.208	2.219	2.230	2.241	2.251	2.262	2.272	2.282	2.293
10	2.303	2.313	2.322	2.332	2.342	2.351	2.361	2.370	2.380	2.389

Logarithms to the Base 10

N	0	1	2	3	4	5	6	7	8	9
10	0000	0043	0086	0128	0170	0212	0253	0294	0334	0374
11	0414	0453	0492	0531	0569	0607	0645	0682	0719	0755
12	0792	0828	0864	0899	0934	0969	1004	1038	1072	1106
13	1139	1173	1206	1239	1271	1303	1335	1367	1399	1430
14	1461	1492	1523	1553	1584	1614	1644	1673	1703	1732
15	1761	1790	1818	1847	1875	1903	1931	1959	1987	2014
16	2041	2068	2095	2122	2148	2175	2201	2227	2253	2279
17	2304	2330	2355	2380	2405	2430	2455	2480	2504	2529
18	2553	2577	2601	2625	2648	2672	2695	2718	2742	2765
19	2788	2810	2833	2856	2878	2900	2923	2945	2967	2989
20	3010	3032	3054	3075	3096	3118	3139	3160	3181	3201
21	3222	3243	3263	3284	3304	3324	3345	3365	3385	3404
22	3424	3444	3464	3483	3502	3522	3541	3560	3579	3598
23	3617	3636	3655	3674	3692	3711	3729	3747	3766	3784
24	3802	3820	3838	3856	3874	3892	3909	3927	3945	3962
25	3979	3997	4014	4031	4048	4065	4082	4099	4116	4133
26	4150	4166	4183	4200	4216	4232	4249	4265	4281	4298
27	4314	4330	4346	4362	4378	4393	4409	4425	4440	4456
28	4472	4487	4502	4518	4533	4548	4564	4579	4594	4609
29	4624	4639	4654	4669	4683	4698	4713	4728	4742	4757
30	4771	4786	4800	4814	4829	4843	4857	4871	4886	4900
31	4914	4928	4942	4955	4969	4983	4997	5011	5024	5038
32	5051	5065	5079	5092	5105	5119	5132	5145	5159	5172
33	5185	5198	5211	5224	5237	5250	5263	5276	5289	5302
34	5315	5328	5340	5353	5366	5378	5391	5403	5416	5428
35	5441	5453	5465	5478	5490	5502	5514	5527	5539	5551
36	5563	5575	5587	5599	5611	5623	5635	5647	5658	5670
37	5682	5694	5705	5717	5729	5740	5752	5763	5775	5786
38	5798	5809	5821	5832	5843	5855	5866	5877	5888	5899
39	5911	5922	5933	5944	5955	5966	5977	5988	5999	6010
40	6021	6031	6042	6053	6064	6075	6085	6096	6107	6117
41	6128	6138	6149	6160	6170	6180	6191	6201	6212	6222
42	6232	6243	6253	6263	6274	6284	6294	6304	6314	6325
43	6335	6345	6355	6365	6375	6385	6395	6405	6415	6425
44	6435	6444	6454	6464	6474	6484	6493	6503	6513	6522
45	6532	6542	6551	6561	6571	6580	6590	6599	6609	6618
46	6628	6637	6646	6656	6665	6675	6684	6693	6702	6712
47	6721	6730	6739	6749	6758	6767	6776	6785	6794	6803
48	6812	6821	6830	6839	6848	6857	6866	6875	6884	6893
49	6902	6911	6920	6928	6937	6946	6955	6964	6972	6981
50	6990	6998	7007	7016	7024	7033	7042	7050	7059	7067
51	7076	7084	7093	7101	7110	7118	7126	7135	7143	7152
52	7160	7168	7177	7185	7193	7202	7210	7218	7226	7235
53	7243	7251	7259	7267	7275	7284	7292	7300	7308	7316
54	7324	7332	7340	7348	7356	7364	7372	7380	7388	7396
N	0	1	2	3	4	5	6	7	8	9

Logarithms to the Base 10 (cont.)

N	0	1	2	3	4	5	6	7	8	9
55	7404	7412	7419	7427	7435	7443	7451	7459	7466	7474
56	7482	7490	7497	7505	7513	7520	7528	7536	7543	7551
57	7559	7566	7574	7582	7589	7597	7604	7612	7619	7627
58	7634	7642	7649	7657	7664	7672	7679	7686	7694	7701
59	7709	7716	7723	7731	7738	7745	7752	7760	7767	7774
60	7782	7789	7796	7803	7810	7818	7825	7832	7839	7846
61	7853	7860	7868	7875	7882	7889	7896	7903	7910	7917
62	7924	7931	7938	7945	7952	7959	7966	7973	7980	7987
63	7993	8000	8007	8014	8021	8028	8035	8041	8048	8055
64	8062	8069	8075	8082	8089	8096	8102	8109	8116	8122
65	8129	8136	8142	8149	8156	8162	8169	8176	8182	8189
66	8195	8202	8209	8215	8222	8228	8235	8241	8248	8254
67	8261	8267	8274	8280	8287	8293	8299	8306	8312	8319
68	8325	8331	8338	8344	8351	8357	8363	8370	8376	8382
69	8388	8395	8401	8407	8414	8420	8426	8432	8439	8445
70	8451	8457	8463	8470	8476	8482	8488	8494	8500	8506
71	8513	8519	8525	8531	8537	8543	8549	8555	8561	8567
72	8573	8579	8585	8591	8597	8603	8609	8615	8621	8627
73	8633	8639	8645	8651	8657	8663	8669	8675	8681	8686
74	8692	8698	8704	8710	8716	8722	8727	8733	8739	8745
75	8751	8756	8762	8768	8774	8779	8785	8791	8797	8802
76	8808	8814	8820	8825	8831	8837	8842	8848	8854	8859
77	8865	8871	8876	8882	8887	8893	8899	8904	8910	8915
78	8921	8927	8932	8938	8943	8949	8954	8960	8965	8971
79	8976	8982	8987	8993	8998	9004	9009	9015	9020	9025
80	9031	9036	9042	9047	9053	9058	9063	9069	9074	9079
81	9085	9090	9096	9101	9106	9112	9117	9122	9128	9133
82	9138	9143	9149	9154	9159	9165	9170	9175	9180	9186
83	9191	9196	9201	9206	9212	9217	9222	9227	9232	9238
84	9243	9248	9253	9258	9263	9269	9274	9279	9284	9289
85	9294	9299	9304	9309	9315	9320	9325	9330	9335	9340
86	9345	9350	9355	9360	9365	9370	9375	9380	9385	9390
87	9395	9400	9405	9410	9415	9420	9425	9430	9435	9440
88	9445	9450	9455	9460	9465	9469	9474	9479	9484	9489
89	9494	9499	9504	9509	9513	9518	9523	9528	9533	9538
90	9542	9547	9552	9557	9562	9566	9571	9576	9581	9586
91	9590	9595	9600	9605	9609	9614	9619	9624	9628	9633
92	9638	9643	9647	9652	9657	9661	9666	9671	9675	9680
93	9685	9689	9694	9699	9703	9708	9713	9717	9722	9727
94	9731	9736	9741	9745	9750	9754	9759	9763	9768	9773
95	9777	9782	9786	9791	9795	9800	9805	9809	9814	9818
96	9823	9827	9832	9836	9841	9845	9850	9854	9859	9863
97	9868	9872	9877	9881	9886	9890	9894	9899	9903	9908
98	9912	9917	9921	9926	9930	9934	9939	9943	9948	9952
99	9956	9961	9965	9969	9974	9978	9983	9987	9991	9996
N	0	1	2	3	4	5	6	7	8	9

INDEX

SYMBOLS USED: *fig*, figure; *ftn*, footnote; *prb*, problem; *tbl*, table.